T0226192

Lecture Notes in Computer Science 10929

Commenced Publication in 1973
Founding and Former Series Editors:
Gerhard Goos, Juris Hartmanis, and Jan van Leeuwen

More information about this series at http://www.springer.com/series/7407

Olaf Beyersdorff · Christoph M. Wintersteiger (Eds.)

Theory and Applications of Satisfiability Testing – SAT 2018

21st International Conference, SAT 2018
Held as Part of the Federated Logic Conference, FloC 2018
Oxford, UK, July 9–12, 2018
Proceedings

 Springer

Editors
Olaf Beyersdorff
Friedrich Schiller University Jena
Jena
Germany

Christoph M. Wintersteiger ⓘ
Microsoft Research Ltd.
Cambridge
UK

ISSN 0302-9743   ISSN 1611-3349 (electronic)
Lecture Notes in Computer Science
ISBN 978-3-319-94143-1   ISBN 978-3-319-94144-8 (eBook)
https://doi.org/10.1007/978-3-319-94144-8

Library of Congress Control Number: 2018947336

LNCS Sublibrary: SL1 – Theoretical Computer Science and General Issues

Printed on acid-free paper

This Springer imprint is published by the registered company Springer International Publishing AG
part of Springer Nature
The registered company address is: Gewerbestrasse 11, 6330 Cham, Switzerland

Preface

This volume contains the papers presented at SAT 2018, the 21st International Conference on Theory and Applications of Satisfiability Testing, held during July 6–19, 2018 in Oxford, UK. SAT 2018 was part of the Federated Logic Conference (FLoC) 2018 and was hosted by the University of Oxford.

The International Conference on Theory and Applications of Satisfiability Testing (SAT) is the premier annual meeting for researchers focusing on the theory and applications of the propositional satisfiability problem, broadly construed. Aside from plain propositional satisfiability, the scope of the meeting includes Boolean optimization, including MaxSAT and pseudo-Boolean (PB) constraints, quantified Boolean formulas (QBF), satisfiability modulo theories (SMT), and constraint programming (CP) for problems with clear connections to Boolean reasoning.

Many hard combinatorial problems can be tackled using SAT-based techniques, including problems that arise in formal verification, artificial intelligence, operations research, computational biology, cryptology, data mining, machine learning, mathematics, etc. Indeed, the theoretical and practical advances in SAT research over the past 25 years have contributed to making SAT technology an indispensable tool in a variety of domains.

SAT 2018 welcomed scientific contributions addressing different aspects of SAT interpreted in a broad sense, including (but not restricted to) theoretical advances (such as exact algorithms, proof complexity, and other complexity issues), practical search algorithms, knowledge compilation, implementation-level details of SAT solvers and SAT-based systems, problem encodings and reformulations, applications (including both novel application domains and improvements to existing approaches), as well as case studies and reports on findings based on rigorous experimentation.

A total of 58 papers were submitted to SAT 2018, comprising 35 regular papers, 17 short papers, and six tool papers. Each submission was reviewed by four Program Committee members and their selected external reviewers. The review process included an author response period, during which the authors of submitted papers were given the opportunity to respond to the initial reviews for their submissions. To reach a final decision, a Program Committee discussion period followed the author response period. External reviewers supporting the Program Committee were also invited to participate directly in the discussions for the papers they reviewed. This year, most submissions further received a meta-review, summarizing the discussion that occurred after the author response and an explanation of the final recommendation. In the end, the committee decided to accept a total of 26 papers; 20 regular, four short, and two tool papers.

The Program Committee singled out the following two submissions for the Best Paper Award and the Best Student Paper Award, respectively:

– Tobias Friedrich and Ralf Rothenberger: "Sharpness of the Satisfiability Threshold for Non-Uniform Random k-SAT"

- Dimitris Achlioptas, Zayd Hammoudeh, and Panos Theodoropoulos: "Fast Sampling of Perfectly Uniform Satisfying Assignments"

In addition to presentations on the accepted papers, the scientific program of SAT included three invited talks:

- Marijn Heule (University of Texas at Austin, US): "Computable Short Proofs"
- Rahul Santhanam (University of Oxford, UK): "Modelling SAT"
- Christoph Scholl (Albert Ludwigs University Freiburg, Germany): "Dependency Quantified Boolean Formulas: An Overview of Solution Methods and Applications"

Two additional keynote and plenary talks, as well as a public lecture, were held jointly with other conferences of FLoC:

- Shafi Goldwasser (MIT): "Pseudo-deterministic Algorithms and proofs"
- Peter O'Hearn (Facebook): "Continuous Reasoning for Big Code"
- A public lecture by Stuart Russell (University of California at Berkeley, USA) on "Unifying Logic and Probability: The BLOG Language"

SAT, together with the other constituent conferences of FLoC, hosted various associated events. In particular, the following four workshops were held July 7–8, (co-) affiliated with SAT:

- Pragmatics of SAT
  Organized by Matti Järvisalo and Daniel Le Berre
- Quantified Boolean Formulas and Beyond
  Organized by Hubie Chen, Florian Lonsing, and Martina Seidl
- Proof Complexity
  Organized by Olaf Beyersdorff, Leroy Chew, and Jan Johannse
- Cross-Fertilization Between CSP and SAT
  Organized by Alexander Ivrii and Yehuda Naveh

As in previous years, the results of several competitive events were announced at SAT:

- SAT Competition 2018
  Organized by Marijn Heule, Matti Järvisalo, and Martin Suda
- MaxSAT Evaluation 2018
  Organized by Fahiem Bacchus, Matti Järvisalo, and Ruben Martins
- Sparkle SAT Challenge 2018
  Organized by Chuan Luo and Holger H. Hoos
- QBFEVAL 2018
  Organized by Luca Pulina and Martina Seidl

We thank everyone who contributed to making SAT 2018 a success. We are indebted to the Program Committee members and the external reviewers, who dedicated their time to review and evaluate the submissions to the conference. We thank the authors of all submitted papers for their contributions, the SAT association for their guidance and support in organizing the conference, the EasyChair conference management system for facilitating the submission and selection of papers, as well as the assembly of these proceedings. We wish to thank the workshop chair, Martina Seidl, and all the organizers of the SAT affiliated workshops and competitions. Special thanks go to the organizers of FLoC, in particular to Moshe Vardi, Daniel Kroening, and Marta Kwiatkowska, for coordinating the various conferences and taking care of the local arrangements.

We gratefully thank the sponsors of SAT 2018: the SAT Association, for providing travel support for students attending the conference, Springer, for sponsoring the best paper awards for SAT 2018, the *Artificial Intelligence* journal, and SATALIA for financial and organizational support for SAT 2018. Thank you.

May 2018   Christoph M. Wintersteiger
                                                      Olaf Beyersdorff

Organization

Program Committee

Gilles Audemard	CRIL, France
Fahiem Bacchus	University of Toronto, Canada
Olaf Beyersdorff	Friedrich Schiller University Jena, Germany
Armin Biere	Johannes Kepler University Linz, Austria
Nikolaj Bjorner	Microsoft, US
Nadia Creignou	Aix-Marseille Université, France
Uwe Egly	Vienna University of Technology, Austria
John Franco	University of Cincinnati, USA
Vijay Ganesh	Waterloo, Canada
Serge Gaspers	UNSW Sydney and Data61, CSIRO, Australia
Marijn Heule	The University of Texas at Austin, USA
Mikolas Janota	INESC-ID/IST, University of Lisbon, Portugal
Matti Järvisalo	University of Helsinki, Finland
George Katsirelos	MIAT, INRA, France
Oliver Kullmann	Swansea University, UK
Massimo Lauria	Sapienza University of Rome, Italy
Daniel Le Berre	CNRS, Université d'Artois, France
Florian Lonsing	Vienna University of Technology, Austria
Ines Lynce	INESC-ID/IST, University of Lisbon, Portugal
Joao Marques-Silva	University of Lisbon, Portugal
Ruben Martins	Carnegie Mellon University, USA
Stefan Mengel	CNRS, CRIL UMR 8188, France
Alexander Nadel	Intel, Israel
Jakob Nordstrom	KTH Royal Institute of Technology, Sweden
Markus N. Rabe	University of California, Berkeley, USA
Martina Seidl	Johannes Kepler University Linz, Austria
Laurent Simon	Labri, Bordeaux Institute of Technology, France
Carsten Sinz	Karlsruhe Institute of Technology, Germany
Friedrich Slivovsky	Vienna University of Technology, Austria
Takehide Soh	Information Science and Technology Center, Kobe University, Japan
Stefan Szeider	Vienna University of Technology, Austria
Jacobo Torán	Universität Ulm, Germany
Ralf Wimmer	Albert-Ludwigs-Universität Freiburg, Germany
Christoph M. Wintersteiger	Microsoft, UK
Xishun Zhao	Institute of Logic and Cognition, Sun Yat-Sen University, China

Additional Reviewers

Aghighi, Meysam
Berg, Jeremias
Berkholz, Christoph
Bliem, Bernhard
Blinkhorn, Joshua
Bonacina, Ilario
Bova, Simone
Capelli, Florent
Cashmore, Michael
Chew, Leroy
Clymo, Judith
Daude, Herve
Davies, Jessica
de Givry, Simon
de Haan, Ronald
Diller, Martin
Durand, Arnaud
Elffers, Jan
Fazekas, Katalin
Ferreira, Margarida
Fichte, Johannes K.
Ge-Ernst, Aile
Giráldez-Cru, Jesus
Gocht, Stephan
Hinde, Luke
Hunter, Paul
Ignatiev, Alexey
Ivrii, Alexander
Jarret, Michael
Kleine Büning, Hans

Klieber, Will
Koshimura, Miyuki
Lampis, Michael
Liang, Jimmy
Manquinho, Vasco
Mitsou, Valia
Montmirail, Valentin
Morgado, Antonio
Nabeshima, Hidetomo
Narodytska, Nina
Nejati, Saeed
Neubauer, Felix
Oetsch, Johannes
Ordyniak, Sebastian
Paxian, Tobias
Peitl, Tomáš
Reimer, Sven
Richoux, Florian
Robere, Robert
Ryvchin, Vadim
Saikko, Paul
Shen, Yuping
Sokolov, Dmitry
Szczepanski, Nicolas
Talebanfard, Navid
Terra-Neves, Miguel
Truchet, Charlotte
van Dijk, Tom
Weaver, Sean
Winterer, Leonore

Invited Talks

Computable Short Proofs

Marijn J. H. Heule

Department of Computer Science, The University of Texas at Austin

The success of satisfiability solving presents us with an interesting peculiarity: modern solvers can frequently handle gigantic formulas while failing miserably on supposedly easy problems. Their poor performance is typically caused by the weakness of their underlying proof system—resolution. To overcome this obstacle, we need solvers that are based on stronger proof systems. Unfortunately, existing strong proof systems— such as extended resolution [1] or Frege systems [2]—do not seem to lend themselves to mechanization.

We present a new proof system that not only generalizes strong existing proof systems but that is also well-suited for mechanization. The proof system is surprisingly strong, even without the introduction of new variables — a key feature of short proofs presented in the proof-complexity literature. Moreover, we introduce a new decision procedure that exploits the strengths of our new proof system and can therefore yield exponential speed-ups compared to state-of-the-art solvers based on resolution.

Our new proof system, called PR (short for Propagation Redundancy), is a *clausal proof system* and closely related to state-of-the-art SAT solving. Informally, a clausal proof system allows the addition of redundant clauses to a formula in conjunctive normal form. Here, a clause is considered redundant if its addition preserves satisfiability. If the repeated addition of clauses allows us finally to add the empty clause— which is, by definition, unsatisfiable—the unsatisfiability of the original formula has been established.

Since the redundancy of clauses is not efficiently decidable in general, clausal proof systems only allow the addition of a clause if it fulfills some efficiently decidable criterion that ensures redundancy. For instance, the popular DRAT proof system [3], which is the de-facto standard in practical SAT solving, only allows the addition of so-called *resolution asymmetric tautologies* [4]. Given a formula and a clause, it can be decided in polynomial time whether the clause is a resolution asymmetric tautology with respect to the formula and therefore the soundness of DRAT proofs can be checked efficiently. Several formally-verified checkers for DRAT proofs are available [5, 6].

We present a new notion of redundancy by introducing a characterization of clause redundancy based on a semantic implication relationship between formulas. By replacing the implication relation in this characterization with a restricted notion of implication that is computable in polynomial time, we then obtain powerful notion of redundancy that is still efficiently decidable. The proof system, which based on this notion of redundancy, turns out to be highly expressive, even without allowing the

---

Based on joint work with Benjamin Kiesl, Armin Biere, and Martina Seidl.

introduction of new variables. This is in contrast to resolution, which is considered relatively weak as long as the introduction of new variables via definitions—as in the stronger proof system of extended resolution—is not allowed. The introduction of new variables, however, has a major drawback—the search space of variables and clauses we could possibly add to a proof is clearly exponential. Finding useful clauses with new variables is therefore hard in practice and resulted only in limited success in the past [7, 8].

In order to capitalize on the strengths of the PR proof system in practice, we enhance conflict-driven clause learning (CDCL) [9]. To do so, we introduce *satisfaction-driven clause learning* (SDCL) [10], a SAT solving paradigm that extends CDCL as follows: If the usual unit propagation does not lead to a conflict, we do not immediately decide for a new variable assignment (as would be the case in CDCL). Instead, we first try to prune the search space of possible truth assignments by learning a so-called PR clause. We demonstrate the strength of SDCL by computing short PR proofs for the famous pigeon hole formulas without new variables.

At this point there exists only an unverified checker to validate PR proofs, written in C. In order to increase the trust in the correctness of PR proofs, we implemented a tool to convert PR proofs into DRAT proofs [11], which in turn can be validated using verified proof checkers. Thanks to various optimizations, the size increase during conversion is rather modest on available PR proofs, thereby making this a useful certification approach in practice.

References

1. Tseitin, G.S.: On the complexity of derivation in propositional calculus. In: Siekmann, J.H., Wrightson, G. (eds.) Automation of Reasoning. Symbolic Computation (Artificial Intelligence). Springer, Heidelberg (1983)
2. Cook, S.A., Reckhow, R.A.: The relative efficiency of propositional proof systems. J. Symb. Log. **44**(1), 36–50 (1979)
3. Wetzler, N., Heule, M.J.H., Hunt, W.A.: DRAT-trim: efficient checking and trimming using expressive clausal proofs. In: Sinz, C., Egly, U. (eds.) SAT 2014. LNCS, vol. 8561, pp. 422–429. Springer, Cham (2014)
4. Järvisalo, M., Heule, M.J.H., Biere, A.: Inprocessing rules. In: Gramlich, B., Miller, D., Sattler, U. (eds) IJCAR 2012. LNCS, vol. 7364, pp. 355–370. Springer, Heidelberg (2012)
5. Cruz-Filipe, L., Heule, M.J.H., Hunt, W.A., Kaufmann, M., Schneider-Kamp, P.: Efficient certified RAT verification. In: de Moura, L. (eds.) CADE 2017. LNCS, vol. 10395, pp. 220–236. Springer, Cham (2017)
6. Lammich, P. Efficient verified (UN)SAT certificate checking. In: de Moura, L.: (eds.) CADE 2017. LNCS, vol. 10395, pp. 237–254. Springer, Cham (2017)
7. Audemard, G., Katsirelos, G., Simon, L.: A restriction of extended resolution for clause learning sat solvers. In: Proceedings of the 24th AAAI Conference on Artificial Intelligence (AAAI 2010). AAAI Press (2010)
8. Manthey, N., Heule, M.J.H., Biere, A.: Automated reencoding of boolean formulas. In: Biere, A., Nahir, A., Vos, T. (eds.) HVC 2012. LNCS, vol. 7857. Springer, Heidelberg (2013)

9. Marques-Silva, J.P., Sakallah, K.A.: GRASP – a new search algorithm for satisfiability. In: Proceedings of the 1996 IEEE/ACM international conference on Computer-aided design, ICCAD 1996, Washington, DC, USA, pp. 220–227. IEEE Computer Society (1996)
10. Heule, M.J.H., Kiesl, B., Seidl, M., Biere, A.: PRuning through satisfaction. In: Strichman, O., Tzoref-Brill, R. (eds.) HVC 2017. LNCS, vol. 10629, pp. 179–194. Springer, Cham (2017)
11. Heule, M.J.H., Biere, A.: What a difference a variable makes. In: Beyer, D., Huisman, M. (eds.) TACAS 2018. LNCS, vol. 10806, pp. 75–92. Springer, Cham (2018)

Modelling SAT

Rahul Santhanam

Department of Computer Science, University of Oxford, UK
rahul.santhanam@cs.ox.ac.uk

Abstract. Satisfiability (SAT) is the canonical NP-complete problem [3, 5]. But how hard is it exactly, and on which instances? There are several existing approaches that aim to analyze and understand the complexity of SAT:

1. Proof complexity [2]: Here the main goal is to show good lower bounds on the sizes of proofs for unsatisfiability of certain "hard" formulas in various proof systems. The importance of this approach owes partly to the fact that lower bounds on proof size can be translated into lower bounds on running time for SAT solvers of interest.
2. Exact algorithms [4]: Here the goal is to get better upper bounds on worst-case running time for SAT algorithms. Assuming NP $\neq$ P, these improved upper bounds will not be polynomial, but they could still improve substantially over the naive brute force search algorithm.
3. Random SAT [1]: Here the goal is to understand the hardness of SAT on random instances, where clauses are picked independently and uniformly at random. The methods of statistical physics turn out to be helpful - physical insights about phase transitions and the structure of solution spaces can be used to quantify the performance of a large class of algorithms.

I broadly discuss these approaches and the relationships between them. I suggest that their complementary perspectives could be useful in developing new models and answering questions that do not seem to be answerable by any individual approach. Indeed SAT serves partly as an excuse to investigate a larger issue: what are good algorithmic models, and what questions should we be asking about them?

Keywords: Satisfiability · Modelling · Proof complexity · Exact algorithms
Phase transitions

References

1. Achlioptas, D.: Random Satisfiability. Handbook of Satisfiability, pp. 245–270 (2009)
2. Beame, P., Pitassi, T.: Propositional proof complexity: past, present, and future. Curr. Trends Theor. Comput. Sci. 42–70 (2001)
3. Cook, S.: The complexity of theorem-proving procedures. In: Proceedings of 3rd Annual ACM Symposium on Theory of Computing (STOC), pp. 151–158 (1971)
4. Dantsin, E., Hirsch, E.: Worst-Case Upper Bounds. Handbook of Satisfiability, pp. 403–424 (2009)
5. Levin, L.: Universal sequential search problems. Problemy Peredachi Informatsii **9**(3), 115–116 (1973)

Contents

Invited Talk

Dependency Quantified Boolean Formulas: An Overview of Solution Methods and Applications

Extended Abstract

Christoph Scholl[✉] and Ralf Wimmer

Albert-Ludwigs-Universität Freiburg, Freiburg im Breisgau, Germany
{scholl,wimmer}@informatik.uni-freiburg.de

Abstract. Dependency quantified Boolean formulas (DQBFs) as a generalization of quantified Boolean formulas (QBFs) have received considerable attention in research during the last years. Here we give an overview of the solution methods developed for DQBF so far. The exposition is complemented with the discussion of various applications that can be handled with DQBF solving.

1 Introduction

Dependency quantified Boolean formulas (DQBFs) [30] have received considerable attention in research during the last years. They are a generalization of ordinary quantified Boolean formulas (QBFs). While the latter have the restriction that every existential variable depends on all universal variables in whose scope it is, DQBFs allow arbitrary dependencies, which are explicitly specified in the formula. This makes DQBFs more expensive to solve than QBFs – for DQBF the decision problem is NEXPTIME-complete [29], for QBF 'only' PSPACE-complete [28]. However, there are practically relevant applications that require the higher expressiveness of DQBF for a natural and tremendously more compact modeling. Among them is the analysis of multi-player games with incomplete information [29], the synthesis of safe controllers [8] and of certain classes of LTL properties [12], and the verification of incomplete combinational and sequential circuits [17,33,39].

Driven by the needs of the applications mentioned above, research on DQBF solving has not only led to fundamental theoretical results on DQBF analyzing which proof calculi for QBF are still sound and/or complete for DQBF [3,5], but also to first solvers like IDQ and HQS [15,16,18,36].

In this work, we give an overview of the solution methods for DQBF developed during the last years as well as of various applications that can be handled with DQBF solving.

This work was partly supported by the German Research Council (DFG) as part of the project "Solving Dependency Quantified Boolean Formulas".

O. Beyersdorff and C. M. Wintersteiger (Eds.): SAT 2018, LNCS 10929, pp. 3–16, 2018.
https://doi.org/10.1007/978-3-319-94144-8_1

2 Notions and Problem Definition

In this section, we provide preliminaries and, in particular, we define dependency quantified Boolean formulas as a generalization of quantified Boolean formulas.

For a finite set V of Boolean variables, $\mathcal{A}(V)$ denotes the set of variable assignments of V, i.e., $\mathcal{A}(V) = \{\nu : V \to \mathbb{B}\}$ with $\mathbb{B} = \{0,1\}$. Given quantifier-free Boolean formulas φ and κ over V and a Boolean variable $v \in V$, $\varphi[\kappa/v]$ denotes the Boolean formula which results from φ by replacing all occurrences of v simultaneously by κ (simultaneous replacement is necessary when κ contains the replaced variable v).

Quantified Boolean formulas are obtained by prefixing Boolean formulas with a 'linearly ordered' quantifier prefix.

Definition 1 (Syntax of QBF). *Let $V = \{x_1, \ldots, x_n, y_1, \ldots, y_m\}$ be a finite set of Boolean variables. A quantified Boolean formula (QBF) ψ over V (in prenex form) is given by $\psi := \forall X_1 \exists Y_1 \ldots \forall X_k \exists Y_k : \varphi$, where $k \geq 1$, $X_1, \ldots, X_k$ is a partition of the universal variables $\{x_1, \ldots, x_n\}$, $Y_1, \ldots, Y_k$ is a partition of the existential variables $\{y_1, \ldots, y_m\}$, $X_i \neq \emptyset$ for $i = 2, \ldots, k$, and $Y_j \neq \emptyset$ for $j = 1, \ldots, k-1$, and φ is a quantifier-free Boolean formula over V, called the matrix of ψ.*

Dependency quantified Boolean formulas are obtained by prefixing Boolean formulas with so-called Henkin quantifiers [21].

Definition 2 (Syntax of DQBF). *Let $V = \{x_1, \ldots, x_n, y_1, \ldots, y_m\}$ be a finite set of Boolean variables. A dependency quantified Boolean formula (DQBF) ψ over V (in prenex form) has the form $\psi := \forall x_1 \ldots \forall x_n \exists y_1(D_{y_1}) \ldots \exists y_m(D_{y_m}) : \varphi$, where $D_{y_i} \subseteq \{x_1, \ldots, x_n\}$ is the dependency set of y_i for $i = 1, \ldots, m$, and φ is a quantifier-free Boolean formula over V, called the matrix of ψ.*

A QBF can be seen as a DQBF where the dependency sets are linearly ordered. A QBF $\psi := \forall X_1 \exists Y_1 \ldots \forall X_k \exists Y_k : \varphi$ is equivalent to the DQBF $\psi := \forall x_1 \ldots \forall x_n \exists y_1(D_{y_1}) \ldots \exists y_m(D_{y_m}) : \varphi$ with $D_{y_i} = \cup_{j=1}^{\ell} X_j$ iff Y_ℓ is the unique set with $y_i \in Y_\ell$, $1 \leq \ell \leq k$, $1 \leq i \leq m$.

We denote the existential variables of a DQBF ψ with $V_\psi^\exists = \{y_1, \ldots, y_m\}$ and its universal variables with $V_\psi^\forall = \{x_1, \ldots, x_n\}$. We often write $\psi = Q : \varphi$ with the quantifier prefix Q and the matrix φ. As the order of the variables in a DQBF quantifier prefix Q does not matter, we can regard it as a set: For instance, $Q \setminus \{v\}$ with a variable $v \in V$ is the prefix which results from Q by removing the variable v together with its quantifier (as well as its dependency set in case v is existential, and all its occurrences in dependency sets if it is universal).

The semantics of a DQBF is typically defined in terms of so-called Skolem functions.

Definition 3 (Semantics of DQBF). *Let ψ be a DQBF as above. It is satisfied if there are functions $s_y : \mathcal{A}(D_y) \to \mathbb{B}$ for $y \in V_\psi^\exists$ such that replacing each existential variable y by (a Boolean expression for) s_y turns φ into a tautology. The functions $(s_y)_{y \in V_\psi^\exists}$ are called Skolem functions for ψ.*

Definition 4 (Equisatisfiability of DQBFs). *Two DQBFs ψ and ψ' are* equisatisfiable *iff they are either both satisfied or both not satisfied.*

Deciding whether a given DQBF is satisfied is NEXPTIME-complete [29], whereas deciding whether a given QBF is satisfied is 'only' PSPACE-complete [28].

3 Overview of Solution Methods

In this section, we give an overview of different solution methods to the DQBF problem. After briefly discussing incomplete solution methods, we present various sound and complete methods. Whereas search-based solvers using the conflict-driven clause learning (CDCL) paradigm [34] seem to outperform other sound and complete solution paradigms for the SAT problem, the situation is not that clear for QBF solving and neither is it for DQBF solving.

3.1 Incomplete Approximations

An obvious approximation approach is an approximation of a DQBF by a QBF formulation whose (implicitly given) dependency sets are supersets of the original dependency sets in the DQBF. If there is no solution to the relaxed problem, then there is also no solution to the original problem (see [33], e.g.). The QBF approximation can even be approximated further by replacing (some or all) universal variables by existential variables which may assume values from the ternary domain $\{0, 1, X\}$ [22,33]. Here X represents an unknown value and is propagated according to standard rules for unknowns [1].

The work of Finkbeiner and Tentrup [14] was the first one to increase the exactness of the obvious QBF approximations by a series of more and more complex QBF formulations.

Balabanov et al. [3] show that resolution together with universal reduction, which is sound and complete for QBF, is no longer complete (but still sound) for DQBF, leading to another incomplete method for DQBF.

3.2 Search-Based Solution

A natural idea for DQBF solving is to generalize successful search-based QBF solvers like DepQBF [25,26]. The problem with QBF solvers applied to DQBF is that the solver assignments to existential variables depend on all universal variables assigned before. That means that an unmodified search-based QBF solver can only respect linearly ordered dependency sets. In [15], for a given DQBF linearly ordered dependency sets are computed that over-approximate the original dependency sets. A search-based QBF solver respecting those linearly ordered dependency sets now has to consider additional constraints: In different paths of the search tree that lead to the same existential variable y_i, but do not differ in the assignments to the variables in D_{y_i}, y_i has to be assigned the same value. In order to enforce those constraints, [15] extends the search-based QDPLL algorithm by learning additional clauses, called Skolem clauses, after assignments to existential variables y_i. The Skolem clauses just record an implication between the current assignments to the variables in D_{y_i} and the current assignment to y_i for transporting this information into other paths of the search tree. Backtracking in case of a conflict has to take these Skolem clauses into account and removes them once they become invalid (which is in contrast to conventional learned clauses that can remain after a conflict, since they are implied).

3.3   Abstraction-Based Solution

In the QBF context there are rather strong solvers like RAReQS [23] and Qesto [24] which work according to the CEGAR paradigm. Fröhlich et al. [16] use a similar idea by solving a series of SAT instantiations. Their solver is based on computing partial universal expansions, which yield over-approximations. If a SAT solver determines that the over-approximation is unsatisfiable, they can conclude the unsatisfiability of the DQBF. In case of satisfiability, they check if the satisfying assignment is consistent with the dependencies of the DQBF; if this is the case, the original DQBF is satisfied. Otherwise, the instantiation is refined using the inconsistent assignment, and the check is repeated. It can be shown that this process finally terminates.

3.4   Fork Resolution

In [32] information fork splitting is proposed as a basic concept for DQBF solving. Information forks are clauses C which can be split into two parts C_1 and C_2 that depend on incomparable dependency sets D_1 and D_2. After splitting C into C_1 and C_2, a fresh existential variable y_j depending on $D_1 \cap D_2$ is introduced, y_j is added to C_1, $\neg y_j$ to C_2. [32] proves that information fork splitting together with resolution and universal reduction forms a sound and complete proof calculus. The proof idea for the completeness is based on the observation that existential variables with incomparable dependency sets occurring in a single clause impede variable elimination by resolution. So information fork splitting is done before. To the best of our knowledge no practical implementation of a DQBF solver is available so far which uses the information fork splitting idea.

3.5   Basic Approach Using Universal Expansion

Universal Expansion for QBF. Already in the QBF context universal expansion has been used as a basic operation to remove universal quantifiers [2,6]. Universal expansion for QBF is defined as follows:

Definition 5 (Universal Expansion for QBF). *For a QBF $\psi = \forall X_1 \exists Y_1 \ldots$ $\forall X_k \exists Y_k : \varphi$, universal expansion of variable $x_i \in X_\ell$ ($1 \leq \ell \leq k$) is defined by*

$$\forall X_1 \exists Y_1 \ldots \forall X_{\ell-1} \exists Y_{\ell-1} \forall X_\ell \setminus \{x_i\} \exists Y_\ell' \forall X_{\ell+1} \exists Y_{\ell+1}' \ldots \forall X_k \exists Y_k' :$$

$$\varphi[0/x_i][y_j^0/y_j \text{ for all } y_j \in Y_h, h \geq \ell] \wedge \varphi[1/x_i][y_j^1/y_j \text{ for all } y_j \in Y_h, h \geq \ell].$$

with $Y_h' = \{y_j^0 \mid y_j \in Y_h\} \cup \{y_j^1 \mid y_j \in Y_h\}$.

Existential variables to the right of x_i (i.e., depending on x_i) result in two copies, all other existential variables are not copied.

Universal Expansion for DQBF. Universal expansion can be easily generalized to DQBF which has been observed, e.g., in [3,9,10,17].

Definition 6 (Universal Expansion for DQBF). *For a DQBF* $\psi = \forall x_1 \ldots$ $\forall x_n \exists y_1(D_{y_1}) \ldots \exists y_m(D_{y_m}) : \varphi$ *with* $Z_{x_i} = \{y_j \in V_\psi^\exists \mid x_i \in D_{y_j}\}$, *universal expansion of variable* $x_i \in V_\psi^\forall$ *is defined by*

$$\left(Q \setminus \left(\{x_i\} \cup \bigcup_{y_j \in Z_{x_i}} \{y_j\} \right) \right) \cup \{\exists y_j^b(D_{y_j} \setminus \{x_i\}) \mid y_j \in Z_{x_i}, b \in \{0,1\}\} :$$

$$\varphi[0/x_i][y_j^0/y_j \text{ for all } y_j \in Z_{x_i}] \wedge \varphi[1/x_i][y_j^1/y_j \text{ for all } y_j \in Z_{x_i}].$$

As for QBFs universal expansion leads to an equisatisfiable formula.

By universally expanding all universal variables both QBFs and DQBFs can be transformed into a SAT problem with a potential exponential increase in variables. Thus, complete universal expansion followed by SAT solving has a double exponential upper bound for the run-time. The upper bound is suboptimal for QBF which is just PSPACE-complete (and can be solved by a simple search-based algorithm in exponential time), whereas for DQBF (which is NEXPTIME-complete) it is unknown whether there is an algorithm with a better worst-case complexity.

3.6 Transformation into QBF

Due to the high cost of complete universal expansion, the solver HQS [18] eliminates universal variables only until a QBF is obtained, which can be solved by an arbitrary QBF solver then.

Transformation into QBF by Universal Expansion. The basic observation underlying the transformation of a DQBF into a QBF is the fact that a DQBF $\psi = \forall x_1 \ldots \forall x_n \exists y_1(D_{y_1}) \ldots \exists y_m(D_{y_m}) : \varphi$ can be written as a QBF if and only if the dependency sets are all comparable, i.e., iff for all $i, j \in \{1, \ldots m\}$ $D_{y_i} \subseteq D_{y_j}$ or $D_{y_j} \subseteq D_{y_i}$ ($\star$). If this condition ($\star$) holds, a linear order of the dependency sets w.r.t. $\subseteq$ can be computed. Such a linear order immediately results in the needed QBF prefix.

For each pair of existential variables y_i and y_j with incomparable dependency sets, either the universal variables in $D_{y_i} \setminus D_{y_j}$ or in $D_{y_j} \setminus D_{y_i}$ have to be eliminated. In [18] finding a minimum set $U \subseteq V^\forall$ of elimination variables leading to a QBF prefix is reduced to a MAXSAT problem. (For each universal variable x a variable m_x is created in the MAXSAT solver such that $m_x = 1$ means that x needs to be eliminated. Soft clauses are used to get an assignment with a minimum number of variables assigned to 1. Hard clauses enforce that for all $y_i, y_j \in V^\exists$, $y_i \neq y_j$, either all variables in $D_{y_i} \setminus D_{y_j}$ or in $D_{y_j} \setminus D_{y_i}$ are eliminated.)

Transformation into QBF by Dependency Elimination. In [37] the method of universal expansions turning a DQBF into a QBF is refined by so-called *dependency elimination*. Dependency elimination is able not only to remove universal variables x_i *completely* from the formula, but also to remove universal variables x_i from *single* dependency sets D_{y_j}, i.e., it works with a finer granularity. Dependency elimination is used with the goal of producing fewer copies of existential variables in the final QBF.

The basic transformation removing a universal variable x_i from a single dependency set D_{y_j} is based on the following theorem:

Theorem 1 (Dependency Elimination). *Assume ψ is a DQBF as in Definition 2 and, w.l.o.g., $x_1 \in D_{y_1}$. Then ψ is equisatisfiable to:*

$$\psi' := \forall x_1 \ldots \forall x_n \, \exists y_1^0 (D_{y_1} \setminus \{x_1\}) \, \exists y_1^1 (D_{y_1} \setminus \{x_1\}) \, \exists y_2(D_{y_2}) \ldots \exists y_m(D_{y_m}) :$$
$$\phi\Big[\big(((\neg x_1 \wedge y_1^0) \vee (x_1 \wedge y_1^1))/y_1\big)\Big].$$

The following example shows that dependency elimination is able to transform a DQBF into an equisatisfiable QBF with much fewer copies of existential variables than needed for universal expansion:

Example 1. The DQBF $\forall x_1 \forall x_2 \exists y_1(x_1) \exists y_2(x_2) \exists y_3(x_1, x_2) \ldots \exists y_n(x_1, x_2) : \varphi$ does not have an equivalent QBF prefix. Therefore the expansion of either x_1 or x_2 is necessary. When universal expansion of x_1 is performed, $y_1, y_3, \ldots, y_n$ are doubled, creating $n-1$ additional existential variables. The universal expansion of x_2 creates copies of $y_2, y_3, \ldots, y_n$.

However, only the dependencies of y_1 on x_1 and of y_2 on x_2 are responsible for the formula not being a QBF. If we eliminate the dependency of y_1 on x_1, e.g., we obtain the formula

$$\forall x_1 \forall x_2 \exists y_1^0(\emptyset) \exists y_1^1(\emptyset) \exists y_2(x_2) \exists y_3(x_1, x_2) \ldots \exists y_n(x_1, x_2) : \varphi[(\neg x_1 \wedge y_1^0) \vee (x_1 \wedge y_1^1)/y_1].$$

This formula can be written as the QBF

$$\exists y_1^0 \exists y_1^1 \forall x_2 \exists y_2 \forall x_1 \exists y_3 \ldots \exists y_n : \varphi[(\neg x_1 \wedge y_1^0) \vee (x_1 \wedge y_1^1)/y_1].$$

Instead of creating $n-1$ additional existential variables, we only have to double y_1 in order to obtain an equisatisfiable QBF.

In order to facilitate the selection of dependencies to eliminate, [37] makes use of the following dependency graph:

Definition 7 (Dependency Graph). *Let ψ be a DQBF as above. The dependency graph $G_\psi = (V_\psi, E_\psi)$ is a directed graph with the set $V_\psi = V$ of variables as nodes and edges*

$$E_\psi = \big\{(x, y) \in V_\psi^\forall \times V_\psi^\exists \,|\, x \in D_y\big\} \,\dot\cup\, \big\{(y, x) \in V_\psi^\exists \times V_\psi^\forall \,|\, x \notin D_y\big\}.$$

G_ψ is a so-called *bipartite tournament graph* [4,11,20]: The nodes can be partitioned into two disjoint sets according to their quantifier and there are only edges that connect variables with different quantifiers – this is the bipartiteness property. Additionally, for each pair $(x, y) \in V_\psi^\forall \times V_\psi^\exists$ there is either an edge from x to y or vice-versa – this property is referred to by the term 'tournament'.

Theorem 2 ([37]). *Let ψ be a DQBF and G_ψ its dependency graph. The graph G_ψ is acyclic iff ψ has an equivalent QBF prefix.*

Eliminating a dependency essentially corresponds to flipping the direction of an edge $(x, y) \in E_\psi \cap (V_\psi^\forall \times V_\psi^\exists)$ from a universal to an existential variable. Our goal is to find a cost-minimal set of edges such that flipping those edges makes the dependency graph acyclic.

The cost of a flipping set $R \subseteq E_\psi \cap (V_\psi^\forall \times V_\psi^\exists)$ corresponds to the number of existential variables in the formula after the dependencies in R have been eliminated. It is given by $\mathrm{cost}(R) := \sum_{y \in V_\psi^\exists} 2^{|R_y|}$. where for $y \in V_\psi^\exists$ $R_y = \{x \in V_\psi^\forall \mid (x, y) \in R\}$. In spite of this non-linear cost function, the computation of a cost-minimal flipping set can be reduced to *integer linear programming with dynamically added constraints* similar to the so-called cutting plane approach [40]. The efficiency of the optimal elimination set computation is significantly increased by integrating *symmetry reduction*. Symmetry reduction is based on the observation that in typical applications the number of different dependency sets is rather small.

Don't-Care Dependencies. Moreover, based on research on dependency schemes [38], we consider in our optimization also dependencies which can be removed 'free of charge' without dependency elimination, since their removal does not change the truth value of the DQBF.

3.7   The Role of Preprocessing

Part of the success of SAT and QBF solving is due to efficient preprocessing of the formula under consideration. The goal of preprocessing is to simplify the formula by reducing/modifying the number of variables, clauses and quantifier alternations, such that it can be solved more efficiently afterwards. For SAT and QBF, efficient and effective preprocessing tools are available like SatELite [13], Coprocessor [27] for SAT and squeezeBF [19], bloqqer [7] for QBF. In [36] we demonstrated that preprocessing is an essential step for DQBF solving as well. Standard preprocessing techniques were generalized and adapted to work with a DQBF solver core. Those techniques include

- using backbones, monotonic variables, and equivalent literals;
- reducing dependency sets based on dependency schemes [38];
- universal reduction, variable elimination by resolution, universal expansion as preprocessing steps;
- blocked clause elimination with hidden and covered literal addition; and

– structure extraction that leads to circuit representations of the matrix instead of a CNF representation.

An important observation made in [36] is that different preprocessing strategies are advisable depending on the DQBF solver core used (e.g., CNF-based vs. circuit-based solvers).

3.8 Computing Skolem Functions

Computing Skolem functions is important both for proof checking of satisfied DQBFs and for various applications such as the ones mentioned in the next section. In [35] we demonstrated how Skolem functions can be obtained from our DQBF solver HQS. The approach computes Skolem functions for the original formula, even taking all preprocessing steps from [36] into account. The computation of Skolem functions can be done with very little overhead compared to the mere solution of the formula.

4   Applications

Here we give three applications that can be formulated as DQBF problems in a natural way. For the first and the third one we can even prove that they are *equivalent* to DQBF which means that each DQBF problem can be translated into the corresponding problem class in polynomial time. Moreover, this means for those applications that they are NEXPTIME-complete as well [17,39].

In all applications mentioned below the translation into DQBF is based on an observation that has been summarized by Rabe [32] as 'DQBF can encode existential quantification over functions'.

4.1   Partial Equivalence Checking for Combinational Circuits

As a first application we look into partial equivalence checking for combinational circuits [33]. As a specification we consider a complete combinational circuit C_{spec}. As an implementation we consider an *incomplete* combinational circuit C_{impl} containing missing parts, so-called 'Black Boxes'. Missing parts may result from abstraction or they are not yet implemented so far. For each Black Box only the interface of the Black Boxes, i.e., their input and output signals, are known, their functionality is completely unknown. The Partial Equivalence Checking (PEC) problem answers the following question:

Definition 8 (Partial Equivalence Checking Problem (PEC)). *Given an incomplete circuit C_{impl} and a (complete) specification C_{spec}, are there implementations of the Black Boxes in C_{impl} such that C_{impl} and C_{spec} become equivalent (i.e., they implement the same Boolean function)?*

Assume that the specification C_{spec} implements a Boolean function $f_{\text{spec}}(x)$ with primary input variables x. For each Black Box BB_i the input signals are denoted by I_i, its output signals by O_i. Let us further assume that the Black Boxes can be sorted topologically (otherwise there are replacements leading to cycles in the combinational circuit), w.l.o.g. in the order $BB_1, \dots, BB_n$. Then the input cone computing the input signals I_i of BB_i represents a vector of Boolean functions $F_i(x, O_1, \dots, O_{i-1})$. The incomplete implementation C_{impl} implements a Boolean function $f_{\text{impl}}(x, O_1, \dots, O_n)$ depending on the primary inputs and the Black Box outputs.

The following DQBF is satisfied iff there is an appropriate implementation of the Black Boxes:

$$\forall x \forall I_1 \dots \forall I_n \exists O_1(I_1) \dots \exists O_m(I_m) : \left(\bigwedge_{i=1}^{n} I_i \equiv F_i \right) \Rightarrow \left(f_{\text{spec}} \equiv f_{\text{impl}} \right).$$

We have to ask that for all valuations of the primary inputs x and all input signals of the Black Boxes $I_1, \dots, I_n$ of the Black Boxes there is a choice for the output signals $O_1, \dots, O_n$ of the Black Boxes such that specification and implementation are equivalent, i.e., $f_{\text{spec}}(x) \equiv f_{\text{impl}}(x, O_1, \dots, O_n)$. However, this is only required for all valuations to the signals that are *consistent with the given circuit*, i.e., only if $\bigwedge_{i=1}^{n} I_i \equiv F_i(x, O_1, \dots, O_{i-1})$ holds. The requirement that the Black Box output signals are only allowed to depend on the Black Box input signals is simply expressed by the dependency sets I_i of the corresponding output signals O_i.

4.2   Controller Synthesis

In [8] SAT- and QBF-based techniques for controller synthesis of safety specifications are considered. A footnote in [8] gives a hint how a simple and elegant DQBF formulation can be used for that purpose as well.

In the controller synthesis problem a sequential circuit with a vector of present state bits s and a vector of next state bits s' is considered. The next state is computed by a transition function $\Lambda(s, x, c)$. Here x are uncontrollable primary inputs and c are controllable inputs which are computed by a controller on the basis of the present state bits and the uncontrollable primary inputs. We consider invariant properties $\text{inv}(s, x)$ which are required to hold at any time. The controller synthesis problem asks whether there is an implementation of the controller such that the resulting sequential circuit satisfies the invariant $\text{inv}(s, x)$ in all states that are reachable from the circuit's initial state(s), given as a predicate init.

The DQBF formulation of controller synthesis is based on the notion of a 'winning set'.

Definition 9. *Let S be the state set of the sequential circuit. A subset $W \subseteq S$ is a* winning set *if all states in W satisfy the invariant and, for all values of the primary inputs, the controller can ensure (by computing appropriate values for the controlled inputs) that the successor state is again in W.*

An appropriate controller can be found iff there is a winning set that includes the initial states of the sequential circuit. This can be formulated as a DQBF. To encode the winning sets, we introduce two existential variables w and w'; w depends on the current state and is supposed to be true for a state s if s is in the winning set. The variable w' depends on the next state variables s' and has the same Skolem function as w (but defined over s' instead of s). To ensure that w and w' have the same semantics the condition $(s \equiv s' \Rightarrow w \equiv w')$ is used. Using those two encodings of the winning set the controller synthesis problem is reduced to the following DQBF [8]:

$$\forall s \, \forall s' \forall x \, \exists w(s) \, \exists w'(s') \exists c(s, x) :$$
$$\big(\mathrm{init}(s) \Rightarrow w\big) \wedge \big(w \Rightarrow \mathrm{inv}(s, x)\big) \wedge \big(s \equiv s' \Rightarrow w \equiv w'\big) \wedge$$
$$\Big(\big(w \wedge (s' \equiv \Lambda(s, x, c))\big)\Big) \Rightarrow w'\Big). \quad (1)$$

The controlled input variables c are allowed to depend on the current state variables s and uncontrolled inputs x only. If the DQBF is satisfied, then the Skolem functions for c provide a suitable controller implementation. (Note that the solver HQS can compute Skolem functions with very little overhead compared to the mere solution of the formula [35].)

4.3   Realizability Checking for Sequential Circuits

The controller synthesis problem can be seen as a special sequential problem with the controller as a single Black Boxes having access to all state bits and all primary circuit inputs. Here we look into a generalization where sequential circuits may contain an arbitrary number of Black Boxes and the exact interface of the Black Boxes, i.e., the signals entering and leaving the Black Boxes, is strictly taken into account [39]. That means that Black Boxes are *not* necessarily able to read all primary inputs and state bits. We confine ourselves to combinational Black Boxes or Black Boxes with bounded memory. The even more general problem considering distributed architectures containing several Black Boxes with *unbounded* memory is undecidable [31].

Black Boxes with *bounded* memory can be reduced to combinational Black Boxes, simply by extracting the memory elements out of the Black Box into the known part of the circuit, such that the incoming and outgoing signals of these memory elements are written and read only by the Black Boxes. Thus, we assume w.l.o.g. sequential circuits with arbitrary combinational Black Boxes in the circuit implementing their transition function.

As in Sect. 4.1, we assume n Black Boxes $\mathrm{BB}_1, \ldots, \mathrm{BB}_n$ with input signals I_i and output signals O_i, respectively. Again, the input cone computing the input signals I_i of BB_i represents a vector of Boolean functions $F_i(x, O_1, \ldots, O_{i-1})$. The transition function depending on the current state variables s, the primary inputs x and the Black Box outputs $O_1, \ldots, O_n$ is given by $\Lambda(s, x, O_1, \ldots, O_n)$. As before, the transition function computes new valuations to the next state variables s'.

We investigate the following problem:

Definition 10. *The* realizability problem for incomplete sequential circuits (RISC) *is defined as follows: Given an incomplete sequential circuit with multiple combinational (or bounded-memory) Black Boxes and an invariant property, are there implementations of the Black Boxes such that in the complete circuit the invariant holds at all times?*

In order to formulate the realizability problem as a DQBF problem, we slightly modify Definition 9 into:

Definition 11. *A subset $W \subseteq S$ is a* winning set *if all states in W satisfy the invariant and, for all values of the primary inputs, the Black Boxes can ensure (by computing appropriate values) that the successor state is again in W.*

Similarly to the controller synthesis problem, a given RISC is realizable iff there is a winning set that includes the initial states of the circuit. This leads us to the following theorem (using the same encoding of the winning set by existential variables w and w' depending on the current state variables s and next state variables s', respectively):

Theorem 3. *Given a RISC as defined above, the following DQBF is satisfied if and only if the RISC is realizable:*

$$\forall s \, \forall s' \forall x \, \forall I_1 \ldots \forall I_n \; \exists w(s) \; \exists w'(s') \; \exists O_1(I_1) \ldots \exists O_n(I_n):$$

$$(\text{init}(s) \Rightarrow w) \wedge (w \Rightarrow \text{inv}(s, x)) \wedge (s \equiv s' \Rightarrow w \equiv w') \wedge$$

$$\left(\left(w \wedge \left[\bigwedge_{i=1}^{n} I_i \equiv F_i(x, O_1, \ldots, O_{i-1}) \right] \wedge (s' \equiv \Lambda(s, x, O_1, \ldots, O_n)) \right) \Rightarrow w' \right).$$

The main difference to (1) consists in the following fact: The requirement that the successor state of a winning state is again a winning state obtains an additional precondition (similar to the DQBF for PEC in Sect. 4.1). The requirement is only needed for signal assignments that are completely consistent with the circuit functionality, i.e., only if the Black Box inputs are assigned consistently with the values computed by their input cones and, of course, the next state variables s' are assigned in accordance with the transition function Λ.

The Black Box outputs O_i of Black Box BB_i are only allowed to depend on the Black Box inputs I_i and, if the DQBF is satisfied, the Skolem functions for O_i provide an appropriate implementation for BB_i.

5   Conclusion and Future Challenges

Dependency quantified Boolean formulas are a powerful formalism for a natural and compact description of various problems. In this paper, we provided an overview of several solution methods for DQBFs.

In the future, the scalability of the solvers has to be further improved and they might be tuned towards specific applications. Further optimizing the single solution methods as well as combining advantages of different solution strategies seems to be an interesting and rewarding task. This should be combined with more powerful preprocessing techniques as well. Moreover, it will be interesting in the future to look into sound but incomplete approximations both disproving *and* proving the satisfiability of DQBFs.

We hope that with the availability of solvers more applications of these techniques will become feasible or will be newly discovered, thereby inspiring further improvements of the solvers – just as it is/was the case for propositional SAT solving and for QBF solving.

Acknowledgment. We are grateful to Bernd Becker, Ruben Becker, Andreas Karrenbauer, Jennifer Nist, Sven Reimer, Matthias Sauer, and Karina Wimmer for heavily contributing to the contents summarized in this paper.

References

1. Abramovici, M., Breuer, M.A., Friedman, A.D.: Digital Systems Testing and Testable Design. Computer Science Press, New York (1990)
2. Ayari, A., Basin, D.: QUBOS: deciding quantified Boolean logic using propositional satisfiability solvers. In: Aagaard, M.D., O'Leary, J.W. (eds.) FMCAD 2002. LNCS, vol. 2517, pp. 187–201. Springer, Heidelberg (2002). https://doi.org/10.1007/3-540-36126-X_12
3. Balabanov, V., Chiang, H.K., Jiang, J.R.: Henkin quantifiers and Boolean formulae: a certification perspective of DQBF. Theoret. Comput. Sci. **523**, 86–100 (2014). https://doi.org/10.1016/j.tcs.2013.12.020
4. Beineke, L.W., Little, C.H.C.: Cycles in bipartite tournaments. J. Comb. Theory Ser. B **32**(2), 140–145 (1982). https://doi.org/10.1016/0095-8956(82)90029-6
5. Beyersdorff, O., Chew, L., Schmidt, R.A., Suda, M.: Lifting QBF resolution calculi to DQBF. In: Creignou, N., Le Berre, D. (eds.) SAT 2016. LNCS, vol. 9710, pp. 490–499. Springer, Cham (2016). https://doi.org/10.1007/978-3-319-40970-2_30
6. Biere, A.: Resolve and expand. In: Hoos, H.H., Mitchell, D.G. (eds.) SAT 2004. LNCS, vol. 3542, pp. 59–70. Springer, Heidelberg (2005). https://doi.org/10.1007/11527695_5
7. Biere, A., Lonsing, F., Seidl, M.: Blocked clause elimination for QBF. In: Bjørner, N., Sofronie-Stokkermans, V. (eds.) CADE 2011. LNCS (LNAI), vol. 6803, pp. 101–115. Springer, Heidelberg (2011). https://doi.org/10.1007/978-3-642-22438-6_10
8. Bloem, R., Könighofer, R., Seidl, M.: SAT-based synthesis methods for safety specs. In: McMillan, K.L., Rival, X. (eds.) VMCAI 2014. LNCS, vol. 8318, pp. 1–20. Springer, Heidelberg (2014). https://doi.org/10.1007/978-3-642-54013-4_1
9. Bubeck, U.: Model-based transformations for quantified Boolean formulas. Ph.D. thesis, University of Paderborn, Germany (2010)
10. Bubeck, U., Büning, H.K.: Dependency quantified horn formulas: models and complexity. In: Biere, A., Gomes, C.P. (eds.) SAT 2006. LNCS, vol. 4121, pp. 198–211. Springer, Heidelberg (2006). https://doi.org/10.1007/11814948_21
11. Cai, M., Deng, X., Zang, W.: A min-max theorem on feedback vertex sets. Math. Oper. Res. **27**(2), 361–371 (2002). https://doi.org/10.1287/moor.27.2.361.328
12. Chatterjee, K., Henzinger, T.A., Otop, J., Pavlogiannis, A.: Distributed synthesis for LTL fragments. In: FMCAD 2013, pp. 18–25. IEEE, October 2013. https://doi.org/10.1109/FMCAD.2013.6679386

13. Eén, N., Biere, A.: Effective preprocessing in SAT through variable and clause elimination. In: Bacchus, F., Walsh, T. (eds.) SAT 2005. LNCS, vol. 3569, pp. 61–75. Springer, Heidelberg (2005). https://doi.org/10.1007/11499107_5

14. Finkbeiner, B., Tentrup, L.: Fast DQBF refutation. In: Sinz, C., Egly, U. (eds.) SAT 2014. LNCS, vol. 8561, pp. 243–251. Springer, Cham (2014). https://doi.org/10.1007/978-3-319-09284-3_19

15. Fröhlich, A., Kovásznai, G., Biere, A.: A DPLL algorithm for solving DQBF. In: International Workshop on Pragmatics of SAT (POS) 2012, Trento, Italy (2012)

16. Fröhlich, A., Kovásznai, G., Biere, A., Veith, H.: iDQ: instantiation-based DQBF solving. In: Le Berre, D. (ed.) International Workshop on Pragmatics of SAT (POS) 2014. EPiC Series, vol. 27, pp. 103–116. EasyChair, Vienna (2014)

17. Gitina, K., Reimer, S., Sauer, M., Wimmer, R., Scholl, C., Becker, B.: Equivalence checking of partial designs using dependency quantified Boolean formulae. In: ICCD 2013, Asheville, NC, USA, pp. 396–403. IEEE CS, October 2013. https://doi.org/10.1109/ICCD.2013.6657071

18. Gitina, K., Wimmer, R., Reimer, S., Sauer, M., Scholl, C., Becker, B.: Solving DQBF through quantifier elimination. In: DATE 2015, Grenoble, France. IEEE, March 2015. https://doi.org/10.7873/date.2015.0098

19. Giunchiglia, E., Marin, P., Narizzano, M.: sQueezeBF: an effective preprocessor for QBFs based on equivalence reasoning. In: Strichman, O., Szeider, S. (eds.) SAT 2010. LNCS, vol. 6175, pp. 85–98. Springer, Heidelberg (2010). https://doi.org/10.1007/978-3-642-14186-7_9

20. Guo, J., Hüffner, F., Moser, H.: Feedback arc set in bipartite tournaments is NP-complete. Inf. Process. Lett. **102**(2–3), 62–65 (2007). https://doi.org/10.1016/j.ipl.2006.11.016

21. Henkin, L.: Some remarks on infinitely long formulas. In: Infinitistic Methods: Proceedings of the 1959 Symposium on Foundations of Mathematics, Warsaw, Panstwowe, pp. 167–183. Pergamon Press, September 1961

22. Jain, A., Boppana, V., Mukherjee, R., Jain, J., Fujita, M., Hsiao, M.S.: Testing, verification, and diagnosis in the presence of unknowns. In: IEEE VLSI Test Symposium (VTS) 2000, Montreal, Canada, pp. 263–270. IEEE Computer Society (2000). https://doi.org/10.1109/VTEST.2000.843854

23. Janota, M., Klieber, W., Marques-Silva, J., Clarke, E.: Solving QBF with counterexample guided refinement. In: Cimatti, A., Sebastiani, R. (eds.) SAT 2012. LNCS, vol. 7317, pp. 114–128. Springer, Heidelberg (2012). https://doi.org/10.1007/978-3-642-31612-8_10

24. Janota, M., Marques-Silva, J.: Solving QBF by clause selection. In: Yang, Q., Wooldridge, M. (eds.) IJCAI 2015, Buenos Aires, Argentina, pp. 325–331. AAAI Press (2015). http://ijcai.org/Abstract/15/052

25. Lonsing, F., Biere, A.: DepQBF: a dependency-aware QBF solver. J. Satisf. Boolean Model. Comput. **7**(2–3), 71–76 (2010)

26. Lonsing, F., Egly, U.: Incremental QBF solving by DepQBF. In: Hong, H., Yap, C. (eds.) ICMS 2014. LNCS, vol. 8592, pp. 307–314. Springer, Heidelberg (2014). https://doi.org/10.1007/978-3-662-44199-2_48

27. Manthey, N.: Coprocessor 2.0 – a flexible CNF simplifier. In: Cimatti, A., Sebastiani, R. (eds.) SAT 2012. LNCS, vol. 7317, pp. 436–441. Springer, Heidelberg (2012). https://doi.org/10.1007/978-3-642-31612-8_34

28. Meyer, A.R., Stockmeyer, L.J.: Word problems requiring exponential time: preliminary report. In: STOC 1973, pp. 1–9. ACM Press (1973). https://doi.org/10.1145/800125.804029

29. Peterson, G., Reif, J., Azhar, S.: Lower bounds for multiplayer non-cooperative games of incomplete information. Comput. Math. Appl. **41**(7–8), 957–992 (2001). https://doi.org/10.1016/S0898-1221(00)00333-3
30. Peterson, G.L., Reif, J.H.: Multiple-person alternation. In: Annual Symposium on Foundations of Computer Science (FOCS), San Juan, Puerto Rico, pp. 348–363. IEEE Computer Society, October 1979. https://doi.org/10.1109/SFCS.1979.25
31. Pnueli, A., Rosner, R.: Distributed reactive systems are hard to synthesize. In: Annual Symposium on Foundations of Computer Science 1990, St. Louis, Missouri, USA, pp. 746–757. IEEE Computer Society, October 1990. https://doi.org/10.1109/FSCS.1990.89597
32. Rabe, M.N.: A resolution-style proof system for DQBF. In: Gaspers, S., Walsh, T. (eds.) SAT 2017. LNCS, vol. 10491, pp. 314–325. Springer, Cham (2017). https://doi.org/10.1007/978-3-319-66263-3_20
33. Scholl, C., Becker, B.: Checking equivalence for partial implementations. In: DAC 2001, Las Vegas, NV, USA, pp. 238–243. ACM Press, June 2001. https://doi.org/10.1145/378239.378471
34. Silva, J.P.M., Sakallah, K.A.: GRASP: a search algorithm for propositional satisfiability. IEEE Trans. Comput. **48**(5), 506–521 (1999). https://doi.org/10.1109/12.769433
35. Wimmer, K., Wimmer, R., Scholl, C., Becker, B.: Skolem functions for DQBF. In: Artho, C., Legay, A., Peled, D. (eds.) ATVA 2016. LNCS, vol. 9938, pp. 395–411. Springer, Cham (2016). https://doi.org/10.1007/978-3-319-46520-3_25
36. Wimmer, R., Gitina, K., Nist, J., Scholl, C., Becker, B.: Preprocessing for DQBF. In: Heule, M., Weaver, S. (eds.) SAT 2015. LNCS, vol. 9340, pp. 173–190. Springer, Cham (2015). https://doi.org/10.1007/978-3-319-24318-4_13
37. Wimmer, R., Karrenbauer, A., Becker, R., Scholl, C., Becker, B.: From DQBF to QBF by dependency elimination. In: Gaspers, S., Walsh, T. (eds.) SAT 2017. LNCS, vol. 10491, pp. 326–343. Springer, Cham (2017). https://doi.org/10.1007/978-3-319-66263-3_21
38. Wimmer, R., Scholl, C., Wimmer, K., Becker, B.: Dependency schemes for DQBF. In: Creignou, N., Le Berre, D. (eds.) SAT 2016. LNCS, vol. 9710, pp. 473–489. Springer, Cham (2016). https://doi.org/10.1007/978-3-319-40970-2_29
39. Wimmer, R., Wimmer, K., Scholl, C., Becker, B.: Analysis of incomplete circuits using dependency quantified Boolean formulas. In: Reis, A.I., Drechsler, R. (eds.) Advanced Logic Synthesis, pp. 151–168. Springer, Cham (2018). https://doi.org/10.1007/978-3-319-67295-3_7
40. Wolsey, L.A.: Integer Programming. Wiley-Interscience, New York (1998)

Maximum Satisfiability

Approximately Propagation Complete and Conflict Propagating Constraint Encodings

Rüdiger Ehlers[(✉)] and Francisco Palau Romero

University of Bremen, Bremen, Germany
ruediger.ehlers@uni-bremen.de

Abstract. The effective use of satisfiability (SAT) solvers requires problem encodings that make good use of the reasoning techniques employed in such solvers, such as unit propagation and clause learning. Propagation completeness has been proposed as a useful property for constraint encodings as it maximizes the utility of unit propagation. Experimental results on using encodings with this property in the context of satisfiability modulo theory (SMT) solving have however remained inconclusive, as such encodings are typically very large, which increases the bookkeeping work of solvers.

In this paper, we introduce approximate propagation completeness and approximate conflict propagation as novel SAT encoding property notions. While approximate propagation completeness is a generalization of classical propagation completeness, (approximate) conflict propagation is a new concept for reasoning about how early conflicts can be detected by a SAT solver. Both notions together span a hierarchy of encoding quality choices, with classical propagation completeness as a special case. We show how to compute approximately propagation complete and conflict propagating constraint encodings with a minimal number of clauses using a reduction to MaxSAT. To evaluate the effect of such encodings, we give results on applying them in a case study.

1 Introduction

Satisfiability (SAT) solvers have become an important tool for the solution of NP-hard practical problems. In order to utilize them, the practical problem to be solved needs to be encoded as a satisfiability problem instance, which is then passed to an off-the-shelf SAT solver. The way in which this encoding is performed has a huge influence on the solver computation times. Hence, the effective use of SAT solvers requires encodings that keep the workload of the solvers as small as possible. Capturing how an encoding needs to look like to have this property is however not a simple task. While it is commonly agreed on that problem-specific knowledge should be made use of, only few general guidelines for efficient encodings are known [7].

A good encoding should keep the numbers of variables *and* the number of clauses as small as possible, while allowing the solver to make most use of clause

© Springer International Publishing AG, part of Springer Nature 2018
O. Beyersdorff and C. M. Wintersteiger (Eds.): SAT 2018, LNCS 10929, pp. 19–36, 2018.
https://doi.org/10.1007/978-3-319-94144-8_2

learning and unit propagation, which are reasoning steps that are (part of) the foundation of modern *CDCL-style* SAT solving [14]. While the effect of clause learning depends on the variable selection scheme for branching employed by the solver, and hence is hard to predict, how an encoding makes most use of unit propagation is better studied. For instance, the class of *cardinality constraints* has many known encodings [1,3], and it is frequently suggested that encodings that are *propagation complete* should be preferred [7,13]. An encoding of a constraint is propagation complete (also known under the name *generalized arc-consistent* [13]) if every literal implied by some partial valuation and the encoded constraint is detected by the unit propagation mechanism of the SAT solver. A constraint encoding with this property reduces the number of times in which costly backtracking is performed until a satisfying assignment is found or the SAT instance is found to be unsatisfiable.

Propagation completeness is of interest for all types of constraints that appear in practically relevant SAT problems, so having an automated way to make the encoding of a constraint type that appears as a *building block* in real-world problems propagation complete is likely to be useful. Brain et al. [9] presented an approach to rewrite SAT encodings to make them propagation complete. They apply their approach to multiple building blocks commonly found in problems from the field of formal verification and modify the SAT-based satisfiability modulo theory (SMT) solver CVC4 to use the computed encodings for bitvector arithmetic operations. Their experiments show that the change increased solver performance somewhat, but made limited overall difference.

This result is surprising. If propagation complete encodings enable a SAT solver to make most use of unit propagation and reduce the number of conflicts during the solution process, then SAT solving times should decrease when using such encodings. A contributing factor to this lack of substantial improvement is that propagation complete encodings are often much larger than minimally-sized encodings. As an example, a three-bit multiplier (with five output bits) can be encoded with 45 clauses, but a propagation complete encoding needs at least 304 clauses. As a consequence, the bookkeeping effort of the solver is higher for propagation complete encodings, which reduces solving efficiency [17]. This observation gives rise to the question if there is a way to balance encoding size and the *propagation quality* of an encoding to get some of the benefits of propagation complete encodings but still keep the burden to the solver by the additional clauses low.

In this paper, we present such an approach to balance the propagation quality of a constraint encoding into conjunctive normal form (CNF) and its size. We define the novel notions of *approximate propagation completeness* and *approximate conflict propagation*. The former is a generalization of propagation completeness, and we say that a CNF formula ψ is approximately propagation complete for a quality level of $c \in \mathbb{N}$ if for every partial valuation to the variables in ψ that can be completed to a satisfying assignment and that implies at least c other variable values, one of them need to be derivable from ψ by unit propagation. Approximate conflict propagation is concerned with how early conflicts are detected. We say that a CNF constraint encoding ψ is approximately conflict propagating with a quality level of $c \in \mathbb{N}$ if every partial valuation that cannot be completed to one that satisfies ψ and for which the values of at most c variables are not set in the partial valuation leads to unit propagation (or induces a conflict) in ψ.

Approximate propagation completeness and approximate conflict propagation both target making the most use of the unit propagation capabilities of solvers. While approximate propagation completeness deals with *satisfiable* partial valuations, i.e., those that can be extended to satisfying assignments, approximate conflict propagation deals with *unsatisfiable* partial valuations. Together these concepts allow to reason about the *propagation quality* of CNF encodings in a relatively fine-grained way.

To evaluate the two new concepts, we present an approach to compute approximately propagation complete and approximately conflict propagating encodings with a minimal number of clauses. The approach starts from a representation of the constraint to be encoded as a binary decision diagram (BDD) and enumerates all shortest clauses implied by the BDD. Every minimal CNF encoding consists of a subset of these clauses. We then compute clause selection requirements for the solution based on the desired propagation quality levels. The resulting requirement set is then processed by a (partial) MaxSAT solver [20] to find a smallest encoding. The approach supports finding minimal propagation complete encodings and minimal arbitrary CNF encodings as special cases.

We apply the approach to a wide variety of constraints, including the ones already used by Brain et al. [9]. We show that their approach can sometimes produce encodings with a clause cardinality that is higher than necessary, and that for a good number of constraints, the various propagation quality level combinations for our new propagation quality notions give rise to many different (minimal) encodings with vastly different sizes. Our approach is also very competitive in terms of computation time when using the MaxSAT solver LMHS [25] in combination with the integer linear programming (ILP) solver CPLEX as backend. To gain some intuition on how efficient the SAT solving process with the new encodings is, we compare them on some *integer factoring* benchmarks.

1.1   Related Work

Brain et al. [9] introduced abstract satisfaction as a theoretical foundation to reason about propagation strength of encodings. They use a modified SAT solver to generate propagation complete encodings and then minimize their sizes by removing redundant clauses using a procedure proposed by Bordeaux and Marques-Silva [8]. As we show in Sect. 5, this approach does not guarantee a minimal number of clauses (but guarantees that no clause from the encoding can be removed without making it propagation incomplete or incorrect), whereas the new algorithm given in Sect. 4 does. Brain et al. also give a variant of their approach in which auxiliary variables are added to the SAT instance, which can substantially reduce the encoding size. This makes the encoding propagation incomplete, however, except when assuming that the SAT solver never branches on the auxiliary variables.

Inala et al. [17] used *syntax-guided program synthesis* to automatically compute propagation complete encodings that improve the efficiency of SMT-based verification. In contrast to the work by Brain et al., their approach synthesizes code to generate encodings rather than computing the encodings directly. The code can then be used for multiple concretizations of the constraint type (e.g., for varying bit vector widths when encoding an addition operation).

Bordeaux and Marques-Silva [8] already solved the problem of generating propagation complete encodings earlier than the aforementioned works. They show that when starting from a CNF encoding that should be made propagation complete, by restricting the search for clauses to be added to so-called *empowering clauses*, the computation time can be substantially reduced. However, their approach requires the use of a quantified Boolean formula (QBF) solver, whereas the later approach by Brain et al. [9] only requires a modified SAT solver. Bordeaux and Marques-Silva also showed that there are constraint classes that require exponentially-sized propagation complete encodings, which further motivates the study of approximate versions of this notion in the present work.

Gwynne and Kullmann [15,16] define two hierarchies for specifying how easy a SAT solver can detect implied literals and conflicts, which we also do in this paper. Their notions are however based on how often a solver has to branch in the optimal case to detect implied literals or conflicts. In contrast, our notions base on how *many* variable values are implied (for approximate propagation completeness) and how many variables do not have values assigned in a partial valuation (for approximate conflict propagation). Hence, our encoded constraints do not rely on the solver to branch on the right variables. This also allows us to automate the process of finding encodings with a minimal number of clauses for a wide variety of constraints, while the experimental evaluation of their work [15] focused on few cases for which encodings in the levels of their hierarchies were available.

Minimal propagation complete CNF encodings typically have more clauses than minimal arbitrary CNF encodings. Many of the additional clauses can also be found automatically by SAT solvers that perform *preprocessing* or *inprocessing* [18] through techniques such as variants of hyper resolution [5]. Due to the high computational cost incurred by them, they are typically only used to a limited extent. Some clauses that are important for approximate propagation completeness and conflict propagation can therefore not be found or are found very late by these techniques. Furthermore, our approach computes minimal encodings from scratch, which be structured in completely different ways that expert-made encodings.

Manthey et al. [22] proposed a technique which can be used for inprocessing or preprocessing and that is based on introducing auxiliary variables that can make the CNF encoding smaller (where they count the number of clauses plus number of variables). They show that SAT solving performance improves on many benchmarks with their approach. While the introduction of additional variables could in principle break the propagation completeness, their approach does not suffer from this problem as they only undo clause distribution in a way that one of the variable phases occurs in two-literal clauses only. Their positive experimental results therefore do not give insight into the practical importance of using propagation complete encodings.

Babka et al. [4] study the theoretical properties of the problem of making a CNF encoding propagation complete. They identify the complexity classes of the different variants of this problem.

Kučera et al. [19] give lower bounds on the minimal number of clauses needed to encode so-called *"at most one"* or *"exactly one"* constraints. In contrast to the work in this paper, their result generalizes to an arbitrary size, also holds for the case that auxiliary variables are used, but is restricted to this particular constraint type.

2 Preliminaries

Given a set of variables $\mathcal{V}$ and Boolean formula ψ over $\mathcal{V}$, the *satisfiability* (SAT) problem is to find a satisfying assignment to the formula if one exists, or to deduce that no such assignment exists, in which case the formula is called *unsatisfiable*. Boolean formulas to be checked for satisfiability are also called *SAT instances* and are assumed to be given in *conjunctive normal form* (CNF) in the following. Such instances are conjunctions of *clauses*, which are in turn disjunctions of *literals*, i.e., from $\mathcal{L}(\mathcal{V}) = \mathcal{V} \cup \{\neg v \mid v \in \mathcal{V}\}$.

A *search-based SAT solver* maintains and manipulates a *partial valuation* to $\mathcal{V}$. A partial valuation $p : \mathcal{V} \rightharpoonup \{\textbf{false}, \textbf{true}\}$ is a partial function from the variables to **false** and **true**. We say that p is consistent with some other partial valuation p' if $p(v) = p'(v)$ holds for all variables v in the domain of p'. A *completion* of p is a full assignment to $\mathcal{V}$ that is consistent with p. We say that p satisfies some literal l over a variable v if $p(v)$ is defined and p can be completed to a full valuation that satisfies l. Likewise, p *falsifies* some literal l if $p(v)$ is defined and no completion of p satisfies l. We say that p' *implies* a literal $l \in \mathcal{L}(\mathcal{V})$ if every completion of p' that satisfies ψ also satisfies l. With a slight abuse of terminology, for some fixed set of clauses, we say that a partial valuation is *satisfiable* if it can be extended to a satisfying assignment, and it is *unsatisfiable* otherwise. We say that a clause c subsumes another clause c' if the literals of c are a subset of the literals of c'. If a CNF formula has two clauses c and c' such that c subsumes c', then c' can be removed without changing the encoded constraint.

During the search process, partial valuations $p : \mathcal{V} \rightharpoonup \{\textbf{false}, \textbf{true}\}$ are extended by the solver (1) by performing *decisions*, where the domain of p is extended by one variable, and (2) by *unit propagation*, where for some clause $l_1 \vee \ldots \vee l_n$ in the formula, there exists an $i \in \{1, \ldots, n\}$ such that the variable of l_i is not in the domain of p, but p falsifies all literals l_j with $i \neq j$.

Given some (sub-)set of variables $\mathcal{V}'$, a *constraint* over $\mathcal{V}'$ is a subset of valuations to $\mathcal{V}'$ that models the satisfaction of the constraint. A CNF encoding

ψ of such a constraint is a set of clauses over $\mathcal{V}'$ that are together satisfied by exactly the valuations in the subset to be encoded. A CNF encoding of a constraint over a variable set $\mathcal{V}'$ is *propagation complete* if for every partial valuation p to $\mathcal{V}'$ and every literal $l \in \mathcal{L}(\mathcal{V}')$, if p implies l, then there exists a clause $l_1 \vee \ldots \vee l_n$ in the encoding for which all literals in the clause except for at most one are falsified by p.

Example 1. As an example, we consider the following CNF encoding:

$$(x_1 \vee x_2) \wedge (\neg x_1 \vee \neg x_2) \wedge (\neg x_2 \vee x_3) \wedge (x_2 \vee \neg x_3) \wedge (x_4 \vee x_2 \vee x_3)$$

This constraint encoding is not propagation complete, as the partial valuation $p = \{x_4 \mapsto \textbf{false}\}$ does not give rise to unit propagation, but in every satisfying valuation, x_1 needs to have a **false** value if x_4 has a **false** value. Shortening the last clause to $(x_2 \vee x_4)$ does not change the set of satisfying valuations and makes it propagation complete. The new clause enables the SAT solver to extend the partial valuation p to $p' = \{x_4 \mapsto \textbf{false}, x_2 \mapsto \textbf{true}\}$ by unit propagation, from where unit propagation can then deduce that x_1 must have a **false** value. Adding the clause $(\neg v_1 \vee x_4)$ instead of changing the clause $(v_4 \vee x_2 \vee x_3)$ would also make the encoding propagation complete.

Partial Maximum Satisfiability (MaxSAT) solvers take a CNF formula in which some clauses are *soft*. The solver searches for a variable assignment that satisfies all the remaining *hard* clauses and maximizes the number of soft clauses that are satisfied by the assignment.

   Binary decision diagrams (BDDs) [10] are compact representations of Boolean formulas over some set of Boolean variables. They are internally represented as directed acyclic graphs and every path through the BDD to a designated **true** node represents one or multiple satisfying assignments of the formula. We will not need their details in the following, and refer to [11] for a well-accessible introduction. BDDs support the usual Boolean operators such as disjuction, complementation, and universal or existential abstraction of a variable. For instance, given a BDD F and a variable v, computing $\exists v.F$ yields a BDD that maps all valuations to the variables to **true** for which the value for v can be set to either **false** or **true** such that the resulting valuation is a model of F.

3   Approximate Propagation Completeness and Conflict Propagation

Propagation complete encodings enable search-based SAT solvers to deduce implied literals by unit propagation. By definition, *every* partial valuation that implies a literal over a variable that does not have a defined value in the partial valuation must give rise to unit propagation by the solver. We want to weaken this requirement in a way that enables SAT practitioners to better balance propagation quality and the encoding size of a constraint. This is done in two ways:

- We separate the consideration of satisfiable and unsatisfiable partial valuations, and
- we relax the requirement that every partial valuation that could give rise to unit propagation should do so, but rather that only partial valuations that enable the solver to make much progress need to so do.

These ideas are implemented in the following two *propagation quality notions* for CNF encodings:

Definition 1 (Approximate Propagation Completeness). *Given a CNF encoding ψ over some set of variables $\mathcal{V}$ and some constant $n \in \mathbb{N}$, we say that ψ is* approximately propagation complete *with a quality level of n if for all satisfiable partial valuations $p : \mathcal{V} \rightharpoonup \{\text{false}, \text{true}\}$ for which n different literals are implied by p and ψ and for which the variables for these literals are not in the domain of p, at least one of them can be derived from p and ψ by unit propagation.*

Definition 2 (Approximate Conflict Propagation). *Given a CNF encoding ψ over some set of variables $\mathcal{V}$ and some constant $n \in \mathbb{N}$, we say that ψ is* approximately conflict propagating *with a quality level of n if for all unsatisfiable partial valuations $p \in \mathcal{V} \rightharpoonup \{\text{false}, \text{true}\}$ for which p is defined on at least $|\mathcal{V}| - n$ variables, there exists a clause in ψ all of whose literals except for at most one are falsified by p.*

In both definitions, we only care about situations in which at least *one* clause should lead to unit propagation (or a conflict in case of approximate conflict propagation). The definition however induces stronger propagation quality requirements through repeated application. If, for instance, ψ is an approximately propagation complete encoding with a quality level of 2 and p is a partial valuation that implies four new literals, then at least one clause needs to lead to the derivation of an extended partial valuation p' by unit propagation. As p' then still implies three literals, by the fact that ψ has an approximate propagation completeness quality level of 2, another clause must give rise to unit propagation. By this line of reasoning, an *n-approximate propagation complete encoding* can never leave more than $n - 1$ implied literals of a partial valuation undetected by unit propagation.

The quality level for approximate conflict propagation states how early conflicts induced by a constraint need to be detected by the unit propagation capabilities of the solver. In contrast to approximate propagation completeness, higher values are better as they mean that more variables can be unassigned in a partial valuation that already violates the constraint and where unit propagation should lead to an extension of the partial valuation or the detection of the conflict.

The requirement that p needs to be defined on at least $|\mathcal{V}| - n$ variables for a partial valuation to be of interest in the approximate conflict propagation definition could also be replaced by considering all partial valuations for which at *most* $|\mathcal{V}| - n$ variables are undefined. This definition would also make sense and requires that unsatisfiable partial valuations in which few variables have values assigned need to give rise to unit propagation until at most n variables are left unassigned. However, this definition would not ensure that conflicts are actually detected by the solver in such a case, while our definition does, which we find more natural. To see that our definition indeed ensures this, note that if a partial valuation p meets the requirements given in the definition, then either a conflict is detected or unit propagation should be able to extend p to a partial valuation p' that is defined on more variables and that still cannot be extended to a satisfying assignment. Hence, the definition can be applied again (repeatedly), and eventually a conflict is found by the solver. Thus, the "at least" in Definition 2 is reasonable, even though in this way, higher quality levels for approximate conflict propagation are better, whereas for approximate propagation completeness, lower numbers are better.

While these definitions consider satisfiable and unsatisfiable partial valuations separately, there is still a connection between them:

Proposition 1. *If a CNF encoding ψ over some set of variables $\mathcal{V}$ is approximately propagation complete with a quality level of 1, then it is (approximately) conflict propagating with a quality level of $|\mathcal{V}|$.*

Proof. Let p be an unsatisfiable partial valuation. We can transform p to an unsatisfiable partial valuation p' to which p is consistent and for which removing any variable of the domain of p' would make it satisfiable. This transformation only requires removing variables from the domain of p until no more variables can be removed without making it satisfiable. Let us now remove an arbitrary variable v' from the domain of p' and let the resulting valuation be called p''. We have that p'' implies the literal $l = \neg v'$ if $p(v') = \textbf{true}$ or $l = v'$ if $p(v') = \textbf{false}$. Since ψ is approximately propagation complete with a quality level of 1, l needs to be deduced by unit propagation from p''. The last clause used in the propagation has, by the definition of unit propagation, all literals instead of l falsified for the partial valuation p''. Since l is in conflict with p' by construction, we have $p''(v) = p'(v)$ for all variables v in the domain of p'', and we have that p is an extension of p', we obtain that p falsifies a clause of ψ. □

A corollary of this proposition is that approximately propagation complete encodings with a quality level of 1 are exactly the same as propagation complete encodings. While the two definitions differ for unsatisfiable partial valuations, they both imply that the unsatisfiability of a partial valuation needs to be detectable by unit propagation. For the approximate notion, this follows from the proposition above. For the older non-approximate propagation completeness definition, this follows from the fact that such partial valuations imply all literals in $\mathcal{L}(\mathcal{V})$, which in turn need to lead to at least one clause in the CNF encoding to be falsified.

4   Computing Minimal Approximately Optimal Encodings

In this section, we present an approach to compute approximately propagation complete and conflict propagating encodings of minimal size. Both of these concepts are parameterized by quality levels, so our procedure will read a *propagation quality level tuple* (q_p, q_c), where q_p denotes the quality level for approximate propagation completeness, and q_c is the quality level for approximate conflict propagation. Applying the approach with the quality level tuple $(1, |\mathcal{V}|)$ thus yields the smallest propagation complete encoding, whereas applying the approach with $(|\mathcal{V}|, 1)$ gives the overall smallest possible CNF encoding for a constraint (as every encoding is automatically approximately conflict propagating with a quality level of 1 by the definition of this propagation quality notion). To avoid the occurrence of the set of variables in propagation quality tuples in the following, we use the ∞ symbol to denote all numbers $\geq |\mathcal{V}|$, as the propagation quality level definitions do not lead to differences for values greater than or equal to the number of variables in a constraint.

The main idea of our approach is that after enumerating all clauses that *could* occur in a minimal CNF encoding with the specified quality level, we can compute requirements on the selection of a subset of clauses that ensure (1) the completeness of the encoding (i.e., that it accepts the correct set of full variable valuations), (2) the desired quality level for approximate propagation completeness, and (3) the desired quality level for approximate conflict propagation. These requirements are then encoded into a (partial) MaxSAT instance whose optimal solution represents a minimally-sized CNF encoding.

The following proposition gives rise to a procedure to efficiently enumerate the set of clauses that *can* occur in a minimal CNF encoding.

Proposition 2. *Let ψ be a CNF encoding of a constraint with propagation quality level tuple (q_p, q_c). If a literal can be removed from a clause in ψ without changing the set of satisfying variable valuations, then removing the literal does not degrade the propagation quality levels of ψ.*

Proof. If a partial valuation p leads to unit propagation for ψ, then it still does so after removing a literal from a clause *except* if the removed literal would be propagated. This can only happen for unsatisfiable partial valuations. For both propagation quality notions, such a case cannot reduce the quality level. □

It follows that without loss of generality, we can assume smallest encodings to only use clauses that are as short as possible (in the sense that removing a literal from the clause would lead to a clause that some allowed variable valuation of the constraint to be encoded violates). Note that Bordeaux and Marques-Silva [8] already proved that when enumerating clauses for a constraint ordered by length, a longer clause can never make a shorter one redundant. Their result is not applicable here, as we later only select a subset of the enumerated *candidate clauses* to be contained in the computed encoding, while the application of their result requires that a clause remains a part of the CNF encoding once it has been found.

Enumerating all shortest clauses can be done in multiple ways. In our implementation for which we report experimental results in the next section, we start with a binary decision diagram description of the constraint and encode the search for a clause into a SAT instance by letting the SAT solver guess a partial valuation and which nodes in the BDD are reachable for the partial valuation. The **true** node of the BDD must not be reachable so that the valuation represents a possible clause in an encoding. We use a cardinality constraint with a ladder encoding [24] to count how many variables are set in the partial valuation, and iteratively search for partial valuations with the smallest possible domain size. Whenever one is found, a clause is added to the SAT instance that excludes all extensions of the partial valuation, and the search continues in an incremental manner.

After a *candidate* set of clauses $C = \{c_1, \ldots, c_m\}$ for the constraint to be encoded is computed, we employ a MaxSAT solver to find a minimal encoding with the desired propagation quality level tuple (q_p, q_c). The MaxSAT instance ϕ_M has the variables $x_1, \ldots, x_m$, i.e., one variable per clause in C, and we compute clauses for ϕ_M that ensure that the selected subset of C is (1) complete enough to encode the correct constraint, (2) approximately propagation complete with a quality level of q_p, and (3) approximately conflict propagating with a quality level of q_c. In the remainder of this section, we describe how to compute these clauses for ϕ_M.

4.1 Ensuring Encoding Correctness

All clauses in C can be part of a correct encoding of the constraint. The final set of selected clauses needs to be large enough not to allow spurious satisfying assignments, however. To achieve this, we recursively enumerate all (partial) assignments p while keeping track of the set of clauses C' not (yet) satisfied by the partial assignment. Whenever a complete assignment is reached and $C' \subseteq C$ is the set of clauses not satisfied by p, the (hard) MaxSAT clause $\bigvee_{c_i \in C'} x_i$ is added to ϕ_M. As optimizations for this process, a recursion step is aborted whenever C' becomes empty, and p is never extended by values for variables that do not occur in C'.

It is possible that a MaxSAT clause found late in this process subsumes a clause found earlier. We use a simple clause database structure that sorts clauses by length and removes subsumed clauses to avoid generating unnecessarily large MaxSAT instances. Storing all clauses with ordered literals increases the efficiency of the approach.

4.2 Encoding Approximate Propagation Completeness

By the definition of approximate propagation completeness, we need to ensure that for every partial valuation that implies at least q_p literals for variables that do not have values in the partial valuation, the final encoded CNF formula includes one clause that gives rise to unit propagation for the partial valuation.

We start by building a BDD that represents all partial valuations for which at least q_p new literals are implied. Algorithm 1 shows how this is done. In order to encode partial valuations, we use the auxiliary variable set $\mathcal{D} = \{d_v \mid v \in \mathcal{V}\}$ to encode which variables have *defined* values in a partial valuation. The algorithm iterates over all variables $v \in \mathcal{V}$ (line 3) and finds the set of partial valuations

Algorithm 1. Procedure to compute the satisfiable partial valuations that imply at least q_p many literals

1:   $F \leftarrow$ satisfying assignments of the constraint to be encoded over $\mathcal{V}$
2:   $X[0] \leftarrow F$
3:   **for** $v \in \mathcal{V}$ **do**
4:      $I = \neg d_v \wedge (\exists v.F) \wedge (\neg \forall v.F)$
5:      **for** $v' \in \mathcal{V} \setminus \{v\}$ **do**
6:        $I = I \wedge ((\forall v'.I) \vee d_{v'})$
7:      $Y \leftarrow X$
8:      $X[0] \leftarrow Y[0] \wedge \neg I$
9:      **for** $i \in \{1, \ldots, |Y|\}$ **do**
10:       $X[i] \leftarrow (Y[i] \wedge \neg I) \vee (Y[i-1] \wedge I)$
11:     $X[|Y|] \leftarrow Y[|Y| - 1] \wedge I$
12: **return** $\bigvee_{i=q_p}^{|\mathcal{V}|} X[i]$

that imply v or $\neg v$ (line 4). We only consider partial valuations that can be extended to satisfying assignments, so only one of them can be implied. If some other variable v' does not have a defined value in a partial valuation, then the valuation can only induce a literal over v if the value of v' does not matter for implying v (lines 5 to 6). The resulting partial valuation set is stored into the variable I in the algorithm.

After I is computed, the algorithm updates a partitioning of the partial valuations by how many literals over the variables already considered in the outer loop are induced by the respective partial valuation (lines 7 to 11). Finally, the algorithm returns the partial valuations that induce at least q_p literals.

For every partial valuation in the resulting BDD, we compute the subset $C' \subseteq C$ of clauses in C that give rise to unit propagation over the valuation, as for the final CNF encoding of the constraint to be approximately propagation complete with a quality level of q_p, one of them needs to be contained. We then add $\bigvee_{c_i \in C'} x_i$ as a hard clause to the MaxSAT instance ϕ_M and update the BDD with the remaining partial valuations to be considered by taking its conjunction with $\neg \bigwedge_{c_i \in C'} \bigvee_{l \in c_i} \bigwedge_{l' \in c_i, l \neq l'} m(l)$, where $m(\neg v) = d_v \wedge v$ and $m(v) = d_v \wedge \neg v$ for every variable $v \in \mathcal{V}$. In this way, all partial valuations that are guaranteed to lead to unit propagation whenever at least one of the clauses in C' is contained in the CNF encoding are removed from the BDD. We do the same with all clause subsets C' found in the procedure given in the previous subsection, as this further reduces the number of partial valuations to be considered. As before, we use a special clause database for ϕ_M to remove subsumed clauses.

4.3 Encoding Approximate Conflict Propagation

Next, we take care of partial valuations that cannot be completed to satisfying assignments and for which at most q_c variables in $\mathcal{V}$ do not have an assigned value. Algorithm 2 describes how to compute this set of partial valuations. After

Algorithm 2. Procedure to compute the unsatisfiable partial valuations for which at least q_c variables do not have a assigned values.

1: $X[0] \leftarrow \neg$(satisfying assignments of the constraint to be encoded over $\mathcal{V}$)
2: **for** $v \in \mathcal{V}$ **do**
3:     $Y \leftarrow X$
4:     **for** $i \in \{1, \ldots, |X|\}$ **do**
5:         $T \leftarrow (\forall v. Y[i]) \wedge \neg d_v$
6:         **if** $i > 1$ **then**
7:             $T \leftarrow T \vee (Y[i-1] \wedge d_v)$
8:         $X[i] \leftarrow T$
9:     $X[|Y|+1] \leftarrow Y[|Y|] \wedge d_v$
10: **return** $\bigvee_{i=|\mathcal{V}|-q_c}^{|\mathcal{V}|} X[i]$

it has been computed, the process is exactly the same as in the previous subsection: for every partial valuation that the BDD maps to **true**, the subset of clauses C' that lead to a conflict or unit propagation for this partial valuation is computed, and the (hard) MaxSAT clause $\bigvee_{c_i \in C'} x_i$ is added to the MaxSAT instance. The BDD for the remaining partial valuations is also updated in the same way and the clauses in the MaxSAT clause database generated by the procedures in the preceding two subsections are used to reduce the number of partial valuations to be considered before enumerating them.

4.4   MaxSAT Solving

All constraints in ϕ_M so far are *hard* constraints, i.e., need to be fulfilled by all satisfying variable valuations of the MaxSAT instance. To request the solver to minimize the number of clauses in the CNF encoding, we add the *soft clauses* $\neg x_1, \ldots, \neg x_m$. Maximizing the number of satisfied soft clauses then exactly corresponds to minimizing the number of clauses in the final constraint encoding.

5   Experiments

We implemented the approximate propagation complete and conflict propagating CNF encoding procedure in C++ using the BDD library CuDD [26]. Unlike the tool GenPCE for computing propagation complete encodings with the approach by Brain et al. [9], our new optic tool[1] does not start with a constraint encoding in CNF, but supports arbitrary Boolean formulas as constraints, which makes their specification easier. For the following experiments, we use lingeling bbc 9230380 [6] as SAT solver for enumerating the candidate clauses for the encoding, and apply the MaxSAT solver LMHS [25] in the version from the end of March 2018, using minisat [12] as backend solver. LMHS performs calls to an integer linear programming solver, for which we employ IBM CPLEX V12.8.0. We use the following benchmark sets:

---

[1] Available at https://github.com/progirep/optic.

1. Full adders with and without carry bit output
2. Multipliers of various input and output bit widths
3. Square and square root functions
4. Unsigned less-than-or-equal comparison functions
5. All-different constraints, including Cook's encoding [23]
6. Bipartite matching problems, where for some set of nodes $V_1 = V_2 = \{1, \ldots, n\}$ for $n \in \mathbb{N}$ and (random) $E \subseteq V_1 \times V_2$, we have one Boolean variable v_e for every $e \in E$ and allow all edge subset choices encoded into $\{v_e\}_{e \in E}$ for which every node in V_1 is predecessor node for exactly one chosen edge and every node in V_2 is the successor node of exactly one chosen edge
7. All other benchmarks from the work of Brain et al. [9] not contained in the benchmark families above.

We apply our implementation with various propagation quality tuples and compare the resulting encoding sizes and computation times with the propagation complete encodings generated by the GenPCE tool and the time it took that tool to compute them. As baseline, we also compare against a simple BDD-based tool that enumerates some shortest possible clauses in a CNF encoding until the encoding is complete, without guaranteeing any propagation quality. All benchmarks were executed on a computer with AMD Opteron 2220 processors with 2.8 GHz clock rate, running an x64 version of Linux. The memory limit for every execution was 8 GB, all executions where single-threaded, and we imposed a time limit of 30 min.

Table 1 shows an excerpt of the results. We tested the propagation quality of all generated encodings using an additional tester tool, and report the results in the table as well. In many cases, optic generated encodings of higher propagation quality levels than requested when this was possible without increasing the number of clauses in the encoding. A few other observations can be made:

1. The GenPCE tool did not always compute propagation complete encodings with a minimal number of clauses (e.g., all-different constraints for 3 objects, some bipartite matching problems), but did so quite often.
2. There are a many cases (e.g., addition with a large number of bits, multiplication, all-different constraints) for which the encodings for different propagation quality tuples have different sizes.
3. Encodings that are fully conflict propagating but for which approximate propagation completeness was not requested are typically small.
4. The GenPCE tool is often slower than optic for computing propagation complete encodings, but the scalability of optic and GenPCE are quite similar. However, optic also times out in a few cases in which GenPCE does not. Computing minimally-sized encodings (as optic does but not GenPCE) appears to be much harder in these cases.

Table 1. Number of computed clauses, computation times (in seconds) and measured propagation quality for a selection of benchmarks and a selection of target propagation quality tuples. A × entry in the measured propagation quality tuples means that the quality level could not be determined due to a timeout (set to 60 s) of the respective measuring tool. We compare against a greedy encoding into CNF (by enumerating shortest possible clauses using BDDs) and against the tool **GenPCE** that computes propagation complete encodings. Computation times are given in seconds. The output of the greedy tool is used as input to **GenPCE** (as that tool requires the input to already be in CNF form, with possibly some variables that need to be existentially quantified away). The $(1,\infty)$ quality tuple refers to propagation completeness, which **GenPCE** also guarantees. The $(\infty,1)$ quality tuple refers to plain minimally-sized encodings. Encodings of quality level (∞,∞) are (fully) conflict propagating, but not necessarily propagation complete.

Benchmark	Greedy	(∞,∞)	$(1,\infty)$	$(2,\infty)$	$(3,\infty)$	$(3,3)$	$(\infty,1)$	GenPCE
All-Difference Cook 3 obj.	36 2.64 (11,10)	timeout	timeout	timeout	timeout	timeout	17 20.74 (11,11)	33 0.00 (1,∞)
All-Difference 3 obj.	26 0.00 (7,6)	15 1.15 (4,∞)	24 0.24 (1,∞)	20 0.35 (2,∞)	15 0.39 (3,∞)	15 0.34 (3,∞)	12 0.10 (4,6)	26 memout
All-Difference Cook 4 obj.	125 6.33 (×,23)	timeout	timeout	timeout	timeout	timeout	28 74.80 (8,12)	203 0.00 (1,∞)
All-Difference 4 obj.	86 0.10 (13,12)	timeout	timeout	timeout	timeout	timeout	timeout	timeout
cvc-add15to4.cnf	65534 1633.63 (×,∞)	144 2.31 (4,∞)	1536 1.64 (1,∞)	808 8.30 (2,∞)	512 3.64 (3,∞)	500 47.51 (3,4)	122 10.49 (6,8)	1536 19.81 (1,∞)
cvc-add3-bw3.cnf	824 1.74 (6,8)	30 0.23 (6,∞)	76 0.10 (1,∞)	62 0.16 (2,∞)	51 15.18 (3,∞)	51 6.90 (3,∞)	32 0.41 (6,4)	76 0.00 (1,∞)
cvc-add3-carry2-gadget.cnf	824 0.88 (6,8)	144 1.46 (4,∞)	1152 1.22 (1,∞)	624 2.97 (2,∞)	416 2.11 (3,∞)	416 2.66 (3,∞)	126 1.02 (6,8)	1152 16.98 (1,∞)
cvc-add3-opt-bw3.cnf	62 0.07 (2,∞)	8 0.25 (3,∞)	254 2.72 (1,∞)	254 16.03 (1,∞)	254 32.34 (3,∞)	254 17.04 (3,∞)	158 20.98 (7,5)	254 0.55 (1,∞)
cvc-add7to3.cnf	254 0.23 (1,∞)	16 0.31 (7,6)	254 0.31 (1,∞)	9 0.25 (2,∞)	8 0.30 (3,∞)	8 0.32 (3,∞)	8 0.30 (3,∞)	254 0.00 (1,∞)
cvc-mult-const-const3-2n-2.cnf	12 0.22 (5,∞)	15 1.70 (8,∞)	11 0.31 (1,∞)	22 0.38 (2,∞)	19 0.38 (3,∞)	19 0.73 (3,6)	13 0.40 (8,7)	11 0.00 (1,∞)
cvc-mult-const-const5-2n-3.cnf	27 0.08 (7,6)	36 0.28 (5,∞)	24 0.36 (1,∞)	20 2.41 (2,∞)	20 3.03 (3,∞)	20 3.58 (3,6)	14 3.07 (8,6)	24 0.00 (1,∞)
cvc-mult-const-const7-2n-3.cnf	30 0.25 (7,6)	60 1.39 (6,∞)	32 1.25 (1,∞)	72 0.28 (2,∞)	56 0.28 (3,∞)	56 0.40 (3,∞)	32 0.39 (5,6)	32 0.09 (1,∞)
cvc-plus-3.cnf	48 0.06 (6,4)	6 0.13 (1,∞)	96 0.12 (1,∞)	243 1.64 (2,∞)	158 3.08 (3,∞)	158 5.16 (3,∞)	54 6.35 (7,9)	96 5.53 (1,∞)
cvc-plus-4.cnf	120 0.21 (8,7)	158 0.39 (1,∞)	336 1.42 (1,∞)					336 0.00 (1,∞)
cvc-slt-gadget.cnf	6 0.08 (1,∞)	6 0.07 (1,∞)	6 0.09 (1,∞)	6 0.08 (1,∞)	6 0.09 (1,∞)	6 0.13 (1,∞)	6 0.11 (1,∞)	6 0.00 (1,∞)
cvc-ult-6.cnf	158 0.28 (1,∞)	18 0.06 (5,∞)	158 0.47 (1,∞)	158 0.26 (1,∞)	158 0.36 (1,∞)	158 0.49 (1,∞)	158 0.42 (1,∞)	158 38.42 (1,∞)
cvc-ult-gadget.cnf	6 0.06 (1,∞)	42 0.15 (6,∞)	6 0.07 (1,∞)	6 0.07 (1,∞)	6 0.07 (1,∞)	6 0.07 (1,∞)	6 0.06 (1,∞)	6 0.20 (1,∞)
Adder, 2 bits, carry output	23 0.04 (3,∞)	66 2.11 (8,∞)	30 0.05 (1,∞)	27 0.07 (2,∞)	22 0.06 (3,∞)	22 0.06 (3,4)	17 0.05 (5,4)	30 0.00 (1,∞)
Adder, 3 bits, carry output	61 0.00 (6,4)	60 1.30 (6,∞)	102 0.31 (1,∞)	93 0.15 (2,∞)	65 0.68 (3,∞)	65 19.23 (3,∞)	34 0.68 (7,7)	102 0.20 (1,∞)
Adder, 4 bits, carry output	149 0.09 (9,8)	90 76.98 (9,∞)	342 2.16 (1,∞)	303 2.48 (2,∞)	188 3.00 (3,∞)	158 273.06 (3,∞)	56 42.89 (9,10)	342 14.57 (1,∞)
Adder, 4 bits	120 0.06 (8,7)	84 24.39 (8,∞)	336 1.33 (1,∞)	243 1.37 (2,∞)	158 3.28 (3,∞)	timeout	54 4.62 (7,9)	336 5.77 (1,∞)
Adder, 5 bits, carry output	343 0.38 (11,11)	timeout	1086 64.51 (1,∞)	948 76.16 (2,∞)	560 106.05 (3,∞)	timeout	80 376.57 (11,13)	1086 848.67 (1,∞)
Adder, 5 bits	272 0.36 (10,10)	timeout	1080 17.00 (1,∞)	762 20.37 (2,∞)	446 32.59 (3,∞)	timeout	78 136.79 (8,12)	1080 353.66 (1,∞)
Multiplier, 2 to 2 bits	12 0.00 (2,∞)	12 0.06 (2,∞)	12 0.06 (1,∞)	12 0.06 (2,∞)	12 0.06 (2,∞)	12 0.05 (2,∞)	12 0.06 (2,∞)	12 0.00 (1,∞)
Multiplier, 2 to 4 bits	24 0.06 (2,∞)	16 0.08 (4,∞)	19 0.09 (1,∞)	17 0.09 (2,∞)	17 0.09 (3,∞)	17 0.08 (3,∞)	16 0.09 (4,∞)	19 0.00 (1,∞)
Multiplier, 3 to 3 bits	31 0.05 (5,4)	29 0.06 (5,∞)	71 0.07 (1,∞)	41 0.13 (2,∞)	39 0.06 (3,∞)	39 0.05 (3,∞)	29 0.09 (6,5)	71 0.00 (1,∞)
Multiplier, 3 to 5 bits	98 0.07 (8,7)	timeout	304 3.63 (1,∞)	165 3.95 (2,∞)	124 10.12 (3,∞)	123 6.01 (3,5)	45 25.33 (8,7)	305 2.10 (1,∞)
Multiplier, 4 to 5 bits	212 0.05 (9,9)	timeout	2274 112.48 (1,∞)	1106 121.12 (2,∞)	timeout	timeout	timeout	2276 40.92 (1,∞)
Multiplier, 5 to 7 bits	1327 0.57 (×,×)	timeout	timeout	timeout	timeout	timeout	timeout	timeout
Square root, 4 bits	10 0.04 (3,∞)	8 0.11 (2,∞)	11 0.30 (1,∞)	8 0.08 (2,∞)	14 0.36 (3,∞)	8 0.08 (2,∞)	8 0.10 (2,∞)	11 0.00 (1,∞)
Square root, 5 bits	14 0.30 (3,∞)	14 0.39 (3,∞)	18 0.84 (1,∞)	15 0.41 (2,∞)	24 0.87 (3,∞)	14 0.37 (3,∞)	14 0.41 (3,∞)	18 0.00 (1,∞)
Square root, 6 bits	28 0.65 (5,5)	22 0.87 (5,∞)	41 2.83 (1,∞)	24 0.86 (2,∞)	45 2.79 (3,∞)	24 0.80 (3,∞)	22 0.86 (5,6)	41 0.08 (1,∞)
Square root, 7 bits	41 2.28 (7,7)	35 2.64 (7,∞)	71 6.63 (1,∞)	51 2.78 (2,∞)	103 8.13 (3,∞)	45 2.92 (3,∞)	35 2.77 (7,7)	71 0.08 (1,∞)
Square root, 8 bits	74 5.03 (9,8)	54 7.23 (8,∞)	209 6.63 (1,∞)	124 6.73 (2,∞)		103 8.61 (3,∞)	52 6.03 (9,7)	209 3.78 (1,∞)
Square function, 4 bits	52 0.07 (8,6)	timeout	56 0.62 (1,∞)	44 0.65 (2,∞)	335 0.93 (3,∞)	335 0.69 (3,∞)	26 0.32 (9,8)	56 0.07 (1,∞)
Square function, 5 bits	109 0.08 (11,10)	timeout	44 1.01 (1,∞)	184 752.77 (2,∞)	timeout	135 1443.43 (3,∞)	42 54.10 (12,10)	273 3.37 (1,∞)
Square function, 6 bits	228 0.25 (×,13)	timeout	273 254.83 (1,∞)	timeout	timeout	timeout	timeout	1701 1195.27 (1,∞)
Square function, 7 bits	492 0.41 (×,×)	timeout	timeout	timeout	timeout	timeout	timeout	timeout
Bipatite matching, 3 obj's, ex. 12	9 0.05 (1,∞)	7 0.08 (3,∞)	8 0.08 (1,∞)	8 0.06 (1,∞)	7 0.07 (3,∞)	7 0.07 (3,∞)	7 0.08 (3,∞)	9 0.00 (1,∞)
Bipatite matching, 4 obj's, ex. 59	16 0.07 (6,4)	11 0.19 (4,∞)	14 0.08 (1,∞)	14 0.08 (1,∞)	11 0.11 (3,∞)	11 0.20 (3,∞)	10 0.05 (4,6)	16 0.00 (1,∞)

We can see that using BDDs for enumerating candidate clauses and generating MaxSAT clauses is not a major bottleneck, as `optic` frequently outperforms `GenPCE` on propagation-complete encodings. We also tested how long computing good encodings for the four-object all-different constraints takes. For the $(1, \infty)$ and $(2, \infty)$ propagation quality levels, it was determined after 18.1 and 28.4 h that 200 and 156 clauses are needed, respectively, of which 11.2 and 621.3 min were spent solving the MaxSAT problems. In the $(3, \infty)$ case, the overall computation took 141.7 h (120 clauses).

5.1   Case Study: Integer Factoring

To evaluate the effect of the encodings on propagation quality, we apply them in an integer factoring case study. We generated 5 numbers that are products of two primes each. For each number c and each $2 \leq n_1 \leq \lceil \log_2(\sqrt{c+1}) \rceil$, we then computed SAT instances for finding a factoring for which the first number has n_1 bits with the most significant bit set. We compose the SAT instance of encodings for full adders with 1, 2, 3, or 4 input bits, and multipliers with 1, 2, or 3 input bits. We use the propagation quality tuples $(3, 3)$, (∞, ∞), $(\infty, 1)$, and $(1, \infty)$, for which the latter three refer to best possible conflict propagation, minimal encoding size, and classical propagation completeness. For the $(3, 3)$ case, we were unable to generate a minimal encoding for the four-bit full adder and had to use a suboptimal (possibly too large) encoding that we obtained by using the MaxSAT solver `maxino-2015-k16` [2] that can output suboptimal solutions found early. We aborted its computation after 400 min. We also compare the solving performance against using the *greedy* encoding introduced as comparison baseline in Table 1.

Figure 1 shows the cactus plot for the computation times of the SAT solvers `lingeling bbc 9230380` [6] and `MapleSAT_LRB` [21] as representatives for modern solvers with and without advanced *inprocessing*, respectively. It can be seen that for shorter time limits, minimally sized encodings without propagation quality guarantees are inferior in overall performance. Above 600 s of computation time for every benchmark, `MapleSAT` however works quite well with minimally-sized encodings, but the difference between the encodings is quite small for high time limits anyway. `Lingeling` works best with encodings of higher propagation quality. Encodings that only enforce conflict propagation seem to be particularly well suited for easier benchmarks.

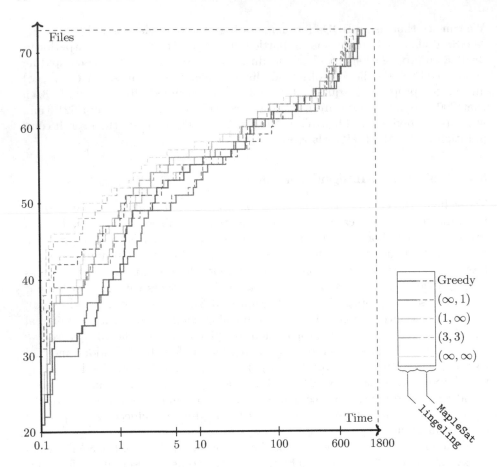

Fig. 1. Cactus plot for the integer factoring case study. Time is given in seconds, the overall number of files is 73. The legend describes the line styles for the studied combinations of solvers and propagation quality tuples.

6   Conclusion

We presented two new propagation quality notions and described an approach to compute minimally-sized CNF encodings from constraint descriptions. Our approach reduces the problem to solving a single MaxSAT instance, and the experiments show that many constraints found in practice give rise to differently sized encodings for different propagation quality level combinations.

In contrast to the work by Brain et al. [9], we based our experimental case study on problems encoded from scratch rather than modifying an existing SMT solver, as the techniques used in SMT solvers are highly tuned to work in concert, and hence replacing a single element has side-effects that cannot be tuned out in an experimental evaluation.

Studying the precise effect of *inprocessing* [18] on what constraint encodings can be used most efficiently is left for future work. For instance, clauses that solvers with inprocessing could automatically derive could be left out, which may give rise to very different minimally-sized encodings.

Acknowledgements. This work was supported by DFG grant EH 481/1-1 and the Institutional Strategy of the University of Bremen, funded by the German Excellence Initiative. The authors want to thank Armin Biere for early feedback on the propagation quality notions defined in this work and Erika Abraham for proposing MaxSAT solvers as reasoning backend.

References

1. Abío, I., Nieuwenhuis, R., Oliveras, A., Rodríguez-Carbonell, E.: A parametric approach for smaller and better encodings of cardinality constraints. In: Schulte, C. (ed.) CP 2013. LNCS, vol. 8124, pp. 80–96. Springer, Heidelberg (2013). https://doi.org/10.1007/978-3-642-40627-0_9

2. Alviano, M., Dodaro, C., Ricca, F.: A MaxSAT algorithm using cardinality constraints of bounded size. In: Twenty-Fourth International Joint Conference on Artificial Intelligence, IJCAI 2015, Buenos Aires, Argentina, 25–31 July 2015, pp. 2677–2683 (2015). http://ijcai.org/Abstract/15/379

3. Asín, R., Nieuwenhuis, R., Oliveras, A., Rodríguez-Carbonell, E.: Cardinality networks: a theoretical and empirical study. Constraints **16**(2), 195–221 (2011). https://doi.org/10.1007/s10601-010-9105-0

4. Babka, M., Balyo, T., Cepek, O., Gurský, S., Kucera, P., Vlcek, V.: Complexity issues related to propagation completeness. Artif. Intell. **203**, 19–34 (2013). https://doi.org/10.1016/j.artint.2013.07.006

5. Bacchus, F., Winter, J.: Effective preprocessing with hyper-resolution and equality reduction. In: Giunchiglia, E., Tacchella, A. (eds.) SAT 2003. LNCS, vol. 2919, pp. 341–355. Springer, Heidelberg (2004). https://doi.org/10.1007/978-3-540-24605-3_26

6. Biere, A.: Lingeling, Plingeling, PicoSAT and PrecoSAT at SAT race 2010. FMV Report Series Technical report 10/1, Johannes Kepler University, Linz, Austria (2010)

7. Bjork, M.: Successful SAT encoding techniques. J. Satisfiability Boolean Model. Comput. **7**, 189–201 (2009)

8. Bordeaux, L., Marques-Silva, J.: Knowledge compilation with empowerment. In: Bieliková, M., Friedrich, G., Gottlob, G., Katzenbeisser, S., Turán, G. (eds.) SOFSEM 2012. LNCS, vol. 7147, pp. 612–624. Springer, Heidelberg (2012). https://doi.org/10.1007/978-3-642-27660-6_50

9. Brain, M., Hadarean, L., Kroening, D., Martins, R.: Automatic generation of propagation complete SAT encodings. In: Jobstmann, B., Leino, K.R.M. (eds.) VMCAI 2016. LNCS, vol. 9583, pp. 536–556. Springer, Heidelberg (2016). https://doi.org/10.1007/978-3-662-49122-5_26

10. Bryant, R.E.: Graph-based algorithms for Boolean function manipulation. IEEE Trans. Comput. **35**(8), 677–691 (1986). https://doi.org/10.1109/TC.1986.1676819

11. Clarke, E.M., Grumberg, O., Peled, D.A.: Model checking. MIT Press (2001). http://books.google.de/books?id=Nmc4wEaLXFEC

12. Eén, N., Sörensson, N.: An extensible SAT-solver. In: Giunchiglia, E., Tacchella, A. (eds.) SAT 2003. LNCS, vol. 2919, pp. 502–518. Springer, Heidelberg (2004). https://doi.org/10.1007/978-3-540-24605-3_37

13. Eén, N., Sörensson, N.: Translating pseudo-boolean constraints into SAT. JSAT 2(1–4), 1–26 (2006). http://jsat.ewi.tudelft.nl/content/volume2/JSAT2_1_Een.pdf

14. Franco, J., Martin, J.: A history of satisfiability. In: Biere, A., Heule, M.J.H., van Maaren, H., Walsh, T. (eds.) Handbook of Satisfiability, Frontiers in Artificial Intelligence and Applications, vol. 185, chap. 1, pp. 3–74. IOS Press (2009)

15. Gwynne, M., Kullmann, O.: Towards a theory of good SAT representations. CoRR abs/1302.4421 (2013). http://arxiv.org/abs/1302.4421

16. Gwynne, M., Kullmann, O.: Generalising unit-refutation completeness and SLUR via nested input resolution. J. Autom. Reasoning 52(1), 31–65 (2014). https://doi.org/10.1007/s10817-013-9275-8

17. Inala, J.P., Singh, R., Solar-Lezama, A.: Synthesis of domain specific CNF encoders for bit-vector solvers. In: Creignou, N., Le Berre, D. (eds.) SAT 2016. LNCS, vol. 9710, pp. 302–320. Springer, Cham (2016). https://doi.org/10.1007/978-3-319-40970-2_19

18. Järvisalo, M., Heule, M.J.H., Biere, A.: Inprocessing rules. In: Gramlich, B., Miller, D., Sattler, U. (eds.) IJCAR 2012. LNCS (LNAI), vol. 7364, pp. 355–370. Springer, Heidelberg (2012). https://doi.org/10.1007/978-3-642-31365-3_28

19. Kučera, P., Savický, P., Vorel, V.: A lower bound on CNF encodings of the at-most-one constraint. In: Gaspers, S., Walsh, T. (eds.) SAT 2017. LNCS, vol. 10491, pp. 412–428. Springer, Cham (2017). https://doi.org/10.1007/978-3-319-66263-3_26

20. Li, C.M., Manyà, F.: MaxSAT, hard and soft constraints. In: Biere, A., Heule, M.J.H., van Maaren, H., Walsh, T. (eds.) Handbook of Satisfiability, Frontiers in Artificial Intelligence and Applications, vol. 185, chap. 19, pp. 613–631. IOS Press (2009)

21. Liang, J.H., Ganesh, V., Poupart, P., Czarnecki, K.: Learning rate based branching heuristic for SAT solvers. In: Creignou, N., Le Berre, D. (eds.) SAT 2016. LNCS, vol. 9710, pp. 123–140. Springer, Cham (2016). https://doi.org/10.1007/978-3-319-40970-2_9

22. Manthey, N., Heule, M.J.H., Biere, A.: Automated reencoding of boolean formulas. In: Biere, A., Nahir, A., Vos, T. (eds.) HVC 2012. LNCS, vol. 7857, pp. 102–117. Springer, Heidelberg (2013). https://doi.org/10.1007/978-3-642-39611-3_14

23. Petke, J.: Bridging Constraint Satisfaction and Boolean Satisfiability. AIFTA. Springer, Cham (2015). https://doi.org/10.1007/978-3-319-21810-6

24. Roussel, O., Manquinho, V.: Pseudo-boolean and cardinality constraints. In: Biere, A., Heule, M.J.H., van Maaren, H., Walsh, T. (eds.) Handbook of Satisfiability, Frontiers in Artificial Intelligence and Applications, vol. 185, chap. 22, pp. 695–733. IOS Press (2009)

25. Saikko, P., Berg, J., Järvisalo, M.: LMHS: A SAT-IP hybrid MaxSAT solver. In: Creignou, N., Le Berre, D. (eds.) SAT 2016. LNCS, vol. 9710, pp. 539–546. Springer, Cham (2016). https://doi.org/10.1007/978-3-319-40970-2_34

26. Somenzi, F.: CUDD: CU Decision Diagram package release 3.0.0 (2015)

Dynamic Polynomial Watchdog Encoding for Solving Weighted MaxSAT

Tobias Paxian$^{(\boxtimes)}$, Sven Reimer$^{(\boxtimes)}$, and Bernd Becker$^{(\boxtimes)}$

Albert-Ludwigs-Universität Freiburg, Georges-Köhler-Allee 051,
79110 Freiburg, Germany
{paxiant,reimer,becker}@informatik.uni-freiburg.de

Abstract. In this paper we present a novel pseudo-Boolean (PB) constraint encoding for solving the weighted MaxSAT problem with iterative SAT-based methods based on the Polynomial Watchdog (PW) CNF encoding. The watchdog of the PW encoding indicates whether the bound of the PB constraint holds. In our approach, we lift this static watchdog concept to a dynamic one allowing an incremental convergence to the optimal result. Consequently, we formulate and implement a SAT-based algorithm for our new Dynamic Polynomial Watchdog (DPW) encoding which can be applied for solving the MaxSAT problem. Furthermore, we introduce three fundamental optimizations of the PW encoding also suited for the original version leading to significantly less encoding size. Our experimental results show that our encoding and algorithm is competitive with state-of-the-art encodings as utilized in QMaxSAT (2nd place in last MaxSAT Evaluation 2017). Our encoding dominates two of the QMaxSAT encodings, and at the same time is able to solve unique instances. We integrated our new encoding into QMaxSAT and adapt the heuristic to choose between the only remaining encoding of QMaxSAT and our approach. This combined version solves 19 (4%) more instances in overall 30% less run time on the benchmark set of the MaxSAT Evaluation 2017. Compared to each encoding of QMaxSAT used in the evaluation, our encoding leads to an algorithm that is on average at least $2X$ faster.

1 Introduction

MaxSAT and its variations are SAT-related optimization problems seeking for a truth assignment to a Boolean formula in Conjunctive Normal Form (CNF) such that the satisfiability of the formula is maximized. Maximizing the satisfiability in a pure MaxSAT problem is to maximize the number of simultaneously satisfied clauses in the CNF. In the weighted MaxSAT variation for each clause a positive integer weight is appended and hence, the maximization of the formula is yielded if the accumulated weights of the satisfied clauses are maximized.

This work is partially supported by the DFG project "Algebraic Fault Attacks" (funding id PO 1220/7-1, BE 1176 20/1, KR 1907/6-1).

O. Beyersdorff and C. M. Wintersteiger (Eds.): SAT 2018, LNCS 10929, pp. 37–53, 2018.
https://doi.org/10.1007/978-3-319-94144-8_3

There exists a wide range of different solving techniques [1] such as branch and bound algorithms [2], iterative SAT solving [3], unsat core based techniques [4] and ILP solver [5], to name a few. A very successful approach is iterative SAT-based solving. The core idea is to adjust the bounds for the maximized result by iterative (and incremental) SAT solver calls. One possibility to do so is a direct encoding of Pseudo-Boolean (PB) constraints of the maximization objective into the SAT instance, such that the truth assignment of the whole formula directly represents the result of the maximization. By forcing the current optimization result to be larger than the last one found, this approach runs iteratively towards the optimum. The recent MaxSAT Evaluation [6] indicates that this technique can be successfully employed for unweighted and weighted MaxSAT, as the iterative SAT-based solver QMaxSAT [7] demonstrates. QMaxSAT adopts four different variations of the totalizer network [8–11] and one adder network [12] as PB encoding. A simple heuristic selects between these encodings.

In this paper we introduce a new encoding and algorithm based on the Polynomial Watchdog (PW) encoding [13]. The PW is an efficient encoding for PB constraints, though it is not designed to be employed in an iterative MaxSAT approach. Hence, we modified the original encoding by replacing the static watchdog of [13] by a dynamic one allowing to adjust the optimization goal. Based on this encoding, we provide a complete algorithm for deciding the weighted MaxSAT problem. Additionally, we introduce three fundamental optimizations/heuristics leading to significantly smaller PW encodings.

To demonstrate the effectiveness we adjoin our new encoding to the QMaxSAT solver. Experimental results on the benchmark set of the Evaluation 2017 [6] show that our encoding leads to an algorithm that is (1) competitive in solved instances and (2) on average $2X$ faster than existing ones. In particular, our approach is clearly superior to [8,11], especially for weighted MaxSAT instances with large clause weights. Moreover, our approach solves complementary instances with the employed adder network [12] and thus, we adjust the heuristics for deciding the used network leading to 4% more solved instances on the benchmark set of [6].

The remaining paper is structured as follows: We present related work on totalizer networks and the weighted MaxSAT application in Sect. 2. In Sect. 3 we introduce the weighted MaxSAT problem as well as the totalizer and PW encoding. In Sect. 4 we present our proposed dynamic PW encoding and algorithm for iterative MaxSAT solving and propose further optimizations in Sect. 5. Finally, we demonstrate the applicability of our new encoding and the optimizations in Sect. 6 and conclude the paper in Sect. 7.

2   Related Work

Since the original version of the totalizer network is published in 2003 [8], many different variations have been investigated since then. Some of them are also employed in context of (weighted) MaxSAT. In particular, the iterative MaxSAT solver QMaxSAT [7] uses many different variations of the totalizer network. Namely, the original totalizer sorting network [8], weighted or generalized totalizer networks [9], mixed radix weighted totalizer [10], and modulo totalizer [11].

The original totalizer is well suited for unweighted instances. However, for weighted instances a naïve implementation does not scale in the encoding size.

The generalized totalizer [9] allows a more direct encoding of weighed inputs. This encoding is integrated into the recursive rules constructing the totalizer. The mixed radix weighted totalizer [10] is an extension of the generalized totalizer combined with the concept of mixed radix base [14].

The modulo totalizer [11] was initially developed for unweighted MaxSAT instances. It reduces the number of used clauses for the encoding by counting fulfilled clauses with modulo operations. As our experimental results show, the modulo totalizer still has scaling issues for a large sum of weights.

Our encoding is based on the Polynomial Watchdog (PW) encoding [13] which also uses totalizer sorting networks. Essentially the PW encoding employs multiple totalizer networks to perform an addition with carry on the sorted outputs. The sorting network based encoding of minisat+ described in [15] has similarities, the differences to the PW encoding are described in detail in [13]. In particular, minisat+ introduces additional logic to observe the exact bounds of the current constraint. Whereas the PW encoding utilizes additional inputs to control and observe the current bounds. Hence, we employ the PW encoding as the additional inputs are easier to manipulate for our dynamic approach.

Apart from the totalizer, other encodings for PB constraints are successfully employed for mapping the MaxSAT constraints. E.g., QMaxSAT uses an adder network [12]. This type of network is better suited for a large sum of input values than totalizer networks as adder have linear complexity in encoding size – in contrast to at least $\mathcal{O}(n \log n)$ for sorting networks.

Other encoding schemes are investigated in [15], where adder, sorting network [16] and BDD [17] implementations are compared. A BDD preserves generalized arc consistency (GAC) for PB constraints, if it can be constructed [15] – in contrast to sorting networks and adders in general. However, the enhanced encoding scheme of the Local Polynomial Watchdog (LPW) in [13] preserves GAC at the cost of encoding complexity. Another GAC preserving encoding is presented in [18] which employs a different kind of sorting networks.

All mentioned totalizer modifications only adjust the recursive rules of the totalizer. In contrast, our proposed encoding utilizes the standard totalizer and modifies the cascading version of [13].

3 Preliminaries

In this section we introduce the foundations of MaxSAT and in particular iterative PB encoding based approaches. Furthermore, we introduce the totalizer network [8] and the Polynomial Watchdog (PW) encoding [13] which are the fundamental encodings utilized in our approach.

3.1 MaxSAT

First, we introduce some basic terminologies which will be used within this paper. The input of MaxSAT problems is a Boolean formula in *Conjunctive Normal Form (CNF)*. A CNF is a conjunction of *clauses*, where a clause is a disjunction of *literals*. A clause which contains one literal is called *unit* (clause). In the following, we adopt the commonly used notation that clauses are sets of literals and a CNF is a set of clauses. A *SAT solver* decides the satisfiability problem, i.e. whether a Boolean formula φ in CNF is satisfiable. In this case, the solver returns a satisfying assignment for all variables, which is also called *model* of φ.

MaxSAT is a SAT-related optimization problem seeking for an assignment *maximizing* the number of simultaneously satisfied clauses of a CNF formula φ. The *partial* MaxSAT problem consists also of so-called *hard* clauses, which *must* be satisfied. All other clauses are called *soft* clauses. Thus, a MaxSAT formula can be formulated as follows: $\varphi = S \cup H$, where S denotes the set of soft clauses and H the set of hard clauses. The *weighted* (partial) MaxSAT problem is a generalization, where each soft clause is denoted with an integer weight w_j. The optimization goal is to maximize the accumulated weight of satisfied soft clauses.

A common approach for solving the MaxSAT problem is the *iterative SAT-based* algorithm [3] which incrementally employs a SAT solver. To do so, a *pseudo-Boolean (PB) constraint* C is directly encoded into CNF, where C is defined as $\Sigma_j a_j \cdot x_j \vartriangleright M$ over Boolean variables x_j and positive integers a_j and M. $\vartriangleright$ is one of the relational operators $\vartriangleright \in \{=, >, \geq, <, \leq\}$. By using only constraints of the form $\Sigma_j a_j \cdot x_j \geq M$, the MaxSAT problem can be reduced to the question of finding a maximum value M^* still satisfying C. To do so, the soft clauses are directly connected to the PB constraint network, where x_j is true if and only if the soft clause s_j is true and the weight w_j is connected to a_j of C.

There are various methods and schemes for the encoding of PB constraints as discussed in Sect. 2. State-of-the-art iterative MaxSAT solvers like QMaxSAT [7] use various and customized CNF encodings. For instance, the QMaxSAT version used in the MaxSAT evaluation effectively employs three different encodings: totalizer network [8], modulo totalizer network [11] and Warners adder network [12].

Regardless of the employed encoding, the iterative approach works as follows: A PB constraint network is encoded as described above and added as hard clauses to the original CNF. For each soft clause $s_j \in S$ a so-called *relaxation* literal r_j is introduced: $s'_j = s_j \vee \overline{r_j}$ (i.e. setting r_j forces s_j to be true) and connected to the PB encoding. Let S' be the set of all modified clauses s'_j, then a SAT solver decides the $\varphi' = S' \cup H$. The returned model allows to determine the current sum of satisfied weights M. The CNF is iteratively modified by adding a constraint demanding a larger optimization result than M. The SAT solver is called incrementally with this new constraint. The whole procedure is repeated until the SAT solver returns "unsatisfiable", i.e. the last added constraint represents a result which is just larger than the optimal result. Hence, the result of the last satisfiable SAT solver call corresponds to the optimization result M^* of the MaxSAT instance.

3.2 Totalizer Network

The *totalizer network* as introduced in [8] is a unary sorting network Φ : $\{0,1\}^n \to \{0,1\}^n$, arranging a binary input vector such that the output vector is sorted in descending order. E.g., the input vector $\langle 1,0,1,1,0 \rangle$ will be processed as follows: $\Phi(\langle 1,0,1,1,0 \rangle) = \langle 1,1,1,0,0 \rangle$. The output vector V represents a natural number v in unary representation: if the ith entry of the output vector V is one, the unary representation matches $v \geq i$.

The totalizer sorting network allows an efficient propagation of output values on CNF. The network divides the input vector recursively into two parts until the resulting vector consists only of one element which is sorted by definition. Two unary sorted vectors are merged together by the formula Ψ. Let $U = \langle u_1, \ldots, u_k \rangle$ and $V = \langle v_1, \ldots, v_l \rangle$ be sorted vectors corresponding to unary representations of natural numbers u and v, respectively. Ψ assures that the resulting vector $W = \langle w_1, \ldots, w_{k+l} \rangle$ is the unary representation of w with $w = u + v$.

The totalizer encoding consists of two mirrored parts $D_1(a,b) = (\overline{u_a} \vee \overline{v_b} \vee w_{a+b})$ and $D_2(a,b) = (u_{a+1} \vee v_{b+1} \vee \overline{w_{a+b+1}})$, where $0 \leq a \leq k$ and $0 \leq b \leq l$. By definition $u_0 = v_0 = w_0 = 1$ and $u_{k+1} = v_{l+1} = w_{k+l+1} = 0$ holds. As stated in [8] the resulting formula $\Psi(W = U \oplus V)$ represents the relation $w = u + v$:

$$\Psi(W = U \oplus V) = \bigwedge_{a=0}^{k} \bigwedge_{b=0}^{l} D_1(a,b) \wedge D_2(a,b) \tag{1}$$

Note, by using only the encoding D_1 we guarantee $w \geq a + b$, hence, we are able to set an upper bound for w. Likewise D_2 guarantees $w \leq a + b$, i.e. a lower bound for w is represented. Note that in each case the other direction does not hold. The input vector is split up recursively in two equally sized parts U and V connected by Ψ until U or V has a size of one, which is sorted by definition. The complete encoding of all Ψ parts is called Φ.

3.3 Polynomial Watchdog Encoding Scheme

In this section we briefly introduce the *Polynomial Watchdog (PW)* encoding scheme for PB constraints C. For more details the interested reader is referred to [13]. The PW encoding uses the functions Ψ and Φ of the totalizer in order to represent one PB constraint as depicted in Fig. 1. A *Global Polynomial Watchdog (GPW)* is introduced allowing to efficiently detect a violation of M in C, with C: $\Sigma_j a_j \cdot x_j < M$. To do so, a so-called *watchdog* ω is introduced. Essentially, whenever C is falsified the literal ω will be assigned to true. I.e. a lower bound is defined and only the D_1 clauses of Ψ are needed. The GPW is defined as

$$GPW(C) = PW(C) \wedge \overline{\omega} \tag{2}$$

which guarantees the previous mentioned property.

The PW is a binary addition with carry of the weights. The coefficients of the constraints are split into their binary representation, the bits with the same weight 2^i are added to one totalizer Φ called *top bucket* TB_i with weight 2^i. The most significant bit position of all coefficients equals the number of top buckets p. These top buckets are connected appropriately representing the carry of the unary addition, where two buckets with weight 2^i and 2^{i+1} are merged applying Ψ. To do so, only every second output of the bucket of weight 2^i and every output of the 2^{i+1} bucket has to be connected into a so-called *bottom bucket* BB_{i+1} with weight 2^{i+1}. Generally, the first top buckets TB_0 and TB_1 are merged resulting in BB_1 and for all other buckets TB_{i+1} and BB_i are merged into BB_{i+1}. The bottom bucket with the largest index p is also called *last bucket*. The naming of top, bottom and last bucket is motivated by the graphical representation as seen in Fig. 1. Each output ω_m of the last bucket represents a weight which is a multiple $m = \lceil \frac{M}{2^p} \rceil$ of 2^p. Since only every second output of the low ordered buckets are used, the actual satisfied weight M is represented by $m \cdot 2^p$ minus a tare sum t of size $0 \leq t < 2^p$. This tare is added to the TB_i's using its binary representation as tare variables T_i with weight 2^i for $0 \leq i < p$. I.e. $t = \Sigma_{T_i=1} 2^i = 2^p - (M \bmod 2^p)$. Hence M can be reformulated as:

$$M = m \cdot 2^p - \Sigma_{T_i=1} 2^i \tag{3}$$

In summary, GPW adds $\overline{\omega_m}$ to guarantee a solution smaller than $m \cdot 2^p$. The exact target weight of the PB constraint C is achieved by calculating the tare values T_i a priori[1] according to Eq. 3. Consequently, the constraint $\overline{\omega_m}$ guarantees that any solution with weight $\geq M$ instantly results in a conflict.

Example 1. Given the constraint $C : 2x_1 + 3x_2 + 5x_3 + 7x_4 < 11$, the a_j values are separated due to their binary representation. As $\lfloor \log_2 7 \rfloor = 2$ holds, the largest bucket size is 2^2 and hence $p = 2$. The position of the watchdog can be achieved by: $m = \lceil \frac{M}{2^p} \rceil = \lceil \frac{11}{2^2} \rceil = 3$. The tare values can be calculated with: $t = \Sigma_{T_i=1} 2^i = 2^p - (M \bmod 2^p) = 2^2 - (11 \bmod 2^2) = 1$, i.e. the binary representation of the tare values is 1_{10} leading to: $T_0 = 1, T_1 = 0$. We can check our result by applying them into Eq. 3: $M = m \cdot 2^p - \Sigma_{T_i=1} 2^i = 3 \cdot 2^2 - (1 \cdot 2^0 + 0 \cdot 2^1) = 11$. This leads to the following PW encoding in Fig. 1. Note, the blue dashed lines in this figure are not present in the actual encoding, since $T_1 = 0$.

In [13] it is stated that merging two totalizers of size n requires $\mathcal{O}(n^2)$ clauses and for the whole totalizer $\mathcal{O}(n^2 \log(n))$ clauses are required. The complete PW encoding complexity is given by $\mathcal{O}(n^2 \log(n) \log(a_{max}))$ clauses, where a_{max} is the largest integer weight of all clauses of the MaxSAT problem.

However, as already stated in [8] the number of encoded clauses of the whole totalizer can be bounded by $\mathcal{O}(n^2)$ and not $\mathcal{O}(n^2 \log(n))$. Since we have no doubt about the remaining argumentation of [13], we conclude that the number of clauses of the (G)PW encoding is actually in $\mathcal{O}(n^2 \log(a_{max}))$.

---

[1] Note that tare values of zero do not have any influence. Hence, only the tare bits $T_i = 1$ are added to TB_i, and are also directly set to 1.

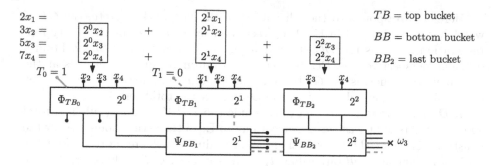

Fig. 1. Polynomial watchdog encoding scheme (Color figure online)

Note, in [13] the concept of a *Local Polynomial Watchdog (LPW)* is introduced which preserves GAC. The complexity is given by $\mathcal{O}(n^3 \log n \log(a_{max}))$. Likewise, we propose that also this complexity has to be corrected to $\mathcal{O}(n^3 \log(a_{max}))$. Still, we do not apply the LPW since the encoding size is not feasible for our application. (G)PW in contrast does not maintain GAC.

4 Dynamic PW Encoding and GPW Algorithm

In this section we introduce our new dynamic PW encoding scheme. We state details of the encoding adjustments and the employment in an iterative MaxSAT solver. The principle of our approach is to lift the static watchdog as described in [13] to a dynamic version, which allows to set a lower bound M of the PB constraint dynamically for optimizing this bound.

The remainder of this section is as follows: we introduce the adjustments of the original PW encoding in Sect. 4.1 and present a complete algorithm to solve the MaxSAT problem based on this encoding in Sect. 4.2.

4.1 Dynamic PW Encoding

In order to employ the PW encoding for representing MaxSAT constraints, we need to allow different watchdog positions and consequently lift the concept of statically a-priori set tare values and watchdog positions to a dynamic one. Thus, we call our modification the *Dynamic Polynomial Watchdog (DPW)* encoding.

As mentioned in Sect. 3.1 the MaxSAT problem can be reduced to find the optimal M^* in a constraint C with $\Sigma_j a_j \cdot x_j \geq M$, where all x_j's and a_j's are appropriately connected to all r_j's and w_j's of the MaxSAT problem leading to $\Sigma_j w_j \cdot r_j \geq M$ in our approach. Note, r_j implies "s_j is satisfiable" but not vice versa. However, the r_j's still represent a lower bound, we will revisit this issue later. In order to increment M, we need to adjust the watchdog and tare values appropriately. Section 4.2 state further details on this adjustment.

Analogously to [13] we define the *Dynamic Global Polynomial Watchdog (DGPW)* as follows:

$$DGPW_i(C) = DPW(C) \wedge \omega_i \tag{4}$$

Here, i corresponds to the ith output of the last bucket. Note, the GPW watchdog as introduced in Eq. 2 represents an *upper bound* for the unary representation, whereas the $DGPW$ watchdog in Eq. 4 is a *lower bound*. From a different perspective: the $DGPW$ requires a minimum number of ones at the output vector, whereas the GPW demands a minimum number of zeros. Thus, our DPW encoding only employs the D_2 part of the totalizer.

If $DGPW_m$ is satisfied, we can conclude the DPW encoding fulfills a total sum of weights of at least $m \cdot 2^p$. According to Eq. 3 the actual bound M for the constraint is achieved by subtracting the tare values which are set to 1. Note, the DPW encoding adds *all* tare values T_i to the top buckets TB_i. Further, the tare variables are not set to a precomputed value. We rather allow the SAT solver to alter the logic value of these variables. Hence, one crucial part of the algorithm is to efficiently determine exact values for the tares.

Example 2. Reconsider Example 1: The DPW additionally adds the tares T_0 and T_1, i.e. the blue dashed lines are part of the DPW encoding. By using D_2 we change the operator of the underlying constraint C from $<$ to $\geq$. If we use the same watchdog position 3 applying $DGPW_3$ and fix the tare values as in Example 1, the represented constraint is $C : 2x_1 + 3x_2 + 5x_3 + 7x_4 \geq 11$.

Based on this encoding we formulate an algorithm leading to the optimization result of the original MaxSAT formulation.

4.2   Dynamic GPW Algorithm

Our algorithm is separated in two phases. First, we apply a *Coarse Convergence (CC)* and finalize with a *Fine Convergence (FC)* as follows:

CC: The watchdog position is increased until the formula is unsatisfiable.
FC: Refines the result of CC by adjusting the tare variables appropriately.

Coarse Convergence (CC). Algorithm 1 gives an overview of the CC phase. It takes the complete CNF of the MaxSAT problem and DPW encoding as input and returns the maximum position m^* of the last bucket for which the constraint is still satisfied.

First we perform an initial SAT solver call in line 2 without additional (watchdog) constraints returning an initial watchdog position. We increment the watchdog connected to the last bucket until the formula is not satisfiable anymore (cf. lines 3–13): Therefore, we calculate the next watchdog position based on the current model (cf. GETLASTSATPOS in line 4) seeking for the last position of the last bucket for which the model is set to true. Since we know that this position represents a lower bound of our solution, we increment it and add the resulting watchdog as an assumption for the next SAT solver call (cf. GETWATCHDOG in lines 5 and 6). If the result is satisfiable we add the last assumption as unit to the CNF allowing the solver to simplify the CNF representation. This is repeated until the solver returns "unsatisfiable", i.e. we found the maximum position m^*.

Algorithm 1. Coarse Convergence

1: **procedure** CoarseConvergence(CNF)
2:     ⟨result, model⟩ ← Solve(CNF);
3:     **while** true **do**
4:         position ← GetLastSatPos(model);
5:         assumption ← GetWatchdog(position+1));
6:         ⟨result, model⟩ ← Solve(CNF + assumption);
7:         **if** result = SAT **then**
8:             CNF ← AddUnitClause(assumption);
9:         **else**
10:            CNF ← AddUnitClause(GetWatchdog(position));
11:            **return** position;   ▷ Return last SAT position
12:        **end if**
13:    **end while**
14: **end procedure**

By doing so, we have determined the first part of Eq. 3 leading to $M^* = m^* \cdot 2^p - \Sigma_{T_i=1} 2^i$. Note, up to this point, no constraints are assumed for the tare variables T_i. Hence, the current model of the tare values does not correspond to the optimal solution. We have to adjust these values as stated in the next subsection. However, the result of the CC phase states the possible solution interval for our searched optimal bound M^* as follows: $(m^* - 1) \cdot 2^p < M^* \leq m^* \cdot 2^p$.

We further optimize the GetLastSatPos function of line 4: As mentioned in Sect. 3.1, s'_j is satisfied if r_j is set to true. Nevertheless, r_j might be false and simultaneously s_j is satisfied, too. This result cannot be seen by our GPW encoding since only the r_j values are connected. Thus, we check each and every soft clause if s_j is satisfied regardless of the value of r_j. Finally, we add up the weight of every *actual* satisfied soft clause resulting in an actual current optimal weight $\hat{M}$ for which holds: $\hat{M} \geq M$. By $\hat{m} = \lceil \frac{\hat{M}}{2^p} \rceil$ we obtain the watchdog position $\hat{m}$ of this optimum. Note, in this case we can immediately add another unit clause to the CNF. As $\hat{M}$ might be larger than M we also might skip output positions of the last bucket and hence, we can add the unit corresponding to the position $\hat{m}$ before calling the SAT solver in line 6 of Algorithm 1. Note that we actually do not need this additional "by-chance" concept, if we would consider appropriate constraints representing the relation $\overline{r_j} \rightarrow \overline{s_j}$, where $\overline{s_j}$ indicates that every literal of the soft clause s_j is falsified. By adding these constraints, r_j would be true iff the soft clause s_j is falsified. However, by adding these constraints, we would lose the potential of this by-chance mechanism as all literals of s_j would be immediately set to false whenever r_j is set to true.

Fine Convergence (FC). Once, we found the coarse solution interval, we seek for the exact result by adjusting the tare variables as part of the Fine Convergence (FC) phase. The general idea is the same as in the CC phase: we force the SAT solver to find a better solution than the current one. Algorithm 2 summarizes our approach. In addition to the modified CNF resulting from Algorithm 1, the procedure also gets the last model from a satisfying SAT solver call from the CC phase as an input.

Algorithm 2. Fine Convergence

1: **procedure** FINECONVERGENCE(CNF, model)
2:     **for** $(n = p - 1; \ n \geq 0; \ n = n - 1)$ **do**
3:         **if** model$[T_n]$ = false **then**
4:             CNF $\leftarrow$ ADDUNITCLAUSE(Negate(T_n));
5:         **else**
6:             assumption $\leftarrow$ Negate(T_n);
7:             $\langle$result, model$\rangle$ $\leftarrow$ SOLVE(CNF + assumption);
8:             **if** result = SAT **then**
9:                 CNF $\leftarrow$ ADDUNITCLAUSE(Negate(T_n));
10:            **else**
11:                CNF $\leftarrow$ ADDUNITCLAUSE(T_n));
12:            **end if**
13:        **end if**
14:    **end for**
15: **end procedure**

First, consider the following observation: In Eq. 3 we have defined the upper bound $(m \cdot 2^p)$ as reference point. Instead, we can also define the value of M using the lower bound $(m - 1) \cdot 2^p + 1$ as reference:

$$M = (m - 1) \cdot 2^p + 1 + \Sigma_{T_i=0} 2^i \qquad (5)$$

As Eq. 5 indicates the tare weights for which the corresponding tare is not satisfied is added to the lower bound. Hence, the tare is only a remainder of the sum of all satisfied weights relating to the lower bound. Consequently, in order to maximize the value of M, we have to maximize $\Sigma_{T_i=0} 2^i$.

Algorithm 2 iterates over all tare variables, from the most significant T_{p-1} to the least significant T_0. If the current tare T_n is already 0, we can add the corresponding unit to our CNF (cf. lines 3 and 4). Otherwise, we have to check whether we can set T_n to zero (and thus maximize the sum of weights) by adding the assumption $T_n = false$ to the solver (cf. lines 6 and 7). If the SAT solver returns satisfiable, we proceed as in line 4 by adding the corresponding unit. Otherwise, it is ensured that the PB constraint is always violated with $T_n = false$, and hence we can fix this tare to true. Note, the last property only holds, if we iterate from the most to the least significant tare position. By doing so, essentially a binary search of the maximum possible tare weight is performed.

As in the coarse convergence, we can explicitly calculate the actual current weight $\hat{M}$. By doing so, we may skip tare positions: If $\hat{M}$ implies that the most significant open tare position T_n must be set to zero, we can directly add the corresponding unit (cf. line 4) and proceed with the next tare position.

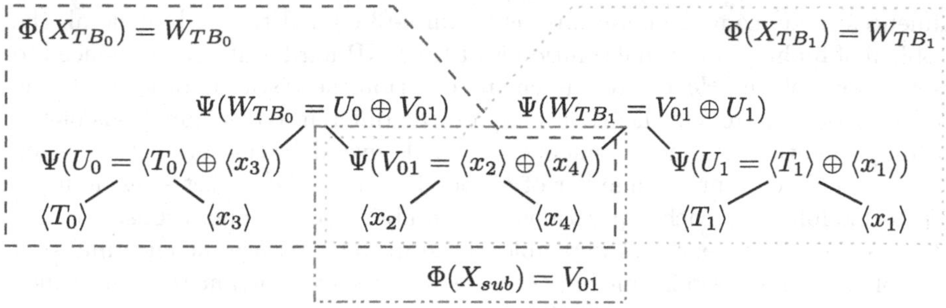

$$\Phi(X_{TB_0}) = W_{TB_0} \qquad\qquad\qquad \Phi(X_{TB_1}) = W_{TB_1}$$

$$\Psi(W_{TB_0} = U_0 \oplus V_{01}) \qquad \Psi(W_{TB_1} = V_{01} \oplus U_1)$$

$$\Psi(U_0 = \langle T_0 \rangle \oplus \langle x_3 \rangle) \qquad \Psi(V_{01} = \langle x_2 \rangle \oplus \langle x_4 \rangle) \qquad \Psi(U_1 = \langle T_1 \rangle \oplus \langle x_1 \rangle)$$

$$\langle T_0 \rangle \qquad \langle x_3 \rangle \qquad\qquad \langle x_2 \rangle \qquad \langle x_4 \rangle \qquad\qquad \langle T_1 \rangle \qquad \langle x_1 \rangle$$

$$\Phi(X_{sub}) = V_{01}$$

Fig. 2. Caching of adder: Φ applied to TB_0 and TB_1 of Example 1. The sorted vector V_{01} can be used for the encoding of TB_0 and TB_1.

5 PW Encoding Optimizations

In this section we state three fundamental optimizations on the PW encoding. In Sect. 5.1, we describe a technique which allows reusing already encoded parts of the network. Note, this optimization concept is already mentioned in [13] as future work. In Sect. 5.2, we propose an approach for determining the cone-of-influence of encoded outputs of the network. By doing so, we are able to incrementally build the needed parts of the encoding, leading to a significant reduction of the encoding size. In Sect. 5.3, we present a concept to set any weight as lower or upper bound. Applying this to the original PW encoding allows to use any operator for the PB constraint directly without converting the formula. Note, all concepts can be utilized in the classical PW encoding and moreover do not affect each other. I.e., utilizing one of the following techniques does not (negatively) influence the efficiency of one of the others, in general.

5.1 Caching of Adder

The PW encoding consists of many shared sub-formulae Ψ among different buckets which actually need be encoded only once [13] as shown in Example 3.

Example 3. Considering Example 1, both TB_0 and TB_1 contain an identical subset of input variables $X_{sub} = \{x_2, x_4\}$. This subset can be encoded once as $\Phi(X_{sub})$ and reused for TB_0 and TB_1 as shown in Fig. 2. The dashed and dotted boxes indicate involved buckets, and the dash-dotted box the shared part.

There is another subset $X_{sub2} = \{x_2, x_3\}$ of TB_0 and TB_2 in Example 1, which could be proceeded likewise. Note, if we try to share both subsets X_{sub} and X_{sub2} at the same time, x_4 of TB_0 will be encoded twice. This additional encoding will degrade the outcome of the method and should therefore be avoided.

The core idea of the Adder Caching (AC) is to reuse the encoding of this shared parts whenever such a sub-formulae is identified. Unfortunately, merging buckets/sub-formulae is quite sophisticated, since caching of one sub-formula influences the upcoming operations, as Example 3 demonstrates. In general, the problem of finding an optimal solution is at least NP-hard as it can be reduced to a set cover problem. Hence, we implemented various heuristics in order to decide which parts to share. Our heuristics rely on the different cost estimations of one caching operation. As measurements, we tried several static parameters which can be calculated a priori: number of encoded clauses, the bucket sizes, number of possible follow up cache operations and number of cache operations, to name a few. Although all heuristics were able to significantly reduce the encoding size, none of them has a significant impact in terms of solved instances or run time.

Instead, we developed a heuristic which is not as effective in terms of encoding size, but has a significant impact on run time as experimental results show. We collocate the sorter inputs according to their corresponding soft clause weight, such that for each weight w, there is a list of corresponding inputs with weight w. Then, for one weight w exactly one totalizer Φ is encoded, i.e. if there are n soft clauses with identical weight w, the totalizer for sorting these n soft clauses is only built once. This sub-formula is shared over all buckets needed for the binary representation of weight w. This is repeated for all weights, and finally the top bucket are constructed by connecting the built sub-formulae using Ψ. In contrast to all other heuristics, multiple cache steps are considered at once as *all* involved adders are merged. We assume that this is one reason why the other heuristics are not as effective and suggest future work on this topic.

5.2   Cone-of-Influence Encoding

As the methodology in the previous section reduces the cost of top buckets, we develop a technique mainly reducing the encoding size of the bottom buckets. Note, the encoding of bottom buckets usually dominates the whole PW encoding as these buckets have more inputs than top buckets. We apply this technique within the CC phase, where the watchdog position is incremented.

The cone-of-influence (COI) encoding converts only needed parts of the (D)PW into CNF. Consider a standard $GPW(C) = PW(C) \wedge \overline{\omega_i}$. We observed that only the output ω_i needs to be encoded for deciding GPW as the encoding of Ψ guarantees *at most* $i - 1$ ones at the output. By encoding only ω_i it is ensured to create at most $\mathcal{O}(n)$ clauses for the last bucket with n inputs, instead of $\mathcal{O}(n^2)$. Depending on the position of ω_i in the output vector W, even more encoding size can be saved: positions close to the borders of W lead to smaller encodings than in the middle as the upcoming Example 4 will motivate.

We introduce a binary tree, which represents the recursive construction of the totalizer encoding including the information whether a variable is already encoded. Hence, each node represents a snapshot of the current Ψ encoding for all partially sorted outputs. Before encoding a specific (output) variable, the tree nodes directly indicate which variables and clauses have to be added and which are already encoded. We traverse the tree from the root to the leaves collecting all clauses within the cone-of-influence of the demanded variable considering also introduced helper variables. Note, the tree also represents and considers the carry inputs of the bottom buckets where only every second output is encoded.

$$\Psi(U = U' \oplus \langle x_3 \rangle)$$
$$U = \langle *, u_2, u_3 \rangle$$

$$\Psi(W = U \oplus V)$$
$$W = \langle *, *, *, w_4, * \rangle$$

$$\Psi(U' = \langle x_1 \rangle \oplus \langle x_2 \rangle)$$
$$U = \langle u_1', u_2' \rangle$$

$$\Psi(V = \langle x_4 \rangle \oplus \langle x_5 \rangle)$$
$$V = \langle v_1, v_2 \rangle$$

u_1'	x_3	$\overline{u_1}$			
u_1'		$\overline{u_2}$	u_2'	x_3	$\overline{u_2}$
u_2'		$\overline{u_3}$		x_3	$\overline{u_3}$

u_1	v_1	$\overline{w_1}$						
u_1	v_2	$\overline{w_2}$	u_2	v_1	$\overline{w_2}$			
u_1		$\overline{w_3}$	u_2	v_2	$\overline{w_3}$	u_3	v_1	$\overline{w_3}$
u_2		$\overline{w_4}$	u_3	v_2	$\overline{w_4}$			
u_3		$\overline{w_5}$						

x_1	x_2	$\overline{u_1'}$		
x_1		$\overline{u_2'}$	x_2	$\overline{u_2'}$

x_4	x_5	$\overline{v_1}$		
x_4		$\overline{v_2}$	x_5	$\overline{v_2}$

v_1	$\overline{w_4}$
v_2	$\overline{w_5}$

Fig. 3. Binary tree representing the cone-of-influence of w_4.

Example 4. Figure 3 illustrates our generated tree. Each node is represented by a table indicating the clauses created by the corresponding Ψ. The ith row in the table shows the needed clauses for the ith output entry according to Ψ. Consider that only w_4 (as seen at the very left) has to be encoded as it might be the current watchdog. By applying our cone-of-influence technique, we only need to consider the 12 highlighted clauses, whereas 11 clauses (in gray) could be saved for the encoding. A $*$ in the vector indicates that the corresponding variable (row of table) is not encoded. Moreover, assume that we may only consider w_5: in this case we just need to encode the very last row of each table.

5.3 Exact Bound Encoding

We explicitly enforce specific values as a lower or upper bound, where we need to encode D_2 for the lower and D_1 for the upper bound. Note, for an exact upper bound $<M$, we need constraints of the form $\overline{r_j} \Rightarrow \overline{s_j}$ and set ω_i to false as in Sect. 3.3. In both cases, we have to adjust the tare values by utilizing Eq. 3.

We utilize the Exact Bound (EB) encoding for setting the next sum of weights in our CC phase. Instead of incrementing the next watchdog position, we explicitly add sufficient assumptions to enforce the weight $\hat{M} + 1$ for the next solver call. By doing so, instances are easier to solve for the SAT solver (since we add more and specific assumptions leading to the solution), but the number of solver calls is increased in the worst case. However, as experimental results show, we actually often converge faster to the optimum, also due to weights satisfied by soft clauses by chance (cf. computation of $\hat{M}$ in Sect. 4.2).

We also employ a restricted version of the exact upper bound encoding by setting the first unsatisfied watchdog of the CC phase to $\overline{\omega_{m^*+1}}$. Note, we do not restrict the tare values or add $\overline{r_j} \Rightarrow \overline{s_j}$ constraints. By doing so, we implicitly forbid assignments of the relaxation literals leading to a weight $>m^* \cdot 2^p$ and thus guiding the SAT solver. We only employ the restricted upper bound in combination with the COI encoding of Sect. 5.2 in order to avoid the additional encoding of $\mathcal{O}(n^2)$ D_1 clauses for the last bucket.

6   Experimental Results

We implemented the new encoding scheme and algorithm in C++ as extension of the QMaxSAT solver [7] as used in the MaxSAT Evaluation 2017 [6].

All experiments were run on one Intel Xeon E5-2650v2 core at 2.60 GHz, with 64 GB of main memory and Ubuntu Linux 16.04 in 64-bit mode as operating system. We aborted all experiments whose computation time exceeded 3,600 CPU seconds or which required more than 32 GB of memory as in the evaluation of 2017. We also used the benchmark set of [6] consisting of 767 weighted instances.

Table 1 shows the efficiency of our optimizations (Sect. 5) of the (D)GPW. As results show, AC and COF have significant impact on either the number of encoded variables or clauses leading to a encoding size of 50% compared to Plain. Whereas, EB is not designed to decrease the encoding size, but still has a significant impact on the run time. Compared to Plain, DGPW is about $2X$ faster and has almost $3X$ less encoding size in number of clauses and variables wrt. the commonly solved instances. Moreover, 28 more instances could be solved.

Table 2 compares the number of solver calls and run time for the two solving phases as well as SAT solver result for the DPGW[2]. In total only 21 SAT solver calls are needed on average, most of them are satisfiable calls in the CC phase. The comparison between the average and median time shows that for easy instances the most time is spent in the CC phase and for hard instances it is the FC phase. The same holds for satisfiable and unsatisfiable solver calls.

In a second experiment, we compare DGPW with QMaxSAT. Table 3 shows the results on the QMaxSAT encodings. The AutoQD heuristic chooses between adder and DGPW as (modulo) totalizer are inferior to our encoding (cf. Fig. 4a). Our heuristic chooses DGPW if either the sum of weights is small ($<400,000$) or large ($>2,000,000,000$). In all other cases the adder is chosen. The original QMaxSAT AutoQ heuristic also chooses (modulo) totalizers for a small sum of weights ($<2^{17}$) as they usually outperform adders in this case. In addition, our empirical results show that DGPW dominates the adder for huge weights. Table 3 is composed as Table 1 comparing two neighboring columns wrt. the commonly solved instances. As expected, the adder needs two orders of magnitudes less clauses due to linear encoding complexity, and the totalizer needs two orders of magnitudes more clauses due to the naive encoding of weights. Figure 4 underlines our results for the individual encodings in a scatter and cacti plot.

Notably, our results are comparable to the MaxSAT evaluation where AutoQ also solved 503 instances with an average run time of 385.18 s [6]. DGPW is competitive in the number of solved instances wrt. the other networks (470 vs. 228, 329, and 491). Remarkably, DGPW is at least $2X$ faster than every other network for the commonly solved instances. The new VBS solves 31 more instances in 60% of the run time, whereas our basic AutoQD solves 19 more instances than the evaluation version of QMaxSAT in overall 70% of the run

---

[2] The run time difference to Table 1 is caused by the time needed for the encoding and the remaining part of our algorithm (e.g. analyzing the SAT solver model).

Table 1. Comparison of *DGPW* without extensions (Plain) with adder caching (AC), cone-of-influence (COI), exact bound (EB) and a combined (DGPW) version using all optimizations. First, the average run time and the total number of solved instances are given. Two neighboring columns compare an optimization with Plain opposing the average run time, median run time and encoding size wrt. the commonly (com.) solved instances. The encoding size is given by the average number of clauses "#cl" (in millions) and variables "#var".

		Plain	AC	Plain	COI	Plain	EB	Plain	DGPW
	#instances	442	453	442	455	442	451	442	470
	Avg. time	323.00	269.63	323.00	313.93	323.00	297.47	323.00	264.16
	Med time	20.56	12.52	20.56	23.26	20.56	19.71	20.56	12.05
com.	#instances	430	430	439	439	439	439	429	429
	Avg. time	302.89	207.07	302.51	288.84	303.08	251.59	306.73	164.60
	Med time	16.55	10.04	18.48	17.98	18.48	16.48	17.22	7.60
	Avg. #cl	29.19	21.32	28.62	14.66	28.59	28.62	29.26	9.78
	Avg. #var	96,803	53,639	95,324	80,295	95,298	95,512	97,000	38,336

Table 2. DGPW divided by Coarse Convergence (CC), Fine Convergence (FC), satisfiable (SAT) and unsatisfiable (UNSAT) solver calls. For each phase the average/median number of the solver calls and solving time in seconds are given.

	CC		FC		SAT		UNSAT	
	Avg.	Med	Avg.	Med	Avg.	Med	Avg.	Med
Solver calls	14.33	9.00	6.69	3.00	19.29	12.00	1.72	2.00
Solving time	120.30	4.93	139.20	1.62	117.73	6.17	141.77	0.87

Table 3. Comparing DGPW with totalizer (Tot), modulo totalizer (ModT) and adder (Add). The virtual best solver (VBS) of the QMaxSAT with integrated DGPW (VBSQD) and without (VBSQ) is depicted. Finally, results are shown for the original heuristic of QMaxSAT (AutoQ) and our adopted one (AutoQD).

		Tot	DGPW	ModT	DGPW	Add	DGPW	VBSQ	VBSQD	AutoQ	AutoQD
	#instances	228	470	329	470	491	470	504	535	503	522
	Avg. time	326.29	264.16	372.49	264.16	430.39	264.16	381.04	301.44	408.30	334.44
	Med time	33.12	12.05	21.60	12.05	29.59	12.05	26.96	19.01	28.68	23.34
com.	#instances	225	225	313	313	428	428	504	504	497	497
	Avg. time	303.45	142.78	330.64	141.23	388.37	163.16	381.04	228.60	380.95	268.35
	Med time	32.12	5.01	18.48	3.69	17.39	7.53	26.96	13.55	27.62	19.80
	Avg. #cl	28.57	0.60	11.61	2.88	0.17	16.45	1.92	12.37	0.33	6.70
	Avg. #var	63,920	11,901	108,006	23,341	45,768	57,404	60,796	56,062	51,328	48,890

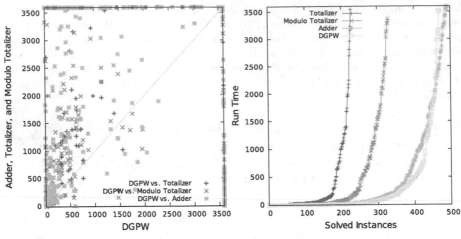

(a) Run time comparison per instance   (b) Solved instances for a given run time

Fig. 4. Comparing different QMaxSAT encodings (adder, totalizer and modulo totalizer) with our newly introduced *DGPW* encoding

time, both wrt. the commonly solved instances. Thus, there is still some potential for our heuristic, however we conserve most of the combined capacity.

7   Conclusions

In this paper, we presented a new encoding scheme for mapping the weighted MaxSAT problem to the Polynomial Watchdog (PW) encoding. To do so, we extended the original encoding by the support of dynamic watchdogs and tare variables. Furthermore, we introduced three optimizations for the PW encoding.

As experimental results show, our optimizations lead to $3X$ smaller encoding sizes and $2X$ faster run times on average compared with the original PW encoding in [13]. Furthermore, we showed the applicability of our new encoding scheme while achieving a speed-up of more than $2X$ compared to the competitors. Finally, we integrated our encoding into a state-of-the-art MaxSAT solver and implemented a prototypical heuristic for deciding the encoding used.

The presented encoding could also be used to handle PB constraints in other solvers like Open-WBO [19]. As future work, we want to investigate the minimization of bucket sizes, which is usually the bottleneck of the PW encoding, by multiplying or dividing the weights with a common factor, and thus changing the binary representation. Moreover, we plan to investigate further heuristics for the adder caching as there is even more potential for the LPW presented in [13].

References

1. Menai, M.E.B., Al-Yahya, T.N.: A taxonomy of exact methods for partial Max-SAT. J. Comput. Sci. Technol. **28**(2), 232–246 (2013)
2. Hansen, P., Jaumard, B.: Algorithms for the maximum satisfiability problem. Computing **44**(4), 279–303 (1990)
3. Zhang, H., Shen, H., Manya, F.: Exact algorithms for MAX-SAT. Electron. Notes Theor. Comput. Sci. **86**(1), 190–203 (2003)
4. Marques-Sila, J., Planes, J.: Algorithms for maximum satisfiability using unsatisfiable cores. In: Gulati, K. (ed.) Advanced Techniques in Logic Synthesis, Optimizations and Applications, pp. 171–182. Springer, New York (2011). https://doi.org/10.1007/978-1-4419-7518-8_10
5. Ansótegui, C., Gabàs, J.: Solving (weighted) partial MaxSAT with ILP. In: CPAIOR, pp. 403–409 (2013)
6. Twelfth MaxSAT evaluation (2017). http://mse17.cs.helsinki.fi/index.html
7. Koshimura, M., Zhang, T., Fujita, H., Hasegawa, R.: QMaxSAT: a partial MaxSAT solver system description. J. Satisf. Boolean Model. Comput. **8**, 95–100 (2012)
8. Bailleux, O., Boufkhad, Y.: Efficient CNF encoding of Boolean cardinality constraints. In: Rossi, F. (ed.) CP 2003. LNCS, vol. 2833, pp. 108–122. Springer, Heidelberg (2003). https://doi.org/10.1007/978-3-540-45193-8_8
9. Joshi, S., Martins, R., Manquinho, V.: Generalized totalizer encoding for pseudo-Boolean constraints. In: Pesant, G. (ed.) CP 2015. LNCS, vol. 9255, pp. 200–209. Springer, Cham (2015). https://doi.org/10.1007/978-3-319-23219-5_15
10. Uemura, N., Fujita, H., Koshimura, M., Zha, A.: A sat encoding of pseudo-Boolean constraints based on mixed radix, March 2017
11. Ogawa, T., Liu, Y., Hasegawa, R., Koshimura, M., Fujita, H.: Modulo based CNF encoding of cardinality constraints and its application to MaxSAT solvers. In: 2013 IEEE 25th International Conference on Tools with Artificial Intelligence (ICTAI), pp. 9–17. IEEE (2013)
12. Warners, J.P.: A linear-time transformation of linear inequalities into conjunctive normal form. Inf. Process. Lett. **68**(2), 63–69 (1998)
13. Bailleux, O., Boufkhad, Y., Roussel, O.: New encodings of pseudo-Boolean constraints into CNF. In: Kullmann, O. (ed.) SAT 2009. LNCS, vol. 5584, pp. 181–194. Springer, Heidelberg (2009). https://doi.org/10.1007/978-3-642-02777-2_19
14. Codish, M., Fekete, Y., Fuhs, C., Schneider-Kamp, P.: Optimal base encodings for pseudo-Boolean constraints. In: Abdulla, P.A., Leino, K.R.M. (eds.) TACAS 2011. LNCS, vol. 6605, pp. 189–204. Springer, Heidelberg (2011). https://doi.org/10.1007/978-3-642-19835-9_16
15. Eén, N., Sörensson, N.: Translating pseudo-Boolean constraints into SAT. J. Satisf. Boolean Model. Comput. **2**, 1–26 (2006)
16. Batcher, K.E.: Sorting networks and their applications. In: AFIPS Spring Joint Computing Conference, pp. 307–314. ACM (1968)
17. Bryant, R.E.: Graph-based algorithms for Boolean function manipulation. IEEE Trans. Comput. **100**(8), 677–691 (1986)
18. Manthey, N., Philipp, T., Steinke, P.: A more compact translation of pseudo-Boolean constraints into CNF such that generalized arc consistency is maintained. In: Lutz, C., Thielscher, M. (eds.) KI 2014. LNCS (LNAI), vol. 8736, pp. 123–134. Springer, Cham (2014). https://doi.org/10.1007/978-3-319-11206-0_13
19. Martins, R., Manquinho, V., Lynce, I.: Open-WBO: a modular MaxSAT solver'. In: Sinz, C., Egly, U. (eds.) SAT 2014. LNCS, vol. 8561, pp. 438–445. Springer, Cham (2014). https://doi.org/10.1007/978-3-319-09284-3_33

Solving MaxSAT with Bit-Vector Optimization

Alexander Nadel[(✉)]

Intel Corporation, P.O. Box 1659, 31015 Haifa, Israel
alexander.nadel@intel.com

Abstract. We explore the relationships between two closely related optimization problems: MaxSAT and Optimization Modulo Bit-Vectors (OBV). Given a bit-vector or a propositional formula F and a *target bit-vector* T, Unweighted Partial MaxSAT maximizes *the number of satisfied bits* in T, while OBV maximizes the *value* of T. We propose a new OBV-based Unweighted Partial MaxSAT algorithm. Our resulting solver–**Mrs. Beaver**–outscores the state-of-the-art solvers when run with the settings of the Incomplete-60-Second-Timeout Track of MaxSAT Evaluation 2017. **Mrs. Beaver** is the first MaxSAT algorithm designed to be incremental in the following sense: it can be re-used across multiple invocations with different hard assumptions and target bit-vectors. We provide experimental evidence showing that enabling incrementality in MaxSAT significantly improves the performance of a MaxSAT-based Boolean Multilevel Optimization (BMO) algorithm when solving a new, critical industrial BMO application: cleaning-up weak design rule violations during the Physical Design stage of Computer-Aided-Design.

1 Introduction

Modern SAT solvers [9,30,44] can be applied to solve various optimization problems in the domain of propositional and bit-vector logic. One such well-known problem is Weighted MaxSAT [23,24][1]. A Weighted MaxSAT instance comprises a set of *hard* satisfiable propositional clauses H (H may also contain bit-vector constraints, reducible to propositional clauses) and a set of weighted *soft* constraints $T = \{t_{n-1}, t_{n-2}, \ldots, t_0\}$, where each constraint t_i is associated with a strictly positive integer weight w_i. To solve such an instance, the solver is required to return an assignment which satisfies H and maximizes the function $\sum_{i=0}^{n-1} t_i * w_i$, comprising the overall weight of the satisfied soft constraints. For the rest of the paper, for convenience and without restricting generality, we assume that every soft constraint is a unit clause.[2] Thus, T can be thought of as a bit-vector, where t_0 is its Least Significant Bit (LSB) and t_{n-1} is its Most

---

[1] For the rest of the paper, MaxSAT refers to Partial MaxSAT, where arbitrary hard constraints are allowed.

[2] An arbitrary soft constraint t_i, reducible to a set of propositional clauses F, can be transformed to a unit clause s', where s' is a fresh variable, by adding the clause $\neg s' \vee c$ to H, for each clause $c \in F$.

© Springer International Publishing AG, part of Springer Nature 2018
O. Beyersdorff and C. M. Wintersteiger (Eds.): SAT 2018, LNCS 10929, pp. 54–72, 2018.
https://doi.org/10.1007/978-3-319-94144-8_4

Significant Bit (MSB). We call T the *target bit-vector*, or, simply, the *target* and every $t_i \in T$ a *target bit*.

Various optimization problems can be expressed as a sub-class of Weighted MaxSAT. Unweighted MaxSAT comprises a restriction of Weighted MaxSAT to problems where all the weights are equal to 1. Essentially, in Unweighted MaxSAT, one has to maximize *the number of satisfied target bits* or, in another words, minimize the number of unsatisfied target bits.

Optimization Modulo Bit-Vectors (OBV), also known as Lexicographic SAT (LEXSAT), is another optimization problem, recently studied in [10,11,25,35, 40,41]. In OBV, the *value* of T has to be maximized (where T is interpreted as an unsigned integer). To reduce OBV to Weighted MaxSAT, one can simply assign every target bit t_i the weight 2^i, thus ensuring that the weight of any bit $t_{i \neq 0}$ is greater than the overall weight of the bits less significant than i. The first OBV algorithm to be implemented, νZ [10,11], solved OBV by applying this very reduction. However, it was shown in [35] that dedicated SAT-based OBV algorithms are substantially more efficient.

Can one then take the opposite route, that is, reduce MaxSAT to OBV? Our answer is affirmative.

We propose a new OBV-based Unweighted MaxSAT algorithm, called **Mrs. Beaver**. **Mrs. Beaver** is composed of two stages: the *incomplete* stage, followed by the *complete* stage. The basic version of the algorithm, applied at the incomplete stage, invokes an OBV algorithm with the *original target* to approximate an Unweighted MaxSAT solution. We propose several enhancements to the basic algorithm in order to find a better approximation faster. The basic version of the complete stage invokes an OBV algorithm whose target comprises the *sum of the bits of the original target* starting with the approximate solution, generated by the incomplete stage.

At its core, **Mrs. Beaver** is purely SAT-based. It re-uses a single incremental SAT instance across all the SAT invocations. Performance-wise, **Mrs. Beaver** is especially useful in the context of incomplete solving. It outperforms the state-of-the-art Unweighted MaxSAT solvers when run with the settings of the Incomplete-60-Second-Timeout Track of the MaxSAT Evaluation 2017.

Unlike the state-of-the-art Unweighted MaxSAT algorithms, **Mrs. Beaver** was designed to be incremental in the following sense: it can always be reused with different hard assumptions and targets. We demonstrate that incrementality in MaxSAT is useful in the context of the Boolean Multilevel Optimization problem (BMO) [26]. BMO can be thought of as the following generalization of Unweighted MaxSAT: instead of a target bit-vector T, there are multiple target bit-vectors $T_{m-1}, T_{m-2}, \ldots, T_0$. The goal is to maximize the number of satisfied bits in each of the targets, where satisfying one bit of T_i is preferred to satisfying all the bits in $T_{i-1}, T_{i-2}, \ldots, T_0$. Note that when $m = 1$, BMO is essentially identical to Unweighted MaxSAT, while if every target has only one bit, BMO is identical to OBV. BMO can be solved with iterative invocations of an Unweighted MaxSAT solver [26]. We show that, on benchmarks generated by a critical industrial problem we encountered at Intel and have described in the following paragraph, an incremental **Mrs. Beaver**-based solution is 6 times faster than a non-incremental `Mrs. Beaver`-based one, and 10 times faster than a non-incremental Unweighted MaxSAT-based algorithm which applies the best state-of-the-art Unweighted MaxSAT solver. In addition, while our incremental `Mrs. Beaver`-based algorithm is 1.2 times slower than the best dedicated BMO solver, it uses 50 times less memory (2 Gb vs. 100 Gb on average).

As part of the Physical Design stage of Computer-Aided-Design (CAD), one has to solve the problem of placing and routing all the devices, while making sure that the resulting layout satisfies so-called hard design rules that originate in the manufacturing requirements. This problem can be solved by reducing it to bit-vector logic and applying some dedicated algorithms [32,33]. In practice, however, there also exist *soft* design rules, whose satisfaction is not necessary but desirable. A failure to satisfy a soft rule increases the manufacturing cost. The soft design rules are divided into classes according to the actual cost of their violation, such that satisfying a design rule of a certain class i is more important that satisfying all the design rules of lower classes. The problem of satisfying the soft design rules after completing the process of place & route under the hard design rules is immediately reducible to BMO.

In the text that follows, Sect. 2 contains preliminaries. Section 3 discusses two desirable properties of SAT-based optimization algorithms: responsiveness and incrementality, while Sect. 4 reviews OBV algorithms in light of these two properties. Section 5 introduces our new Unweighted MaxSAT algorithm– `Mrs. Beaver`. Section 6 discusses how to apply `Mrs. Beaver` to solve BMO. Section 7 analyzes the experimental results, and Sect. 8 sums up our work and conclusions.

2   Preliminaries

2.1   SAT Solving

A SAT solver [9,30,44] receives a propositional formula F in Conjunctive Normal Form (CNF) and returns a satisfying assignment (also called a *model*) if one exists. In *incremental SAT solving under assumptions* [19,34,37], the user may invoke the SAT solver multiple times, each time with a different set of *assumptions*, where each assumption is a literal, and, possibly, additional clauses. The solver then checks the satisfiability of all the clauses provided so far while enforcing the values of the current set of assumptions only.

Modern SAT solvers apply phase saving [20,42,46] as their polarity selection heuristic. In phase saving, once a variable is picked by the variable decision heuristic, the literal is chosen according to its latest value, where the values are normally initialized with 0.

2.2   State-of-the-Art Unweighted MaxSAT Solvers

Unweighted MaxSAT is an active area of research as can be seen from the ever-improving results in the MaxSAT Evaluations held since 2006 [2]. We briefly summarize the state-of-the-art in Unweighted MaxSAT solving, based on the MaxSAT Evaluation 2017 results [1].

Since 2011, the MaxSAT Evaluations have had two types of categories: *complete* and *incomplete*. Complete solvers look for a solution that guarantees the absolute optimum, given a relatively generous time-out. Incomplete solvers on the other hand seek to find a good solution (that is, a solution in which there are as few as possible unsatisfied target bits), given a small time-out. Incomplete solving can be useful in applications where time resources are limited and good enough solutions are sufficient.

The winner in the complete category of MaxSAT Evaluation 2017 [1] is `Open-WBO-RES` [38], closely followed by `MaxHS` [14]. `Open-WBO-RES` is a strategy implemented within the framework of the SAT-based `Open-WBO` solver. `Open-WBO-RES` applies unsatisfiable core [5,17,36] analysis-based algorithms [21, 27–29] with resolution-based partitioning [38]. `MaxHS` combines SAT and Mixed Integer Programming (MIP) [14–16].

There were two incomplete categories in MaxSAT Evaluation 2017, based on time-outs of 60 and 300 seconds. An `Open-WBO` strategy–`Open-WBO-LSU` (based on the linear search SAT-UNSAT algorithm (LSU) [6])–won the 60-second category, followed by `MaxHS` and the `MaxRoster` [47] solver. `MaxRoster` won the 300-second category, followed by `Open-WBO-LSU` and `MaxHS`. `MaxRoster` is a SAT-based solver that switches dynamically between different MaxSAT strategies [47].

2.3   Totalizer Encoding

Given a target bit-vector $T = \{t_{n-1}, t_{n-2}, \ldots, t_0\}$ and a model μ, let $unsBits(T, \mu) = \sum_{i=0}^{n-1} \neg\mu(t_i)$ be the number of unsatisfied target bits in μ. We drop μ and use simply $unsBits(T)$, when allowed by the context.

Our algorithms need a way to *a*) efficiently create a bit-vector representing the number of unsatisfied target bits, while at the same time *b*) imposing an upper bound on the number of unsatisfied target bits, or, in other words, asserting the *cardinality constraint* $unsBits(T) \leq b$ for a given b.

To that end, we apply totalizer encoding [4]. The totalizer encodes the sum of the bits in a bit-vector in *unary representation*, which is known to be much more efficient than binary representation in terms of propagation power [3,4].

Given a bit-vector $S = \{s_{n-1}, s_{n-2}, \ldots, s_0\}$, the totalizer is a binary tree whose top-most node–$tot(S)$–is a bit-vector of width n, representing the sum of S's bits in unary representation; that is, we have $tot(S)_i = 1$ iff $\sum_{j=0}^{n-1} s_j \geq i$. The totalizer encoding requires $O(n * log(n))$ variables and $O(n^2)$ clauses.

The totalizer encoding is substantially more efficient if a (low) upper bound–b–on the number of unsatisfied bits in S is known [13]. In that case, the order of the number of clauses goes down to $O(n * b)$. This is because the width of all the nodes longer than b (including the top node $tot(S)$) can be cut down to b. To impose the cardinality constraint $\sum_{i=0}^{n-1} s_i \leq b$, one has to add one additional bit $tot(S)_b$ to every bit-vector longer than b (in the original totalizer encoding), and set $tot(S)_b$ to 0.

We denote by $tot(S, \leq b)$ the bit-vector of width $b + 1$, representing the totalizer's top node, which encodes the sum of all the bits in S in unary representation, where the cardinality constraint $\sum_{i=0}^{n-1} s_i \leq b$ is asserted.

3   Responsiveness and Incrementality

This section discusses two desirable properties of SAT-based optimization algorithms: responsiveness and incrementality. We also briefly review existing MaxSAT algorithms with respect to these properties.

3.1   Responsiveness

By *responsiveness*, we mean the ability of the solver to keep generating and outputting better and better solutions during the solving process. Such a property can be useful in various applications when the time resources are limited. Responsiveness is essential for incomplete MaxSAT solving.

3.2   Incrementality

By *incrementality*, we mean the ability of the solver to stay alive and handle many queries, as incremental SAT solving under assumptions does.

Incrementality in Current MaxSAT Algorithms. Unfortunately, we are unaware of any state-of-the-art MaxSAT solver that does not become invalid after a single invocation. The only attempt at incremental MaxSAT solving was made in [43], where Open-WBO was modified so that one could re-use the solver and add hard and soft clauses between invocations. Unfortunately, the proposed algorithm has not been integrated into the main Open-WBO release. In any event, however, the ability merely to *add* clauses is not sufficient for implementing a BMO algorithm, based on incremental MaxSAT.

Incrementality Under Soft Assumptions. For our application, we need a more generic incremental API, where each invocation has its own target bit-vector. In other words, we want to be able to change the set of soft clauses completely between invocations. Our Unweighted MaxSAT solver–Mrs. Beaver–meets that requirement.

Incrementality Under Hard and Soft Assumptions. We believe that, in addition to the ability to change the target, MaxSAT users would benefit if the solvers could handle *hard assumptions*, which hold only for one particular invocation of the solver (like incremental SAT solving). Given such an API, one could de-activate irrelevant clauses and alternate between MaxSAT and pure SAT calls (where pure SAT calls would have an empty target). Our application does not require hard assumptions, but hard assumptions could make it possible to use MaxSAT across other applications. For example, one could then integrate Unweighted MaxSAT into the IC3 [12] (aka PDR [18]) algorithm for incremental SAT-based model checking for maximizing the number of state elements that are assigned don't care values in satisfiable queries (based on dual-rail encoding).

Hard assumptions do not appear in the pseudo-code of the algorithms proposed in this paper. However, adding hard assumptions β to our algorithms is trivial, since, at their core, our algorithms are SAT-based. One simply has to add the assumptions in β to the list of hard assumptions for every SAT invocation.

4   Optimization Modulo Bit-Vectors (OBV) Algorithms

This section reviews existing OBV algorithms in light of their performance, responsiveness, and incrementality. We needed to analyze these properties in order to choose the underlying OBV algorithm for Mrs. Beaver.

As we mentioned in Sect. 1, the first OBV solver was νZ [10,11]. νZ applied a straightforward reduction to Weighted MaxSAT. However, this approach does not scale [35]. Two dedicated OBV algorithms were proposed in [35]: OBV-WA and OBV-BS.

OBV-WA can be thought of as a linear search for the maximal model, starting with the highest possible value of the target and working towards 0. The algorithm stops at the first satisfying assignment. OBV-WA is an incremental algorithm implemented inside a SAT solver. OBV-WA can be quite efficient, but, unfortunately, it is not responsive as it finds only one (best) model. Thus it cannot serve as the underlying building block for Mrs. Beaver.

OBV-BS is depicted in Algorithm 1. Essentially, the algorithm implements a binary search over the possible values of a target T. The algorithm receives a CNF formula F and the target T.[3] It maintains the current model μ, initialized with an arbitrary model to F at line 3, and a partial assignment α, which is empty at the beginning. The main loop of the algorithm (starting at line 6) goes over all the bits of target T starting from the MSB t_{n-1} down to t_0. Each iteration extends α with either t_i or $\neg t_i$, where t_i is preferred over $\neg t_i$ iff there exists a model where t_i is assigned 1 while bits higher than i have already been assigned in previous iterations. *Phase saving optimization*, shown in lines 2 and 10, sets the phase saving array for the target bits with 1's before every SAT invocation, thus encouraging the solver to prefer a higher value for T. Phase saving optimization improves the performance of the algorithm. OBV-BS is incremental, since it is based on incremental SAT solving. It is also quite responsive, since it keeps finding better models throughout its execution.

Independently, OBV-BS, without phase saving optimization, was also suggested in [25] in the context of solving the LEXSAT problem, which is, essentially, identical to OBV.

BINARY [40,41] is another OBV algorithm. BINARY can be thought of as a partial integration of OBV-WA into OBV-BS. BINARY applies OBV-BS, where, for every SAT solver iteration inside the main loop, it adds the upper half of the bits, that is, $\{t_i, t_{i+1}, \ldots, t_{\lceil i/2 \rceil}\}$, as assumptions, rather than only the current bit $\{t_i\}$. If the invocation is satisfiable, the solver can update i to bit number $\lceil i/2 \rceil$. Otherwise, it halves the number of satisfied assumptions and stays at iteration i.

---

[3] If the original formula F is a bit-vector formula; it is preprocessed and translated to CNF first.

Algorithm 1. OBV-BS

1: **function** SOLVE(CNF Formula F, Target $T = \{t_{n-1}, t_{n-2}, \ldots, t_0\}$)
2:     Set the phase saving values of $\{t_{n-1}, t_{n-2}, \ldots, t_0\}$ to 1
3:     $\mu := \text{SAT}$
4:     **if** SAT solver returned UNSAT **then return** UNSAT
5:     $\alpha := \{\}$
6:     **for** $i \leftarrow n - 1$ **downto** 0 **step** 1 **do**
7:         **if** $t_i \in \mu$ **then** $\triangleright\ t_i \in \mu \equiv t_i = 1$ in μ
8:             $\alpha := \alpha \cup \{t_i\}$
9:         **else**
10:             Set the phase saving values of $\{t_{n-1}, t_{n-2}, \ldots, t_0\}$ to 1
11:             $\tau := \text{SATUNDERASSUMPTIONS}(\alpha \cup \{t_i\})$
12:             **if** SAT solver returned SAT **then** $\mu := \tau$ **else** $\alpha := \alpha \cup \{\neg t_i\}$
13:     **return** μ

In addition, BINARY skips the first SAT invocation. BINARY was reported to be faster than OBV-BS in [40,41]. However, BINARY is less responsive than OBV-BS, since it apparently increases the number of unsatisfiable queries.

All things considered, we picked OBV-BS as the baseline algorithm for Mrs. Beaver, since it combines good performance, responsiveness, and incrementality. Note that OBV-BS can easily be updated to handle user-given hard assumptions β by adding β's literals to the assumption list for every SAT invocation.

5 Mrs. Beaver: An Unweighted MaxSAT Algorithm

This section introduces our new Unweighted MaxSAT algorithm Mrs. Beaver. The high-level algorithm is shown in Algorithm 2. It receives a satisfiable CNF formula F, the target bit-vector T, the incrementality mode *incrMode* and the search mode *searchMode*. Algorithm 2 outputs a model μ which minimizes $unsBits(T, \mu)$. *incrMode* lets the user decide whether the algorithm should be incremental, and how it should operate in incremental mode. *searchMode* determines the behavior of the algorithm at the complete stage, as will be explained later in Sect. 5.1.

Assume for now that *incrMode* = *none*, that is, that the algorithm is *not* incremental, and that *searchMode* = *SU*. First, for the *incomplete* stage of the algorithm, Mrs. Beaver invokes a preprocessor, Mrs. Beaver-Inc (described in Sect. 5.2), designed to quickly find a model μ with as low $unsBits(T, \mu)$ as possible. Then, during the *complete* stage, the algorithm invokes OBV-BS to minimize a new target $T' := tot(\neg T, \leq unsBits(T, \mu) - 1)$, comprising the sum of unsatisfied target bits starting with the value $unsBits(T, \mu) - 1$. If the latter invocation is satisfiable with the model μ', Mrs. Beaver returns μ'. Otherwise, there is no better model than μ, hence μ is returned.

It is imperative for performance to count the number of unsatisfied target bits towards 0, rather than the number of satisfied target bits towards n.

This is because creating a totalizer with an upper-bound on the sum is substantially more efficient than creating one with a lower-bound on the sum.

Our algorithm reuses the same SAT solver instance across all the calls, hence all learning is re-used. As we mentioned, when $incrMode = none$, the algorithm is *not* incremental. This is because the totalizer encoding asserts a cardinality constraint which is, apparently, not inferred by F. Section 5.4 describes Mrs. Beaver's behavior in incremental modes.

5.1 Mrs. Beaver and Linear Search

Mrs. Beaver is closely related to the linear search SAT-UNSAT (LSU) and UNSAT-SAT (LUS) algorithms [6].

LSU starts by finding a solution μ using a SAT solver. It then enters the *SAT-UNSAT loop*, which adds a cardinality constraint ensuring that the next solution will have strictly fewer unsatisfied target bits than $unsBits(T, \mu)$ after which it invokes a SAT solver. The algorithm updates μ with any newly found solution and terminates when the SAT solver returns UNSAT. It is guaranteed that the latest solution is an optimal one.

LUS keeps a lower bound l (initialized to 0) for the number of unsatisfied target bits for which no solution exists. LUS operates in an *UNSAT-SAT loop* which runs a SAT solver assuming that $unsBits(T) = l$. If the solver returns UNSAT, LUS updates l to $l + 1$ and proceeds to the next iteration of the loop. If the solver finds a solution μ, LUS terminates, in which case μ is guaranteed to be an optimal solution.

Note that the *complete* stage of Mrs. Beaver can behave as either the SAT-UNSAT loop (when $searchMode = SU$) or the UNSAT-SAT loop (when $searchMode = US$). In the latter case, the solver reverses the bits of T, so as to start falsifying T from the LSB towards the MSB. Thus it is the usage of the incomplete preprocessor–Mrs. Beaver-Inc–that differentiates between the linear search algorithms and Mrs. Beaver. Specifically, the difference between LSU and Mrs. Beaver in SU mode is that LSU uses a single SAT invocation for the incomplete stage, while Mrs. Beaver applies Mrs. Beaver-Inc. The difference between Mrs. Beaver in US mode and the LUS algorithm is that the former finds an upper bound on the number of unsatisfied target bits using Mrs. Beaver-Inc, while the latter may use a single SAT invocation to find an upper bound (if incremental weakening [28] is applied).

5.2 Mrs. Beaver-Inc: The Incomplete Preprocessor

Mrs. Beaver-Inc is designed to quickly find improving models. Our basic idea is to run OBV-BS over the target T to gradually reduce the number of unsatisfied target bits for the *current order* of T's literals. We realized that, to find tighter lower bounds faster, the following two optimizations would be useful:

1. Change OBV-BS so as to satisfy more target bits, even if the resulting algorithm no longer solves the OBV problem. We present such an algorithm–UMS-OBV-BS–next.

Algorithm 2. Mrs. Beaver

1: **function** SOLVE(CNF Formula F, Target $T = \{t_{n-1}, t_{n-2}, \ldots, t_0\}$, $incrMode \in \{none, full, maxPreserving\}$, $searchMode \in \{SU, US\}$)

Require: F is satisfiable

2:     $\mu :=$ Mrs. Beaver-Inc(F,T)

3:     **if** $unsBits(T, \mu) = 0$ **then return** μ

4:     **if** $incrMode = none$ **then**

5:         $T' := tot(\neg T, \leq unsBits(T, \mu) - 1)$

6:     **else if** $incrMode = maxPreserving$ **then**

7:         $T' := tot(\neg T, \leq unsBits(T, \mu))$

8:     **else**   ▷ $incrMode = full$

9:         $T' := tot(\neg T, \leq unsBits(T, \mu) - 1)$ with a fresh selector; see text

10:     **if** $searchMode = US$ **then** $T :=$ reverse(T)

11:     $\mu' :=$ OBV $-$ BS($F, \neg T'$)   ▷ Maximizing $\neg T' \equiv$ minimizing T'

12:     **if** OBV-BS solver returned SAT **then return** μ' **else return** μ

---

2. Run *several* iterations of UMS-OBV-BS and/or OBV-BS, where the target bits are the same, but their order changes (by reversing or shuffling). Changing the order of the target bits increases the chances of encountering a MaxSAT-friendly order. Below we assume that any algorithm that changes the order of the bits in the target T recreates the original T just before it finishes.

From OBV-BS to UMS-OBV-BS. We propose modifying OBV-BS to increase the chances of satisfying more target bits as follows: after a new model μ is encountered, the algorithm pushes all the target bits assigned 1 towards the most significant bit, so as to fix the value of such bits to 1 for the rest of the algorithm. Algorithm 3 shows a function that transforms OBV-BS to UMS-OBV-BS. It is designed to be invoked immediately after Algorithm 1 finds a new model for bit i at line 11. Note that UMS-OBV-BS no longer solves the OBV problem.

UMS-OBV-BS maintains an index k, initialized with the current index i minus 1. It visits every bit whose value has not been set and swaps any newly satisfied bits with t_k, where, when a satisfied bit is discovered, k is decreased by 1.

Algorithm 3. UMS-OBV-BS

1: **function** MODIFYING OBV-BS TO UMS-OBV-BS

Require: Invoke this function immediately after line 11 of Alg. 1

2:     $k := i - 1$

3:     **for** $j \leftarrow i - 1$ **downto** 0 **step** 1 **do**

4:         **if** $\mu(t_j) = 1$ **then**

5:             Swap the bits t_k and t_j

6:             $k := k - 1$

7:     **return** μ

The Preprocessor. The generic scheme of our preprocessor `Mrs. Beaver-Inc` is shown in Algorithm 4. It allows some freedom, the concrete heuristics being regulated by several user-given parameters discussed below. `Mrs. Beaver-Inc` receives a CNF formula F and the target T. It operates in a loop which runs for a user-given number of iterations. Each iteration invokes either `UMS-OBV-BS` or `OBV-BS`. The returned model μ is stored after the initial iteration and updated whenever a better model is found. After each iteration, T is either reversed or shuffled. The algorithm is regulated by the following 3 user-given parameters:

1. `ALG`: the inner algorithm, applied at line 3. It can either be *a*) `UMS-OBV-BS` or *b*) `OBV-BS` or *c*) `Mixed-OBV`, which is an alternation between `UMS-OBV-BS` and `OBV-BS`. If `ALG` is either plain `OBV-BS` or plain `UMS-OBV-BS`, the target is reversed at line 5 after each odd iteration and randomly shuffled after each even iteration. If `ALG` is `Mixed-OBV`, then `UMS-OBV-BS` is applied at iterations $i : i\%4 \in \{0,1\}$, while `OBV-BS` is applied at iterations $i : i\%4 \in \{2,3\}$. The target is reversed after iterations $i : i\%4 \in \{1,2,3\}$ (note that reversing T after iteration $i : i\%4 = 3$ recreates the original order) and shuffled after iteration $i : i\%4 = 3$.
2. *itNum*: the number of iterations.
3. *obvConfThr*: a threshold on the number of conflicts for each invocation of SAT-under-assumptions to find the satisfiability status of a single bit inside `UMS-OBV-BS` and `OBV-BS` (line 11 in Algorithm 1). Since `Mrs. Beaver-Inc` is incomplete, we found it useful to stop the solver when a threshold on the number of conflicts is reached in order not to get stuck with difficult bits. An unsolved target bit is assigned 0 by the algorithm.

Algorithm 4. `Mrs. Beaver-Inc`

1: **function** `Mrs. Beaver-Inc`(CNF Formula F, Target $T = \{t_{n-1}, t_{n-2}, \ldots, t_0\}$)
2:     **for** $i \leftarrow 1$ **to** *itNum* **step** 1 **do**　　　　　▷ *itNum* is a user-given threshold
3:         $\mu' := $ `UMS − OBV − BS`(F, T) or `OBV − BS`(F, T)
4:         **if** μ doesn't exist or *unsBits*$(T, \mu') < $ *unsBits*(T, μ) **then** $\mu := \mu'$
5:         $T := $ `reverse(T)` or `shuffle(T)`

5.3   Responsiveness

`Mrs. Beaver-Inc` is quite responsive. Not only can each invocation of `OBV-BS`/`UMS-OBV-BS` update the best model, the best model can also be updated by the inner iterations of `OBV-BS`/`UMS-OBV-BS`. Hence, the main algorithm `Mrs. Beaver` is responsive at the incomplete stage. At the complete stage `Mrs. Beaver` is responsive only in the *SU* mode.

5.4   Incrementality

Recall that `Mrs. Beaver` can operate in non-incremental mode, fully incremental mode, or maximization-preserving incremental mode (described below), depending on the user-given value *incrMode* $\in \{none, full, maxPreserving\}$.

Full Incrementality. An algorithm is fully incremental if all the learned clauses are inferred by the input formula F. To make Mrs. Beaver fully incremental, we need to eliminate any clauses created by the totalizer. To that end, we simply add a fresh selector variable s to every clause generated by the totalizer and add $\neg s$ as a hard assumption to OBV-BS, applied at line 11 of Algorithm 2. One can also add the unit clause s after Mrs. Beaver is completed to remove all the clauses, generated by the totalizer.

Maximization-Preserving Incrementality. An invocation of incremental Unweighted MaxSAT is *maximization-preserving* if it asserts that the number of unsatisfied bits in the current target T cannot be higher than the number $unsBits(T, \mu)$, found by the algorithm. As we shall see, a maximization-preserving incremental Unweighted MaxSAT solution is useful in the context of BMO solving.

Algorithm 2, in the mode $incrMode = none$, is *almost* maximization-preserving, except than the totalizer, created at line 5, asserts that the number of unsatisfied bits is strictly lower than $unsBits(T, \mu)$. This might cause the formula to become unsatisfiable if the actual minimum happens to be $unsBits(T, \mu)$. To fix this problem for the maximization-preserving mode $incrMode = maxPreserving$, we simply provide the totalizer the number $unsBits(T, \mu)$ as the upper bound. Note that this might result in a certain performance degradation.

6   Applying Mrs. Beaver to Solve BMO

Recall that in BMO, instead of a target bit-vector T, there are multiple target bit-vectors $T_{m-1}, T_{m-2}, \dots, T_0$. The goal is to maximize the number of satisfied bits in each of the targets, where satisfying one bit of T_i is preferred to satisfying all the bits in $T_{i-1}, T_{i-2}, \dots, T_0$.

One way to solve BMO, proposed in [26], is by reducing the problem to Weighted MaxSAT by concatenating the bits of all the target bit-vectors into one target bit-vector, and assigning each bit $t_i^0 \in t_0$ the weight $w^0 = 1$, and each bit $t_i^l \in T_{l>0}$ the weight $w^l = 1 + \sum_{k=0}^{l-1} w^k * |T_k|$. However, as we shall see, such a solution does not scale.

Algorithm 5 shows our BMO algorithm–Oh Mrs. Beaver–which adapts the iterative MaxSAT-based BMO algorithm from [26] to apply an *incremental* Unweighted MaxSAT solver underneath. Oh Mrs. Beaver takes full advantage of Mrs. Beaver's functionality in maximization-preserving mode. Oh Mrs. Beaver simply goes over all the targets from the most important one towards the least important one, and applies Mrs. Beaver in maximization-preserving mode to each target. In this way it guarantees that the optimal solution for each target T_i is asserted after invocation i is completed.

Oh Mrs. Beaver invokes Mrs. Beaver as the underlying building block, but it could, in principle, use any maximization-preserving incremental Unweighted MaxSAT algorithm that allows the user to change the target. Unfortunately, no such algorithm exists in the literature. It is possible, however, to use a fresh *non-incremental* Unweighted MaxSAT solver for every iteration i of Algorithm 5 for the current target T_i. For that to work, one has to assert the cardinality constraint $\sum_{i=0}^{n-1} \neg T_i \leq unsBits(T_i)$ after each iteration i. Unfortunately, cardinality constraints are not part of the standard MaxSAT format. To evaluate the performance of non-incremental instantiations of Oh Mrs. Beaver with the different state-of-the-art Unweighted MaxSAT solvers, we encoded the cardinality constraints into clauses using the totalizer encoding.

Algorithm 5. Oh Mrs. Beaver

1: **function** SOLVE(CNF Formula F, Targets $T_{m-1}, T_{m-2}, \ldots, T_0$)
Require: F is satisfiable
2:     **for** $i \leftarrow n - 1$ **to** 0 **step** 1 **do**
3:         $\mu := $ Mrs. Beaver $(F, T_i, maxPreserving)$

---

7 Experimental Results

This section studies the performance of our algorithms. Section 7.1 analyzes the performance of Unweighted MaxSAT solvers run with settings that mimic the MaxSAT Evaluation 2017. Section 7.2 examines the performance of MaxSAT and BMO solvers on benchmarks we generated from our industrial application.

The benchmarks we generated, as well as the detailed results, are available in [31]. Unless specified differently, the experiments were executed on machines with 32 Gb of memory running Intel® Xeon® processors with 3 GHz CPU frequency. The time is always shown in seconds.

7.1 Unweighted MaxSAT: MaxSAT Evaluation 2017

In this section we compare the performance of Mrs. Beaver to that of the winners of the MaxSAT Evaluation. In addition, we study the impact of Mrs. Beaver's search mode (*SU* vs. *US*) and Mrs. Beaver-Inc's three parameters, introduced in Sect. 5.2, on the performance of Mrs. Beaver.

We denote by {ALG, *itNum*, *obvConfThr*, *searchMode*} a configuration of Mrs. Beaver, where the search mode *searchMode* is either *SU* or *US* and the incomplete preprocessor applies the algorithm ALG $\in$ {OBV−BS, UMS−OBV−BS, Mixed−OBV} using *itNum* iterations and the conflict threshold *obvConfThr* in OBV-BS and/or UMS-OBV-BS. We denote by the configurations $\{-, 1, 0, SU\}$ and $\{-, 1, 0, US\}$ the implementations of LSU and LUS, respectively, in the framework of Mrs. Beaver (i.e., a conflict threshold of 0 per bit means that the incomplete stage of Mrs. Beaver invokes SAT instead of Mrs. Beaver-Inc).

Recall that the MaxSAT Evaluation had two incomplete categories, with 60-second and 300-second timeouts, respectively, and one complete category with a 3600-second timeout.

Incomplete Categories. We compared the performance of different configurations of Mrs. Beaver to that of the leading incomplete solvers MaxHS, Open-WBO-LSU, and MaxRoster. Based on preliminary experiments, we picked the following configuration as the baseline for Mrs. Beaver when the timeout is 60 second: $\{\texttt{Mixed} - \texttt{OBV}, 10^5, 10^4, SU\}$. To provide evidence that the baseline configuration is indeed the best one and to study the impact of the parameters, we also provide results for "neighbor" configurations, constructed by changing one of the parameters of the baseline configuration, and that of our linear search implementations ($\{-, 1, 0, SU\}$ and $\{-, 1, 0, US\}$).

Mimicking the MaxSAT Evaluation, our primary ranking criteria was average *score*. *score* for instance i and solver S is the ratio between the number of unsatisfied target bits in the best solution found by any of the solvers and the number of unsatisfied target bits in the best solution found by solver S. *score* is 0 if no solution was found by S. It holds that *score* $\in [0, 1]$.

Consider the upper part of Table 1 which displays the results for the 60-second timeout. Each row compares the results of a single configuration of Mrs. Beaver, shown in the first column, to those of the other solvers. Following the presentation style used in the MaxSAT Evaluation, we provide *score*, the number of solved instances (in columns titled #S) and the number of times each algorithm found the best solution (in columns titled #B). The best score in each row is highlighted. The table is sorted according to the score of the Mrs. Beaver configuration.

The best result was achieved by the baseline configuration $\{\texttt{Mixed} - \texttt{OBV}, 10^5, 10^4, SU\}$. It outperforms all the other solvers, including our LSU implementation ($\{-, 1, 0, SU\}$) and the LSU implementation in the MaxSAT evaluation winner Open-WBO-LSU. Mrs. Beaver's performance is slightly better in the SU (SAT-UNSAT) mode.

Concerning the parameters of Mrs. Beaver-Inc, changing the conflict threshold or the number of iterations led to a mild deterioration of the score. Alternating between OBV-BS and UMS-OBV-BS yielded the best results. Using only UMS-OBV-BS ($\{\texttt{UMS} - \texttt{OBV} - \texttt{BS}, \infty, 10^5\}$) was insufficient for outscoring the other solvers.

The bottom part of Table 1 shows the results for the 300-second timeout. We found in preliminary experiments that the best-performing Mrs. Beaver configuration for the 300-second timeout is $\{\texttt{Mixed} - \texttt{OBV}, 10^4, 10^5, SU\}$; it is slightly different from that used for the 60-second timeout. As in the MaxSAT Evaluation, MaxRoster emerges as the best solver in this category. Mrs. Beaver comes out as the second best.

Complete Category Based on preliminary experiments, we picked the following configuration as the baseline for Mrs. Beaver for complete solving: $\{\texttt{UMS} - \texttt{OBV} - \texttt{BS}, 1, 10^3, US\}$. It applies one iteration of the preprocessor using UMS-OBV-BS and a relatively low conflict threshold of 10^3 as well as the US mode at the complete stage.

Table 1. MaxSAT evaluation: incomplete categories

Mrs. Beaver Conf.	Mrs. Beaver			Open-WBO-LSU			MaxHS			MaxRoster		
	Score	#S	#B	Score	#S	#B	Score	#S	#B	Score	#S	#B
60-Second Timeout												
$\{$Mixed-OBV, $10^5, 10^4, SU\}$	0.81792	182	56	0.73498	178	54	0.73017	192	37	0.67423	145	89
$\{$Mixed-OBV, $10^4, 10^4, SU\}$	0.81787	182	55	0.73498	178	54	0.73017	192	37	0.67423	145	89
$\{$Mixed-OBV, $10^5, 10^4, US\}$	0.81756	182	56	0.73498	178	54	0.73017	192	37	0.67423	145	89
$\{$Mixed-OBV, $10^6, 10^4, SU\}$	0.81748	182	56	0.73498	178	54	0.73017	192	37	0.67423	145	89
$\{$Mixed-OBV, $10^5, 10^5, SU\}$	0.81378	182	64	0.73902	178	50	0.73122	192	37	0.67094	145	87
$\{$Mixed-OBV, $10^5, 10^3, SU\}$	0.79008	181	49	0.74207	178	59	0.73654	192	38	0.67593	145	89
$\{$OBV-BS, $10^5, 10^4, SU\}$	0.7855	182	55	0.74274	178	57	0.73764	192	38	0.67765	145	92
$\{-, 1, 0, SU\}$	0.74531	182	65	0.75043	178	53	0.74577	192	45	0.68675	145	91
$\{$UMS-OBV-BS, $10^5, 10^4, SU\}$	0.73236	181	39	0.74611	178	59	0.74121	192	44	0.67835	145	90
$\{-, 1, 0, US\}$	0.57173	182	8	0.77117	178	77	0.76302	192	47	0.69246	145	96
300-Second Timeout												
$\{$Mixed-OBV, $10^4, 10^5, SU\}$	0.77807	183	39	0.71429	182	43	0.75285	194	55	0.87118	182	126
$\{$Mixed-OBV, $10^4, 10^5, US\}$	0.77806	183	39	0.71429	182	43	0.75285	194	55	0.87118	182	126
$\{$Mixed-OBV, $10^5, 10^5, SU\}$	0.77774	183	40	0.71424	182	43	0.75279	194	55	0.87112	182	126
$\{$Mixed-OBV, $10^3, 10^5, SU\}$	0.77563	183	39	0.71429	182	43	0.75285	194	55	0.87118	182	126
$\{$Mixed-OBV, $10^4, 10^4, SU\}$	0.77259	183	32	0.71354	182	44	0.75276	194	56	0.87191	182	128
$\{$Mixed-OBV, $10^4, 10^6, SU\}$	0.75761	183	40	0.71442	182	44	0.75277	194	55	0.8715	182	127
$\{$OBV-BS, $10^5, 10^4, SU\}$	0.72725	183	32	0.71503	182	44	0.75782	194	57	0.8768	182	129
$\{-, 1, 0, SU\}$	0.71329	184	36	0.7232	182	46	0.76508	194	62	0.88502	182	133
$\{$UMS-OBV-BS, $10^5, 10^4, SU\}$	0.70314	183	32	0.71608	182	43	0.75515	194	60	0.87329	182	128
$\{-, 1, 0, US\}$	0.50183	184	3	0.72414	182	48	0.76543	194	62	0.88605	182	135

Consider Table 2. It displays the number of solved instances and the overall run-time of Open-WBO-RES, MaxHS, Open-WBO-LSU, our linear search implementations and several Mrs. Beaver configurations (MaxRoster cannot be applied for complete solving). The algorithms are sorted according to their performance.

Unsurprisingly, unlike in the incomplete categories, Mrs. Beaver did not perform as well as the leading solvers, Open-WBO-RES and MaxHS. Apparently, the reason is that Mrs. Beaver's top-performing complete algorithm relies merely on US linear search. Applying SU instead of US at the complete stage or changing the underlying algorithm at the incomplete stage results in a performance deterioration. Notably, the preprocessor allows Mrs. Beaver to solve 18 more instance as compared to the LUS implementation in our framework ($\{$UMS $-$ OBV $-$ BS, $1, 10^3, US\}$ vs. $\{-, 1, 0, US\}$).

Table 2. MaxSAT evaluation: complete category

Results			Results (Continued)		
Solver	#S	Overall Time	Solver	#S	Overall Time
MaxHS	655	927384.89	$\{$UMS-OBV-BS, $1, 10^3, SU\}$	547	1251925.21
Open-WBO-RES	654	880493.49	$\{$OBV-BS, $1, 10^3, US\}$	546	1281167
$\{$UMS-OBV-BS, $1, 10^3, US\}$	572	1176926.57	$\{$OBV-BS, $1, 10^3, SU\}$	543	1292191.23
Open-WBO-LSU	554	1209114.62	$\{-, 1, 0, SU\}$	541	1258922.31
$\{-, 1, 0, US\}$	554	1214299.29			

7.2   Industrial Instances

For the experiments in this section, we generated 9 Weighted MaxSAT benchmarks that encode the industrial task of cleaning up soft design rules in Intel designs. We used the straightforward translation from BMO to Weighted MaxSAT, described in Sect. 6, for generating the benchmarks. The number of variables in the benchmarks ranges from 4,367,381 to 8,220,593, while the number of clauses ranges from 12,960,427 to 26,676,683. The number of target bit-vectors (before applying the reduction from BMO to Weighted MaxSAT) is 44 in every benchmark.

The main goal of our experiments was to study the impact of enabling incrementality in MaxSAT solving on the performance of the Unweighted MaxSAT-based BMO algorithm in the context of our application. We compared Oh Mrs. Beaver against the non-incremental Unweighted MaxSAT flow with the following underlying solvers: Open-WBO-RES, Open-WBO-LSU, MaxHS, and Mrs. Beaver. We used the best performing configuration in the complete category for both Oh Mrs. Beaver and the non-incremental Mrs. Beaver. We used a timeout of 1800 seconds for Oh Mrs. Beaver and a timeout of 600 seconds for each invocation of a non-incremental solver.

The results are shown in Table 3. For each solver, the table shows the number of completed invocations (out of 44 incremental invocations, one for each target bit-vector) and the time. The last row shows the average number of completed invocations and the average time for each solver. Oh Mrs. Beaver is clearly the best solver. It solved all the benchmarks, being 6 times faster than the Mrs. Beaver-based non-incremental flow and 10 times faster than the MaxHS-based non-incremental flow. Neither Open-WBO-RES nor Open-WBO-LSU scaled to our instances. All in all, it pays off to apply *incremental* Unweighted MaxSAT solving to solve industrial instances of BMO.

In addition, for comparison, we ran the following Weighted MaxSAT solvers: Open-WBO, MaxHS, MaxRoster, Clasp [22] and Sat4j [6]. It turned out that only Sat4j was able to process our benchmarks successfully, since in our benchmarks the weight can be as high as 10^{129}, while the maximal weight in the MaxSAT format is restricted to 2^{63}. Sat4j timed-out on all the instances.

Finally, we translated our benchmarks to the BMO format, used during the Lion9 Challenge [7] (the only BMO competition ever held), and ran the following three dedicated BMO solvers, comprising all the participants in the challenge: Sat4j [6], Open-WBO-SU, and Open-WBO-MSU3 (the latter two solvers were implemented in the Open-WBO framework). The results are shown in Table 4. Initially, when we invoked the BMO solvers on our standard machines (with 32 Gb of memory), all three solvers failed to solve a single instance: Sat4j timed-out and both versions of Open-WBO came back with memory-outs. As a follow-up experiment, we ran Oh Mrs. Beaver, Open-WBO-SU, and Open-WBO-MSU3 on a machine having 512 Gb of memory. It turned out that both versions of Open-WBO slightly outperformed Oh Mrs. Beaver (by 1.2 times on average), but they used 50 times more memory (2 Gb vs. 100 Gb on average). Both Open-WBO-SU and Open-WBO-MSU3 reduce BMO to iterative MaxSAT invocations, similarly to Oh

Mrs. Beaver (using different MaxSAT algorithms underneath: LSU in the case of Open-WBO-SU and MSU3 [27] in the case of Open-WBO-MSU3). As for the huge difference in memory usage, it is difficult to determine the reason for this, as the source code of both Open-WBO-SU and Open-WBO-MSU3 is unavailable.

Table 3. Evaluation on industrial BMO instances

#	Incr. Calls	Oh Mrs. Beaver		Open-WBO-RES		Open-WBO-LSU		MaxHS		Mrs. Beaver	
		Solved	Time	Solved	Time	Solved	Time	Solved	Time	Solved	Time
1	44	44	310	0	26400	0	26400	44	3060	44	1881
2	44	44	452	0	26400	0	26400	44	4343	44	2890
3	44	44	346	0	26400	0	26400	44	4141	44	2740
4	44	44	366	0	26400	0	26400	44	3404	44	2120
5	44	44	197	0	26400	0	26400	44	3635	44	2271
6	44	44	229	44	8883	44	8895	44	1998	44	1163
7	44	44	282	44	11483	44	11483	44	2336	44	1406
8	44	44	325	11	25828	8	25937	44	2737	44	1667
9	44	44	459	0	26400	0	26400	44	3159	44	1998
Avrg	44	44	330	11	22733	11	22746	44	3201	44	2015

Table 4. Evaluation of BMO solvers on industrial BMO instances

#	Standard Settings (32Gb)			512Gb and 1.2Ghz CPU frequency					
	Sat4j Res	Open-WBO-SU Res	Open-WBO-MSU3 Res	Oh Mrs. Beaver		Open-WBO-SU		Open-WBO-MSU3	
				Time	Mem (Mb)	Time	Mem (Mb)	Time	Mem (Mb)
1	Timeout	Memout	Memout	413	1917	316	96881	333	96917
2	Timeout	Memout	Memout	630	2710	469	133578	451	133577
3	Timeout	Memout	Memout	469	2594	451	127032	298	127121
4	Timeout	Memout	Memout	356	2164	371	108399	367	108449
5	Timeout	Memout	Memout	438	2287	360	114650	463	114673
6	Timeout	Memout	Memout	275	1289	183	67374	235	67274
7	Timeout	Memout	Memout	298	1514	219	76320	249	74618
8	Timeout	Memout	Memout	349	1737	319	89581	296	89605
9	Timeout	Memout	Memout	437	2019	366	102081	359	102104
Avrg				407	2026	339	101766	339	101593

8   Conclusion

We explored how Unweighted MaxSAT solving can benefit from the recently introduced Optimization Modulo Bit-Vectors (OBV) algorithms. We proposed a new OBV-based Unweighted MaxSAT algorithm–Mrs. Beaver. Mrs. Beaver outscored the top solvers when run with the settings of the Incomplete-60-Second-Timeout Track of MaxSAT Evaluation 2017. Unlike the existing state-of-the-art algorithms, Mrs. Beaver was designed to be incremental in the sense that it can be reapplied with a different set of hard assumptions and soft clauses. We demonstrated that enabling incrementality in MaxSAT significantly improves the performance a MaxSAT-based BMO algorithm applied for solving a new critical industrial BMO application: cleaning-up weak design rule violations during the Physical Design stage of Computer-Aided-Design at Intel.

Acknowledgments. We are grateful to Ruben Martins for his valuable advice concerning different aspects of MaxSAT research. We thank Jessica Davies and Fahiem Bacchus for their help with MaxHS, and Vadim Ryvchin for suggesting the use of a conflict threshold in OBV-BS.

References

1. Ansotegui, C., Bacchus, F., Järvisalo, M., Martins, R. (eds.): MaxSAT Evaluation 2017: Solver and Benchmark Descriptions, vol. B-2017-2. Department of Computer Science Series of Publications B. University of Helsinki (2017)
2. Argelich, J., Li, C.M., Manyà, F., Planes, J.: The first and second max-sat evaluations. JSAT **4**(2–4), 251–278 (2008)
3. Bailleux, O.: On the CNF encoding of cardinality constraints and beyond. CoRR, abs/1012.3853 (2010)
4. Bailleux, O., Boufkhad, Y.: Efficient CNF encoding of Boolean cardinality constraints. In: Rossi, F. (ed.) CP 2003. LNCS, vol. 2833, pp. 108–122. Springer, Heidelberg (2003). https://doi.org/10.1007/978-3-540-45193-8_8
5. Belov, A., Heule, M., Marques-Silva, J.: MUS extraction using clausal proofs. In: Sinz and Egly [45], pp. 48–57 (2014)
6. Le Berre, D., Parrain, A.: The SAT4J library, release 2.2. JSAT **7**(2–3), 59–64 (2010)
7. Le Berre, D., Roussel, O., Delmas, R., Marmion, M.-E.: LION 9 challenge. http://www.lifl.fr/LION9/challenge.php
8. Biere, A., Gomes, C.P. (eds.): SAT 2006. LNCS, vol. 4121. Springer, Heidelberg (2006). https://doi.org/10.1007/11814948
9. Biere, A., Heule, M., van Maaren, H., Walsh, T. (eds.): Handbook of Satisfiability. Frontiers in Artificial Intelligence and Applications, vol. 185. IOS Press (2009)
10. Bjørner, N., Phan, A.-D.: νZ - maximal satisfaction with Z3. In: Kutsia, T., Voronkov, A. (eds.) 6th International Symposium on Symbolic Computation in Software Science, SCSS 2014. EPiC Series, Gammarth, La Marsa, Tunisia, 7–8 December 2014, vol. 30, pp. 1–9. EasyChair (2014)
11. Bjørner, N., Phan, A.-D., Fleckenstein, L.: νZ - an optimizing SMT solver. In: Baier, C., Tinelli, C. (eds.) TACAS 2015. LNCS, vol. 9035, pp. 194–199. Springer, Heidelberg (2015). https://doi.org/10.1007/978-3-662-46681-0_14
12. Bradley, A.R.: SAT-based model checking without unrolling. In: Jhala, R., Schmidt, D. (eds.) VMCAI 2011. LNCS, vol. 6538, pp. 70–87. Springer, Heidelberg (2011). https://doi.org/10.1007/978-3-642-18275-4_7
13. Büttner, M., Rintanen, J.: Satisfiability planning with constraints on the number of actions. In: Biundo, S., Myers, K.L., Rajan, K. (eds.) Proceedings of the Fifteenth International Conference on Automated Planning and Scheduling (ICAPS 2005), Monterey, California, USA, 5–10 June 2005, pp. 292–299. AAAI (2005)
14. Davies, J.: Solving MAXSAT by decoupling optimization and satisfaction. Dissertation, University of Toronto (2013)
15. Davies, J., Bacchus, F.: Exploiting the power of MIP solvers in MAXSAT. In: Järvisalo, M., Van Gelder, A. (eds.) SAT 2013. LNCS, vol. 7962, pp. 166–181. Springer, Heidelberg (2013). https://doi.org/10.1007/978-3-642-39071-5_13
16. Davies, J., Bacchus, F.: Postponing optimization to speed up MAXSAT solving. In: Schulte, C. (ed.) CP 2013. LNCS, vol. 8124, pp. 247–262. Springer, Heidelberg (2013). https://doi.org/10.1007/978-3-642-40627-0_21

17. Dershowitz, N., Hanna, Z., Nadel, A.: A scalable algorithm for minimal unsatisfiable core extraction. In: Biere and Gomes [8], pp. 36–41 (2006)
18. Eén, N., Mishchenko, A., Brayton, R.K.: Efficient implementation of property directed reachability. In: Bjesse, P., Slobodová, A. (eds.) International Conference on Formal Methods in Computer-Aided Design, FMCAD 2011, Austin, TX, USA, 30 October–02 November 2011, pp. 125–134. FMCAD Inc. (2011)
19. Eén, N., Sörensson, N.: An extensible SAT-solver. In: Giunchiglia, E., Tacchella, A. (eds.) SAT 2003. LNCS, vol. 2919, pp. 502–518. Springer, Heidelberg (2004). https://doi.org/10.1007/978-3-540-24605-3_37
20. Frost, D., Dechter, R.: In search of the best constraint satisfaction search. In: AAAI, pp. 301–306 (1994)
21. Fu, Z., Malik, S.: On solving the partial MAX-SAT problem. In: Biere and Gomes [8], pp. 252–265 (2006)
22. Gebser, M., Kaufmann, B., Schaub, T.: Conflict-driven answer set solving: from theory to practice. Artif. Intell. **187**, 52–89 (2012)
23. Hansen, P., Jaumard, B.: Algorithms for the maximum satisfiability problem. Computing **44**(4), 279–303 (1990)
24. Karp, R.M.: Reducibility among combinatorial problems. In: Miller, R.E., Thatcher, J.W. (eds.) Proceedings of a Symposium on the Complexity of Computer Computations. The IBM Research Symposia Series, 20–22 March 1972, IBM Thomas J. Watson Research Center, Yorktown Heights, New York, pp. 85–103. Plenum Press, New York (1972)
25. Knuth, D.E.: The Art of Computer Programming, Volume 4, Fascicle 6: Satisfiability. Addison Wesley, Boston (2015)
26. Marques-Silva, J., Argelich, J., Graça, A., Lynce, I.: Boolean lexicographic optimization: algorithms and applications. Ann. Math. Artif. Intell. **62**(3–4), 317–343 (2011)
27. Marques-Silva, J., Planes, J.: On using unsatisfiability for solving maximum satisfiability. CoRR, abs/0712.1097 (2007)
28. Martins, R., Joshi, S., Manquinho, V.M., Lynce, I.: Incremental cardinality constraints for MaxSAT. In: O'Sullivan [39], pp. 531–548 (2014)
29. Morgado, A., Dodaro, C., Marques-Silva, J.: Core-guided MaxSAT with soft cardinality constraints. In: O'Sullivan [39], pp. 564–573 (2014)
30. Moskewicz, M.W., Madigan, C.F., Zhao, Y., Zhang, L., Malik, S.: Chaff: engineering an efficient SAT solver. In: Proceedings of the 38th Design Automation Conference, DAC 2001, Las Vegas, NV, USA, 18–22 June 2001, pp. 530–535. ACM (2001)
31. Nadel, A.: Solving MaxSAT with bit-vector optimization: benchmarks and detailed results. https://goo.gl/VS1WJg
32. Nadel, A.: Routing under constraints. In: Piskac, R., Talupur, M. (eds.) 2016 Formal Methods in Computer-Aided Design, FMCAD 2016, Mountain View, CA, USA, 3–6 October 2016, pp. 125–132. IEEE (2016)
33. Nadel, A.: A correct-by-decision solution for simultaneous place and route. In: Majumdar, R., Kunčak, V. (eds.) CAV 2017. LNCS, vol. 10427, pp. 436–452. Springer, Cham (2017). https://doi.org/10.1007/978-3-319-63390-9_23
34. Nadel, A., Ryvchin, V.: Efficient SAT solving under assumptions. In: Cimatti, A., Sebastiani, R. (eds.) SAT 2012. LNCS, vol. 7317, pp. 242–255. Springer, Heidelberg (2012). https://doi.org/10.1007/978-3-642-31612-8_19
35. Nadel, A., Ryvchin, V.: Bit-vector optimization. In: Chechik, M., Raskin, J.-F. (eds.) TACAS 2016. LNCS, vol. 9636, pp. 851–867. Springer, Heidelberg (2016). https://doi.org/10.1007/978-3-662-49674-9_53

36. Nadel, A., Ryvchin, V., Strichman, O.: Accelerated deletion-based extraction of minimal unsatisfiable cores. JSAT **9**, 27–51 (2014)
37. Nadel, A., Ryvchin, V., Strichman, O.: Ultimately incremental SAT. In: Sinz and Egly [45], pp. 206–218 (2014)
38. Neves, M., Martins, R., Janota, M., Lynce, I., Manquinho, V.: Exploiting resolution-based representations for MaxSAT solving. In: Heule, M., Weaver, S. (eds.) SAT 2015. LNCS, vol. 9340, pp. 272–286. Springer, Cham (2015). https://doi.org/10.1007/978-3-319-24318-4_20
39. O'Sullivan, B. (ed.): CP 2014. LNCS, vol. 8656. Springer, Cham (2014). https://doi.org/10.1007/978-3-319-10428-7
40. Petkovska, A.: Exploiting satisfiability solvers for efficient logic synthesis. Dissertation, École Polytechnique Fédérale de Lausanne (2017)
41. Petkovska, A., Mishchenko, A., Soeken, M., De Micheli, G., Brayton, R.K., Ienne, P.: Fast generation of lexicographic satisfiable assignments: enabling canonicity in sat-based applications. In: Liu, F. (ed.) Proceedings of the 35th International Conference on Computer-Aided Design, ICCAD 2016, Austin, TX, USA, 7–10 November 2016, p. 4. ACM (2016)
42. Pipatsrisawat, K., Darwiche, A.: A lightweight component caching scheme for satisfiability solvers. In: Marques-Silva, J., Sakallah, K.A. (eds.) SAT 2007. LNCS, vol. 4501, pp. 294–299. Springer, Heidelberg (2007). https://doi.org/10.1007/978-3-540-72788-0_28
43. Si, X., Zhang, X., Manquinho, V., Janota, M., Ignatiev, A., Naik, M.: On incremental core-guided MaxSAT solving. In: Rueher, M. (ed.) CP 2016. LNCS, vol. 9892, pp. 473–482. Springer, Cham (2016). https://doi.org/10.1007/978-3-319-44953-1_30
44. Marques-Silva, J.P., Sakallah, K.A.: GRASP: a search algorithm for propositional satisfiability. IEEE Trans. Comput. **48**(5), 506–521 (1999)
45. Sinz, C., Egly, U. (eds.): SAT 2014. LNCS, vol. 8561. Springer, Cham (2014). https://doi.org/10.1007/978-3-319-09284-3
46. Shtrichman, O.: Tuning SAT checkers for bounded model checking. In: Emerson, E.A., Sistla, A.P. (eds.) CAV 2000. LNCS, vol. 1855, pp. 480–494. Springer, Heidelberg (2000). https://doi.org/10.1007/10722167_36
47. Sugawara, T.: MaxRoster: solver description. In: Ansotegui et al. [1], p. 12 (2017)

Conflict Driven Clause Learning

Using Combinatorial Benchmarks to Probe the Reasoning Power of Pseudo-Boolean Solvers

Jan Elffers[1], Jesús Giráldez-Cru[1], Jakob Nordström[1(✉)], and Marc Vinyals[2]

[1] KTH Royal Institute of Technology, Stockholm, Sweden
{elffers,giraldez,jakobn}@kth.se
[2] Tata Institute of Fundamental Research, Mumbai, India
marc.vinyals@tifr.res.in

Abstract. We study cdcl-cuttingplanes, Open-WBO, and Sat4j, three successful solvers from the Pseudo-Boolean Competition 2016, and evaluate them by performing experiments on crafted benchmarks designed to be trivial for the cutting planes (CP) proof system underlying pseudo-Boolean (PB) proof search but yet potentially tricky for PB solvers. Our experiments demonstrate severe shortcomings in state-of-the-art PB solving techniques. Although our benchmarks have linear-size tree-like CP proofs, and are thus extremely easy in theory, the solvers often perform quite badly even for very small instances. We believe this shows that solvers need to employ stronger rules of cutting planes reasoning. Even some instances that lack not only Boolean but also real-valued solutions are very hard in practice, which indicates that PB solvers need to get better not only at Boolean reasoning but also at linear programming. Taken together, our results point to several crucial challenges to be overcome in the quest for more efficient pseudo-Boolean solvers, and we expect that a further study of our benchmarks could shed more light on the potential and limitations of current state-of-the-art PB solving.

1 Introduction

In its most general form, a *pseudo-Boolean function* maps sets of Boolean values to a real number. Such functions have been studied since the 1960s in the context of operations research and 0-1 programming, yielding an extensive body of work as surveyed, e.g., in [5]. In this paper we consider the special case of *linear pseudo-Boolean constraints* $\sum_i a_i \ell_i \geq A$ encoded as integer linear combinations of literals ℓ_i (i.e., Boolean variables or negations of such variables). In the decision problem *pseudo-Boolean solving (PBS)* one asks whether a collection of such constraints is feasible or not. In *pseudo-Boolean optimization (PBO)* the task is to compute the best value of an objective function (written as a linear constraint) subject to other linear constraints, a formalism that captures problems in many different fields. Clearly, PBO can be reduced to PBS by iteratively computing solutions and adding constraints forcing the value of the objective function to improve. In the current work we focus on the decision problem PBS.

© Springer International Publishing AG, part of Springer Nature 2018
O. Beyersdorff and C. M. Wintersteiger (Eds.): SAT 2018, LNCS 10929, pp. 75–93, 2018.
https://doi.org/10.1007/978-3-319-94144-8_5

Pseudo-Boolean constraints are more expressive than conjunctive normal form (CNF) formulas, but are close enough that techniques for CNF SAT solving can be harnessed to attack pseudo-Boolean problems. The connection to integer linear programming (ILP) and, in particular 0-1 programming, makes it natural to also borrow insights from these areas.

Work on applying SAT-based methods in pseudo-Boolean solving seems to have started in the mid-1990s inspired by Barth [1,2] and developed in different directions. One line of research has focused on inference methods based on *cutting planes (CP)* [7,9,14], including works by Chai and Kuehlman [6], Sheini and Sakallah [33], and Dixon et al. [10]. In this context it was reported that focusing on the restricted form of *cardinality constraints* $\sum_i \ell_i \geq A$ can be more effective than dealing with general linear constraints [6,33], and according to [11] a very competitive approach can be to simply translate pseudo-Boolean constraints to CNF and use a *conflict-driven clause learning (CDCL)* SAT solver [3,25,28]. A different path was pursued by Manquinho and Marques-Silva, who devised ways of learning and backtracking non-chronologically using branch-and-bound search [22,23]. Needless to say, this brief historical overview is very far from complete—see, e.g., the excellent survey [31] for more details.

Current state-of-the-art pseudo-Boolean solvers building on the first line of work discussed above include *Open-WBO* [26,29], which reduces the problem to CNF [19] and applies CDCL search, and *Sat4j* [21,32], which uses cutting planes inference rules. These two solvers performed very well in the decision track in the *Pseudo-Boolean Competition 2016* together with the relatively new solver *cdcl-cuttingplanes* [12],[1] which, as the name suggests, also implements conflict-driven search in cutting planes.

1.1 Our Investigations and Conclusions

The survey [31] mentioned above ends on the optimistic note that *"[t]rade-offs between inference power and inference speed are often made in current algorithms and the right balance is still sought"* but that *"we can expect that, once the right balance is found, pseudo-Boolean solvers will become a major tool in problem solving."* From a theoretical point of view, there are strong reasons to concur with this—pseudo-Boolean (PB) solvers are based on an exponentially stronger method than CDCL solvers and so should have the potential to vastly outperform them. Intriguingly, in practice the opposite more often seems to be the case.

We approach this disconnect between theory and practice by studying the performance of the three PB solvers *cdcl-cuttingplanes*, *Open-WBO*, and *Sat4j* on the kind of PBS decision problems where they came out on top in the

---

[1] An updated version of this solver with the new name *RoundingSat* is described in [13].

Pseudo-Boolean Competition 2016. We consider the benchmarks in [36][2] as well as other crafted benchmarks that were specifically designed to be very easy for the cutting planes proof system underlying pseudo-Boolean SAT solving but to be potentially tricky to handle for PB solvers (not in the sense of being "obfuscated" in any way, but in the sense that the instances seem to require inherently pseudo-Boolean reasoning to be efficiently solvable). Since our starting point is proof complexity, our focus is on unsatisfiable benchmarks. We report results from fairly extensive experiments intended to highlight strong and weak points of these solvers when run on our benchmark set, and present some empirical conclusions as well as hypotheses that we hope will stimulate follow-up research.

Before briefly discussing our findings, we want to stress that we do not claim to provide the final word in this matter, but rather our purpose is to initiate a new line of research. By necessity, our set of benchmarks is limited, and the instances are quite particular in that they have been designed to have certain combinatorial properties. Nevertheless, we believe that this work shows that an in-depth study of pseudo-Boolean solver performance on such benchmarks can provide interesting insights. In contrast to industrial benchmarks, here we can have a detailed understanding of the properties of the instances, including, in particular, the fact that they can be solved efficiently in principle. In addition, the possibility to scale their size allows us to draw conclusions about asymptotic behaviour rather than just observing isolated data points.

The Need for Stronger Boolean Reasoning. The most obvious conclusion from our work is that the cutting planes-based reasoning in pseudo-Boolean solvers needs to be strengthened significantly. As mentioned above, our benchmarks have been designed to have short CP proofs, and most often these proofs are even tree-like (meaning that they can be found without learning from conflicts). However, in many cases the solvers struggle hopelessly even for very small instances. To explain by an analogy to CDCL solving, this is as if state-of-the-art CDCL solvers would fail completely for small formulas with linear-size DPLL proofs!

We consider the most plausible explanation for the poor performance to be that the PB solvers do not exploit the full power of the *division* rule in cutting planes, using only a limited form of division as in *cdcl-cuttingplanes* or substituting it altogether by the simpler *saturation* rule as in *Sat4j*. This hypothesis is strengthened by the observation that *cdcl-cuttingplanes* is consistently performing better than *Sat4j* in cases when use of the division rule seems to be crucial from a theoretical point of view.

Looking at the results from a different angle, it is well known that PB solvers such as *Sat4j* performs well on *pigeonhole principle (PHP) formulas*, and the results from the Pseudo-Boolean Competition 2016 show that this is also the

---

[2] It should be noted that [36] is closely related to the current work in that both papers are motivated by similar concerns, namely understanding the power and limitations of pseudo-Boolean reasoning. A key difference, though, is that the instances studied in [36] are designed to be potentially *hard* for the subsystems of cutting planes implemented by PB solvers, whereas in this paper we choose parameter settings so that almost all instances are theoretically *very easy*.

case for so-called *subset cardinality formulas* [27,34,35]. However, the seemingly equally simple *even colouring formulas* [24] appear very hard in practice. On closer inspection, one difference here is that PHP formulas do not have even rational solutions—there is no way to fit $n + 1$ pigeons into n holes even if the pigeons can be sliced—and the same holds for subset cardinality formulas, whereas even colouring formulas are satisfied by assigning every variable value $\frac{1}{2}$.

This raises the question whether hardness correlates with the existence of rational solutions, which we will refer to in what follows as the *rational hypothesis*. Clearly, rational solutions alone do not imply hardness—any 2-CNF formula without unit clauses is satisfied by assigning all variables value $\frac{1}{2}$, yet this does not make such a formula hard—but any formula *without* rational solutions has short proofs that PB solvers can find in theory [36], so we can ask if they can also find such proofs in practice.

Although there are families of formulas where the lack of rational solutions seems to help, the answer from our experiments to whether solvers can *always* efficiently decide rationally infeasible 0-1 integer linear programs is negative— we find examples of instances without rational solutions that are very hard in practice. But when we then go further and study for which instances we can help the solvers to run fast by, e.g., dropping a heavy hint in the form of a good fixed variable order (while keeping other settings at default values), an intriguing pattern emerges—for most of our benchmarks it holds that the solvers can be made efficient *if the instances have small strong backdoors[3] to pseudo-Boolean formulas without rational solutions.* There is of course a selection bias in the benchmarks we study, but we nevertheless find this refined version of the rational hypothesis quite intriguing and hope it can stimulate further study.

The Need for Stronger LP Reasoning. The refined rational hypothesis just discussed cannot explain all our findings, however, especially since solvers cannot always count on getting helpful hints. For some of our benchmark families—in particular, encodings of the *dominating set* problem on hexagonal grids—the formulas are extremely challenging even when the corresponding linear program has no rational solutions.[4] For these instances it can be shown that the method of reasoning used in *Sat4j* and *cdcl-cuttingplanes* can in principle derive extra constraints by simple addition [36], i.e., without any Boolean reasoning, and with these constraints added the formulas become trivial also in practice. The solvers do not find these constraints on their own, though, and we have not been able to coax them into doing so even by trying out different helpful variable orderings. Instead, the solvers get stuck exploring parts of the search space where even the LP is infeasible. This shows that PB solvers need to strengthen not only their Boolean reasoning but also their linear programming capabilities.

---

[3] A *strong backdoor* for an instance F to a family $\mathcal{F}$ of (easy) instances—in this case, instances without rational solutions—is a set of variables in F such that any assignment ρ to these variables yield a restricted instance $F\restriction_\rho$ that is in $\mathcal{F}$.

[4] It might be worth pointing out that for an instance to lack rational solutions is the same as saying that the linear programming relaxation is infeasible, and so such instances can be shown unsatisfiable in polynomial time simply by solving the LP.

The Need to Become more Competitive with CDCL and MIP Solvers. For formulas that are provably exponentially hard for resolution but easy for cutting planes we see, not surprisingly, that *cdcl-cuttingplanes* and *Sat4j* outperform *Open-WBO*. However, for instances that are inherently pseudo-Boolean in nature, but where resolution can nevertheless efficiently simulate PB reasoning if given a natural CNF encoding, we see that most often *cdcl-cuttingplanes* and *Sat4j* are orders of magnitude slower than *Open-WBO*. It is also often the case, though, that the roles become reversed if the formula is randomly shuffled before being fed to the solvers. And it is also often true that if we force *cdcl-cuttingplanes* to use a good fixed decision order, then its performance matches that of *Open-WBO*, but if left to its own devices *cdcl-cuttingplanes* will deviate from this decision order. This raises the question of whether the way *Open-WBO* encodes pseudo-Boolean constraints into CNF helps it to find and stick to a good variable order when the formula is presented in such a way as to suggest such a good order.

Since pseudo-Boolean solving is closely related to integer linear programming, it is also natural to compare PB solvers to mixed integer linear (MIP) solvers. We have run experiments with the MIP solver *Gurobi* [15] on our combinatorial benchmarks and can observe that it is consistently better than all the PB solvers studied. It should be emphasized that this is perhaps not so suprising—many of our formulas have been constructed to be hard for CDCL but trivial for tree-like cutting planes, and this means that they are by definition amenable to branch-and-bound techniques. Furthermore, *Gurobi* solves an LP relaxation at every node in its search tree, and so will immediately detect the rationally infeasible instances that turn out to be hard for PB solvers. Thus, for the benchmarks considered in this paper *Gurobi* is playing on home turf. Still, we can see no good reason why PB solvers should be so bad for formulas that are dead-easy for tree-like CP. And it certainly would seem well worth it to take a long, hard look at MIP solving techniques and see what can be ported to PB solvers.

Our findings might seem depressing in that they are mostly bad news for state-of-the-art pseudo-Boolean solving. However, we would rather view our work as pointing forward to some crucial challenges that need to be overcome. We hope that our combinatorial formulas can be valuable as challenge benchmarks in the quest to develop more efficient PB solvers, which could then fulfil the vision of [31] and assume their rightful place as *"major tools in problem solving."*

1.2   Organization of this Paper

We describe our experimental set-up in Sect. 2 and discuss our benchmarks in Sect. 3. Section 4 contains an analysis of our results, and some concluding remarks are presented in Sect. 5.

2   Experimental Set-up

We have conducted an experimental evaluation using the pseudo-Boolean solvers *cdcl-cuttingplanes* [12] *Open-WBO* [26,29], and *Sat4j* [21,32]. These were the top three solvers in the Pseudo-Boolean Competition 2016 [30] in the category DEC-SMALLINT-LIN ("no optimization, small integers, linear constraints, SAT+UNSAT answers"), and we ran the solvers on such benchmarks as described in more detail in Sect. 3. Since our benchmarks are inspired by proof complexity, where one studies the complexity of certifying unsatisfiability, we focused almost exclusively on UNSAT instances. For our experiments on shuffled instances we randomly shuffled the variable indices, literal polarities, and the order of the constraints as well as variables within the constraints.

We used the versions of *cdcl-cuttingplanes* and *Open-WBO* submitted to the PB Competition 2016 and a slightly later version of *Sat4j* from November 4, 2016. By default *Sat4j* runs two subsolvers in parallel (Resolution and Cutting-Planes) and returns the answer of the first of them solving the problem. This gives *Sat4j* an advantage, since it gets double the amount of CPU time compared to the other solvers, but it only makes our point stronger when it performs poorly. Since we are particularly interested in analysing cutting planes-based solvers we included the standalone solver *Sat4jCP* in our experiments, but we only show results when they differ from *Sat4j* (i.e., when the formula was decided by the Resolution subsolver). For the *cdcl-cuttingplanes* experiments with fixed decision orders we used a version from April 19, 2017, since the competition version had no support for fixed orders. For our mixed integer programming experiments we used the solver *Gurobi* [15] version 7.5.2 restricted to a single thread.

We ran our experiments on a cluster with a set-up of 6 AMD Opteron 6238 (2.6 GHz) cores and 16 GB of memory. The timeout for the experiments was 3000 s otherwise stated.

3   Description of Benchmarks

All our benchmarks were designed to be very easy for the cutting planes (CP) proof system, so that the experiments would measure proof search quality for instances where CP-based solvers should in principle be able to perform well.

The well-known *pigeonhole principle (PHP) formula* claims that $n+1$ pigeons can be placed into n holes with only one pigeon per hole, encoded as *pigeon axioms* $\sum_{j \in [n]} x_{i,j} \geq 1$ and *hole axioms* $\sum_{i \in [n+1]} x_{i,j} \leq 1$ for $i \in [n+1]$, $j \in [n]$. We also consider more complicated versions by introducing *emergency exits* as follows. We generate k disjoint PHP instances over variables $x_{i,j}^\ell$, where for each $\ell \in [k]$ we allow some pigeon(s) i^* to "take the emergency exit" by changing the pigeon axiom to $y^\ell + \sum_{j \in [n]} x_{i^*,j}^\ell \geq 1$, where y^ℓ, $\ell \in [k]$, are new variables. However, a special constraint $\sum_{\ell=1}^{k} y^\ell \leq k - 1$ enforces that at most $k - 1$ emergency exits are taken in total. We study two variants where either (a) one particular pigeon per PHP instance can take the emergency exit or (b) all pigeons in an instance can do so. All these versions of PHP are rationally unsatisfiable.

A *subset cardinality (SC) formula* [27,34,35] is generated from a 0/1 $n \times n$ matrix $A = (a_{i,j})$ with 4 ones in every row and column, except that one row and column contains 5 ones. Writing $R_i = \{j \mid a_{i,j} = 1\}$ and $C_j = \{i \mid a_{i,j} = 1\}$ to denote the positions of 1s in row i and column j, respectively, the formula obtained from A consists of the constraints $\sum_{j \in R_i} x_{i,j} \geq |R_i|/2$ and $\sum_{i \in C_j} x_{i,j} \leq |C_i|/2$ for $i, j \in [n]$. We use randomly generated matrices and "fixed bandwidth" matrices with a fixed pattern of ones shifted down the diagonal. We also consider a more restrictive version with equality constraints $\sum_{j \in R_i} x_{i,j} = \lceil |R_i|/2 \rceil$ and $\sum_{i \in C_j} x_{i,j} = \lfloor |C_i|/2 \rfloor$. Again, all of these instances are rationally unsatisfiable.

The *even colouring (EC) formula* [24] over a connected graph $G = (V, E)$ with all $v \in V$ of even degree $\deg(v)$ consists of constraints $\sum_{e \in E(v)} x_e = \deg(v)/2$, where $E(v)$ denotes the set of edges incident to v. This formula claims the existence of a black-white edge colouring such that every vertex has the same number of black and white edges, and is unsatisfiable if and only if $|E|$ is odd. We study these formulas for two families of graphs: (a) long, narrow toroidal grids, where every vertex has edges horizontally and vertically to its 4 neighbours, and with one edge split into a degree-2 vertex to get an odd number of edges; (b) random regular graphs of even degree $2d$, splitting an edge if d is even.

The *vertex cover (VC) formula* with constraints $x_u + x_v \geq 1$, $(u, v) \in E$, and $\sum_{v \in V} x_v \leq S$ encodes that a graph $G = (V, E)$ has a size-S vertex cover (i.e., a set such that every edge is incident to some vertex in it). As in [36], we examine long, narrow rectangular toroidal grids with m rows and n columns for $m = O(1)$ even. The minimal vertex cover for such a graph has size $m\lceil n/2 \rceil$. We generate four versions by varying the value of S, where for the first three n is odd: (a) $S = m\lceil n/2 \rceil - 1$, the largest value such that the formula is still unsatisfiable (version *hard*); (b) $S = mn/2$ (version *easy*), which is more obviously unsatisfiable but still has a rational solution with value $\frac{1}{2}$ for all variables; (c) $S = m\lfloor n/2 \rfloor - 1$ (version *norat*), without rational solutions; (d) $S = m\lfloor n/2 \rfloor - 1$ for n even (version *norat-even*), where S is the largest value that makes the formula unsatisfiable both for Boolean and rational values. To obtain slightly harder instances without superfluous edges we also consider such grids with all vertical edges removed, yielding a collection of disjoint cycles, and use the same values of S as above.

The *dominating set (DS) formula* for a graph G consists of constraints $\sum_{u \in \{v\} \cup N(v)} x_u \geq 1$ for all $v \in V$ and $\sum_{v \in V} x_v \leq S$, saying that G has a size-S dominating set (i.e., such that every vertex in G either belongs to or is a neighbour of a vertex in the set). We study these formulas for hexagonal grids as represented in [36] with m rows and n columns, with one dimension fixed while the other scales. Since hexagonal grids are 3-regular any dominating set must have size at least $\lceil |V|/4 \rceil$, and so we choose $S = \lfloor |V|/4 \rfloor$. When $4 \nmid |V|$ the resulting instance is rationally unsatisfiable, but otherwise there is always a rational solution setting all variables to $\frac{1}{4}$, whereas the Boolean satisfiability depends in nontrivial ways on the exact geometry of the grid [36] (in particular, in contrast to the other families some of our dominating set instances are satisfiable).

The *linearized pebbling (LinPeb) formula* of arity d over a directed acyclic graph with a unique sink has variables $v_1, \ldots, v_d$ for each vertex v and consists of the following contradictory constraints (where we let $d' = 2\lfloor (d-1)/2 \rfloor + 1$): (a) for every source vertex v the constraint $2\sum_{i=1}^{d} v_i \geq d'$; (b) for every non-source

Table 1. Overview of benchmarks and results

Formula family	Rational solutions	Small backdoor	Division needed	Performance		
				cdcl-CP	Sat4j	O-WBO
PHP	No	–	–	Easy	Easy	Hard
SC	No	–	–	Easy	Easy	Hard
EC even grid	Yes	Yes	No	Easy	Easy	Hard
EC odd grid	Yes	No	Helpful	Hard[a]	Hard	Hard
EC random	Yes	No(?)	Crucial(?)	Fairly hard	Hard	Hard
VC hard	Yes	No	Helpful	Hard[b]	Hard	Easy[c]
VC easy	Yes	Many	No	Hard[b]	Hard	Easy[c]
VC norat(-even)	No	–	–	Hard[b]	Hard	Easy[c]
DS	Yes	Many	No	Hard[d]	Hard[d]	Easy[c]
LinPeb	Yes	Yes	No	Hard[b]	Hard[e]	Easy

[a] Easy if formula appropriately reordered. [b] Fairly easy for forced order.
[c] Hard if shuffled. [d] Easy if LP-derivable constraints added. [e] Easy for *Sat4jRes*.

vertex w with predecessors u and v the constraint $2\sum_{i=1}^{d} w_i \geq \sum_{i=1}^{d}(u_i + v_i)$;
(c) for the unique sink vertex z the constraint $2\sum_{i=1}^{d} z_i \leq d'$. We study instances generated from so-called pyramid graphs.[5]

4   Experimental Evaluation

In this section we describe and analyse the results of our experiments (summarized in Table 1).[6] As already mentioned, our benchmarks can be scaled in size by varying a parameter, allowing to study the asymptotic behaviour of the solvers as the instance size increases. Our figures illustrate this by plotting performance on the y-axis against the value of the scaling parameter on the x-axis (which, in particular, seems to be a better way of visualizing our data than using so-called cactus plots).

4.1   Pseudo-Boolean Solvers and Boolean Reasoning

Our first conclusion is that PB solvers need to strengthen their reasoning by using stronger rules than saturation and implementing better proof search.

As an example, consider even colouring (EC) formulas, which can be refuted (i.e., proven unsatisfiable) with tree-like cutting planes proofs in linear size using just two applications of division. We could thus expect solvers based on cutting

---

[5] We remark that some linearized pebbling formulas were submitted to the Pseudo-Boolean Competition 2016 under the name `sumineq` (sum inequalities).

[6] By necessity, our discussion is far from exhaustive, but readers can find all our benchmarks and the data from our experiments at http://www.csc.kth.se/~jakobn/publications/CombinatorialBenchmarksPBsolvers.

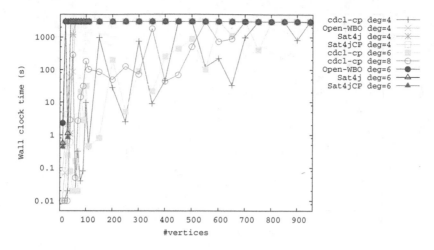

Fig. 1. Solver performance for even colouring formulas on random graphs (#constraints = #vertices $|V|$; #variables = $\deg \cdot |V|/2 + O(1)$).

planes (CP) to run blisteringly fast for EC formulas over any graph, but this is not the case. We have generated formulas from $m \times n$ grids with $m = O(1)$ a small constant, which have short proofs in resolution when encoded in CNF, and random regular graphs, which are exponentially hard for resolution.[7]

For grids with m even the formulas are trivial for both *Sat4j* and *cdcl-cuttingplanes*, but for m odd they are hard. Interestingly, in the latter case *Sat4jRes* performs better than *Sat4jCP*, suggesting that CP-based solvers do not find better proofs than CDCL-based solvers. It is also notable that flipping the vertex (and hence variable) order from column-major to row-major, even though it does not change the formula, helps *cdcl-cuttingplanes* find an efficient proof. This indicates that there is ample room to improve on search heuristics.

To explain the difference between even-row and odd-row grids, we observe that EC formulas always have a rational solution with all variables assigned value $\frac{1}{2}$, but grids with an even number of rows and columns have backdoors of size 1 (namely, either of the edges incident to the degree-2 vertex on the split edge), and as long as the number of rows m is even there are backdoors of size at most $m = O(1)$. If m is odd, however, the backdoor size jumps to $n - O(1)$.

EC formulas on random graphs are very hard for *Open-WBO* and *Sat4j*. They are not easy for *cdcl-cuttingplanes* either, but this solver performs markedly better, and does not seem to be sensitive to the degree of the graph (see Fig. 1). The only short proofs known for these formulas crucially use division [36], and we believe that the superior performance of *cdcl-cuttingplanes* is explained by the frequent (though still limited) use of division in this solver. It is worth noting, though, that for the few instances solved by *Sat4j* the number of conflicts is not

---

[7] Such a lower bound cannot be found in the literature, but is possible to obtain for graphs with good enough expansion using a variation of the techniques in [4].

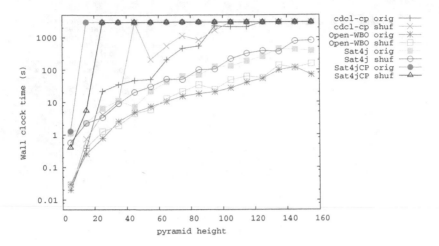

Fig. 2. Performance for arity-5 linearized pebbling formulas on pyramids, with (**shuf**) and without (**orig**) shuffling (#constraints $\approx$ #variables $\approx$ **height**^{2}/2).

too far from *cdcl-cuttingplanes*. The most likely explanation is that *Sat4j* does divide the constraint in the rare case when all coefficients are equal, which is all that is needed in [36]. To be sure whether the above explanations are correct we would need to do proof logging, but for PB solvers there is unfortunately nothing like the *DRAT* format [17,18,37] used for CDCL solvers.

Another formula family where *cdcl-cuttingplanes* performs better than *Sat4j* are linearized pebbling formulas, which are easy for *Sat4jRes* but extremely hard for *Sat4jCP* (see Fig. 2). Interestingly, for these instances *Sat4j* generates constraints with coefficients larger than 10^9 in a matter of seconds, whereas *cdcl-cuttingplanes* keeps all coefficients small. Division is not critically needed for efficient refutations here, but it might be that it is what helps *cdcl-cuttingplanes* keep coefficient sizes down and achieve better performance. *Sat4j* has problems with large coefficients also for vertex cover (VC) and dominating set (DS) instances, where *cdcl-cuttingplanes* performs better, but not for PHP and subset cardinality (SC) formulas, where both solvers are fast.

Let us next review what our data say about the *extended rational hypothesis*, i.e., that instances with small backdoors to rational unsatisfiability should be easy. PHP and SC formulas do not have rational solutions, and the fact that instances are trivial for both *cdcl-cuttingplanes* and *Sat4j* supports the rational hypothesis. PHP formulas with emergency exits were designed to be potentially harder instances that still do not have rational solutions, but they fail to fool *cdcl-cuttingplanes* and hence can be interpreted as circumstantial evidence in favour of the rational hypothesis for this solver (but less so for *Sat4j*).

As mentioned above, *cdcl-cuttingplanes* and *Sat4j* run fast on EC formulas for backdoor size 1 but not larger (random graphs very likely yield instances without small backdoors, though we did not attempt a rigorous proof). This supports the hypothesis, but the fact that *cdcl-cuttingplanes* also runs fast when slightly modifying instances for odd-row grids makes the connection less clear.

The VC instances `norat` and `norat-even` without rational solutions are easier than the `easy` version, which is in turn easier than the `hard` version. This is consistent with the hypothesis since the backdoor size is 0 for `norat(-even)` and 1 for `easy` (rational solutions disappear after branching on any vertex), whereas the smallest backdoor for version `hard` has size $m - 1$ ($m - 1$ vertices in the same column form a backdoor, but any $m - 2$ vertices can be assigned to leave a rational solution). This holds for both grids and collections of cycles.

Linearized pebbling formulas have a size-d backdoor (the d variables associated with the sink). *Sat4jCP* does not run fast on these formulas, but with some tweaking *cdcl-cuttingplanes* can be convinced to perform well. It seems like a stimulating challenge to develop natural heuristics for the CP-based solvers that would make them competitive with *Open-WBO* and *Sat4jRes* for these instances.

Dominating set formulas on hexagonal grids have rational solutions when the total number of vertices is divisible by 4, in which case there is strong empirical evidence for backdoors of size 3 (obtained by considering any vertex and two of its neighbours). Somewhat annoyingly, we have not been able to always make *cdcl-cuttingplanes* run fast for such instances, however, so as of now our experimental results for these formulas do *not* support the rational hypothesis.

4.2 Pseudo-Boolean Solvers and Linear Programming

To support the claim that PB solvers also need better linear programming capabilities, we again consider dominating set instances on hexagonal grids. These are very challenging for both *Sat4j* and *cdcl-cuttingplanes*. They are manageable for *Open-WBO* when the fixed dimension is small but quickly become very hard as this dimension grows. Also, *Open-WBO* is extremely sensitive to random shuffling, a phenomenon that we discuss further in Sect. 4.3.

Quite intriguingly, all instances become trivial if modified as follows. Recall that we have a greater-or-equal (`GEQ`) constraint $\sum_{u \in \{v\} \cup N(v)} x_u \geq 1$ for each vertex v encoding that it is dominated. Since hexagonal grids are 3-regular, and since the required dominating set size is at most $|V|/4$, it follows that at most one of the vertices in $\{v\} \cup N(v)$ is in the dominating set. These less-or-equal (`LEQ`) constraints can easily be derived using only the addition rule in *cdcl-cuttingplanes* and *Sat4j* [36], and with such constraints added the instances become trivial as shown in Fig. 3. So far we have not been able to get the solvers to realize that these `LEQ` constraints should be derived, though, although elementary linear programming would be sufficient to achieve this.

4.3 Pseudo-Boolean Solvers Versus CDCL and MIP

When comparing cutting planes-based solvers to CDCL solvers we get mixed results. On PHP and subset cardinality formulas both CP-based solvers *Sat4j* and *cdcl-cuttingplanes* perform very well, while *Open-WBO* does very poorly,

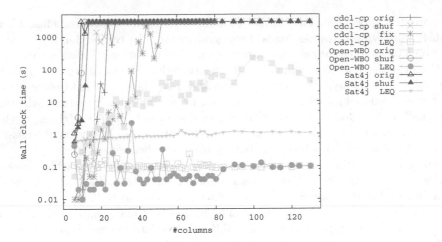

Fig. 3. Performance for dominating set on hex grids with 7 rows with shuffling (**shuf**) and without (**orig**) and also with added **LEQ** constraints as well as for fixed-order (**fix**) *cdcl-cuttingplanes* (#constraints = #variables = $7 \cdot$ **columns** + O(1)).

which nicely matches that these formulas are easy for CP but exponentially hard for resolution. For our other benchmarks we find *Open-WBO* to be surprisingly competitive, but performance is often brittle. In contrast, although all our benchmarks are very easy for CP, on many instances *cdcl-cuttingplanes* and *Sat4j* are quite far from performing well (though *cdcl-cuttingplanes* can often be made to match *Open-WBO* performance by manual intervention such as fixing good variable decision orders).

An example family of benchmarks for which *Open-WBO* shines are vertex cover (VC) formulas. Here the performance of *cdcl-cuttingplanes* and *Sat4j* is quite poor in general, though the former solver is clearly better than the latter. A closer look at the results reveals that the number of conflicts seems to be similar, but since *Sat4j* solves so few instances within the timeout limit it is hard to know for sure whether the differences in running time are due to proof search quality or lower-level implementation details.

Open-WBO performs quite well for almost all our VC instances (except for the **hard** version on collections of cycles), which indicates that the encoding to CNF admits an efficient resolution proof. Since the covering constraints $x_u + x_v \geq 1$ for the edges $(u, v) \in E$ are already disjunctive clauses, the performance is likely to depend on how the vertex cover size constraint $\sum_{v \in V} x_v \leq S$ is encoded into CNF. A key aspect here is that the vertices are listed consecutively when we generate the grid graphs (more precisely, in column-major order). This means that as *Open-WBO* decides on consecutive variable indices, it will explore neighbouring vertices constrained by common covering constraints, and it seems plausible that propagation on the auxiliary variables in the encoding of the size

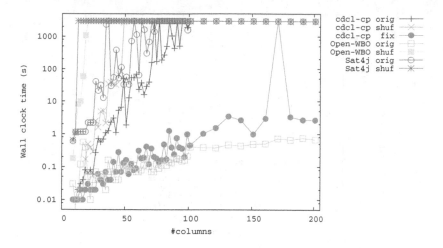

Fig. 4. Performance for vertex cover formulas on grids with 8 rows, version **norat**, with shuffling (**shuf**) and without (**orig**), and also for fixed-order (**fix**) *cdcl-cuttingplanes* (#variables = $|V| = 8 \cdot$ **columns**, #constraints = $2 \cdot |V|$).

constraint helps the solver count efficiently. This hypothesis[8] is supported by the fact that when instances are shuffled the performance degenerates dramatically for *Open-WBO*, but much less so for *cdcl-cuttingplanes*. A further observation is that though *cdcl-cuttingplanes* is rather bad for these instances, it can be made much more efficient by fixing the decision order (namely, branching on vertices in column-major order). This indicates that with a better decision heuristic *cdcl-cuttingplanes* could potentially be competitive with *Open-WBO* here. See Fig. 4 for plots of all the findings above.

As an example of benchmarks that *Sat4j* and *cdcl-cuttingplanes* can solve easily we have pigeonhole principle (PHP) formulas with $n + 1$ pigeons and n pigeonholes. These have CP proofs in size $O(n^2)$ that can be found with $O(n)$ conflicts. While both solvers only need $O(n)$ conflicts, and seem to find the same (essentially optimal) proof, we found that running times scale very differently. A linear regression analysis using the logarithm of the number of constraints indicates running time $O(n^{3.2})$ for *cdcl-cuttingplanes* but $O(n^{5.0})$ for *Sat4j* (see Fig. 5). Interestingly, this turned out to be due to an implementation inefficiency in *Sat4j*, which could be identified and fixed thanks to our experiments, after which running times became more similar. It is not surprising that PHP formulas are very hard for *Open-WBO*, since there is an exponential lower bound for resolution [16] which can also be adapted (using techniques in [4]) to work for other common ways of encoding the at-most-1 pigeonhole constraints into CNF.

---

[8] It would be interesting to verify this by a more in-depth study of *Open-WBO*. However, the PB version of this solver was not open-source at the time of our experiments, and also our main focus in this work is on solvers implementing CP-based reasoning.

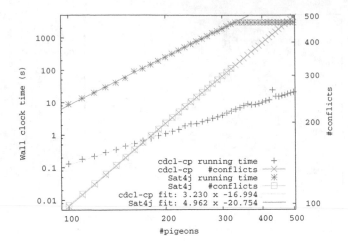

Fig. 5. Running time and #conflicts for *cdcl-cuttingplanes* and *Sat4j* on PHP (#variables = **pigeons**(**pigeons** − 1), #constraints = 2 · **pigeons** − 1).

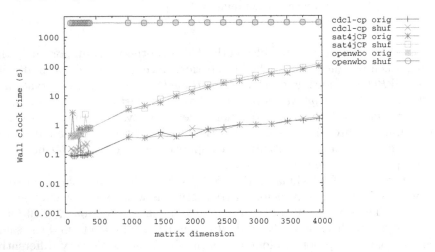

Fig. 6. Performance for subset cardinality formulas on random graphs (#variables = 4 · **dimension** + 1, #constraints = 2 · **dimension**).

PHP formulas with emergency exits are always easy for *cdcl-cuttingplanes* but remain hard for *Open-WBO* independently of the number of emergency exits k. This latter finding is also as expected, since even if the solver chooses $k − 1$ emergency exits in the right way to satisfy $k − 1$ subinstances of PHP, the residual formula is a standard PHP instance which is exponentially hard. Interestingly, *Sat4j* performs well on the version where all pigeons can take the emergency exit, but much worse on the version with only one pigeon per exit (which is more constrained, and could thus have been expected to be easier). We remark that when both *Sat4j* and *cdcl-cuttingplanes* solve these formulas efficiently the number of conflicts seem to grow like $O(kn)$, but when *Sat4j* does not perform well the number of conflicts grows faster. Thus, in contrast to the results for standard PHP, here the proof search quality seems worse in *Sat4j*.

Looking at subset cardinality formulas, they seem to be solved much faster than PHP when plotting running time against the scaling parameter, but this is since the instances are much smaller. Again, we observe that *Sat4j* takes significantly longer than *cdcl-cuttingplanes* as the instance size increases. *Open-WBO* is completely lost, as expected in view of the exponential lower bound for resolution in [27] (see Fig. 6).

Let us finally make a brief comparison to MIP solving. *Gurobi* does remarkably well on all our benchmarks, solving all but the three largest EC instances in under 10 s. On the one hand, this is not too surprising, since all of our instances have tree-like proofs, and hence just branching and backtracking without learning is enough to solve them. Furthermore, the challenging instances that are rationally unsatisfiable will be solved *Gurobi* very quickly, since it also considers linear relaxations of the problem and this is enough to decide unsatisfiability. However, it is hard to avoid the conclusion that one promising approach for strengthening PB solvers would be to incorporate techniques from MIP solving.

5   Concluding Remarks

In this paper we evaluate the three pseudo-Boolean solvers *cdcl-cuttingplanes*, *Open-WBO*, and *Sat4j* on decision problems encoded as linear constraints with small integer coefficients, a kind of problems where these solvers were among the best in the Pseudo-Boolean Competition 2016. The solvers differ in that *Open-WBO* re-encodes the problem into CNF and runs a CDCL solver, thus performing proof search in resolution for the re-encoded instance, whereas *cdcl-cuttingplanes* and *Sat4j* implement conflict-driven search natively with pseudo-Boolean constraints, corresponding to cutting planes (CP) proof search.

We have performed extensive experiments on carefully constructed combinatorial benchmarks to investigate how efficiently these solvers implement their chosen methods of reasoning. Although all of our instances have been specifically designed to be very easy for the cutting planes proof system, the performance of *cdcl-cuttingplanes* and *Sat4j* varies greatly, and is often quite poor. Theoretical as well as empirical evidence points to the conclusion that the reasoning in these solvers needs to be strengthened, in particular, by exploiting the division rule.

For many of the benchmarks studied we can help *Sat4j* and *cdcl-cuttingplanes* run fast by giving advice in the form of a good, fixed variable decision order, or sometimes by reordering variables and constraints. An immediate question is whether the solvers could achieve such good performance on their own by some enhanced heuristic. It can be observed that this phenomenon occurs most often for instances which either do not even have rational solutions—i.e., when the real polytope defined by the linear constraints is in fact empty—or have small backdoor sets such that any assignment to these backdoor variables eliminates all rational solutions. We find this to be a very intriguing connection, and believe it would be interesting to investigate further whether it can be the case more generally that strong solver performance correlates with the existence of small backdoors to rationally unsatisfiable instances.

As expected, *Open-WBO* stands no chance against *cdcl-cuttingplanes* and *Sat4j* when run on instances that are hard for resolution when encoded into CNF. However, when there are efficient proofs in both resolution and cutting planes we see that the CP-based solvers can be orders of magnitude slower. Curiously, if *cdcl-cuttingplanes* is helped by being given a good variable order on such instances, then the performance is competitive with *Open-WBO*, but when left to its own devices *cdcl-cuttingplanes* does not choose this order. This raises the question whether the encoding to CNF that is used helps *Open-WBO* find a good variable order and stick with it. It should be noted, though, that *Open-WBO* is very sensitive to permutations of the input, so the encoding to CNF is only good when the constraints in the initial pseudo-Boolean instance are presented in a helpful order. The CP-based solvers appear much more robust in this regard.

Finally, we observe that for the instances considered in this paper all three PB solvers that we study are clearly outperformed by the general-purpose mixed integer programming solver *Gurobi*. At first sight this is slightly disappointing, since PB solvers working on 0/1-valued problems should be able to exploit techniques not available to MIP solvers, but a big part of the explanation is probably that our benchmarks have been constructed to be easy for tree-like CP, and so they are by design amenable to branch-and-bound techniques. But another reason is likely to be that *Gurobi* solves linear programming relaxations of the problem during the search, which makes it run fast on instances that lack rational solutions but are apparently very challenging for PB solvers. We believe that there would be great potential for improvement by incorporating such linear programming reasoning in PB solvers. It would also be interesting to find benchmarks that are easy for conflict-driven pseudo-Boolean search, at least in theory, but not for MIP or CDCL, i.e., instances that are hard for resolution and tree-like cutting planes but easy for general, DAG-like cutting planes.

Taken together, our results can be viewed as a concrete set of challenges to be overcome in order to construct more efficient pseudo-Boolean solvers. It is also our belief that a further study of crafted benchmarks like the ones in this paper has the potential to shed valuable light on the inner workings of PB solvers.

Acknowledgements. We are most grateful to Daniel Le Berre for long and patient explanations of the inner workings of pseudo-Boolean solvers, and to João Marques-Silva for helping us get an overview of relevant references for pseudo-Boolean solving. We also want to thank Ruben Martins for sharing an executable of *Open-WBO* with us and answering questions about the solver. Finally, we are thankful for the many detailed comments from the *SAT 2018* anonymous reviewers, which helped to improve this paper considerably.

Our computational experiments were performed on resources provided by the Swedish National Infrastructure for Computing (SNIC). Many of our benchmarks were generated using the tool CNFgen [8,20], for which we gratefully acknowledge Massimo Lauria.

The fourth author performed part of this work while at KTH Royal Institute of Technology. All authors were funded by the European Research Council under the European Union's Seventh Framework Programme (FP7/2007–2013)/ERC grant agreement no. 279611. The third author was also supported by Swedish Research Council grants 621-2012-5645 and 2016-00782, and the fourth author by the Prof. R Narasimhan Foundation.

References

1. Barth, P.: Linear 0-1 inequalities and extended clauses. Technical report MPI-I-94-216, Max-Planck-Institut für Informatik, April 1994. Preliminary version in LPAR 1993

2. Barth, P.: A Davis-Putnam based enumeration algorithm for linear pseudo-Boolean optimization. Technical report MPI-I-95-2-003, Max-Planck-Institut für Informatik, January 1995

3. Bayardo Jr., R.J., Schrag, R.: Using CSP look-back techniques to solve real-world SAT instances. In: Proceedings of the 14th National Conference on Artificial Intelligence (AAAI 1997), pp. 203–208, July 1997

4. Ben-Sasson, E., Wigderson, A.: Short proofs are narrow—resolution made simple. J. ACM 48(2), 149–169 (2001). Preliminary version in STOC 1999

5. Boros, E., Hammer, P.L.: Pseudo-Boolean optimization. Discrete Appl. Math. 123(1–3), 155–225 (2002)

6. Chai, D., Kuehlmann, A.: A fast pseudo-Boolean constraint solver. IEEE Trans. Comput. Aided Des. Integr. Circ. Syst. 24(3), 305–317 (2005). Preliminary version in DAC 2003

7. Chvátal, V.: Edmonds polytopes and a hierarchy of combinatorial problems. Discrete Math. 4(1), 305–337 (1973)

8. CNFgen: Combinatorial benchmarks for SAT solvers. https://github.com/MassimoLauria/cnfgen

9. Cook, W., Coullard, C.R., Turán, G.: On the complexity of cutting-plane proofs. Discrete Appl. Math. 18(1), 25–38 (1987)

10. Dixon, H.E., Ginsberg, M.L., Hofer, D.K., Luks, E.M., Parkes, A.J.: Generalizing Boolean satisfiability III: implementation. J. Artif. Intell. Res. 23, 441–531 (2005)

11. Eén, N., Sörensson, N.: Translating pseudo-Boolean constraints into SAT. J. Satisf. Boolean Model. Comput. 2(1–4), 1–26 (2006)

12. Elffers, J.: cdcl-cuttingplanes: a conflict-driven pseudo-Boolean solver (2016). Submitted to the Pseudo-Boolean Competition 2016

13. Elffers, J., Nordström, J.: Divide and conquer: towards faster pseudo-Boolean solving. In: Proceedings of the 27th International Joint Conference on Artificial Intelligence (IJCAI-ECAI 2018), July 2018 (to appear)

14. Gomory, R.E.: An algorithm for integer solutions of linear programs. In: Graves, R., Wolfe, P. (eds.) Recent Advances in Mathematical Programming, pp. 269–302. McGraw-Hill, New York (1963)

15. Gurobi optimizer. http://www.gurobi.com/

16. Haken, A.: The intractability of resolution. Theor. Comput. Sci. 39(2–3), 297–308 (1985)

17. Heule, M., Hunt Jr., W.A., Wetzler, N.: Trimming while checking clausal proofs. In: Proceedings of the 13th International Conference on Formal Methods in Computer-Aided Design (FMCAD 2013), pp. 181–188, October 2013
18. Heule, M.J.H., Hunt, W.A., Wetzler, N.: Verifying refutations with extended resolution. In: Bonacina, M.P. (ed.) CADE 2013. LNCS (LNAI), vol. 7898, pp. 345–359. Springer, Heidelberg (2013). https://doi.org/10.1007/978-3-642-38574-2_24
19. Joshi, S., Martins, R., Manquinho, V.: Generalized totalizer encoding for pseudo-Boolean constraints. In: Pesant, G. (ed.) CP 2015. LNCS, vol. 9255, pp. 200–209. Springer, Cham (2015). https://doi.org/10.1007/978-3-319-23219-5_15
20. Lauria, M., Elffers, J., Nordström, J., Vinyals, M.: CNFgen: a generator of crafted benchmarks. In: Gaspers, S., Walsh, T. (eds.) SAT 2017. LNCS, vol. 10491, pp. 464–473. Springer, Cham (2017). https://doi.org/10.1007/978-3-319-66263-3_30
21. Le Berre, D., Parrain, A.: The Sat4j library, release 2.2. J. Satisf. Boolean Model. Comput. **7**, 59–64 (2010)
22. Manquinho, V.M., Marques-Silva, J.: On using cutting planes in pseudo-Boolean optimization. J. Satisf. Boolean Model. Comput. **2**, 209–219 (2006). Preliminary version in SAT 2005
23. Manquinho, V.M., Marques-Silva, J.P.: Integration of lower bound estimates in pseudo-Boolean optimization. In: 16th IEEE International Conference on Tools with Artificial Intelligence (ICTAI 2004), pp. 742–748, November 2004
24. Markström, K.: Locality and hard SAT-instances. J. Satisf. Boolean Model. Comput. **2**(1–4), 221–227 (2006)
25. Marques-Silva, J.P., Sakallah, K.A.: GRASP: a search algorithm for propositional satisfiability. IEEE Trans. Comput. 48(5), 506–521 (1999). Preliminary version in ICCAD 1996
26. Martins, R., Manquinho, V., Lynce, I.: Open-WBO: a modular MaxSAT solver. In: Sinz, C., Egly, U. (eds.) SAT 2014. LNCS, vol. 8561, pp. 438–445. Springer, Cham (2014). https://doi.org/10.1007/978-3-319-09284-3_33
27. Mikša, M., Nordström, J.: Long proofs of (seemingly) simple formulas. In: Sinz, C., Egly, U. (eds.) SAT 2014. LNCS, vol. 8561, pp. 121–137. Springer, Cham (2014). https://doi.org/10.1007/978-3-319-09284-3_10
28. Moskewicz, M.W., Madigan, C.F., Zhao, Y., Zhang, L., Malik, S.: Chaff: engineering an efficient SAT solver. In: Proceedings of the 38th Design Automation Conference (DAC 2001), pp. 530–535, June 2001
29. Open-WBO: An open source version of the MaxSAT solver WBO. http://sat.inesc-id.pt/open-wbo/
30. Pseudo-Boolean competition 2016, July 2016. http://www.cril.univ-artois.fr/PB16/
31. Roussel, O., Manquinho, V.M.: Pseudo-Boolean and cardinality constraints. In: Biere, A., Heule, M.J.H., van Maaren, H., Walsh, T. (eds.) Handbook of Satisfiability, Chap. 22. Frontiers in Artificial Intelligence and Applications, vol. 185, pp. 695–733. IOS Press, Amsterdam (2009)
32. Sat4j: The Boolean satisfaction and optimization library in Java. http://www.sat4j.org/
33. Sheini, H.M., Sakallah, K.A.: Pueblo: a hybrid pseudo-Boolean SAT solver. J. Satisf. Boolean Model. Comput. **2**(1–4), 165–189 (2006). Preliminary version in DATE 2005
34. Spence, I.: sgen1: a generator of small but difficult satisfiability benchmarks. J. Exp. Algorithmics **15**, 1.2:1–1.2:15 (2010)

35. Van Gelder, A., Spence, I.: Zero-one designs produce small hard SAT instances. In: Strichman, O., Szeider, S. (eds.) SAT 2010. LNCS, vol. 6175, pp. 388–397. Springer, Heidelberg (2010). https://doi.org/10.1007/978-3-642-14186-7_37
36. Vinyals, M., Elffers, J., Giráldez-Cru, J., Gocht, S., Nordström, J.: In between resolution and cutting planes: a study of proof systems for pseudo-Boolean SAT solving, July 2018 (to appear)
37. Wetzler, N., Heule, M.J.H., Hunt, W.A.: DRAT-trim: efficient checking and trimming using expressive clausal proofs. In: Sinz, C., Egly, U. (eds.) SAT 2014. LNCS, vol. 8561, pp. 422–429. Springer, Cham (2014). https://doi.org/10.1007/978-3-319-09284-3_31

Machine Learning-Based Restart Policy for CDCL SAT Solvers

Jia Hui Liang[1]([✉]), Chanseok Oh[2], Minu Mathew[3], Ciza Thomas[3],
Chunxiao Li[1], and Vijay Ganesh[1]

[1] University of Waterloo, Waterloo, Canada
jliang@gsd.uwaterloo.ca
[2] Google, New York, USA
[3] College of Engineering, Thiruvananthapuram, India

Abstract. Restarts are a critically important heuristic in most modern conflict-driven clause-learning (CDCL) SAT solvers. The precise reason as to why and how restarts enable CDCL solvers to scale efficiently remains obscure. In this paper we address this question, and provide some answers that enabled us to design a new effective machine learning-based restart policy. Specifically, we provide evidence that restarts improve the quality of learnt clauses as measured by one of best known clause quality metrics, namely, literal block distance (LBD). More precisely, we show that more frequent restarts decrease the LBD of learnt clauses, which in turn improves solver performance. We also note that too many restarts can be harmful because of the computational overhead of rebuilding the search tree from scratch too frequently. With this trade-off in mind, between that of learning better clauses vs. the computational overhead of rebuilding the search tree, we introduce a new machine learning-based restart policy that predicts the quality of the next learnt clause based on the history of previously learnt clauses. The restart policy erases the solver's search tree during its run, if it predicts that the quality of the next learnt clause is below some dynamic threshold that is determined by the solver's history on the given input. Our machine learning-based restart policy is based on two observations gleaned from our study of LBDs of learnt clauses. First, we discover that high LBD percentiles can be approximated with z-scores of the normal distribution. Second, we find that LBDs, viewed as a sequence, are correlated and hence the LBDs of past learnt clauses can be used to predict the LBD of future ones. With these observations in place, and techniques to exploit them, our new restart policy is shown to be effective over a large benchmark from the SAT Competition 2014 to 2017.

1 Introduction

The Boolean satisfiability problem is a fundamental problem in computer science: given a Boolean formula in conjunctive normal form, does there exist an assignment to the Boolean variables such that the formula evaluates to true? Boolean satisfiability is the quintessential NP-complete [13] problem, and hence

© Springer International Publishing AG, part of Springer Nature 2018
O. Beyersdorff and C. M. Wintersteiger (Eds.): SAT 2018, LNCS 10929, pp. 94–110, 2018.
https://doi.org/10.1007/978-3-319-94144-8_6

one might prematurely conjecture that Boolean SAT solvers cannot scale. Yet modern SAT solvers routinely solve instances with millions of Boolean variables. In practice, many practitioners reduce a variety of NP problems to the Boolean satisfiability problem, and simply call a modern SAT solver as a black box to efficiently find a solution to their problem instance [11,12,19]. For precisely this reason, SAT solving has become an important tool for many industrial applications. Through decades of research, the SAT community has built surprisingly effective backtracking solvers called conflict-driven clause-learning (CDCL) [23] SAT solvers that are based on just a handful of key principles [18]: conflict-driven branching, efficient propagation, conflict analysis, preprocessing/inprocessing, and restarts.

Like all backtracking search, the run of a CDCL SAT solver can be visualized as a search tree where each distinct variable is a node with two outgoing edges marked true and false (denoting value assignments to the variable) respectively. The solver frequently restarts, that is, it discards the current search tree and begins anew (but does not throw away the learnt clauses and the variable activities). Although this may seem counterproductive, SAT solvers that restart frequently are significantly faster empirically than solvers that opt not to restart. The connection between restarts and performance is not entirely clear, although researchers have proposed a variety of hypotheses such as exploiting variance in the runtime distribution [14,22] (similar to certain kinds of randomized algorithms). For various reasons however, we find these hypotheses do not explain the power of restarts in the CDCL SAT solver setting. In this paper, we take inspiration from Hamadi et al. who claim that the purpose of restarts is to compact the assignment stack [16]. We then further show that a compact stack tends to improve the quality of clauses learnt where we define quality in terms of the well-known metric *literal block distance* (LBD). Despite the search tree being discarded by a restart, learnt clauses are preserved so learning higher quality clauses continues to reap benefits across restarts. By learning higher quality clauses, the solver tends to find a solution quicker. However, restarting too often incurs a high overhead of constantly rebuilding the search tree. So it is imperative to balance the restart frequency to improve the LBD but avoid excessive overhead.

Based on the above-mentioned analysis, we designed a new restart policy called *machine learning-based restart* (MLR) that triggers a restart when the LBD of the next learnt clause is above a certain threshold. The motivation for this policy is that rather than investing computation into learning a low quality clause, the solver should invest that time in rebuilding the search tree instead in the hopes of learning a better clause. This restart policy is based on two key observations that we made by analyzing CDCL solvers over a large benchmark: First, we observed that recent LBDs are correlated with the next LBD. We introduce a machine learning-based technique exploiting this observation to predict the LBD of the next learnt clause. Second, we observed that the right tail of the LBD distribution is similar to the right tail of the normal distribution. We exploit this observation to set a meaningful LBD threshold for MLR based on percentiles. MLR is then shown to be competitive vis-a-vis the current state-of-the-art restart policy implemented as part of the Glucose solver [4].

Contributions: We make the following contributions in this paper:

1. We provide experimental support for the hypothesis that restarts "compact the assignment stack" as stated by the authors of ManySAT [16] (see Sect. 4). We then add to this hypothesis, and go on to show that a compact assignment stack correlates with learning lower LBD clauses (see Sect. 4.2). Lastly, learning clauses with lower LBD is shown to correlate with better solving time (see Sect. 4.3). Additionally we provide analytical explanations as to why we discount some previously proposed hypotheses that attempt to explain the power of restarts in practice (see Sect. 3).
2. We propose a method to set thresholds for the quality of a LBD of a clause. We experimentally show that the right tail of the LBD distribution closely matches a normal distribution, hence high percentiles can be accurately predicted by simply computing the mean and standard deviation. See Sect. 5.1 for details.
3. We show that LBDs viewed as a sequence are correlated. This is a crucial observation that we back by experimental data. The fact that LBDs viewed as a sequence are correlated enables us to take the LBDs of recent learnt clauses and predict the LBD of the next clause. See Sect. 5.2 for details.
4. Building on all the above-mentioned experimentally-verified observations, we introduce a new restart policy called *machine learning-based restart* (MLR) that is competitive vis-a-vis the current state-of-the-art restart policy on a comprehensive benchmark from the SAT Competition 2014 to 2017 instances. See Sect. 6 for details.

2   Background

We assume the reader is familiar with the Boolean satisfiability problem and SAT solver research literature [6].

LBD Clause Quality Metric: It has been shown, through the lens of proof complexity, that clause-learning SAT solvers (under perfect non-deterministic branching and restarts, and asserting clause learning schemes) are exponentially more powerful than CDCL SAT solvers without clause learning [26]. However, the memory requirement to store all the learnt clauses is too high for many instances since the number of conflicts grows very rapidly. To overcome this issue, all modern SAT solvers routinely delete some clauses to manage the memory usage. The most popular metric to measure the quality of a clause is called *literal block distance* (LBD) [3], defined as the number of distinct decision levels of the variables in the clause. Intuitively, a clause with low LBD prunes more search space than a clause with higher LBD. Hence clauses with high LBD typically are the ones prioritized for deletion. Although LBD was originally proposed for the purpose of clause deletion, it has since proven useful in other contexts where there is need to measure the quality of learnt clauses such as sharing clauses in parallel SAT solvers and restarts. Another measure of quality of a learnt clause is its length. To the best of our knowledge, we are not aware of any other universally accepted clause quality metrics at the time of writing of this paper.

In this paper we will often look at LBDs as a sequence. At any time during the search where i conflicts have occurred, we use the term "previous" LBD to refer to the LBD of the clause learnt at the ith conflict and "next" LBD to refer to the LBD of the clause learnt at the $(i + 1)^{th}$ conflict.

Overview of Restarts in CDCL SAT Solvers: Informally, a restart heuristic in the context of CDCL SAT solver can be defined as a method that discards parts of the solver state at certain points in time during its run. CDCL solvers restart by discarding their "current" partial assignment and starting the search over, but all other aspects of solver state (namely, the learnt clauses, variable activities, and variable phases) are preserved. Although restarts may appear unintuitive, the fact that learnt clauses are preserved means that solver continues to make progress. Restarts are implemented in practically all modern CDCL solvers because it is well known that frequent restarts greatly improve solver performance in practice.

3   Prior Hypotheses on "The Power of Restarts"

In this section, we discuss prior hypotheses on the power of restarts in the DPLL and local search setting and their connection to restarts in the CDCL setting.

Heavy-tailed Distribution and Las Vegas Algorithm Hypotheses: From the perspective of Las Vegas algorithms, some researchers have proposed that restarts in CDCL SAT solvers take advantage of the variance in solving time [14,22]. For a given input, the running time of a Las Vegas algorithm is characterized by a probability distribution, that is, depending on random chance the algorithm can terminate quickly or slowly relatively speaking. A solver can get unlucky and have an uncharacteristically long running time, in which case, a restart gives the solver a second chance of getting a short run-time [22]. More specifically, a heavy-tailed distribution was observed for various satisfiable instances on randomized DPLL solvers [14]. However, this explanation does not lift to restarts in modern CDCL solvers. First, most modern CDCL solvers are not Las Vegas algorithms, that is, they are deterministic algorithms, and hence restarts cannot take advantage of variance in the solving time like in Las Vegas algorithms. Second, the optimal restart policy for Las Vegas algorithms has a restart interval greater than the expected solving time of the input [22], so hard instances should restart very infrequently. However in practice, even difficult instances with high solving time benefit from very frequent restarts in CDCL solvers. Third, the definition of restarts in the context of Las Vegas algorithms differs significantly from the restarts implemented in CDCL solvers. In Las Vegas algorithms, the restarts are equivalent to starting a new process, that is, the algorithm starts an independent run from scratch. Restarts in CDCL are only partial, the assignment stack is erased but everything else preserved (i.e., learnt clauses, saved phases, activity, etc.). Since the phases are saved, the CDCL SAT solver reassigns variables to the same value across restart boundaries [27]. As the authors of ManySAT [16] note: "Contrary to the common belief, restarts are not used to eliminate the heavy tailed phenomena since after restarting SAT solvers dive in the part of the search space that they just left." Fourth, the heavy-tailed phenomena was found to be true only for satisfiable instances, and yet empirically restarts are known to be even more relevant for unsatisfiable instances.

Escaping Local Minima Hypothesis: Another explanation for restarts comes from the context of optimization. Many optimization algorithms (in particular, local search algorithms), get stuck in the local minima. Since local search only makes small moves at a time, it is unlikely to move out of a deep local minimum. The explanation is that restarts allow the optimization algorithm to escape the local minimum by randomly moving to another spot in the solution space. Certain local-search based SAT solvers (that aim to minimize the number of unsatisfied clauses) do use restarts for this very purpose [17,28]. However, restarts in CDCL do not behave in the same manner. Instead of setting the assignment of variables to random values like in local search, rather CDCL solvers revisit the same (or nearby) search space of assignments even after restarts since the variable activities and phases are preserved across restart boundaries [27].

As we show in Sect. 4, our hypothesis for the power of restarts is indeed consistent with the "escaping local minima" hypothesis. However, restarts enable CDCL solvers to escape local minima in a way that works differently from local search algorithms. Specifically, CDCL solvers with restarts enabled escape local minima by jumping to a nearby space to learn "better clauses", while local search algorithms escape local minima by randomly jumping to a different part of the search space.

4   "Restarts Enable Learning Better Clauses" Hypothesis

In this section, we propose that restarts enable a CDCL solver to learn better clauses. To justify our hypothesis, we start by examining the claim by Hamadi et al. [16] stating that "In SAT, restarts policies are used to compact the assignment stack and improve the order of assumptions." Recall that in CDCL SAT solvers, the only thing that changes during a restart is the assignment stack, and hence the benefits of restarts should be observable on the assignment stack. In this paper, we show that this claim matches reality, that is, restarting frequently correlates with a smaller assignment stack. We then go one step further, and show that a compact assignment stack leads to better clause learning. That is, the solver ends up learning clauses with lower LBD, thereby supporting our hypothesis, and this in turn improves the solver performance.

Restarts do incur a cost though [27], for otherwise restart after every conflict would be the optimal policy for all inputs. After a solver restarts, it needs to make many decisions and propagations to rebuild the assignment stack from scratch. We call this the *rebuild time*. More precisely, whenever a solver performs a restart, we note the current time and the assignment stack size x right before the restart.

Then the rebuild time is the time taken until either the solver encounters a new conflict or the new assignment stack size exceeds x through a series of decisions and propagations. Since we want to isolate the benefit of restart, we need to discount the cost of rebuilding. We define *effective time* to be the solving time minus the rebuild times of every restart.

4.1 Confirming the "Compacting the Assignment Stack" Claim

We ran the Glucose 4.1 SAT solver [5] [1] with various frequencies of restarting to show that indeed restarts do compact the assignment stack. For all experiments in this paper, Glucose was run with the argument "-no-adapt" to prevent it from changing heuristics. For each instance in the SAT Competition 2017 main track, we ran Glucose 4.1 with a timeout of 3 h of effective time on StarExec, a platform purposefully designed for evaluating SAT solvers [29]. The StarExec platform uses the Intel Xeon CPU E5-2609 at 2.40 GHz with 10240 KB cache and 24 GB of main memory, running on Red Hat Enterprise Linux Server release 7.2, and Linux kernel 3.10.0-514.16.1.el7.x86_64.

At every conflict, the assignment stack size is logged before backtracking occurs then the solver restarts after the conflict is analyzed (i.e., a uniform restart policy that restarts after every 1 conflict). We then ran the solver again on the same instance except the restart interval is doubled (i.e., a uniform restart policy that restarts after every 2 conflicts). We continue running the solver again and again, doubling the restart interval each time (i.e., a uniform restart policy that restarts after every 2^k conflicts) until the restart interval is so large that the solver never restarts before termination. For each instance, we construct a scatter plot, where the x-axis is the restart interval and the y-axis is the average assignment stack size for that restart policy on that instance, see Fig. 1a for an example. We then compute the Spearman correlation between the two axes, a positive correlation denotes that smaller restart intervals correlate with smaller assignment stack size, that is evidence that frequent restarts compacts the assignment stack. We plot the Spearman correlations of all 350 instances in Fig. 1b. 91.7% of the instances have a positive correlation coefficient. In conclusion, our experiments support the claim by Hamadi et al. [16] "restarts policies are used to compact the assignment stack."

It is important to note that this result is contingent on the branching heuristic implemented by the solver. If the branching heuristic is a static ordering, then the solver picks the decision variables in the same order after every restart and rebuilds the same assignment stack, hence the assignment stack does not get compacted. In our previous work [21], we showed that VSIDS-like branching heuristics "focus" on a small subset of logically related variables at any point in time. We believe a "focused" branching heuristic will see the compacting of assignment stack since a restart erases the assignment stack so a "focused" branching heuristic can reconstruct the assignment stack with only the subset of variables it is focused on.

---

[1] Glucose is a popular and competitive CDCL SAT solver often used in experiments because of its efficacy and simplicity (http://www.labri.fr/perso/lsimon/glucose/).

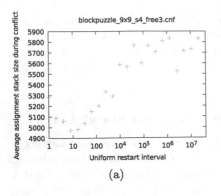
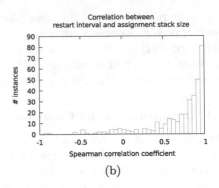

(a)   (b)

Fig. 1. (a) Scatter plot for a given instance showing increasing assignment stack size as the restarts become less frequent. (b) Histogram showing the distribution of Spearman correlations between the restart interval and the average assignment stack size for all 350 instances. The median correlation is 0.839.

4.2 Learning Better Clauses

We hypothesize that compacting the assignment stack generally leads to better learnt clauses, and that this is one of the benefits of restarts in SAT solvers in practice. Note that the clause learning schemes construct the learnt clause from a subset of variables on the assignment stack. Hence, a smaller assignment stack should lead to a learnt clause with smaller LBD than otherwise. To show this experimentally, we repeat the previous experiment where we ran Glucose 4.1 with the uniform restart policy restarting every 2^k conflicts for various parameters of k. At each conflict, we log the assignment stack size before backtracking and the LBD of the newly learnt clause. For each instance, we draw a scatter plot, where the x-axis is the average assignment stack size and the y-axis is the average LBD of learnt clauses, see Fig. 2a. We compute the Spearman correlation between the two axes and plot these correlations in a histogram, see Fig. 2b. 73.1% of the instances have a positive correlation coefficient.

4.3 Solving Instances Faster

Although lower LBD is widely believed to be a sign of good quality clause, we empirically show that indeed lower LBD generally correlates with better effective time. This experiment is a repeat of the last two experiments, with the exception that the x-axis is the average learnt clause LBD and the y-axis is the effective time, see Fig. 3a for an example. As usual, we compute the Spearman correlation between the two axes, discarding instances that timeout, and plot these correlations in a histogram, see Fig. 2b. 77.8% of the instances have a positive correlation coefficient. As expected, learning lower LBD clauses tend to improve solver performance.

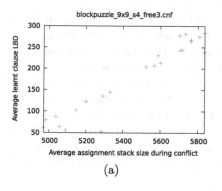

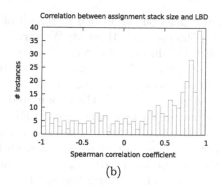

(a)   (b)

Fig. 2. (a) Scatter plot for a given instance showing increasing assignment stack size correlates with increasing LBD of learnt clauses. (b) Histogram showing the distribution of Spearman correlations between the average assignment stack size and the average LBD of learnt clauses for all 350 instances. The median correlation is 0.607.

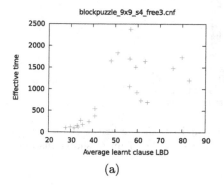

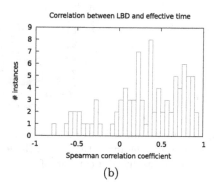

(a)   (b)

Fig. 3. (a) Scatter plot for a given instance showing increasing average learnt clause LBD correlates with increasing effective time. (b) Histogram showing the distribution of Spearman correlations between the average learnt clause LBD and effective time for all 90 instances without timeouts. The median correlation is 0.366.

4.4   Clause Length

If the previous experiments replaced LBD with clause length, then the median Spearman correlation between the average assignment stack size and average learnt clause length is 0.822 and the median Spearman correlation between the average learnt clause length and effective time is 0.08.

4.5   Low LBD in Core Proof

We hypothesize that lower LBD clauses are preferable for unsatisfiable instances because they are more likely to be a core learnt clause, that is, a learnt clause that is actually used in the derivation of the final empty clause. We performed the following experiment to support our hypothesis. We ran Glucose with no clause deletion on all 350 instances of the SAT Competition 2017 main track with 5000 seconds timeout. We turned off clause deletion because the deletion policy in Glucose inherently biases towards low LBD clauses by deleting learnt clauses with higher LBDs. We used DRAT-trim [30] to extract the core proof from the output of Glucose, i.e., the subset of clauses used in the derivation of the empty clause. We then computed the ratio between the mean LBD of the core learnt clauses and the mean LBD of all the learnt clauses. Lastly we plotted the ratios in a histogram, see Fig. 4. For the 57 instances for which core DRAT proofs were generated successfully, all but one instance has a ratio below 1. In other words, lower LBD clauses are more likely to be used in deriving the empty clause than clauses with higher LBD.

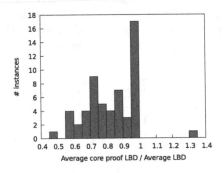

Fig. 4. Histogram for the ratio between the mean LBD of the learnt clauses in the core proof and the mean LBD of all the learnt clauses for the 57 unsatisfiable instances DRAT-trim produced a core proof.

4.6   New Restart Policy

Based on the above observations, we designed a new machine learning-based restart policy that is competitive with the state-of-the-art policies. As shown earlier, empirically restarts reduce LBD, hence we design a restart policy that tries to avoid high LBDs by restarting. Intuitively, our restart policy does the following: restart if the next learnt clause has an LBD in the 99.9^{th} percentile. Implementing this policy requires new techniques to answer the two following questions: is an LBD in the 99.9^{th} percentile and what is the LBD of the next learnt clause. We designed techniques to estimate answers to these two questions. The answer for the first question is the normal distribution is a good approximation for the right tail of the LBD distribution. The answer for the second question is to use machine learning to predict the LBD of the next clause.

5   A Machine Learning-Based Restart Policy

In this section, we describe our machine learning-based restart policy. We first start by answering the two questions posed in the last subsection regarding LBD percentile and predicting LBD of the next clause.

5.1   LBD Percentile

Given the LBD of a clause, it is unclear a priori how to label it as "good" or "bad". Some heuristics set a constant threshold and any LBDs above this threshold are considered bad. For example, Plingeling [7] considers learnt clauses with LBD greater 7 to be bad, and these clauses are not shared with the other workers. COMiniSatPS considers learnt clauses with LBD greater than 8 to be bad, and hence these clauses are readily deleted [25]. The state-of-the-art Glucose restart policy [4] on the other hand uses the mean LBD multiplied by a fixed constant as a threshold. The problem with using a fixed constant or the mean times a constant for thresholds is that we do not have a priori estimate of how many clauses exceed this threshold, and these thresholds seem arbitrary. Using arbitrary thresholds makes it harder to reason about solver heuristics, and in this context, the efficacy of restart policies.

We instead propose that for any given input it is more appropriate to use dynamic threshold that is computed based on the history of the CDCL solver's run on that input. At any point in time during the run of the solver, the dynamic threshold is computed as the 99.9^{th} percentile of LBDs of the learnt clauses seen during the run so far. Before we estimate whether an LBD is in the 99.9^{th} percentile, the first step is to analyze the distribution of LBDs seen in practice. In this experiment, the Glucose solver was run on all 350 instances in SAT Competition 2017 main track for 30 min and the LBDs of all the learnt clauses were recorded. Figure 5 shows the histogram of LBDs for 4 representative instances. As can be seen from the distributions of these representative instances, either their LBD distribution is close to normal or a right-skewed one.

Even though the right-skew distribution is not normal, the high percentiles can still be approximated by the normal distribution since the right tail is close to the normal curve. We conducted the following experiment to support this claim. For each instance, we computed the mean and variance of the LBD distribution to draw a normal distribution with the same mean and variance. We used the normal distribution to predict the LBD x at the 99.9^{th} percentile. We then checked the recorded LBD distribution to see the actual percentile of x. Figure 6 is a histogram of all the actual percentiles. Even in the worst case, the predicted 99.9^{th} percentile turned out to be the 97.1^{th} percentile. Hence for this benchmark the prediction of the 99.9^{th} percentile using the normal distribution has an error of less than 3 percentiles. Additionally, only 6 of the 350 instances predicted an LBD that was in the 100th percentile and all 6 of these instances solved in less than 130 conflicts hence the prediction was made with very little data.

These figures were created by analyzing the LBD distribution at the end of a 30 min run of Glucose, and we note the results are similar before the 30 min is up. Hence the 99.9^{th} percentile of LBDs can be approximated as the 99.9^{th} percentile of $norm(\mu, \sigma^2)$. The mean μ and variance σ^2 are estimated by the sample mean and sample variance of all the LBDs seen thus far, which is computed incrementally so the computational overhead is low. The 99.9^{th} percentile of the normal distribution maps to the z-score of 3.08, that is, an LBD is estimated to be in the 99.9^{th} percentile if it is greater than $\mu + 3.08 \times \sigma$.

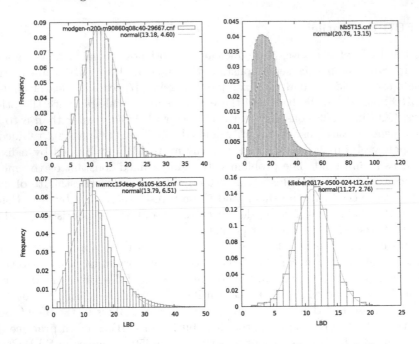

Fig. 5. Histogram of LBDs of 4 instances. A normal distribution with the same mean and variance is overlaid on top for comparison.

5.2 LBD of Next Clause

Since at any point during the run of a solver, the LBD of the "next learnt" clause is unknown, we propose the use of machine learning to predict that LBD instead. This requires finding good features that correlate with the next LBD. We hypothesize that LBDs of recent past learnt clauses correlate with the LBD of the next learnt clause.

In this experiment, Glucose was run on all 350 instances of the 2017 Competition main track and the LBDs of all the learnt clauses were recorded. Let n be the number of LBDs recorded for an instance. A table with two columns of length $n - 1$ are created. For each row i in this two column table, the first column contains the LBD of the ith conflict and the second column contains the LBD of the $(i + 1)^{th}$ conflict. Intuitively, after the solver finishes resolving the ith conflict, the ith learnt clause is the "previous" learnt clause represented by the first column. Correspondingly, the "next" learnt clause is the $(i + 1)^{th}$ learnt clause represented by the second column. For each instance that took more than 100 conflicts to solve, we computed the Pearson correlation between the first and second column and plot all these correlations in a histogram, see Fig. 7.

Our results show that the "previous LBD" is correlated with the "next LBD" which supports the idea that recent LBDs are good features to predict the next LBD via machine learning. In addition, all the correlations are positive, meaning that if the previous LBD is high (resp. low) then the next LBD is expected to

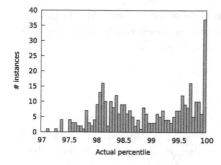

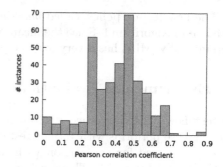

Fig. 6. Histogram of the actual percentiles of the LBD predicted to be the 99.9^{th} percentile using a normal distribution.

Fig. 7. Histogram of the Pearson correlation between the "previous" and "next" LBD for the instances in the SAT Competition 2017 main track benchmark.

be high (resp. low). Perhaps this explains why the Glucose restart policy [4] is effective. Additionally, we note that for the average instance, the LBD of the learnt clause after a restart is smaller than the LBD of the learnt clause right before that restart 74% of the time, showing the effect of restarts on LBD.

This paper proposes learning the function $f(l_{-1}, l_{-2}, l_{-3}, l_{-1} \times l_{-2}, l_{-1} \times l_{-3}, l_{-2} \times l_{-3}) = l_{next}$ where l_{-i} is the LBD of the learnt clause from i conflicts ago and $l_{-i} \times l_{-j}$ are products of previous LBDs to incorporate their feature interaction, and l_{next} is the LBD of the next clause. This function is approximated using linear regression where θ_i are coefficients to be trained by the machine learning algorithm:

$\tilde{f}(l_{-1}, l_{-2}, l_{-3}, l_{-1} \times l_{-2}, l_{-1} \times l_{-3}, l_{-2} \times l_{-3}) = \theta_0 + \theta_1 \times l_{-1} + \theta_2 \times l_{-2} + \theta_3 \times l_{-3} + \theta_4 \times l_{-1} \times l_{-2} + \theta_5 \times l_{-1} \times l_{-3} + \theta_6 \times l_{-2} \times l_{-3}$

Since LBDs are streamed in as conflicts occur, an online algorithm that can incrementally adjust the θ_i coefficients cheaply is required. We use the state-of-the-art Adam algorithm [20] from machine learning literature because it scales well with the number of dimensions, is computationally efficient, and converges quickly for many problems. The Adam algorithm is in the family of stochastic gradient descent algorithms that adjusts the coefficients to minimize the squared error, where the error is the difference between the linear function's prediction and the actual next LBD. The algorithm computes the gradient of the squared error function and adjusts the coefficients in the opposite direction of the gradient to minimize the squared error function. For the parameters of Adam, we use the values recommended by the original authors [20].

The coefficients θ_i are all initialized to 0 at the start of the search. Whenever a new clause is learnt, one iteration of Adam is applied with the LBDs of the three previous learnt clauses and their pairwise products as features and the LBD of the new clause as the target. The coefficients θ_i are adjusted in the process. When BCP reaches a fixed point without a conflict, the function $\tilde{f}$ is queried with the current set of coefficients θ_i to predict the LBD of the next clause. If the prediction exceeds the sample mean plus 3.08 standard deviations (i.e., approximately the 99.9^{th} percentile), a restart is triggered.

The new restart policy, called *machine learning-based restart* (MLR) policy, is shown in Algorithm 1. Since the mean, variance, and coefficients are computed incrementally, MLR has a very low computational overhead.

6 Experimental Evaluation

To test how MLR performs, we conducted an experimental evaluation to see how Glucose performs with various restart policies. Two state-of-the-art restart policies are used for comparison with MLR: Glucose (named after the solver itself) [4] and Luby [22]. The benchmark consists of all instances in the application and hard combinatorial tracks from the SAT Competition 2014 to 2017 totaling 1411 unique instances. The Glucose solver with various restart policies were run over the benchmark on StarExec. For each instance, the solver was given 5000 s of CPU time and 8 GB of RAM. The results of the experiment are shown in Fig. 8. The source code of MLR and further analysis of the experimental results are available on our website [1].

The results show that MLR is in between the two state-of-the-art policies of Glucose restart and Luby restart. For this large benchmark, MLR solves 19 instances more than Luby and 20 instances fewer than Glucose. Additionally, the learned coefficients in MLR $\sigma_1, \sigma_2, \sigma_3$ corresponding to the coefficients of

---

Algorithm 1. Pseudocode for the new restart policy MLR.

```
 1: function INITIALIZE                                        ▷ Called once at the start of search.
 2:     α ← 0.001, ε ← 0.00000001, β₁ ← 0.9, β₂ ← 0.999                    ▷ Adam parameters.
 3:     conflicts ← 0, conflictsSinceLastRestart ← 0
 4:     t ← 0                                                      ▷ Number of training examples.
 5:     prevLbd₃ ← 0, prevLbd₂ ← 0, prevLbd₁ ← 0     ▷ LBD of clause learnt 3/2/1 conflicts ago.
 6:     μ ← 0, m2 ← 0                          ▷ For computing sample mean and variance of LBDs seen.
 7:     for v in 0..|FeatureVector()| − 1 do                  ▷ Initialize θ, m, v to be vectors of zeros.
 8:         θᵢ ← 0, mᵢ ← 0, vᵢ ← 0                     ▷ Coefficients of linear function and Adam internals.
 9: function FEATUREVECTOR
10:     return    [1, prevLbd₁, prevLbd₂, prevLbd₃, prevLbd₁  ×  prevLbd₂, prevLbd₁  ×
         prevLbd₃, prevLbd₂ × prevLbd₃]
11: function AFTERCONFLICT(LearntClause)             ▷ Update the coefficients θ using Adam.
12:     conflicts ← conflicts + 1, conflictsSinceLastRestart ← conflictsSinceLastRestart + 1
13:     nextLbd ← LBD(LearntClause)
14:     δ ← nextLbd − μ, μ ← μ + δ/conflicts, Δ ← nextLbd − μ, m2 ← m2 + δ × Δ
15:     if conflicts > 3 then                                    ▷ Apply one iteration of Adam.
16:         t ← t + 1
17:         features ← FeatureVector()
18:         predict ← θ · features
19:         error ← predict − nextLbd
20:         g ← error × features
21:         m ← β₁ × m + (1 − β₁) × g, v ← β₂ × v + (1 − β₂) × g × g
22:         m̂ ← m/(1 − β₁ᵗ), v̂ ← v/(1 − β₂ᵗ)
23:         θ ← θ − α × m̂/(√v̂ + ε)
24:     prevLbd₃ ← prevLbd₂, prevLbd₂ ← prevLbd₁, prevLbd₁ ← nextLbd
25: function AFTERBCP(IsConflict)
26:     if ¬IsConflict ∧ conflicts > 3 ∧ conflictsSinceLastRestart > 0 then
27:         σ ← √(m2/(conflicts − 1))
28:         if θ · FeatureVector() > μ + 3.08σ then    ▷ Estimate if next LBD in 99.9th percentile.
29:             conflictsSinceLastRestart ← 0, Restart()
```

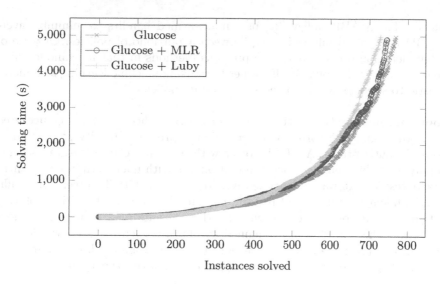

Fig. 8. Cactus plot of two state-of-the-art restart policies and MLR. A point (x, y) is interpreted as x instances have solving time less than y seconds for the given restart policy. Being further right means more instances are solved, further down means instances are solved faster.

the features representing recent past LBDS are nonnegative 91% of the time at the end of the run. This reinforces the notion that previous LBDs are positively correlated with the next LBD.

7 Related Work

Restart policies come in two flavors: static and dynamic. Static restart policies predetermine when to restart before the search begins. The state-of-the-art for static is the Luby [22] restart heuristic which is theoretically proven to be an optimal universal restart policy for Las Vegas algorithms. Dynamic restart policies determine when to restart on-the-fly during the run of the solver, typically by analyzing solver statistics. The state-of-the-art for dynamic is the restart policy proposed by Glucose [4] that keeps a short-term and a long-term average of LBDs. The short-term is the average of the last 50 LBDs and the long-term is the average of all the LBDs encountered since the start of the search. If the short-term exceeds the long-term by a constant factor then a restart is triggered. Hence the Glucose policy triggers a restart when the recent LBDs are high on average whereas the restart policy proposed in this paper restarts when the predicted LBD of the next clause is high. Biere et al. [8] propose a variation of the Glucose restart where an exponential moving average is used to compute the short-term and long-term averages. Haim and Walsh [15] introduced a machine learning-based technique to select a restart policy from a portfolio after 2100 conflicts. The MABR policy [24] uses multi-armed bandits to minimize average LBD by dynamically switching between a portfolio of policies. Our use of machine learning differs from these previous methods in that machine learning is part of the restart policy itself, rather than using machine learning as a meta-heuristic to select between a fixed set of restart policies.

Proof-complexity theoretic Considerations: Theorists have conjectured that restarts give the solver more power in a proof-complexity sense than a solver without restarts. A CDCL solver with asserting clause learning scheme can polynomially simulate general resolution [26] with nondeterministic branching and restarts. It was independently shown that a CDCL solver with sufficiently random branching and restarts can simulate bounded-width resolution [2]. It remains an open question whether these results hold if the solvers does not restart. This question has remained stubbornly open for over two decades now. We refer the reader to the excellent articles by Buss et al. on attempts at understanding the power of restarts via proof-complexity theory [9,10].

8 Conclusion

We showed that restarts positively impact the clause learning of CDCL solvers by decreasing the LBD of learnt clauses (thus improving their quality) compared to no restarts. However restarting too frequently is computationally expensive. We propose a new restart policy called MLR that tries to find the right balance in this trade-off. We use z-scores of the normal distribution to efficiently approximate the high percentiles of the LBD distribution. Additionally, we use machine learning to predict the LBD of the next clause, given the previous 3 LBDs and their pairwise products. Experimentally, the new restart policy is competitive with the current state-of-the-art.

References

1. MLR source code. https://sites.google.com/a/gsd.uwaterloo.ca/maplesat/mlr
2. Atserias, A., Fichte, J.K., Thurley, M.: Clause-learning algorithms with many restarts and bounded-width resolution. In: Kullmann, O. (ed.) SAT 2009. LNCS, vol. 5584, pp. 114–127. Springer, Heidelberg (2009)
3. Audemard, G., Simon, L.: Predicting learnt clauses quality in modern SAT solvers. In: Proceedings of the 21st International Joint Conference on Artificial Intelligence, IJCAI 2009, pp. 399–404. Morgan Kaufmann Publishers Inc., San Francisco (2009)
4. Audemard, G., Simon, L.: Refining restarts strategies for SAT and UNSAT. In: Milano, M. (ed.) CP 2012. LNCS, pp. 118–126. Springer, Heidelberg (2012). https://doi.org/10.1007/978-3-642-33558-7_11
5. Audemard, G., Simon, L.: Glucose and syrup in the SAT17. In: Proceedings of SAT Competition 2017: Solver and Benchmark Descriptions, pp. 16–17 (2017)
6. Biere, A., Biere, A., Heule, M., van Maaren, H., Walsh, T.: Handbook of Satisfiability: Volume 185 Frontiers in Artificial Intelligence and Applications. IOS Press, Amsterdam (2009)

7. Biere, A.: Lingeling, plingeling and treengeling entering the SAT competition 2013. In: Proceedings of SAT Competition 2013: Solver and Benchmark Descriptions, pp. 51–52 (2013)

8. Biere, A., Fröhlich, A.: Evaluating CDCL variable scoring schemes. In: Heule, M., Weaver, S. (eds.) SAT 2015. LNCS, vol. 9340, pp. 405–422. Springer, Cham (2015). https://doi.org/10.1007/978-3-319-24318-4_29

9. Bonet, M.L., Buss, S., Johannsen, J.: Improved separations of regular resolution from clause learning proof systems. J. Artif. Intell. Res. **49**, 669–703 (2014). https://doi.org/10.1613/jair.4260

10. Buss, S.R., Kolodziejczyk, L.A.: Small stone in pool. Logical Methods Comput. Sci. 10(2) (2014)

11. Cadar, C., Ganesh, V., Pawlowski, P.M., Dill, D.L., Engler, D.R.: EXE: automatically generating inputs of death. In: Proceedings of the 13th ACM Conference on Computer and Communications Security, CCS 2006, pp. 322–335. ACM, New York (2006)

12. Clarke, E., Biere, A., Raimi, R., Zhu, Y.: Bounded model checking using satisfiability solving. Formal Methods Syst. Des. **19**(1), 7–34 (2001)

13. Cook, S.A.: The complexity of theorem-proving procedures. In: Proceedings of the Third Annual ACM Symposium on Theory of Computing, STOC 1971, pp. 151–158. ACM, New York (1971). http://doi.acm.org/10.1145/800157.805047

14. Gomes, C.P., Selman, B., Crato, N., Kautz, H.: Heavy-tailed phenomena in satisfiability and constraint satisfaction problems. J. Autom. Reasoning **24**(1–2), 67–100 (2000)

15. Haim, S., Walsh, T.: Restart strategy selection using machine learning techniques. In: Kullmann, O. (ed.) SAT 2009. LNCS, vol. 5584, pp. 312–325. Springer, Heidelberg (2009)

16. Hamadi, Y., Jabbour, S., Sais, L.: ManySAT: a parallel SAT solver. J. Satisfiability **6**, 245–262 (2008)

17. Hirsch, E.A., Kojevnikov, A.: UnitWalk: a new SAT solver that uses local search guided by unit clause elimination. Ann. Math. Artif. Intell. **43**(1), 91–111 (2005)

18. Katebi, H., Sakallah, K.A., Marques-Silva, J.P.: Empirical study of the anatomy of modern SAT solvers. In: Sakallah, K.A., Simon, L. (eds.) SAT 2011. LNCS, vol. 6695, pp. 343–356. Springer, Heidelberg (2011). https://doi.org/10.1007/978-3-642-21581-0_27

19. Kautz, H., Selman, B.: Planning as satisfiability. In: Proceedings of the 10th European Conference on Artificial Intelligence, ECAI 1992, pp. 359–363. Wiley, New York (1992)

20. Kingma, D.P., Ba, J.: Adam: A Method for Stochastic Optimization. CoRR abs/1412.6980 (2014). http://arxiv.org/abs/1412.6980

21. Liang, J.H., Ganesh, V., Zulkoski, E., Zaman, A., Czarnecki, K.: Understanding VSIDS branching heuristics in conflict-driven clause-learning SAT solvers. In: Piterman, N. (ed.) HVC 2015. LNCS, vol. 9434, pp. 225–241. Springer, Cham (2015). https://doi.org/10.1007/978-3-319-26287-1_14

22. Luby, M., Sinclair, A., Zuckerman, D.: Optimal speedup of Las Vegas algorithms. Inf. Process. Lett. **47**(4), 173–180 (1993)

23. Marques-Silva, J.P., Sakallah, K.A.: GRASP-a new search algorithm for satisfiability. In: Proceedings of the 1996 IEEE/ACM International Conference on Computer-Aided Design, ICCAD 1996, pp. 220–227. IEEE Computer Society, Washington, DC (1996)

24. Nejati, S., Liang, J.H., Gebotys, C., Czarnecki, K., Ganesh, V.: Adaptive restart and CEGAR-based solver for inverting cryptographic hash functions. In: Paskevich, A., Wies, T. (eds.) VSTTE 2017. LNCS, vol. 10712, pp. 120–131. Springer, Cham (2017). https://doi.org/10.1007/978-3-319-72308-2_8
25. Oh, C.: COMiniSatPS the chandrasekhar limit and GHackCOMSPS. In: Proceedings of SAT Competition 2017: Solver and Benchmark Descriptions, pp. 12–13 (2017)
26. Pipatsrisawat, K., Darwiche, A.: On the power of clause-learning SAT solvers with restarts. In: Gent, I.P. (ed.) CP 2009. LNCS, vol. 5732, pp. 654–668. Springer, Heidelberg (2009). https://doi.org/10.1007/978-3-642-04244-7_51
27. Ramos, A., van der Tak, P., Heule, M.J.H.: Between restarts and backjumps. In: Sakallah, K.A., Simon, L. (eds.) SAT 2011. LNCS, vol. 6695, pp. 216–229. Springer, Heidelberg (2011)
28. Saitta, L., Sebag, M.: Phase Transitions in Machine Learning, pp. 767–773. Springer, Boston (2010)
29. Stump, A., Sutcliffe, G., Tinelli, C.: StarExec: a cross-community infrastructure for logic solving. In: Demri, S., Kapur, D., Weidenbach, C. (eds.) IJCAR 2014. LNCS (LNAI), vol. 8562, pp. 367–373. Springer, Cham (2014)
30. Wetzler, N., Heule, M.J.H., Hunt, W.A.: DRAT-trim: efficient checking and trimming using expressive clausal proofs. In: Sinz, C., Egly, U. (eds.) SAT 2014. LNCS, vol. 8561, pp. 422–429. Springer, Cham (2014). https://doi.org/10.1007/978-3-319-09284-3_31

Chronological Backtracking

Alexander Nadel and Vadim Ryvchin[(✉)]

Intel Corporation, P.O. Box 1659, 31015 Haifa, Israel
{alexander.nadel,vadim.ryvchin}@intel.com

Abstract. Non-Chronological Backtracking (NCB) has been implemented in every modern CDCL SAT solver since the original CDCL solver GRASP. NCB's importance has never been questioned. This paper argues that NCB is not always helpful. We show how one can implement the alternative to NCB–Chronological Backtracking (CB)–in a modern SAT solver. We demonstrate that CB improves the performance of the winner of the latest SAT Competition, Maple_LCM_Dist, and the winner of the latest MaxSAT Evaluation Open-WBO.

1 Introduction

Conflict-Driven Clause Learning (CDCL) SAT solving has been extremely useful ever since its the original implementation in the GRASP solver over 20 years ago [13], as it enabled solving real-world instances of intractable problems [2]. The algorithmic components of the original GRASP algorithms have been meticulously studied and modified over the years with the one notable exception of Non-Chronological Backtracking (NCB). NCB has always been perceived as an unquestionably beneficial technique whose impact is difficult to isolate, since it is entangled with other CDCL algorithms. NCB's contribution went unstudied even in [6]–a paper which aimed at isolating and studying the performance of fundamental CDCL algorithms. In this paper, we show how to implement the alternative to NCB–Chronological Backtracking (CB)–in a modern SAT solver.

Recall the CDCL algorithm. Whenever Boolean Constraint Propagation (BCP) discovers a falsified *conflicting clause* β, the solver learns a new *conflict clause* σ. Let the *conflict decision level* cl be the highest decision level in the conflicting clause β.[1] The new clause σ must contain one variable v assigned at cl (the 1UIP variable). Let the *second highest decision level* s be the highest decision level of σ's literals lower than cl ($s = 0$ for a unit clause). Let the *backtrack level* bl be the level the solver backtracks to just after recording σ and before flipping v.

Non-Chronological Backtracking (NCB) always backtracks to the second highest decision level (that is, in NCB, $bl = s$). The idea behind NCB is to improve the solver's locality by removing variables irrelevant for conflict analysis

---

[1] In the standard algorithm, cl is always equal to the current decision level, but, as we shall see, that is not the case for CB.

© Springer International Publishing AG, part of Springer Nature 2018
O. Beyersdorff and C. M. Wintersteiger (Eds.): SAT 2018, LNCS 10929, pp. 111–121, 2018.
https://doi.org/10.1007/978-3-319-94144-8_7

from the assignment trail. NCB's predecessor is conflict-directed backjumping, proposed in the context of the Constraint Satisfaction Problem (CSP) [11].

Let *Chronological Backtracking (CB)* be a backtracking algorithm which always backtracks to the decision level immediately preceding the conflict decision level cl (that is, in CB, $bl = cl - 1$). In our proposed implementation, after CB is carried out, v is flipped and propagated (exactly as in the NCB case), and then the solver goes on to the next decision or continues the conflict analysis loop.

Implementing CB is a non-trivial task as it changes some of the indisputable invariants of modern SAT solving algorithms. In particular, the decision level of the variables in the assignment trail is no longer monotonously increasing. Moreover, the solver may learn a conflict clause whose highest decision level is higher than the current decision level. Yet, as we shall see, implementing CB requires only few short modifications to the solver.

To understand why CB can be useful consider the following example. Let $F = S \wedge T$ be a propositional formula in Conjunctive Normal Form (CNF), where S is a long satisfiable CNF formula (for example, assume that S has 10^7 variables), $T \equiv (c \vee \neg b) \wedge (c \vee b)$, and $V(S) \cap V(T) = \emptyset$, where $V(H)$ comprises the set of H's variables. Consider Minisat's [3] execution, given F. The solver is likely to start by assigning the variables in $V(S)$ (since S's variables are likely to have higher scores), satisfying S, and then getting to satisfying T. Assume that the solver has satisfied S and is about to take the next decision. Minisat will pick the literal $\neg c$ as the next decision, since the variable c has a higher index than b and 0 is always preferred as the first polarity. The solver will then learn a new *unit* conflict clause (c) and backtrack to decision level 0 as part of the NCB algorithm. After backtracking, the solver will satisfy S again from the very beginning and then discover that the formula is satisfied. Note that the solver is not expected to encounter any conflicts while satisfying S for the second time because of the phase saving heuristic [4,10,14] which re-assigns the same polarity to every assigned variable. Yet, it will have to re-assign all the 10^7 variables in $V(S)$ and propagate after each assignment. In contrast, a CB-based solver will satisfy F immediately after satisfying S without needing to backtrack and satisfy S once again.

Our example may look artificial, yet in real-word cases applying NCB might indeed result in useless backtracking (not necessarily to decision level 0) and reassignment of almost the same literals. In addition, NCB is too aggressive: it might remove good decisions from the trail only because they did not contribute to the *latest* conflict resolution. Guided by these two insights, our backtracking algorithm applies CB when the difference between the CB backtrack level and the NCB backtrack level is higher than a user-given threshold T, but only after a user-given number of conflicts C passed since the beginning of solving.

We have integrated CB into the SAT Competition 2017 [5] winner, `Maple_LCM_Dist` [7], and MaxSAT Evaluation 2017 [1] winner `Open-WBO` [9] (code available in [8]). As a result, `Maple_LCM_Dist` solves 3 more SAT Competition benchmarks; the improvement on unsatisfiable instances is consistent. `Open-WBO` solves 5 more MaxSAT Evaluation benchmarks and becomes much faster on 10 families.

In the text that follows, Sect. 2 provides CB's implementation details, Sect. 3 presents the experimental results, and Sect. 4 concludes our work.

2   Chronological Backtracking

We show how CB can be integrated into a modern CDCL solver [12] starting with an example. Consider the input formula, comprising 9 clauses $c_1 \dots c_9$, shown on the left-hand side in Fig. 1. We will walk through a potential execution of a CDCL solver using CB, while highlighting the differences between CB and NCB.

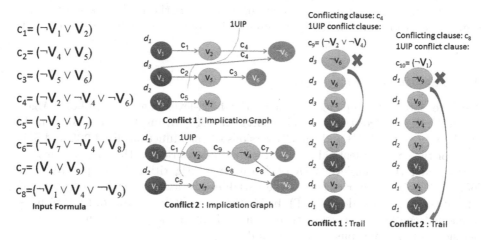

Fig. 1. CB example

Assume the first decision at decision level d_1 is v_1, followed by the implication v_2 in clause c_1 (at the same level d_1). Then, a new decision v_3 implying v_7 in c_5 is carried out at decision level d_2. The next decision (at level d_3) is v_4. It implies v_5 in c_2 and v_6 in c_3, followed by a conflict, as all literals of c_4 are falsified under the current partial assignment. The implication graph and the trail at the time of conflict 1 are shown in Fig. 1. The conflict analysis will then learn a new 1UIP clause $c_9 = (\neg v_2 \lor \neg v_4)$ (resolution between clauses c_2, c_3, c_4).

At this point, a difference between NCB and CB is manifested. NCB would backtrack to the end of level d_1, skipping the irrelevant decision level d_2. We apply CB, which backtracks to the end of the previous decision level d_2. Backtracking to the end of d_2 undoes the assignments of v_6, v_5, v_4. Then, the algorithm asserts the unassigned 1UIP literal $\neg v_4$ and pushes it to the trail.

Our CB implementation marks $\neg v_4$'s decision level as d_1, since d_1 is the second highest level in the newly learned clause; however, $\neg v_4$ is placed into the trail after literals assigned at a higher decision level d_2. Hence, unlike in the NCB case, the decision levels of literals in the trail are not necessary monotonically increasing. It still holds, though, that each literal l implied at clause α is placed in the trail after all the other literals of α.

Let us proceed with our example. The assignment of $\neg v_4$ implies v_9 in c_7. Our algorithm marks the decision level of v_9 as d_1, since it is the highest level in the clause c_7 where v_9 is implied. Then, BCP finds a falsified clause c_8. Our algorithm identifies the decision level of the conflict as d_1, since all the literals in the conflicting clause c_8 were assigned at that level. At that point, CB will backtrack to the end of d_1 *before proceeding with conflict analysis*. Our backtrack algorithm will unassign the variables assigned at d_2, that is, v_3 and v_7, while keeping the variables assigned at d_1 (v_4 and v_9) in the same order. After the backtracking, conflict analysis is invoked. Conflict analysis will learn a new clause $c_{10} = (\neg v_1)$ (resolution between clauses c_1, c_9, c_7, c_8). The algorithm will then backtrack to the decision level $d_0 = d_1 - 1$ (to emphasize: in CB the backtrack level is the previous decision level, determined independently of the newly learned conflict clause).

2.1   Algorithm

Now we show the implementation of the high-level algorithms CDCL (Algorithm 1), BCP (Algorithm 2) and Backtrack (Algorithm 3) with CB. In fact, we show both the NCB and the CB versions of each function. For CDCL and BCP most of the code is identical, except for the lines marked with either $_{ncb}$ or $_{cb}$.

Consider the high-level CDCL algorithm in Algorithm 1. It operates in a loop that finishes after either all the variables are assigned (SAT) or when an empty clause is derived (UNSAT). Inside the loop, BCP is invoked. BCP returns a falsified conflicting clause if there is a conflict. If there is no conflict, a new decision is taken and pushed to the trail.

The first difference between CB and NCB shows up right after a conflict detection. The code between lines 4 – 8 is applied only in the case of CB. If the conflicting clause contains one literal l from the maximal decision level, we let BCP propagating that literal at the second highest decision level in *conflicting_cls*. Otherwise, the solver backtracks to the maximal decision level in the conflicting clause before applying conflict analysis. This is because, as we saw in the example, the conflicting clause may be implied at a decision level earlier than the current level. The conflict analysis function returns the 1UIP variable to be assigned and the conflict clause σ. If σ is empty, the solver returns UNSAT. Assume σ is not empty. The backtrack level bl is calculated differently for NCB and CB. As one might expect, bl comprises the second highest decision level in σ in the case of NCB case and the previous decision level in the case of CB (note that for CB the solver has already backtracked to the maximal decision level in the conflicting clause). Subsequently, the solver backtracks to bl and pushes the 1UIP variable to the trail before continuing to the next iteration of the loop.

Consider now the implementation of BCP in Algorithm 2. BCP operates in a loop as long as there exists at least one unvisited literal in the trail ν. For the first unvisited literal l, BCP goes over all the clauses watched by l. Assume

---

Algorithm 1. CDCL

ν: the trail, stack of decisions and implications
$_{ncb}$: marks the NCB code
$_{cb}$: marks the CB code

Input: CNF formula
Output: SAT or UNSAT

 1: **while** not all variables assigned **do**
 2:    *conflicting_cls* := BCP();
 3:    **if** *conflicting_cls* $\neq$ *null* **then**
 4:      **if** *conflicting_cls* contains one literal from the maximal level **then**
 5:        $_{cb}$ Backtrack(second highest decision level in *conflicting_cls*)
 6:        $_{cb}$ continue
 7:      **else**
 8:        $_{cb}$ Backtrack(maximal level in *conflicting_cls*)
 9:      (*1uip*, σ) := ConflictAnalysis(*conflicting_cls*)
10:      **if** σ is empty **then**
11:        **return** UNSAT
12:      $_{ncb}$ *bl* := second highest decision level in σ (0 for a unit clause)
13:      $_{cb}$ *bl* := current decision level - 1
14:      Backtrack(*bl*)
15:      Push *1uip* to ν
16:    **else**
17:      Decide and push the decision to ν
18: **return** *SAT*

---

a clause β is visited. If β is a unit clause, that is, all β's literals are falsified except for one unassigned literal k, BCP pushes k to the trail. After storing k's implication reason in *reason(k)*, BCP calculates and stores k's implication level *level(k)*. The implication level calculation comprises the only difference between CB and NCB versions of BCP. The current decision level always serves as the implication level for NCB, while the maximal level in β is the implication level for CB. Note that in CB a literal may be implied *not* at the current decision level. As usual, BCP returns the falsified conflicting clause, if such is discovered.

Finally, consider the implementation of Backtrack in Algorithm 3. For the NCB case, given the target decision level bl, Backtrack simply unassigns and pops all the literals from the trail ν, whose decision level is greater than bl. The CB case is different, since literals assigned at different decision levels are interleaved on the trail. When backtracking to decision level bl, Backtrack removes all the literals assigned after bl, but it puts aside all the literals assigned before bl in a queue μ *maintaining their relative order*. Afterwards, μ's literals are returned to the trail in the same order.

2.2 Combining CB and NCB

Our algorithm can easily be modified to heuristically choose whether to use CB or NCB for any given conflict. The decision can be made, for each conflict, in the

Algorithm 2. BCP

dl: current decision level
ν: the trail, stack of decisions and implications
$_{ncb}$: marks the NCB code
$_{cb}$: marks the CB code

BCP()

1: **while** ν contains at least one unvisited literal **do**
2:      $l :=$ first literal in ν, unvisited by BCP
3:      $wcls :=$ clauses watched by l
4:      **for** $\beta \in wcls$ **do**
5:          **if** β is unit **then**
6:              $k :=$ the unassigned literal of β
7:              Push k to the end of ν
8:              $reason(k) := \beta$
9:              $_{ncb}\ level(k) := dl$
10:             $_{cb}\ level(k) :=$ max level in β
11:         **else**
12:             **if** β is falsified **then**
13:                 **return** β
     return $null$

Algorithm 3. Backtrack

dl: current decision level
ν: the trail, stack of decisions and implications
$level\_index(bl + 1)$: the index in ν of $bl + 1$'s decision literal

Backtrack(bl) : **NCB version**
Assume: $bl < dl$

1: **while** ν.size() $\geq level\_index(bl + 1)$ **do**
2:      Unassign ν.back()
3:      Pop from ν

Backtrack(bl) : **CB Version**
Assume: $bl < dl$

1: Create an empty queue μ
2: **while** ν.size() $\geq level\_index(bl + 1)$ **do**
3:      **if** $level(\nu$.back()$) \leq bl$ **then**
4:          Enqueue ν.back() to μ
5:      **else**
6:          Unassign ν.back()
7:      Pop from ν
8: **while** μ is not empty **do**
9:      Push μ.first() to the end of ν
10:     Dequeue from μ

main function in Algorithm 1 by setting the backtrack level to either the second highest decision level in σ for NCB (line 12) or the previous decision level for CB (line 13).

In our implementation, NCB is always applied before C conflicts are recorded since the beginning of the solving process, where C is a user-given threshold. After C conflicts, we apply CB whenever the difference between the CB backtrack level (that is, the previous decision level) and the NCB backtrack level (that is, the second highest decision level in σ) is higher than a user-given threshold T.

We introduced the option of delaying CB for C first conflicts, since backtracking chronologically makes sense only after the solver had some time to aggregate variable scores, which are quite random in the beginning. When the scores are random or close to random, the solver is less likely to proceed with the same decisions after NCB.

3   Experimental Results

We have implemented CB in Maple_LCM_Dist [7], which won the main track of the SAT Competition 2017 [5], and in Open-WBO, which won the complete unweighted track of the MaxSAT Evaluation 2017 [1]. The updated code of both solvers is available in [8]. We study the impact of CB with different values of the two parameters, T and C, in Maple_LCM_Dist and Open-WBO on SAT Competition 2017 and MaxSAT Evaluation 2017 instances, respectively. For all the tests we used machines with 32 GB of memory running Intel® Xeon® processors with 3 GHz CPU frequency. The time-out was set to 1800 s. All the results refer only to benchmarks solved by at least one of the participating solvers.

3.1   SAT Competition

In preliminary experiments, we found that $\{T = 100, C = 4000\}$ is the best configuration for Maple_LCM_Dist. Table 1 shows the summary of run time and unsolved instances of the default Maple_LCM_Dist vs. the best configuration in CB mode, $\{T = 100, C = 4000\}$, as well as "neighbor" configurations $\{T = 100, C = 3000\}$, $\{T = 100, C = 5000\}$, $\{T = 90, C = 4000\}$ and $\{T = 110, C = 4000\}$. Figure 2 and Fig. 3 compare the default Maple_LCM_Dist vs. the overall winner $\{T = 100, C = 4000\}$ on satisfiable and unsatisfiable instances respectively. Several observations are in place.

First, Table 1 shows that $\{T = 100, C = 4000\}$ outperforms the default Maple_LCM_Dist in terms of for both the number of solved instances and the run-time. It solves 3 more benchmarks and is faster by 4536 s.

Second, CB is consistently more effective on unsatisfiable instances. Table 1 demonstrates that the best configuration for unsatisfiable instances $\{T = 100, C = 5000\}$ solves 4 more instances than the default configuration and is faster by 5783 s. The overall winner $\{T = 100, C = 4000\}$ solves 3 more unsatisfiable benchmarks than the default and is faster by 5113 s. Figure 3 shows that CB is beneficial on the vast majority of unsatisfiable instances. Interestingly, we found that there is one family on which CB consistently yields significantly better results: the 27 instances of the g2-T family. On that family, the run-time in CB mode is never worse than that in NCB mode. In addition, CB helps to solve 4 more benchmarks than the default version and causes the solver to be faster by 1.5 times on average.

Finally, although the overall winner is slightly outperformed by the default configuration on satisfiable instances, CB can be tuned for satisfiable instances too. $\{T = 100, C = 3000\}$ solves 2 additional satisfiable instances, while $\{T = 110, C = 4000\}$ solves 1 additional instance faster than the default. We could not pinpoint a family, where CB shows a significant advantage on satisfiable instances.

Table 1. Results of Maple_LCM_Dist on SAT competition 2017 instances

		Base	T = 100			C = 4000	
			C = 3000	C = 4000	C = 5000	T = 90	T = 110
SAT	Unsolved	13	11	13	16	20	12
	Time	50003	53362	50580	59167	59482	47748
UNSAT	Unsolved	6	5	3	2	4	6
	Time	58414	54034	53301	52631	52481	53991
ALL	Unsolved	19	16	16	18	24	18
	Time	108417	107396	103881	111798	111963	101739

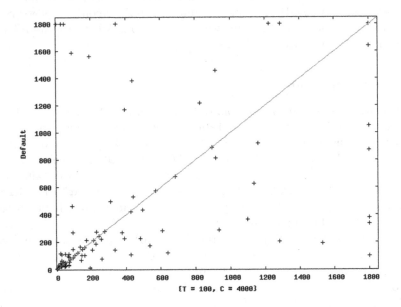

Fig. 2. Maple_LCM_Dist on SAT

3.2 MaxSAT Evaluation

In preliminary experiments, we found that $\{T = 75, C = 250\}$ is the best configuration for Open-WBO with CB. Consider the five left-most columns of Table 2.

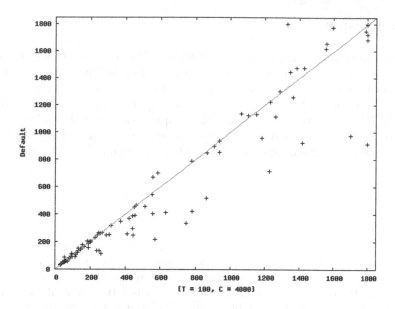

Fig. 3. Maple_LCM_Dist on UNSAT

Table 2. Results of Open-WBO on MaxSAT evaluation 2017 instances

Family	Default		{75, 250}		{75, 0}		{75, 500}		{50, 250}		{100, 250}	
	#S	Time	#S	Time	#S	Time	#S	Time	#S	Time	#S	Time
Grand Total	639	53048	644	50704	642	51370	640	52406	640	53582	643	51022
kbtree	0	3600	2	2756	1	3332	2	2921	2	2771	2	2733
atcoss-sugar	11	2179	12	1812	12	1328	11	2013	11	2004	12	1889
close-solutions	32	2692	33	4235	32	2711	32	2597	33	2589	32	4382
extension-enforcement	7	1963	8	828	7	1975	7	1942	8	1093	8	1306
gen-hyper-tw	5	4348	6	3871	6	3219	5	4057	5	3901	7	3383
treewidth-computation	24	3407	25	2306	24	3661	25	2169	23	4527	24	3778
atcoss-mesat	11	1660	11	605	11	703	11	610	11	674	11	534
min-fill	4	1105	4	413	4	384	4	910	4	244	4	349
packup	35	697	35	253	35	172	35	460	35	252	35	253
scheduling	1	206	1	92	1	153	1	164	1	141	1	130
bcp-syn	21	2535	20	2643	21	2247	21	2642	20	3145	20	2733
mbd	35	1327	34	1982	34	1972	34	2006	35	1275	35	1222
hs-timetabling	1	48	1	317	1	276	1	968	1	396	1	453

They present the number of solved instances and the run-time of the default Open-WBO vs. {T = 75, C = 250} (abbreviated to {75, 250}) over the MaxSAT Evaluation families (complete unweighted track). The second row shows the overall results. CB helps Open-WBO to solve 5 more instances in less time. The subsequent rows of Table 2 show the results for families, where either Open-WBO or {T = 75, C = 250} was significantly faster than the other solver, that is, it either solved more instances or was at least two times as fast. One can see that CB significantly improved the performance of Open-WBO on 10 families, while the performance was significantly deteriorated on 3 families only. The other columns of Table 2 present the results of 4 configurations neighbor to {T = 75, C = 250} for reference.

4   Conclusion

We have shown how to implement Chronological Backtracking (CB) in a modern SAT solver as an alternative to Non-Chronological Backtracking (NCB), which has been commonly used for over two decades. We have integrated CB into the winner of the SAT Competition 2017, Maple_LCM_Dist, and the winner of MaxSAT Evaluation 2017 Open-WBO. CB improves the overall performance of both solvers. In addition, Maple_LCM_Dist becomes consistently faster on unsatisfiable instances, while Open-WBO solves 10 families significantly faster.

References

1. Ansotegui, C., Bacchus, F., Järvisalo, M., Martins, R. (eds.): MaxSAT evaluation 2017: solver and benchmark descriptions, Department of Computer Science Series of Publications B. University of Helsinki, vol. B-2017-2 (2017)
2. Biere, A., Heule, M., van Maaren, H., Walsh, T. (eds.): Handbook of Satisfiability, Volume 185 of Frontiers in Artificial Intelligence and Applications. IOS Press, Amsterdam (2009)
3. Eén, N., Sörensson, N.: An extensible SAT-solver. In: Giunchiglia, E., Tacchella, A. (eds.) SAT 2003. LNCS, vol. 2919, pp. 502–518. Springer, Heidelberg (2004). https://doi.org/10.1007/978-3-540-24605-3_37
4. Frost, D., Dechter, R.: In search of the best constraint satisfaction search. In: AAAI, pp. 301–306 (1994)
5. Heule, M., Järvisalo, M., Balyo, T.: SAT competition (2017). https://baldur.iti.kit.edu/sat-competition-2017/
6. Katebi, H., Sakallah, K.A., Marques-Silva, J.P.: Empirical study of the anatomy of modern SAT solvers. In: Sakallah, K.A., Simon, L. (eds.) SAT 2011. LNCS, vol. 6695, pp. 343–356. Springer, Heidelberg (2011). https://doi.org/10.1007/978-3-642-21581-0_27
7. Luo, M., Li, C.-M., Xiao, F., Manyà, F., Lü, Z.: An effective learnt clause minimization approach for CDCL SAT solvers. In: Sierra, C. (ed.), Proceedings of the Twenty-Sixth International Joint Conference on Artificial Intelligence IJCAI 2017, Melbourne, Australia, 19–25 August 2017, pp. 703–711 (2017). ijcai.org
8. Nadel, A., Ryvchin, V.: Chronological backtracking: solvers. goo.gl/ssukuu
9. Neves, M., Martins, R., Janota, M., Lynce, I., Manquinho, V.: Exploiting resolution-based representations for MaxSAT solving. In: Heule, M., Weaver, S. (eds.) SAT 2015. LNCS, vol. 9340, pp. 272–286. Springer, Cham (2015). https://doi.org/10.1007/978-3-319-24318-4_20
10. Pipatsrisawat, K., Darwiche, A.: A lightweight component caching scheme for satisfiability solvers. In: Marques-Silva, J., Sakallah, K.A. (eds.) SAT 2007. LNCS, vol. 4501, pp. 294–299. Springer, Heidelberg (2007). https://doi.org/10.1007/978-3-540-72788-0_28
11. Prosser, P.: Hybrid algorithms for the constraint satisfaction problem. Comput. Intell. **9**(3), 268–299 (1993)
12. Marques Silva, J.P., Lynce, I., Malik, S.: Conflict-driven clause learning SAT solvers. In: Biere et al. 2, pp. 131–153

13. Marques Silva, J.P., Sakallah, K.A.: GRASP - a new search algorithm for satisfiability. In: ICCAD, pp. 220–227 (1996)
14. Shtrichman, O.: Tuning SAT checkers for bounded model checking. In: Emerson, E.A., Sistla, A.P. (eds.) CAV 2000. LNCS, vol. 1855, pp. 480–494. Springer, Heidelberg (2000). https://doi.org/10.1007/10722167_36

Centrality-Based Improvements to CDCL Heuristics

Sima Jamali and David Mitchell[✉]

Simon Fraser University, Vancouver, Canada
sima_jamali@sfu.ca, mitchell@cs.sfu.ca

Abstract. There are many reasons to think that SAT solvers should be able to exploit formula structure, but no standard techniques in modern CDCL solvers make explicit use of structure. We describe modifications to modern decision and clause-deletion heuristics that exploit formula structure by using variable centrality. We show that these improve the performance of Maple LCM Dist, the winning solver from Main Track of the 2017 SAT Solver competition. In particular, using centrality in clause deletion results in solving 9 more formulas from the 2017 Main Track. We also look at a number of measures of solver performance and learned clause quality, to see how the changes affect solver execution.

1 Introduction

Structure seems important in SAT research. Notions of instance structure are invoked in explaining solver performance, for example on large industrial formulas; many empirical papers relate aspects of solver performance to formula structure; and structure is key in theoretical results on both hardness and tractability.

Despite this, no standard method used in modern CDCL SAT solvers makes direct use of formula structure. (We exclude local structure such as used in resolving two clauses or assigning a unit literal.) The heuristics in CDCL solvers focus on local properties of the computation — what the algorithm has recently done — ignoring overall structure. The VSIDS and LRB decision heuristics give strong preference to variables that have been used many times recently, while clause deletion based on LBD and clause activity selects clauses based on recent use and an indicator of likelihood of being used again soon.

We present modifications to state-of-the-art decision and clause deletion heuristics that take structure into account by using variable betweenness centrality. This measure reflects the number of shortest paths through a variable in the primal graph of the formula. For decision heuristics, we give three different centrality-based modifications that alter VSIDS or LRB variable activities. For the clause deletion heuristic, we replace activity-based deletion with deletion based on clause centrality, a clause quality measure we believe is new.

We demonstrate the effectiveness of the methods by implementing them in Maple LCM Dist, the winning solver from Main Track of the 2017 SAT Solver

© Springer International Publishing AG, part of Springer Nature 2018
O. Beyersdorff and C. M. Wintersteiger (Eds.): SAT 2018, LNCS 10929, pp. 122–131, 2018.
https://doi.org/10.1007/978-3-319-94144-8_8

competition, and running them on the formulas from that track with a 5000 s time-out. All the modifications increased the number of instances solved and reduced the PAR-2 scores. While our methods are simple, to our knowledge this is the first time that explicit structural information has been successfully used to improve the current state-of-the-art CDCL solver on the main current benchmark. We also report a number of other measures of solver performance and learned clause quality, and make some observations about these.

Paper Organization. The remainder of the present section summarizes related work. Section 2 defines betweenness centrality and describes centrality computation. Sections 3 describes our modified decision and deletion heuristics. Section 4 gives the main performance evaluation. Section 5 looks at some execution details, and Sect. 6 makes concluding remarks and mentions work in progress.

Related Work. Several papers have studied the structure of industrial CNF formulas, e.g., [6,13,29]. "Community structure" (CS) has been shown in industrial formulas [11] and CS quality is correlated with solver run time [2,3,24,25]. CS was used in [4] to generate useful learned clauses. In [20,23] CS was used to obtain small witnesses of unsatisfiability. [17] showed that VSIDS tends to choose bridge variables (community connectors), and [14] showed that preferential bumping of bridge variables increased this preference and improved performance of the Glucose SAT solver. [27] described a method that applies large bump values to variables in particular communities. Eigenvalue centrality of variables was studied in [15], and it was shown that CDCL decisions are likely to be central variables. Betweenness centrality was studied in [14], where it was shown that a large fraction of decision variables have high betweenness centrality, and that the performance of Glucose can be improved by preferential bumping of variables with high betweenness centrality. Some features used in learning performance prediction models, as used in SATzilla [30], are structural measures. Lower bounds for CDCL run times on unsatisfiable formulas are implied by resolution lower bounds, and formula structure is central to these [8]. Formulas with bounded treewidth are fixed parameter tractable [1], and also can be efficiently refuted by CDCL with suitable heuristics [5].

2   Centrality Computation

The primal graph of a propositional CNF formula ϕ (also called the variable incidence graph or the variable interaction graph) is the graph $G(\phi) = \langle V, E \rangle$, with V being the set of all variables in ϕ and $(p, q) \in E$ iff there is a clause $C \in \phi$ containing both p and q (either negated or not). The betweenness centrality of a vertex v is defined by $g(v) = \sum_{s \neq v \neq t}(\sigma_{s,t}(v)/\sigma_{s,t})$, where $\sigma_{s,t}$ is the number of shortest s-t paths and $\sigma_{s,t}(v)$ is the number of those that pass through v, normalized to range over $[0,1]$ [12]. The betweenness centrality of a variable v in formula ϕ is the betweenness centrality of v in the primal graph $G(\phi)$.

Exactly computing betweenness centrality involves an all-pairs-shortest-paths computation, and is too expensive for large formulas. We computed approximate centrality values using the NetworkX [22] betweenness_centrality function, with sample size parameter $n/50$, where n is the number of variables. The parameter sets the number of vertices used to compute the approximation.

For some industrial formulas even this approximation takes too long to be useful (under SAT competition conditions), but for many formulas the approximation is fast. With a 300 s time-out, we found approximations for 217 of the 350 formulas from the main track of the 2017 competition. Figure 1 is a histogram of the centrality approximation times for these formulas, showing that a large fraction required less than 70 s.

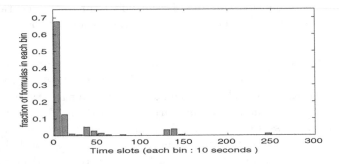

Fig. 1. Histogram of centrality approximation times.

3   Modified Decision and Deletion Heuristics

Decision Heuristics. The VSIDS decision heuristic [21], in several variations, has been dominant in CDCL solvers for well over a decade. Recently, the LRB (Learning-Rate-Based) heuristic [16] was shown to be effective, and winners of recent competitions use a combination of VSIDS and LRB. Both employ variable "activity" values, which are updated frequently to reflect the recent variable usage. The update involves increasing (or "bumping") the activity value for a variable each time it is assigned or appears during the derivation of a new learned clause. A secondary update in MapleSAT [16] and its descendants involves, at each conflict, *reducing* the LRB activity score of each variable that has not been assigned a value since the last restart. Maple LCM Dist uses both VSIDS and LRB, at different times during a run, and LRB activity reduction.

In [14] we reported that increasing the VSIDS bump value for high-centrality variables during an initial period of a run improved the performance of the solver Glucose. This did not help much in solvers using LRB, but motivated further study. As in [14] we define "high-centrality" variables to be the 1/3 fraction of variables with highest centrality values. The modifications reported here are:

HCbump-V: We scale the VSIDS additive bump values for high-centrality variables by a factor greater than 1. In the experiments reported here, the factor is 1.15.

HCbump-L: We periodically scale the LRB activity values of high-centrality variables by a factor greater than 1. In the experiments reported here, we scaled by a factor of 1.2 every 20,000 conflicts.

HCnoReduce: We disable the reduction of LRB scores for "unused variables" that are also high-centrality variables.

Clause Deletion. The Maple LCM Dist clause deletion (or reduction) scheme was inherited from COMiniSatPS, and is as follows [7,19]. The learned clauses are partitioned into three sets called CORE, TIER2 and LOCAL. Two LBD threshold values, t_1, t_2, are used. Learned clauses with LBD less than t_1 are put in CORE. t_1 is initially 3 but changed to 5 if CORE has fewer than 100 clauses after 100,000 conflicts. Clauses with LBD between t_1 and $t_2 = 6$ are put in TIER2. Clauses with LBD more than 6 are put in LOCAL. Clauses in TIER2 that are not used for a long time are moved to LOCAL. Clause deletion is done as follows. Order the clauses in LOCAL by non-decreasing activity. If m is the number of clauses in LOCAL, delete the first $m/2$ clauses that are not reasons for the current assignment. We report on the following modification:

HCdel: Replace ordering of clauses in LOCAL by clause activity with ordering by clause centrality. We define the centrality of a clause to be the mean centrality of the variables occurring in it.

4 Performance Evaluation

We implemented each of our centrality-based heuristics in Maple LCM Dist [19], the solver that took first place in the Main Track of the 2017 SAT Solver Competition [26]. We compared the performance of the modified versions against the default version of Maple LCM Dist by running them on the 350 formulas from the Main Track of the 2017 solver competition, using a 5000 s time-out. Computations were performed on the Cedar compute cluster [9] operated by Compute Canada [10]. The cluster consists of 32-core, 128 GB nodes with Intel "Broadwell" CPUs running at 2.1 GHz.

We allocated 70 s to approximate the variable centralities, based on the cost-benefit trade-off seen in Fig. 1: Additional time to obtain centralities for more formulas grows quickly after this point. If the computation completed, we used the resulting approximation in our modified solver. Otherwise we terminated the computation and ran default Maple LCM Dist. The choice of 70 s is not crucial: Any cut-off between 45 and 300 s gives essentially the same outcome. Centrality values were obtained for 198 of the 350 formulas. Our 5000 s timeout includes the time spent on centrality computation, whether or not the computation succeeded.

Table 1 gives the number of instances solved and the PAR-2 score for each method. All four centrality-based modifications improved the performance of Maple LCM Dist by both measures.

Table 1. Number of formulas solved (out of 350) and PAR-2 score, for default Maple LCM Dist and our four modified versions.

	Maple LCM Dist	HCbump-L	HCbump-V	HCnoReduce	HCdel
Number solved	215	219	218	221	224
PAR-2 score	4421	4382	4375	4381	4242

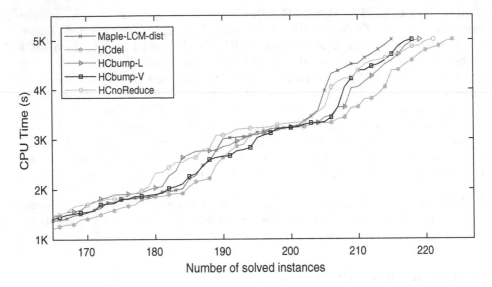

Fig. 2. Cactus plot comparing performance of default Maple LCM Dist and our four modified versions.

Figure 2 gives the "cactus plot" (inverse cumulative distribution function) for the runs. All four modifications result in improved performance. HCdel, which uses centrality-based clause deletion, is the best, and also out-performs default Maple LCM Dist for almost all smaller cut-off time values. The other methods, which modify the decision heuristic, improve on the default for all times longer than 3300 s. The two methods that modify LRB under-perform the default on easy formulas, but catch up at around 3200 s.

Families Affected. It is natural to wonder if these improvements are due to only one or two formula families. They are not. Table 2 shows, for each of our four modified solvers, how many formulas it solved that default Maple LCM Dist did not, and how many families they came from.

5   Performance Details

Reliability. There is an element of happenstance when using a cut-off time. For example, the "best" method would be different with a cut-off of 2800 s, and

Table 2. Number of families involved in formulas solved by our modified solvers by not by default Maple LCM Dist.

Solver	HCdel	HCnoReduce	NCbump-L	HCbumpt-V
Number of formulas	11	10	8	5
Number of families	5	6	5	3

the "worst" would be different with a cut-off of 2200 s. Run-time scatter plots give us an alternate view. Figure 3 gives scatter plots comparing the individual formula run-times for each of our four modified solvers with the default Maple LCM Dist. We observe:

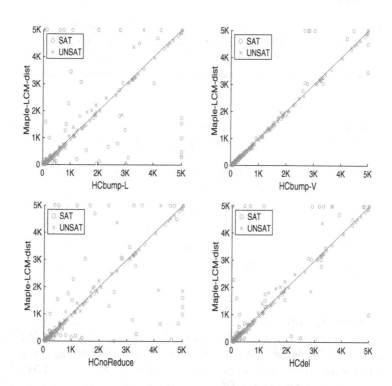

Fig. 3. Comparison of run-times of default Maple LCM Dist with each of our modified versions. Each plot shows all instances that were solved by at least one of the two solvers represented. Satisfiable formulas are denoted with ○, unsatisfiable formulas with ×.

- In each plot many points are lined up just below the main diagonal. These are the formulas without centralities, for which we pay a 70-s penalty.
- The most reliable method is HCdel. It solved the most, was faster on 70% of formulas with centralities and had significant slow-down for only 4 formulas.
- HCbump-V caused the least variation: it solved more formulas, but gave significant speedups on only a few others.

– The two LRB modifications, HCbump-L and HCnoReduce, were very "noisy", speeding up many formulas but also slowing down quite a few.
– It is very interesting that very large differences in run-time were mostly for satisfiable formulas.

Reasoning Rates. Here, we look at measures of the rate of solver reasoning or search (as distinct from, for example, quality of reasoning or total time). Table 3 shows, for each method, the mean decision rate, the mean conflict production rate, the mean unit propagation rate, and the mean Global Learning Rate (GLR). GLR, defined as the ratio of the number of conflicts to the number of decisions, was introduced in [18], where it was observed that decision heuristics producing a higher GLR had reduced solving times. We observe:

– Consistent with the observations in [18], decision heuristic changes that improved performance increased GLR, though only slightly;
– The fastest of our methods, HCdel, did not have a higher GLR, suggesting that it learned or kept "better" clauses, rather than more clauses.

Table 3. Measures of search or reasoning rate for the four solvers. Conflicts, Decisions and Propagations are in thousands of events per second.

Solver	Conflicts	Decisions	Propagations	GLR
Maple LCM Dist	8.25	23.7	1,452	0.623
HCdel	8.43	25.2	1,493	0.623
HCbump-L	8.52	26.3	1,530	0.626
HCbump-V	8.15	23.5	1,432	0.625
HCnoReduce	8.02	21.0	1,420	0.629

Learned Clause Quality. Measures of "clause quality" that have been studied or used in solver heuristics include size, literal block distance (LBD) and activity. Here we add clause centrality to these. Small clauses are good because they eliminate many truth assignments and facilitate propagation. Literal Block Distance is defined relative to a CDCL assignment stack, and is the number of different decision levels for variables in the clause. Small LBD clauses are like short clauses relative to assignments that are near the current one [6]. Clause activity is an analog of VSIDS activity, bumped each time the clause is used in a learned clause derivation [28]. Intuitively, clauses with low centrality connect variables "on the edge of the formula", and a long clause with low centrality connects many such variables, so is likely hard to use.

To see the effect of centrality-based deletion on clause quality, we measured the quality of learned clauses kept in LOCAL for three deletion schemes: Activity based deletion (default Maple LCM Dist); Centrality-based deletion (HCdel);

and LBD-based deletion (implemented in Maple LCM Dist for this study). Table 4 shows the results. Reported numbers are the mean of measurements taken just after each clause deletion phase. We observe:

- Centrality-based deletion keeps better clauses than activity-based deletion, as measured by both size and LBD, and also performs better.
- LBD-based deletion keeps the "best" clauses measured by LBD and size, has the worst performance and keeps the worst clauses measured by centrality.
- Centrality is the only clause quality measure that perfectly predicts ordering of the deletion methods by solving speed.

Table 4. Measures of quality for clauses in the LOCAL clause set, for three deletion schemes. (Centralities are scaled by 10,000).

Deletion method	Clause centrality	Clause LBD	Clause size	Solving time
Activity-based deletion	106	24	56	401
Centrality-based deletion	182	15	36	347
LBD-based deletion	80	9	24	446

6   Discussion

We introduced four centrality-based modifications to standard CDCL decision and deletion heuristics, and implemented these in Maple LCM Dist, first-place solver from the Main Track of the 2017 SAT Solver Competition. All four changes improved the performance on the formulas from this track.

The centrality-based deletion scheme, HCdel, solved the most formulas, produced the smallest PAR-2 scores, and also gave the most reliable speed-ups. This deletion scheme is based on clause centrality, a new measure of clause quality introduced here. We presented other evidence that clause centrality is an interesting clause quality measure, and we believe that further study of this measure will be productive.

The decision heuristic modifications performed less well than HCdel, but confirm the importance of variable centrality, and are interesting because they seem to work for different formulas. For example, among 26 formulas that at least one method solved and at least one did not, there are 12 formulas that are either solved by HCbump-L and no other method, or not solved by HCbump-L but solved by all other methods.

Work in progress includes more in-depth study of the roles of variable and clause centrality in solver execution, and development of a centrality-based restart strategy.

Acknowledgement. This research was supported in part by the Natural Sciences and Engineering Research Council of Canada (NSERC).

References

1. Alekhnovich, M., Razborov, A.A.: Satisfiability, branch-width and Tseitin tautologies. In: 43rd Symposium on Foundations of Computer Science (FOCS 2002), 16–19 November 2002, Vancouver, BC, Canada, pp. 593–603. IEEE (2002)
2. Ansótegui, C., Bonet, M.L., Giráldez-Cru, J., Levy, J.: Community structure in industrial SAT instances (2016). arXiv:1606.03329 [cs.AI]
3. Ansótegui, C., Giráldez-Cru, J., Levy, J.: The community structure of SAT formulas. In: Cimatti, A., Sebastiani, R. (eds.) SAT 2012. LNCS, vol. 7317, pp. 410–423. Springer, Heidelberg (2012). https://doi.org/10.1007/978-3-642-31612-8_31
4. Ansótegui, C., Giráldez-Cru, J., Levy, J., Simon, L.: Using community structure to detect relevant learnt clauses. In: Heule, M., Weaver, S. (eds.) SAT 2015. LNCS, vol. 9340, pp. 238–254. Springer, Cham (2015). https://doi.org/10.1007/978-3-319-24318-4_18
5. Atserias, A., Fichte, J.K., Thurley, M.: Clause-learning algorithms with many restarts and bounded-width resolution. In: Kullmann, O. (ed.) SAT 2009. LNCS, vol. 5584, pp. 114–127. Springer, Heidelberg (2009). https://doi.org/10.1007/978-3-642-02777-2_13
6. Audemard, G., Simon, L.: Predicting learnt clauses quality in modern sat solvers. In: 21st International Joint Conference on Artificial Intelligence, IJCAI 2009, pp. 399–404, San Francisco, CA, USA. Morgan Kaufmann (2009)
7. Balyo, T., Heule, M.J.H., Jarvisalo, M.J.: Proceedings of SAT competition 2016. Technical report, University of Helsinki (2016)
8. Ben-Sasson, E., Wigderson, A.: Short proofs are narrow - resolution made simple. In: Thirty-First Annual ACM Symposium on Theory of Computing, pp. 517–526. ACM (1999)
9. Cedar, A Compute Canada Cluster. https://docs.computecanada.ca/wiki/Cedar
10. Compute Canada: Advanced Research Computing (ARC) Systems. https://www.computecanada.ca/
11. Fortunato, S.: Community detection in graphs. Phys. Rep. **486**(3–5), 75–174 (2010)
12. Freeman, L.C.: A set of measures of centrality based on betweenness. Sociometry **40**(1), 35–41 (1977)
13. Gomes, C.P., Selman, B.: Problem structure in the presence of perturbations. In: 14th National Conference on Artificial Intelligence and Ninth Innovative Applications of Artificial Intelligence Conference, (AAAI 1997, IAAI 1997), 27–31 July 1997, Providence, Rhode Island, pp. 221–226. AAAI (1997)
14. Jamali, S., Mitchell, D.: Improving SAT solver performance with structure-based preferential bumping. In: 3rd Global Conference on Artificial Intelligence (GCAI 2017), Miami, FL, USA, 18–22 October 2017, vol. 50. EPiC, pp. 175–187. EasyChair (2017)
15. Katsirelos, G., Simon, L.: Eigenvector centrality in industrial SAT instances. In: Milano, M. (ed.) CP 2012. LNCS, pp. 348–356. Springer, Heidelberg (2012). https://doi.org/10.1007/978-3-642-33558-7_27
16. Liang, J.H., Ganesh, V., Poupart, P., Czarnecki, K.: Learning rate based branching heuristic for SAT solvers. In: Creignou, N., Le Berre, D. (eds.) SAT 2016. LNCS, vol. 9710, pp. 123–140. Springer, Cham (2016). https://doi.org/10.1007/978-3-319-40970-2_9
17. Liang, J.H., Ganesh, V., Zulkoski, E., Zaman, A., Czarnecki, K.: Understanding VSIDS branching heuristics in conflict-driven clause-learning SAT solvers. In: Piterman, N. (ed.) HVC 2015. LNCS, vol. 9434, pp. 225–241. Springer, Cham (2015). https://doi.org/10.1007/978-3-319-26287-1_14

18. Liang, J.H., Hari Govind, V.K., Poupart, P., Czarnecki, K., Ganesh, V.: An Empirical Study of Branching Heuristics Through the Lens of Global Learning Rate. In: Gaspers, S., Walsh, T. (eds.) SAT 2017. LNCS, vol. 10491, pp. 119–135. Springer, Cham (2017). https://doi.org/10.1007/978-3-319-66263-3_8
19. Luo, M., Li, C.-M., Xiao, F., Manyà, F., Lü, Z.: An effective learnt clause minimization approach for CDCL SAT solvers. In: 26th International Joint Conference on Artificial Intelligence (IJCAI 2017), Melbourne, Australia, 19–25 August 2017, pp. 703–711 (2017). ijcai.org
20. Martins, R., Manquinho, V., Lynce, I.: Community-based partitioning for MaxSAT solving. In: Järvisalo, M., Van Gelder, A. (eds.) SAT 2013. LNCS, vol. 7962, pp. 182–191. Springer, Heidelberg (2013). https://doi.org/10.1007/978-3-642-39071-5_14
21. Moskewicz, M.W., Madigan, C.F., Zhao, Y., Zhang, L., Malik, S.: Chaff: engineering an efficient SAT solver. In: 38th annual Design Automation Conference (DAC 2001), pp. 530–535. ACM (2001)
22. NetworkX, Software for complex networks. https://networkx.github.io/
23. Neves, M., Martins, R., Janota, M., Lynce, I., Manquinho, V.: Exploiting resolution-based representations for MaxSAT solving. In: Heule, M., Weaver, S. (eds.) SAT 2015. LNCS, vol. 9340, pp. 272–286. Springer, Cham (2015). https://doi.org/10.1007/978-3-319-24318-4_20
24. Newsham, Z., Ganesh, V., Fischmeister, S., Audemard, G., Simon, L.: Impact of community structure on SAT solver performance. In: Sinz, C., Egly, U. (eds.) SAT 2014. LNCS, vol. 8561, pp. 252–268. Springer, Cham (2014). https://doi.org/10.1007/978-3-319-09284-3_20
25. Newsham, Z., Lindsay, W., Ganesh, V., Liang, J.H., Fischmeister, S., Czarnecki, K.: SATGraf: visualizing the evolution of SAT formula structure in solvers. In: Heule, M., Weaver, S. (eds.) SAT 2015. LNCS, vol. 9340, pp. 62–70. Springer, Cham (2015). https://doi.org/10.1007/978-3-319-24318-4_6
26. SAT Competition 2017, July 2017. https://baldur.iti.kit.edu/sat-competition-2017/
27. Sonobe, T., Kondoh, S., Inaba, M.: Community branching for parallel portfolio SAT solvers. In: Sinz, C., Egly, U. (eds.) SAT 2014. LNCS, vol. 8561, pp. 188–196. Springer, Cham (2014). https://doi.org/10.1007/978-3-319-09284-3_14
28. Sorensson, N., Een, N.: Minisat v1. 13 - a SAT solver with conflict-clause minimization. In: Poster at the 8th Conference on Theory and Applications of Satisfiability Testing (SAT 2005), St. Andrews, UK, 19–23 June 2005
29. Williams, R., Gomes, C.P., Selman, B.: Backdoors to typical case complexity. In: 18th International Joint Conference on Artifical Intelligence (IJCAI 2003) Acapulco, Mexico, 9–15 August 2003, pp. 1173–1178. Morgan Kaufmann (2003)
30. Xu, L., Hutter, F., Hoos, H.H., Leyton-Brown, K.: SATzilla: portfolio-based algorithm selection for SAT. J. Artif. Intell. Res. **32**, 565–606 (2008)

Model Counting

Fast Sampling of Perfectly Uniform Satisfying Assignments

Dimitris Achlioptas[1,2], Zayd S. Hammoudeh[1(✉)], and Panos Theodoropoulos[2]

[1] Department of Computer Science, University of California, Santa Cruz,
Santa Cruz, CA, USA
{dimitris,zayd}@ucsc.edu
[2] Department of Informatics and Telecommunications, University of Athens,
Athens, Greece
ptheodor@di.uoa.gr

Abstract. We present an algorithm for perfectly uniform sampling of satisfying assignments, based on the exact model counter sharpSAT and reservoir sampling. In experiments across several hundred formulas, our sampler is faster than the state of the art by 10 to over 100,000 times.

1 Introduction

The DPLL [4] procedure forms the foundation of most modern SAT solvers. Its operation can be modeled as the preorder traversal of a rooted, binary tree where the root corresponds to the empty assignment and each edge represents setting some unset variable to 0 or 1, so that each node of the tree corresponds to a distinct partial assignment.

If the residual formula under a node's partial assignment is empty of clauses, or contains the empty clause, the node is a leaf of the tree. Naturally, the leaves corresponding to the former case form a partition of the formula's satisfying assignments (models), each part called a *cylinder* and having size equal to 2^z, where $z \geq 0$ is the number of unassigned variables at the leaf.

Generally, improved SAT solver efficiency is derived by trimming the DPLL search tree. For instance, conflict-driven clause learning (CDCL) amounts to adding new clauses to the formula each time a conflicting assignment is encountered. These added (learned) clauses make it possible to identify partial assignments with no satisfying extensions higher up in the tree.

1.1 Model Counting

Naturally, we can view model counting as the task where each internal node of the aforementioned tree simply adds the number of models of its two children. With this in mind, we see that the aforementioned CDCL optimization carries over, helping identify subtrees devoid of models sooner.

Research supported by NSF grants CCF-1514128, CCF-1733884, an Adobe research grant, and the Greek State Scholarships Foundation (IKY).

O. Beyersdorff and C. M. Wintersteiger (Eds.): SAT 2018, LNCS 10929, pp. 135–147, 2018.
https://doi.org/10.1007/978-3-319-94144-8_9

Despite the similarity with SAT solving, though, certain optimizations are uniquely important to efficient model counting. Specifically, it is very common for different partial assignments to have the same residual formula. While CDCL prevents the repeated analysis of unsatisfiable residual formulas, it does not prevent the reanalysis of previously encountered satisfiable residual formulas. To prevent such reanalysis #SAT solvers, e.g., Cachet [9], try to memoize in a cache the model counts of satisfiable residual formulas. Thus, whenever a node's residual formula is in the cache, the node becomes a leaf in the counting tree. We will refer to the tree whose leaves correspond to the execution of a model counter employing caching as a *compact counting* tree.

Another key optimization stems from the observation that as variables are assigned values, the formula tends to break up into pieces. More precisely, given a formula consider the graph having one vertex per clause and an edge for every pair of clauses that share at least one variable. Routinely, multiple connected components are present in the graph of the input formula. More importantly, as variables are assigned, components split. Trivially, a formula is satisfiable iff all its components are satisfiable. Determining the satisfiability of each component-formula independently can confer dramatic computational benefits [1].

The DPLL-based model counter sharpSAT [11], originally released in 2006 by Thurley and iteratively improved since, is the state-of-the-art exact model counter. It leverages all of the previously discussed optimizations and integrates advanced branch-variable selection heuristics proposed in [10]. Its main advantage over its predecessors stems from its ability to cache more components that are also of greater relevance. It achieves this through a compact encoding of cache entries as well as a cache replacement algorithm that takes into account the current "context", i.e., the recent partial assignments considered. Finally, it includes a novel algorithm for finding failed literals in the course of Boolean Constraint Propagation (BCP), called implicit BCP, which makes a very significant difference in the context of model counting.

Our work builds directly on top of sharpSAT and benefits from all the ideas that make it a fast exact model counter. Our contribution is to leverage this speed in the context of sampling. Generically, i.e., given a model counter as a black box, one can sample a satisfying assignment with $2n$ model counter invocations by repeating the following: pick an arbitrary unset variable v; count the number of models Z_0, Z_1, of the two formulas that result by setting v to $0,1$, respectively; set v to 0 with probability $Z_0/(Z_0 + Z_1)$, otherwise set it to 1.

As we discuss in Sect. 4 it is not hard to improve upon the above by modifying sharpSAT so that, with essentially no overhead, it produces a *single* perfectly uniform sample in the course of its normal, model counting execution. Our main contribution lies in introducing a significantly more sophisticated modification, leveraging a technique known as reservoir sampling, so that with relatively little overhead, it can produce *many* samples. Roughly speaking, the end result is a sampler for which one can largely use the following rule of thumb:

Generating 1,000 perfectly uniform models takes
about 10 times as long as it takes to count the models.

2   Related Work

In digital functional verification design defects are uncovered by exposing the device to a set of test stimuli. These stimuli must satisfy several requirements to ensure adequate verification coverage. One such requirement is that test inputs be diverse, so as to increase the likelihood of finding bugs by testing different corners of the design.

Constrained random verification (CRV) [8] has emerged in recent years as an effective technique to achieve stimuli diversity by employing randomization. In CRV, a verification engineer encodes a set of design constraints as a Boolean formula with potentially hundreds of thousands of variables and clauses. A constraint solver then selects a random set of satisfying assignments from this formula. Efficiently generating these random models, also known as witness, remains a challenge [3].

Current state of the art witness generators, such as UniGen2 [2], use a hash function to partition the set of all witnesses into roughly equal sized groups. Selecting such a group uniformly at random and then a uniformly random element from within the selected group, produces an approximately uniform witness. This approximation of uniformity depends on the variance in the size of the groups in the initial hashing-based partition. In practical applications, this non-uniformity is not a major issue.

Arguably the main drawback of hash-based witness generators is that their total execution time grows *linearly* with the number of samples. Acceleration can be had via parallelization, but at the expense of sacrificing witness independence. Also, by their probabilistic nature, hash-based generators may fail to return the requested number of models.

Our tool SPUR (Satisfying Perfectly Uniformly Random) addresses the problem of generating many samples by combining the efficiencies of sharpSAT with reservoir sampling. This allows us to draw a very large number of samples per traversal of the compact counting tree.

3   Caching and Component Decomposition

Most modern #SAT model counters are DPLL-based, and their execution can be modeled recursively. For example, Algorithm 1 performs model counting with component decomposition and caching similar to sharpSAT [11]. (We have simplified this demonstrative implementation by stripping out efficiency enhancements not directly relevant to this discussion including CDCL, unit clause propagation, non-chronological backtracking, cache compaction, etc.)

The algorithm first looks for F and its count in the cache. If they are not there, then if F is unsatisfiable or empty, model counting is trivial. If F has multiple connected components, the algorithm is applied recursively to each one and the product of the model counts is returned. If F is a non-empty, connected formula not in the cache, then a branching variable is selected, the algorithm is applied to each of the two restricted subformulas, and the sum of the model counts is returned after it has been deposited in the cache along with F.

Algorithm 1. Model counting with component decomposition and caching

```
 1: function COUNTER(F)
 2:     if ISCACHED(F) then
 3:         return CACHEDCOUNT(F)                          ▷ Cache-hit leaf
 4:
 5:     if UNSAT(F) then return 0
 6:     if CLAUSES(F) = ∅ then return 2^|VAR(F)|            ▷ Cylinder leaf
 7:
 8:     C_1, ..., C_k ← COMPONENTDECOMPOSITION(F)   ▷ Component decomposition
 9:     if k > 1 then
10:         for i from 1 to k do
11:             Z_i ← COUNTER(C_i)
12:         Z ← ∏_{i=1}^{k} Z_i
13:         return Z
14:
15:     v ← BRANCHVARIABLE(F)
16:     Z_0 ← COUNTER(F ∧ v = 0)
17:     Z_1 ← COUNTER(F ∧ v = 1)
18:     Z ← Z_0 + Z_1
19:     ADDTOCACHE(F, Z)                 ▷ The count of every satisfiable, connected
20:                                       ▷ subformula ever encountered is cached
21:     return Z
```

4   How to Get One Uniform Sample

It is easy to modify Algorithm 1 so that it returns a *single* uniformly random model of F. All we have to do is: (i) require the algorithm to return one model along with the count (ii) select a uniformly random model whenever we reach a cylinder, and (iii) at each branching node, when the two recursive calls return with two counts and two models, select one of the two models with probability proportional to its count, and store it along with the sum of the two counts in the cache before returning it. In the following, $F(\sigma)$ denotes the restriction of formula F by partial assignment σ and $\text{FREE}(\sigma)$ denotes the variables not assigned a value by σ.

An important observation is that the algorithm does not actually need to select, cache, and return a random model every time it reaches a cylinder. It can instead simply return the partial assignment corresponding to the cylinder. After termination, we can trivially "fill out" the returned cylinder to a complete satisfying assignment. This can be a significant saving as, typically, there are many cylinders, but we only need to return one model. Algorithm 2 employs this idea so that it returns a cylinder (instead of a model), each cylinder having been selected with probability proportional to its size.

The correctness of Algorithm 2 would be entirely obvious in the absence of model caching. With it, for any subformula, F', we only select a model at most once. This is because after selecting a model τ of F' for the first time in line 20 we write τ along with F' in the cache, in line 21, and therefore, if we ever encounter F' again, lines 2, 3 imply we will return τ as a model for F'. Naturally, even though we reuse the same model for a subformula encountered in completely different parts of the tree, no issue of probabilistic dependence arises: since we only return one sample overall, and thus for F', how could it?

It is crucial to note that this fortuitous non-interaction between caching and sampling does *not* hold for multiple samples, since if a subformula appears at several nodes of the counting tree, the sample models associated with these nodes must be independent of one another.

Algorithm 2. Single model sampler

1: **function** ONEMODEL(F, σ)
2:     **if** ISCACHED$(F(\sigma))$ **then**
3:         **return** CACHEDCOUNT$(F(\sigma))$, CACHEDMODEL$(F(\sigma))$
4:
5:     **if** UNSAT$(F(\sigma))$ **then return** $0, -$
6:     **if** CLAUSES$(F(\sigma)) = \emptyset$ **then return** $2^{|\text{FREE}(\sigma)|}, \sigma$
7:
8:     $C_1, \ldots, C_k \leftarrow$ COMPONENTDECOMPOSITION$(F(\sigma))$
9:     **if** $k > 1$ **then**
10:         **for** i from 1 to k **do**
11:             $Z_i, \sigma_i \leftarrow$ ONEMODEL(C_i, σ)
12:         $Z \leftarrow \prod_{i=1}^{k} Z_i$
13:         $\tau \leftarrow \sigma_1, \ldots, \sigma_k$
14:         **return** Z, τ
15:
16:     $v \leftarrow$ BRANCHVARIABLE$(F(\sigma))$
17:     $Z^0, \sigma^0 \leftarrow$ ONEMODEL$(F, \sigma \wedge v = 0)$
18:     $Z^1, \sigma^1 \leftarrow$ ONEMODEL$(F, \sigma \wedge v = 1)$
19:     $Z \leftarrow Z^0 + Z^1$
20:     $\tau \leftarrow \sigma^0$ with probability Z^0/Z, otherwise $\tau \leftarrow \sigma^1$
21:     ADDTOCACHE$(F(\sigma), Z, \tau)$
22:
23:     **return** Z, τ

---

5   How to Get Many Uniform Samples at Once

Consider the set $\mathcal{C}$ which for each leaf σ_j of the compact counting tree comprises a pair (σ_j, c_j), where c_j is the number of satisfying extensions (models) of partial assignment σ_j. The total number of models, Z, therefore equals $\sum_j c_j$. Let $\text{Bin}(n, p)$ denote the Binomial random variable with n trials of probability p.

To sample s models uniformly, independently, and with replacement (u.i.r.), we would like to proceed as follows:

1. Enumerate $\mathcal{C}$, while enjoying full model count caching, as in sharpSAT.
2. Without storing the (huge) set $\mathcal{C}$, produce from it a random set $\mathcal{R}$ comprising pairs $\{(\sigma_i, s_i)\}_{i=1}^t$, for some $1 \leq t \leq s$, such that:
   (a) Each σ_i is a distinct leaf of the compact counting tree.
   (b) $s_1 + \cdots + s_t = s$ (we will eventually generate s_i extensions of σ_i u.i.r.).
   (c) For every leaf σ_j of the compact counting tree and every $1 \leq w \leq s$, the probability that (σ_j, w) appears in $\mathcal{R}$ equals $\Pr[\mathrm{Bin}\,(s, c_j/Z) = w]$.

Given a set $\mathcal{R}$ as above, we can readily sample models corresponding to those pairs (σ_i, s_i) in $\mathcal{R}$ for which either $s_i = 1$ (by invoking $\mathrm{ONEMODEL}(F(\sigma_i))$), or for which $\mathrm{CLAUSES}(F(\sigma_i)) = \emptyset$ (trivially). For each pair (σ_i, s_i) for which $s_i > 1$, we simply run the algorithm again on $F(\sigma_i)$, getting a set $\mathcal{R}'$, etc.

Obviously, the non-trivial part of the above plan is achieving (2c) without storing the (typically huge) set $\mathcal{C}$. We will do this by using a very elegant idea called reservoir sampling [12], which we describe next.

6   Reservoir Sampling

Let A be an arbitrary finite set and assume that we would like to select s elements from A u.i.r. for an arbitrary integer $s \geq 1$. Our task will be complicated by the fact that the (unknown) set A will not be available to us at once. Instead, let $A_1, A_2, \ldots, A_m$ be an arbitrary, unknown partition of A. Without any knowledge of the partition, or even of m, we will be presented with the parts in an arbitrary order. When each part is presented we can select some of its elements to store, but our storage capacity is precisely s, i.e., at any given moment we can only hold up to s elements of A. Can we build a sample as desired?

Reservoir sampling is an elegant solution to this problem that proceeds as follows. Imagine that (somehow) we have already selected s elements u.i.r. from a set B, comprising a multiset S. Given a set C disjoint from B we can produce a sample of s elements selected u.i.r. from $B \cup C$, without access to B, as follows. Note that in Step 3 of Algorithm 3, multiple instances of an element of B in S are considered distinct, i.e., removing one instance leaves the rest unaffected. It is not hard to see that after Step 4 the multiset S will comprise s elements selected u.i.r. from $B \cup C$. Thus, by induction, starting with $B = \emptyset$ and processing the sets $A_1, A_2, \ldots$ one by one (each in the role of C) achieves our goal.

Algorithm 3. Turns a u.i.r. s-sample $S \subseteq B$ to a u.i.r. s-sample of $B \cup C$

1: Generate $q \sim \mathrm{Bin}\,(s, |C|/|B \cup C|)$.
2: Select q elements from C u.i.r.
3: Select q elements from S uniformly, independently, without replacement.
4: Swap the selected elements of S for the selected elements of C.

6.1   Reservoir Sampling in the Context of Model Caching

In our setting, each set A_i amounts to a leaf of the compact counting tree. We would like to build our sample set by (i) traversing this tree exactly as sharpSAT, and (ii) ensuring that every time the traversal moves upwards from a leaf, we hold s models selected u.i.r. from all satisfying extensions of leaves encountered so far. More precisely, instead of actual samples, we would like to hold a random set $\mathcal{R}$ of weighted partial assignments satisfying properties (2a)–(2c) in Sect. 5.

To that end, it will be helpful to introduce the following distribution. Given r bins containing $s_1, \ldots, s_r$ distinct balls, respectively, and $q \geq 0$ consider the experiment of selecting q balls from the bins uniformly, independently, without replacement. Let $\mathbf{q} = (q_1, \ldots, q_r)$ be the (random) number of balls selected from each bin. We will write $\mathbf{q} \sim \mathcal{D}((s_1, \ldots, s_r), q)$. To generate a sample from this distribution, let $b_0 = 0$; for $i \in [r]$, let $b_i = s_1 + \cdots + s_i$, so that $b_1 = s_1$ and $b_r = s_1 + \ldots + s_r := s$. Let $\gamma_1, \gamma_2, \ldots, \gamma_q$ be i.i.d. uniform elements of $[s]$. Initialize q_i to 0 for each $i \in [r]$. For each $i \in [q]$: if $\gamma_i \in (b_{z-1}, b_z]$, then increment q_r by 1.

With this in mind, imagine that we have already processed t leaves so that $Z_t = Z = |A_1| + \cdots |A_t|$ and that the reservoir contains $\mathcal{R} = \{(\sigma_i, s_i)\}_{i=1}^r$, such that $\sum_{i=1}^r s_i = s$. Let σ be the current leaf (partial assignment), let A be the set of σ's satisfying extensions, and let $w = |A|$. To update the reservoir, we first determine the random number, $q \geq 0$, of elements from A to place in our s-sample, as a function of w, Z. Having determined q we draw from $\mathcal{D}((s_1, \ldots, s_r), q)$ to determine how many elements to remove from each set already in the reservoir, by decrementing its weight s_i (if $s_i \leftarrow 0$ we remove $(\sigma_i, 0)$ from the reservoir). Finally, we add (σ, q) to the reservoir to represent the q elements of A.

Note that, in principle, we could have first selected s elements u.i.r. from A and then $0 \leq q \leq s$ among them to merge into the reservoir (again represented as (σ, q)). This viewpoint is useful since, in general, instead of merging into the existing reservoir $0 \leq q \leq s$ elements from a single cylinder of size w, we will need to merge q elements from a set of size w that is the union of $\ell \geq 1$ disjoint sets, each represented by a partial assignment σ_j, such that we have already selected a_j elements from each set, where $\sum_{j=1}^\ell a_j = s$. Indeed, Algorithm 4 below is written with this level of generality in mind, so that our simple single cylinder example above corresponds to merging $\langle w, \{(\sigma, s)\}\rangle$ into the reservoir.

Algorithm 4. Merges $R = \langle Z, \{(\sigma_i, s_i)\}_{i=1}^r \rangle$ with $\langle w, \{(\sigma_j, a_j)\}_{j=1}^\ell \rangle$

1: **function** RESERVOIRUPDATE($R, \langle w, \{(\sigma_j, a_j)\}_{j=1}^\ell \rangle$)
2:     $Z \leftarrow Z + w$
3:     $q \sim \text{Bin}(s, w/Z)$
4:     Generate $(\beta_1, \ldots, \beta_\ell) \sim \mathcal{D}((a_1, \ldots, a_\ell), q)$
5:     Generate $(\gamma_1, \ldots, \gamma_r) \sim \mathcal{D}((s_1, \ldots, s_r), q)$
6:     $R' \leftarrow \langle Z, \{(\sigma_j, \beta_j)\}_{j=1}^\ell \cup \{(\sigma_i, s_i - \gamma_i)\}_{i=1}^r \rangle$
7:     Discard any partial assignment in R' whose weight is 0
8:     **return** R'

---

7   A Complete Algorithm

To sample s models u.i.r. from a formula F, we create an empty reservoir R of capacity s and invoke SPUR$(F, \emptyset, R)$. The call returns the model count of F and modifies R in place to contain pairs $\{(\sigma_i, s_i)\}_{i=1}^{t}$, for some $1 \leq t \leq s$, such that $\sum_{i=1}^{t} s_i = s$. Thus, SPUR partitions the task of generating s samples into t independent, smaller sampling tasks. Specifically, for each $1 \leq i \leq t$, if CLAUSES$(F(\sigma_i)) = \emptyset$, then sampling the s_i models is trivial, while if $s_i = 1$, sampling can be readily achieved by invoking ONEMODEL on $F(\sigma_i)$. If none of the two simple cases occurs, SPUR is called on $F(\sigma_i)$ requesting s_i samples.

Algorithm 5. Counts models and fills up a reservoir with s samples

---

1: **function** SPUR(F, σ, R)
2:     **if** ISCACHED$(F(\sigma))$ **then**
3:         RESERVOIRUPDATE$(R, \langle$CACHEDCOUNT$(F(\sigma)), (\sigma, s)\rangle)$
4:         **return** CACHEDCOUNT$(F(\sigma))$
5:
6:     **if** UNSAT$(F(\sigma))$ **then return** 0
7:     **if** CLAUSES$(F(\sigma)) = \emptyset$ **then**
8:         RESERVOIRUPDATE$(R, \langle 2^{|\text{FREE}(\sigma)|}, (\sigma, s)\rangle)$
9:         **return** $2^{|\text{FREE}(\sigma)|}$
10:
11:     $C_1, \ldots, C_k \leftarrow$ COMPONENTDECOMPOSITION$(F(\sigma))$
12:     **if** $k > 1$ **then**
13:         **for** i from 1 to k **do**
14:             Create a new reservoir R_i of capacity s
15:             $Z_i \leftarrow$ SPUR$(C_i, \emptyset, R_i)$
16:         $w \leftarrow \prod_{i=1}^{k} Z_i$
17:         $A \leftarrow$ STITCH$(\sigma, R_1, R_2, \ldots, R_k)$
18:         RESERVOIRUPDATE$(R, \langle w, A \rangle)$
19:         **return** w
20:
21:     $v \leftarrow$ BRANCHVARIABLE$(F(\sigma))$
22:     $Z_0 \leftarrow$ SPUR$(F, \sigma \wedge v = 0, R)$
23:     $Z_1 \leftarrow$ SPUR$(F, \sigma \wedge v = 1, R)$
24:     ADDTOCACHE$(F, Z_0 + Z_1)$
25:     **return** $Z_0 + Z_1$

If a formula has $k > 1$ components, SPUR is invoked recursively on each component C_i with a new reservoir R_i (also passed by reference). When the recursive calls return, each reservoir R_i comprises some number of partial assignments over the variables in C_i, each with an associated weight (number of samples), so that the sum of the weights equals s. It will be convenient to think of the content of each reservoir R_i as a multiset containing exactly s strings from $\{0, 1, *\}^{\text{Var}(C_i)}$. Under this view, to STITCH together two reservoirs R_1, R_2, we fix an arbitrary permutation of the s strings in, say, R_1, pick a uniformly random permutation of the strings in R_2, and concatenate the first string in R_1 with the first string in R_2, the second string in R_1 with the second string in R_2, etc. To stitch together multiple reservoirs we proceed associatively. The final result is a set $\{(\sigma_j, a_j)\}_{j=1}^{\ell}$, for some $1 \leq \ell \leq s$, such that $\sum_{j=1}^{\ell} a_j = s$.

8   Evaluation and Experiments

We have developed a prototype C++ implementation [6] of SPUR on top of sharpSAT (ver. 5/2/2014) [11]. This necessitated developing multiple new modules as well as extensively modifying several of the original ones.

8.1   Uniformity Verification

Since sharpSAT is an exact model counter, the samples derived from SPUR are perfectly uniform. Since we use reservoir sampling, they are also perfectly independent. As a test of our implementation we selected 55 formulas with model counts ranging from 2 to 97,536 and generated 4 million models of each one.

For each formula F we (i) recorded the number of times each of its $M(F)$ models was selected by SPUR, and (ii) drew 4 million times from the multinomial distribution with $M(F)$ outcomes, corresponding to ideal u.i.r. sampling. We measured the KL-divergence of these two empirical distributions from the multinomial distribution with $M(F)$ outcomes, so that the divergence of the latter provides a yardstick for the former. The ratio of the two distances was close to 1 over all formulas, and the product of the 55 ratios was 0.357.

One of the formulas we considered was case110 with 16,384 models, which was used in the verification of the approximate uniformity of UniGen2 in [2]. Figure 1 plots the output of UniGen2 and SPUR against a background of the ideal multinomial distribution (with mean 244.14...). Each point (x, y) represents the number of models, x, that were generated y times across all 4,000,000 trials.

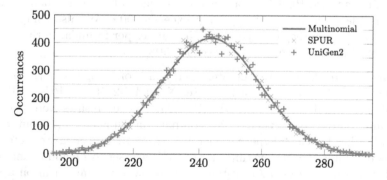

Fig. 1. Uniformity comparison between an ideal uniform sampler, SPUR and UniGen2 on the "case110" benchmark on four million samples.

8.2   Running Time

To demonstrate the empirical performance of SPUR we ran it on several hundred formulas, along with UniGen2 (ver. 9/28/2017), an almost-uniform, almost-i.i.d. SAT witness generator, representing the state of the art prior to our work.

Benchmarks: We considered 586 formulas, varying in size from 14 to over 375,000 variables. They are the union of the 369 formulas used to benchmark UniGen2 in [7] (except for approximately 20 proprietary formulas with suffix _new that are not publicly available) and the 217 formulas used to benchmark sharpSAT in [11]. Of the latter we removed from consideration the 100 formulas in the flat200 graph coloring dataset, since on all of them UniGen2 timed out, while SPUR terminated successfully in a handful of seconds. This left 486 formulas.

An important distinction between the two sets of formulas is that all formulas from [7] come with a *sampling set*, i.e., a relatively small subset, S, of variables. When such a set is given as part of the input, UniGen2 samples (near-)uniformly from the elements of $\{0,1\}^S$ that have at least one satisfying extension (model). For all but 17 of the 369 formulas, the provided set was in fact an *independent support* set, i.e., each of element of $\{0,1\}^S$ was guaranteed to have at most one satisfying extension. Thus for these 352 formulas UniGen2 is, in fact, sampling satisfying assignments, making them fair game for comparison (if anything such formulas slightly favor UniGen2 as we do not include the time required to extend the returned partial assignments to full assignments which, in principle, could be substantial.) None of the 117 formulas used to benchmark sharpSAT come with such a set (since sharpSAT does not support counting the size of projections of the set of models). Of these $486 - 17 = 469$ formulas, 2 are unsatisfiable, while for another 22 UniGen2 crashed or exited with an error. (SPUR did not crash or report an error on any formulas.) Of the remaining 445 formulas, 72 caused both SPUR and UniGen2 to time out. We report on the remaining 373 formulas.

For each formula we generated between 1,000 and 10,000 samples, as originally performed by Chakraborty et al. [2] and report the results in detail. Our main finding is that SPUR is on average more than 400× faster than UniGen2, i.e., the geometric mean[1] of the speedup exceeds 400×. We also compared the two algorithms when they only generate 55 samples per formula. In that setting, the geometric mean of the speedup exceeds 150×.

Experiment Setup: All experiments were performed on a high-performance cluster, where each node consists of two Intel Xeon E5-2650v4 CPUs with up to 10 usable cores and 128 GB of DDR4 DRAM. All our results were generated on the same hardware to ensure a fair comparison. UniGen2's timeout was set to 10 h; all other UniGen2 hyperparameters, e.g., κ, startIteration, etc., were left at their default values. The timeout of SPUR was set to 7 h and its maximum cache size was set to 8 GB. All instances of the two programs run on a single core at a time.

---

[1] The arithmetic mean [of the speedup] is even greater (always). For the aptness of using the geometric mean to report speedup factors see [5].

8.3 Comparison

Table 1 reports the time taken by SPUR and UniGen2 to generate 1,000 samples for a representative subset of the benchmarks. Included in the table is also the speedup factor of SPUR relative to UniGen2, i.e., the ratio of the two execution times. Since sharpSAT represents the execution time floor for SPUR we also provide the ratio between SPUR's execution time and of a *single* execution of sharpSAT. Numbers close to 1 substantiate the heuristic claim "if you can count the models with sharpSAT, you can sample."

Table 1. Time (sec) comparison of SPUR and UniGen2 to generate 1,000 samples.

Benchmark	#Var	#Clause	$\frac{\text{SPUR}}{\text{sharpSAT}}$	UniGen2 (sec)	SPUR (sec)	Speedup
case5	176	518	19.1	633	0.84	753
registerlesSwap	372	1,493	7.0	28,778	0.26	110,684
s953a_3_2	515	1,297	13.4	1,139	1.03	1,105
s1238a_3_2	686	1,850	7.0	610	2.31	264
s1196a_3_2	690	1,805	10.0	516	2.10	245
s832a_15_7	693	2,017	13.5	56	0.81	69
case_1_b12_2	827	2,725	1.4	689	29	23
squaring30	1,031	3,693	3.7	1,079	4.58	235
27	1,509	2,707	1.0	99	0.017	5,823
squaring16	1,627	5,835	1.9	11,053	78	141
squaring7	1,628	5,837	1.4	2,185	38	57
111	2,348	5,479	1.0	163	0.029	5,620
51	3,708	14,594	1.5	714	0.11	6,490
32	3,834	13,594	1.0	181	0.051	3,549
70	4,670	15,864	1.0	196	0.056	3,500
7	6,683	24,816	1.0	173	0.077	2,246
Pollard	7,815	41,258	6.0	181	355	0.51
17	10,090	27,056	1.6	192	0.092	2,086
20	15,475	60,994	2.7	289	2.05	140
reverse	75,641	380,869	6.2	TIMEOUT	2.66	>13,533

Figure 2 compares the time required to generate 1,000 witnesses with SPUR and UniGen2 for the full set of 373 benchmarks. The axes are logarithmic and each mark represents a single formula. Formulas for which a timeout occurred appear along the top or right border, depending on which tool timed out. (For marks corresponding to timeouts, the axis of the tool for which there was a timeout was co-opted to create a histogram of the number of timeouts that occurred.) These complete results can be summarized as follows:

- SPUR was faster than UniGen2 on 371 of the 373 benchmarks.
- On 369 of the 373, SPUR was more than 10× faster.

- On over 2/3 of the benchmarks, it was more than 100× faster.
- The geometric mean of the speedup exceeds 400×.
- On over 70% of the benchmarks, SPUR generated 1,000 samples within at most 10× of a single execution of sharpSAT.
- SPUR was 3 times more likely than UniGen2 to successfully generate witnesses for large formulas, (e.g., >10,000 variables).

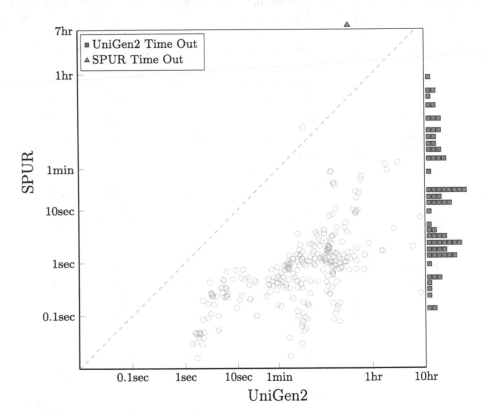

Fig. 2. Comparison of the running time to generate 1,000 samples between UniGen2 and SPUR over 373 formulas.

References

1. Bayardo Jr., R.J., Pehoushek, J.D.: Counting models using connected components. In: Proceedings of the 17th National Conference on Artificial Intelligence and 12th Conference on Innovative Applications of Artificial Intelligence, pp. 157–162. AAAI Press (2000)
2. Chakraborty, S., Fremont, D.J., Meel, K.S., Seshia, S.A., Vardi, M.Y.: On parallel scalable uniform SAT witness generation. In: Baier, C., Tinelli, C. (eds.) TACAS 2015. LNCS, vol. 9035, pp. 304–319. Springer, Heidelberg (2015). https://doi.org/10.1007/978-3-662-46681-0_25

3. Chakraborty, S., Meel, K.S., Vardi, M.Y.: Balancing scalability and uniformity in SAT witness generator. In: Proceedings of the 51st Annual Design Automation Conference, DAC 2014, pp. 60:1–60:6. ACM, New York (2014). http://doi.acm.org/10.1145/2593069.2593097
4. Davis, M., Logemann, G., Loveland, D.: A machine program for theorem-proving. Commun. ACM **5**(7), 394–397 (1962)
5. Fleming, P.J., Wallace, J.J.: How not to lie with statistics: the correct way to summarize benchmark results. Commun. ACM **29**(3), 218–221 (1986). http://doi.acm.org.oca.ucsc.edu/10.1145/5666.5673
6. SPUR source code. https://github.com/ZaydH/spur
7. Meel, K.: Index of UniGen verification benchmarks. https://www.cs.rice.edu/CS/Verification/Projects/UniGen/Benchmarks/
8. Naveh, Y., Rimon, M., Jaeger, I., Katz, Y., Vinov, M., Marcus, E., Shurek, G.: Constraint-based random stimuli generation for hardware verification. In: Proceedings of the 18th Conference on Innovative Applications of Artificial Intelligence, IAAI 2006, vol. 2, pp. 1720–1727. AAAI Press (2006)
9. Sang, T., Bacchus, F., Beame, P., Kautz, H., Pitassi, T.: Combining component caching and clause learning for effective model counting. In: Proceedings of the 7th International Conference on Theory and Applications of Satisfiability Testing, SAT 2004 (2004)
10. Sang, T., Beame, P., Kautz, H.: Heuristics for fast exact model counting. In: Bacchus, F., Walsh, T. (eds.) SAT 2005. LNCS, vol. 3569, pp. 226–240. Springer, Heidelberg (2005). https://doi.org/10.1007/11499107_17
11. Thurley, M.: sharpSAT – counting models with advanced component caching and implicit BCP. In: Biere, A., Gomes, C.P. (eds.) SAT 2006. LNCS, vol. 4121, pp. 424–429. Springer, Heidelberg (2006). https://doi.org/10.1007/11814948_38
12. Vitter, J.S.: Random sampling with a reservoir. ACM Trans. Math. Softw. **11**(1), 37–57 (1985). http://doi.acm.org/10.1145/3147.3165

Fast and Flexible Probabilistic Model Counting

Dimitris Achlioptas[1,2], Zayd Hammoudeh[1], and Panos Theodoropoulos[2(✉)]

[1] Department of Computer Science, University of California,
Santa Cruz, Santa Cruz, CA, USA
{dimitris,zayd}@ucsc.edu
[2] Department of Informatics and Telecommunications,
University of Athens, Athens, Greece
ptheodor@di.uoa.gr

Abstract. We present a probabilistic model counter that can trade off running time with approximation accuracy. As in several previous works, the number of models of a formula is estimated by adding random parity constraints (equations). One key difference with prior works is that the systems of parity equations used correspond to the parity check matrices of Low Density Parity Check (LDPC) error-correcting codes. As a result, the equations tend to be much shorter, often containing fewer than 10 variables each, making the search for models that also satisfy the parity constraints far more tractable. The price paid for computational tractability is that the statistical properties of the basic estimator are not as good as when longer constraints are used. We show how one can deal with this issue and derive rigorous approximation guarantees by performing more solver invocations.

1 Introduction

Given a CNF formula F with n variables, let $S = S(F)$ denote the set of its satisfying assignments (models). One way to estimate $|S|$ is to proceed as follows. For a fixed integer $0 \leq i \leq n$, let $R_i \subseteq \{0,1\}^n$ be a random set such that $\Pr[\sigma \in R_i] = 2^{-i}$ for all $\sigma \in \{0,1\}^n$. Markov's inequality implies that if $|S| < 2^{i-1}$, then $\Pr[S \cap R_i \neq \emptyset] < 1/2$. Therefore, if we select independent random sets $R_i^1, R_i^2, \ldots, R_i^t$ and find that the intersection with S is non-empty for the majority of them, we can declare that $|S| \geq 2^{i-1}$ with confidence $1 - \exp(-\Theta(t))$.

What happens if in the majority of the trials we find the intersection to be empty? Can we similarly draw the conclusion that $|S|$ is unlikely to be much more than 2^i? Unfortunately, no. The informativeness of $S \cap R_i = \emptyset$ depends on significantly more refined statistical properties of the random set R_i than the property that $\Pr[\sigma \in R_i] = 2^{-i}$, i.e., uniformity. For example, imagine that $|S| = 2^i$ and that the distribution of R_i is uniform but such that either

---

Research supported by NSF grants CCF-1514128, CCF-1733884, an Adobe research grant, and the Greek State Scholarships Foundation (IKY).

O. Beyersdorff and C. M. Wintersteiger (Eds.): SAT 2018, LNCS 10929, pp. 148–164, 2018.
https://doi.org/10.1007/978-3-319-94144-8_10

$S \cap R_i = \emptyset$ or $S \cap R_i = S$, always. Then, the number of trials needed to have a reasonable chance of ever witnessing $S \cap R_i \neq \emptyset$ is $\Omega(2^i)$. In other words, with this distribution for R_i, we can not distinguish between an unsatisfiable formula and one with 2^i models.

In the above example, the distribution of the random set R_i is such that the random variable $X = |S \cap R_i|$ exhibits extreme variance, a so-called "lottery phenomenon": it typically equals 0, but with very small probability it is huge. (Nearly) at the other end of the spectrum are distributions for the set R_i that exhibit *pairwise independence*, i.e.,

$$\Pr[\sigma \in R_i \wedge \tau \in R_i] = \Pr[\sigma \in R_i] \cdot \Pr[\tau \in R_i] \quad \text{for every} \quad \sigma \neq \tau \in \{0,1\}^n. \quad (1)$$

To get a feel for (1), fix any $\sigma \in \{0,1\}^n$ and sample R_i. Observe that conditional on $\sigma \in R_i$, the probability that $\tau \in R_i$ must be the same whether τ is at Hamming distance 1 from σ, or at distance, say, $n/2$ (throughout, distance will mean Hamming distance). In other words, the characteristic function of the set R_i must decorrelate in a *single step*!

It is possible to show that Eq. (1) implies that $\Pr[S \cap R_i \neq \emptyset] \geq (\mathbb{E}X)/(1 + \mathbb{E}X)$ and, thus, that if $|S| > 2^i$, then $\Pr[S \cap R_i \neq \emptyset] > 1/2$. Therefore, if, as before, we repeat the experiment t times and find the intersection to be empty in the majority of the trials, now we can declare that $|S| \leq 2^{i+1}$ with confidence $1 - \exp(-\Theta(t))$. Combined with the lower bound argument for $|S|$ outlined earlier, we see that in order to efficiently approximate $|S|$ within a factor of 4 it suffices to have a distribution of sets R_i for which (1) holds and for which checking whether $S \cap R_i = \emptyset$ or not can be done efficiently. Indeed, given such a distribution one can estimate $|S|$ within a $(1 \pm \epsilon)$ factor, for any $\epsilon > 0$, and any desired confidence $1 - \delta$, in $O(\epsilon^{-2} \log(1/\delta))$ trials.

In order to be able to check efficiently whether $S \cap R_i = \emptyset$ we must, at a minimum, be able to represent the random sets R_i compactly, in spite of their exponential size. The key to this is to represent each set R_i *implicitly* as the set of solutions to a system of i random parity (XOR) constraints (linear equations modulo 2). More precisely, for any fixed matrix $A \in \{0,1\}^{i \times n}$, consider the partition (hashing) of $\{0,1\}^n$ induced by the value of $A\sigma \in \{0,1\}^i$. Let

$$R_i = \{\sigma \in \{0,1\}^n : A\sigma = b\} \quad \text{where } b \in \{0,1\}^i \text{ is uniformly random.} \quad (2)$$

Observe that even though the 2^i parts may have dramatically different sizes, the uniformity in the choice of b in (2) implies that $\Pr[\sigma \in R_i] = 2^{-i}$, for every $\sigma \in \{0,1\}^n$, as desired. At the same time, checking whether $S \cap R_i = \emptyset$ or not can be done by converting the i parity constraints to clauses and using a SAT solver, or, more recently, by using a SAT solver supporting parity constraints, e.g., CryptoMiniSat [14].

From the above discussion we see that the only issue left is how the choice of the matrix A affects the variance of the sizes of the different parts and, thus, the variance of $|S \cap R_i|$. To that end, it is not hard to prove that if A is a uniformly random element of $\{0,1\}^{i \times n}$ (equivalently, if each element A_{ij} is set to 0/1 independently with equal probability), then membership in R_i enjoys pairwise independence, i.e., (1) holds. As mentioned above, this is essentially perfect from a statistical point of view. Unfortunately, though, under this distribution for A each parity constraint contains $n/2$ variables, on average, and changing any variable in a parity constraint immediately changes its truth value (whereas in clauses that's not the case, typically, motivating the two watched literals heuristic [11]). As a result, the branching factor of the search for satisfying assignments (models) that also satisfy the parity equations gets rapidly out of hand as the number of variables in the formula increases.

All ideas presented so far, including in particular the choice of a uniformly random matrix $A \in \{0,1\}^{i \times n}$, first appeared in the pioneering theoretical works by Sipser [13], Stockmeyer [15], and Valiant and Vazirani [17]. As we discuss in Sect. 2, there has since been a long line of works aiming to make the approach practical. Specifically, the limitations posed by *long* parity constraints, i.e., those of (average) length $n/2$, was already recognized in the very first works in the area [7,8]. Later works [6,18] tried to remedy the problem by considering parity equations where each constraint includes each variable independently with probability $p < 1/2$. While such sparsity helps the solver in finding elements of $S \cap R$, the statistical properties of the resulting random sets deteriorate rapidly as p decreases. Crucially, in *all* these works, different constraints (parity equations) select their set of variables independently of one another.

In [1] we introduced the idea of using random matrices $A \in \{0,1\}^{i \times n}$ with *dependent* entries, by selecting A uniformly from an ensemble of Low Density Parity Check (LDPC) matrices. A simplest such ensemble comprises all matrices where every row (equation) contains the same number l of ones *and* every column contains the same number $r \geq 3$ of ones. We gave a first mathematical analysis of the statistical properties of the resulting sets R_i and some experimental evidence that their actual statistical properties are probably much better than what is suggested by the mathematical analysis.

A key idea motivating our work here and in [1] is the realization that to prove mathematically rigorous *lower* bounds, the random sets R_i do *not* need to come with any statistical guarantees (besides the trivial requirement of uniformity). The obligation to use distributions $\mathscr{D}_i$ with statistical guarantees exists only for upper bounds and, crucially, only concerns their behavior over sets of size 2^i or greater. When i/n is not tiny we will see that short parity constraints have provably good statistical behavior.

In this paper we present[1] an approximate model counter, called F2, with rigorous guarantees based on these ideas. F2 has three modes of operation, trading accuracy for computation time. To discuss these modes, let us foreshadow that the statistical demerit of a distribution on matrices $A \in \{0,1\}^{i \times n}$ in our context will be captured by a scalar quantity $B = B(i, n) \geq 1$ that increases as the average constraint length decreases, with $B = 1$ corresponding to pairwise independence (and average constraint length $n/2$).

Given any $\delta > 0$, let $q = \ln(1/\delta)$. Given any $\epsilon \in (0, 1/3]$, with probability at least $1 - \delta$, all of the following will occur, in sequence:

---

[1] F2 source code available at https://github.com/ptheod/F2.git.

1. After $O(q + \log_2 n)$ solver invocations, F2 will return a number $\ell \leq \log_2 |S|$ and B.
2. After $O(qB)$ solver invocations, F2 will return a number $u \geq \log_2 |S|$.
3. After $O(qB^2/\epsilon^4)$ solver invocations, F2 will return a number $Z \in (1 \pm \epsilon)|S|$.

Observe that while the bounds $\ell \leq \log_2 |S| \leq u$ are guaranteed (with probability $1 - \delta$), no a priori bound is given for $u - \ell$. In other words, in principle the algorithm may offer very little information on $\log_2 |S|$ at the end of Step 2. As we will see, in practice, this is not the case and, in fact, we expect that in most practical applications Step 3 will be unnecessary. We give a detailed experimental performance of F2 in Sect. 10. The main takeaway is that F2 dramatically extends the range of formulas for which one can get a rigorous model count approximation.

2   Previous Work

The first work on practical approximate model counting using systems of random parity equations was by Gomes et al. [8]. Exactly along the lines outlined in the introduction, they proved that when $A \in \{0,1\}^{i \times n}$ is uniformly random, i.e., when each entry of A is set to 1 independently with probability $p = 1/2$, one can rigorously approximate $\log_2 |S|$ within an additive constant by repeatedly checking if $S \cap R_i = \emptyset$, for various values of i. They further proved that if each entry of A is set to 1 with probability $p < 1/2$ one get a rigorous lower bound, but one which may be arbitrarily far from the truth. In [7], Gomes et al. showed experimentally that it can be possible to achieve good accuracy (without guarantees) using parity constraints of length $k \ll n/2$.

Interest in the subject was rekindled by works of Chakraborty et al. [3] and of Ermon et al. [5]. Specifically, a complete, rigorous, approximate model counter, called ApproxMC, was given in [3] which takes as input any $\delta, \epsilon > 0$, and with probability at least $1 - \delta$ returns a number in the range $(1 \pm \epsilon)|S|$. In [5] an algorithm, called WISH, is given with a similar (δ, ϵ)-guarantee for the more general problem of approximating sums of the form $\sum_{\sigma \in \{0,1\}^n} w(\sigma)$, where w is a non-negative real-valued function over Ω^n, where Ω is a finite domain. Both ApproxMC and WISH also use uniformly random $A \in \{0,1\}^{i \times n}$, so that the resulting parity equations have average length $n/2$, limiting the range of problems they can handle.

ApproxMC uses the satisfiability solver CryptoMiniSat (CMS) [14] which has native support and sophisticated reasoning for parity constraints. CMS can, moreover, take as input a cutoff value $z \geq 1$, so that it will run until it either finds z solutions or determines the number of solutions to be less than z. ApproxMC makes use of this capability in order to target i such that $|S \cap R_i| = \Theta(\delta^{-2})$, instead of i such that $|S \cap R_i| \approx 1$. Our algorithms make similar use of this capability, using several different cutoffs.

The first effort to develop rigorous performance guarantees when $p < 1/2$ was made by Ermon et al. in [6], where an explicit expression was given for the smallest allowed p as a function of $|S|, n, \delta, \epsilon$. The analysis in [6] was recently improved by Zhao et al. in [18] who, among other results, showed that when $\log_2 |S| = \Omega(n)$, one can get rigorous approximation guarantees with $p = O((\log n)/n)$, i.e., average constraint length $O(\log n)$. While, prima facie, this seems a very promising result, we will see that the dependence on the constants involved in the asymptotics is very important in practice. For example, in our experiments we observe that already setting $p = 1/8$ yields results whose accuracy is much worse than those achieved by LDPC constraints.

Finally, in [4] Chakraborty et al. introduced a very nice idea for reducing the number of solver invocations without any compromise in approximation quality. It amounts to using *nested* sequences of random sets $R_1 \supseteq R_2 \supseteq R_3 \supseteq \cdots \supseteq R_n$ in the search for $i \approx \log_2 |S|$. The key insight is that using nested (instead of independent) random sets R_i means that $|S \cap R_i|$ is deterministically non-increasing in i, so that linear search for i can be replaced with binary search, reducing the number of solver invocations from linear to logarithmic in n. We use the same idea in our work.

2.1   Independent Support Sets

A powerful idea for mitigating the severe limitations arising from long parity constraints was proposed by Chakraborty et al. in [2]. It is motivated by the observation that formulas arising in practice often have a small set of variables $I \subseteq V$ such that every value-assignment to the variables in I has at most one extension to a satisfying assignment. Such a set I is called an *independent support set*. Clearly, if $S' \subseteq \{0,1\}^I$ comprises the value assignments to the variables in I that can be extended to satisfying assignments, then $|S| = |S'|$. Thus, given I, we can rethink of model counting as the task of estimating the size of a subset of $\{0,1\}^I$, completely oblivious to the variables in $V - I$. In particular, we can add random parity constraints only over the variables in I, so that even if we use long constraints each constraint has $|I|/2$ instead of $|V|/2$ variables on average. Since independent support sets of small size can often be found in practice [9], this has allowed ApproxMC to scale to certain formulas with thousands of variables.

In our work, independent support sets are also very helpful, but per a rather "dual" reasoning: for any fixed integers i, k, the statistical quality of random sets defined by systems of i parity constraints with k variables each, decreases with the number of variables over which the constraints are taken. Thus, by adding our short constraints over only the variables in an independent support set, we get meaningful results on formulas for which $|I|/2$ is too large (causing CMS and thus ApproxMC to timeout), but for which $|I|/|V|$ is sufficiently large for our short parity constraints to have good statistical properties.

Variable Convention. *In the rest of the paper we will think of the set of variables V of the formula F being considered as being some independent support set of F (potentially the trivial one, corresponding to the set of all variables). Correspondingly, n will refer to the number of variables in that set V.*

3   Our Results

In [1], the first and last authors showed that systems of parity equations based on LDPC codes can be used both to derive a rigorous lower bound for $|S|$ quickly, and to derive a (δ, ϵ)-approximation of $|S|$ with $O(qB^2/\epsilon^4)$ solver invocations, as per Step 3 of F2. The new contributions in this work are the following.

- In Sect. 5 we show how to compute a rigorous *upper* bound for $|S|$ with a number of solver invocations that is *linear* in B. While the bound does not come with any guarantee of being close to $|S|$, in practice it is remarkably accurate. Key to our approach is a large deviations inequality bounding the lower tail of a random variable as a function of the ratio between its second moment and the square of its first moment. Notably, the analogue of this inequality does *not* hold for the upper tail. Recognizing and leveraging this asymmetry is our main intellectual contribution.
- In Sect. 6 we simplify and streamline the analysis of the (δ, ϵ)-approximation algorithm of [1], showing also how to incorporate the idea of nested sampling sets.
- In Sects. 7–9 we refine the analysis of [1] for B, resulting in significantly better bounds for it. Getting such improved bounds is crucial for making our aforementioned upper-bounding algorithm fast in practice (as it is linear in B).
- Finally, we give a publicly available implementation, called F2.

4   First a Lower Bound

To simplify exposition we only discuss lower bounds of the form $|S| \geq 2^i$ for $i \in \mathbb{N}$, deferring the discussion of more precise estimates to Sect. 6. For any distribution $\mathscr{D}$, let $R \sim \mathscr{D}$ denote that random variable R has distribution $\mathscr{D}$.

Definition 1. *Let $\mathscr{D}$ be a distribution on subsets of a set U and let $R \sim \mathscr{D}$. We say that $\mathscr{D}$ is i-uniform if $\Pr[\sigma \in R] = 2^{-i}$ for every $\sigma \in U$.*

Algorithm 1 below follows the scheme presented in the introduction for proving lower bounds, except that instead of asking whether typically $S \cap R \neq \emptyset$, it asks whether typically $|S \cap R| \geq 2$. To do this, $|S \cap R|$ is trimmed to 4 in line 5 (by running CryptoMiniSat with a cutoff of 4), so that the event $Z \geq 2t$ in line 8 can only occur if the intersection had size at least 2 in at least $t/2$ trials.

Theorem 1 ([1]). $\Pr[\textit{The output of Algorithm 1 is incorrect}] \leq e^{-t/8}$.

To get a lower bound for $|S|$ we can invoke Algorithm 1 with $i = 1, 2, \ldots, n$ sequentially and keep the best lower bound returned (if any). To accelerate this linear search we can invoke Algorithm 1 with $i = 1, 2, 4, 8, \ldots$ until the first "Don't know" occurs, say at $i = 2^u$. At that point we can perform binary search in $\{2^{u-1}, \ldots, 2^u - 1\}$, treating every "Don't know" answer as a (conservative) imperative to reduce the interval's upper bound to the midpoint and every "Yes"

Algorithm 1. Given i, t decides if $|S| \geq 2^i$ with error probability $e^{-t/8}$

1: $Z \leftarrow 0$
2: $j \leftarrow 0$
3: **while** $j < t$ and $Z < 2t$ **do**   ▷ The condition $Z < 2t$ is an optimization
4:     Sample $R_j \sim \mathcal{D}_i$ ▷ $\mathcal{D}_i$ can be any i-uniform distribution
5:     $Y_j \leftarrow \min\{4, |S \cap R_j|\}$ ▷ Run CryptoMiniSat with cutoff 4
6:     $Z \leftarrow Z + Y_j$
7:     $j \leftarrow j + 1$
8: **if** $Z \geq 2t$ **then**
9:     **return** "Yes"
10: **else**
11:     **return** "Don't know"

---

answer as an allowance to increase the interval's lower bound to the midpoint. We call this scheme "doubling binary search." In Step 1 of F2 this is further accelerated by invoking Algorithm 1 with a very small number of trials, t, in the course of the doubling-binary search. The result of the search is treated as a "ballpark" estimate and a proper binary search is done in its vicinity, by using for each candidate i the number of iterations suggested by Theorem 1.

5   Then an Upper Bound

As discussed in the introduction, lottery phenomena may cause Algorithm 1 and, thus, Step 1 of F2 to underestimate $\log_2 |S|$ arbitrarily. To account for the possibility of such phenomena we bound the "lumpiness" of the sets $R_i \sim \mathcal{D}_i$ by the quantity defined in (3) below, measuring lumpiness at a scale of M.

Definition 2. *Let $\mathcal{D}$ be any distribution on subsets of $\{0,1\}^n$ and let $R \sim \mathcal{D}$. For any fixed $M \geq 1$, let*

$$\mathrm{Boost}(\mathcal{D}, M) = \max_{\substack{S \subseteq \{0,1\}^n \\ |S| \geq M}} \frac{1}{|S|(|S| - 1)} \sum_{\substack{\sigma, \tau \in S \\ \sigma \neq \tau}} \frac{\Pr[\sigma, \tau \in R]}{\Pr[\sigma \in R] \Pr[\tau \in R]}. \tag{3}$$

To develop intuition for (3) observe that the ratio inside the sum is the factor by which the a priori probability that a truth assignment belongs in R is modified by conditioning on some other truth assignment belonging in R. So, if membership in R is pairwise independent, then $\mathrm{Boost}(\mathcal{D}, \cdot) = 1$. Note also that since $|S| \geq M$ instead of $|S| = M$ in (3), the function $\mathrm{Boost}(\mathcal{D}, \cdot)$ is non-increasing in M. As we will see, the critical quantity for an i-uniform distribution $\mathcal{D}_i$ is $\mathrm{Boost}(\mathcal{D}_i, 2^i)$, i.e., an i-uniform distribution can be useful even if $\mathrm{Boost}(\mathcal{D}_i)$ is huge for sets of size less than 2^i.

To analyze Algorithm 2 we will use the following inequality of Maurer [10].

Algorithm 2. Given $\delta > 0$ and $L \leq |S|$ returns $Z \geq |S|$ with probability $1 - \delta$

1: $\ell \leftarrow \lfloor \log_2 L \rfloor$
2: $\mathscr{D}_\ell \leftarrow$ any ℓ-uniform distribution
3: $B \leftarrow$ any upper bound for $\mathrm{Boost}(\mathscr{D}_\ell, 2^\ell)$
4: $t \leftarrow \lceil 8(B + 1) \ln(1/\delta) \rceil$
5: $Z \leftarrow 0$
6: **for** j from 1 to t **do**
7:     Sample $R_j \sim \mathscr{D}_\ell$
8:     $X_j \leftarrow |S \cap R_j|$ ▷ Run CryptoMiniSat without cutoff
9:     $Z \leftarrow Z + X_j$
10: **return** "$|S| \leq 2^{\ell+1}(Z/t)$"

Lemma 1. *Let* $X_1, \ldots, X_t$ *be non-negative i.i.d. random variables. Let* $Z = \sum_{i=1}^t X_i$. *If* $\mathbb{E}X_1^2/(\mathbb{E}X_1)^2 \leq B$, *then for any* $\alpha \geq 0$,

$$\Pr[Z \leq (1 - \alpha)\mathbb{E}Z] \leq \exp\left(-\frac{\alpha^2 t}{2B}\right).$$

Theorem 2. $\Pr[$*The output of Algorithm 2 is correct*$] \geq 1 - \delta$.

Proof. Let Z be the random variable equal to the value of variable Z in line 9, right before line 10 is executed. If $Z = z$, in order for the output to be wrong it must be that $|S| > 2^{\ell+1}(z/t)$, implying $\mathbb{E}Z = t|S|2^{-\ell} > 2z$ and, therefore, that the event $Z \leq \mathbb{E}Z/2$ occurred. Since Z is the sum of i.i.d. non-negative random variables $X_1, \ldots, X_t$, we can bound $\Pr[Z \leq \mathbb{E}Z/2]$ via Lemma 1.

To bound $\mathbb{E}X_1^2/(\mathbb{E}X_1)^2$, we write $X_1 = \sum_{\sigma \in S} \mathbf{1}_{\sigma \in R_1}$ and observe that

$$\mathbb{E}X_1^2 = \sum_{\sigma, \tau \in S} \Pr[\sigma, \tau \in R_1]$$

$$= \sum_{\sigma \in S} \Pr[\sigma \in R_1] + \sum_{\substack{\sigma, \tau \in S \\ \sigma \neq \tau}} \Pr[\sigma, \tau \in R_1]$$

$$\leq \sum_{\sigma \in S} \Pr[\sigma \in R_1] + 2^{-2i}|S|(|S| - 1)\mathrm{Boost}(\mathscr{D}, |S|)$$

$$\leq \mathbb{E}X_1 + \mathrm{Boost}(\mathscr{D}, |S|)(\mathbb{E}X_1)^2.$$

Since $|S| \geq L \geq 2^\ell$ and $\mathrm{Boost}(\mathscr{D}_\ell, M)$ is non-increasing in M, we see that

$$\frac{\mathbb{E}X_1^2}{(\mathbb{E}X_1)^2} \leq \frac{1}{\mathbb{E}X} + \mathrm{Boost}(\mathscr{D}, |S|) \leq 1 + \mathrm{Boost}(\mathscr{D}_\ell, 2^\ell). \tag{4}$$

Therefore, applying Lemma 1 with $\alpha = 1/2$ and recalling the definitions of B and t in lines 3 and 4 of Algorithm 2, we see that $\Pr[Z \leq \mathbb{E}Z/2] \leq \delta$, as desired.

F2. Given $L \leq |S| \leq U$, $\delta, \theta > 0$ returns $Z \in (1 \pm \delta)|S|$ with probability $1 - \theta$

1: **if** $L < 4/\delta$ **then**
2:     $E \leftarrow$ number of solutions found by CryptoMiniSat ran with cutoff $4/\delta$
3:     **if** $E < 4/\delta$ **then return** E ▷ In this case $|S| = E$

4:
5: $\ell \leftarrow \lfloor \log_2(\delta L/4) \rfloor$
6: $u \leftarrow \lceil \log_2 U \rceil$
7: $B \leftarrow$ Any upper bound for $\max_{\ell \leq i \leq u-2} \text{Boost}(\mathscr{D}_i, 2^i)$

8:
9: $\delta \leftarrow \min\{\delta, 1/3\}$
10: $\xi \leftarrow 8/\delta$
11: $b \leftarrow \lceil \xi + 2(\xi + \xi^2(B-1)) \rceil$ ▷ If $B = 1$, then $b = \lceil 24/\delta \rceil$
12: $t \leftarrow \lceil (2b^2/9) \ln(5/\theta) \rceil$

13:
14: $Z_\ell, Z_{\ell+1}, \ldots, Z_u \leftarrow 0$

15:
16: **for** j from 1 to t **do**
17:     $M \leftarrow$ a uniformly random element of an LDPC ensemble over $\{0,1\}^{u \times n}$
18:     $y \leftarrow$ a uniformly random element of $\{0,1\}^u$
19:     **for** i from ℓ to u **do**
20:         Let M_i, y_i comprise the first i rows of M and y, respectively
21:         $R_{i,j} \leftarrow \{\sigma \in \{0,1\}^n : M_i \sigma = y_i\}$ ▷ Enforce the first i parity constraints
22:         $Y_{i,j} \leftarrow \min\{b, |S \cap R_{i,j}|\}$ ▷ Run CryptoMiniSat with cutoff b
23:         $Z_i \leftarrow Z_i + Y_{i,j}$

24:
25: $j \leftarrow \max\{-1, \max\{\ell \leq i \leq u : Z_i \geq t(1-\delta)(4/\delta)\}\}$
26:
27: **if** $j \neq -1$ **then return** $2^j(Z_j/t)$
28: **else return** "Fail"

6   Finally a $(1 \pm \delta)|S|$ Approximation

Given any bounds $L \leq |S| \leq U$, for example derived by using Algorithms 1 and 2, algorithm F2 below yields a rigorous approximation of $|S|$ within $1 \pm \delta$ with a number of solver invocations proportional to B^2/δ^4, where

$$B = \max_{\ell \leq i \leq u-2} \text{Boost}(\mathscr{D}_i, 2^i),$$

where $\ell \approx \log_2(\delta L)$ and $u \approx \log_2 u$. (If $B = 1$, the iterations drop to $O(\delta^{-2})$.)

Theorem 3. $\Pr[\text{F2 returns } Z \in (1 \pm \delta)|S|] \geq 1 - \theta$.

To prove Theorem 3 we will need the following tools.

Lemma 2 (Hoeffding's Inequality). *If* $Z = Y_1 + \cdots + Y_t$, *where* $0 \leq Y_i \leq b$ *are independent random variables, then for any* $w \geq 0$,

$$\Pr[Z/t \geq \mathbb{E}Z/t + w] \leq e^{-2t(w/b)^2} \quad and \quad \Pr[Z/t \leq \mathbb{E}Z/t - w] \leq e^{-2t(w/b)^2}. \quad (5)$$

Lemma 3 ($[1]$). *Let $X \geq 0$ be an arbitrary integer-valued random variable. Write $\mathbb{E}X = \mu$ and $\mathrm{Var}(X) = \sigma^2$. For some integer $b \geq 0$, define the random variable $Y = \min\{X, b\}$. For any $\lambda > 0$, if $b \geq \mu + \lambda\sigma^2$, then $\mathbb{E}Y \geq \mathbb{E}X - 1/\lambda$.*

Lemma 4 ($[1]$). *Let $\mathscr{D}$ be any i-uniform distribution on subsets of $\{0,1\}^n$. For any fixed set $S \subseteq \{0,1\}^n$, if $R \sim \mathscr{D}$ and $X = |S \cap R|$, then $\mathrm{Var}(X) \leq \mathbb{E}X + (\mathrm{Boost}(\mathscr{D}, |S|) - 1)(\mathbb{E}X)^2$.*

Proof. If $|S| < 4/\delta$, the algorithm returns exactly $|S|$ and exits. Otherwise, the value ℓ defined in line 5 is non-negative and $q := \lfloor \log_2(\delta|S|/4) \rfloor \geq \ell$ since $L \leq |S|$.

Let $A_i = Z_i/t$. We will establish the following propositions:

(a) $\Pr[A_q 2^q \notin (1 \pm \delta)|S|] \leq 2e^{-9t/(2b^2)}$.
(b) $\Pr[A_{q+1} 2^{q+1} \notin (1 \pm \delta)|S|] \leq 2e^{-9t/(2b^2)}$.
(c) If $A_q 2^q \in (1 \pm \delta)|S|$, then $j \geq q$ in line 25 (deterministically).
(d) $\Pr[j \geq q + 2] \leq e^{-8t/b^2}$.

Given propositions (a)–(d) the theorem follows readily. If $A_{q+k} 2^{q+k}$ is in the range $(1 \pm \delta)|S|$ for $k \in \{0,1\}$ but for $k \geq 2$ it is less than $(1 - \delta)(4/\delta)$, then the algorithm will report either $A_q 2^q$ or $A_{q+1} 2^{q+1}$, both of which are in $(1 \pm \delta)|S|$. Thus, the probability that the algorithm does not report a number in $(1 \pm \delta)|S|$ is at most $2 \cdot 2e^{-9t/(2b^2)} + e^{-8t/b^2}$ which, by our choice of t, is less than θ.

To establish propositions (a)–(d) we start by noting the following facts:

(i) $R_{i,j}$ is sampled from an i-uniform distribution for every i, j.
(ii) The sets $R_{i,1}, \ldots, R_{i,t}$ are independent for every i.
(iii) $R_{\ell,j} \supseteq R_{\ell+1,j} \supseteq \cdots \supseteq R_{u-1,j} \supseteq R_{u,j}$ for every j.

Now, fix any $i = q+k$, where $k \geq 0$. Let $X_{i,j} = |S \cap R_{i,j}|$ and write $\mathbb{E}X_{i,j} = \mu_i$, $\mathrm{Var}(X_{i,j}) = \sigma_i^2$. By fact (ii), Z_i is the sum of t independent random variables $0 \leq Y_{i,j} \leq b$. Since $\mathbb{E}Z_i/t \leq \mu_i$, Hoeffding's inequality implies that for all $i \geq q$,

$$\Pr[Z_i/t \geq (1 + \delta)\mu_i] \leq \exp\left(-2t\left(\frac{\delta\mu_i}{b}\right)^2\right). \tag{6}$$

To bound $\Pr[Z_i/t \geq (1 - \delta)\mu_i]$ for $k \in \{0,1\}$ we first observe that $|S| \geq 2^{q+1}$, since $\delta \leq 2$. Since $\mathrm{Boost}(\mathscr{D}, M)$ is non-increasing in M and $q \leq u - 2$ we see that

$$\max_{k \in \{0,1\}} \mathrm{Boost}(\mathscr{D}_{q+k}, |S|) \leq \max\{\mathrm{Boost}(\mathscr{D}_q, 2^{q+1}), \mathrm{Boost}(\mathscr{D}_{q+1}, 2^{q+1})\}$$

$$\leq \max\{\mathrm{Boost}(\mathscr{D}_q, 2^q), \mathrm{Boost}(\mathscr{D}_{q+1}, 2^{q+1})\}$$

$$\leq \max_{\ell \leq i \leq u-2} \mathrm{Boost}(\mathscr{D}_i, 2^i)$$

$$\leq B. \tag{7}$$

Fact (i) implies that X_{ij} satisfies the conditions of Lemma 4. Therefore, for $i \in \{q, q+1\}$, Lemma 4 combined with (7) implies $\sigma_i^2 \leq \mu_i + (B - 1)\mu_i^2$. Since

$\mu_i < 8/\delta$ for all $i \geq q$ while $\xi = 8/\delta$, we see that $b = \lceil \xi + 2(\xi + \xi^2(B-1)) \rceil \geq \mu_i + 2\sigma_i^2$. Thus, for $i \in \{q, q+1\}$ the random variables $X_{i,j}, Y_{i,j}$ satisfy the conditions of Lemma 3 with $\lambda = 2$, implying $\mathbb{E}Y_{i,j} \geq \mathbb{E}X_{i,j} - 1/2$. Therefore, $\mathbb{E}Z_i/t \geq \mu_i - 1/2$ for $i \in \{q, q+1\}$ so that Hoeffding's inequality implies

$$\Pr[Z_i/t \leq (1-\delta)\mu_i] \leq \exp\left(-2t\left(\frac{\delta\mu_i - 1/2}{b}\right)^2\right). \tag{8}$$

To establish propositions (a) and (d) observe that $\mu_{q+k} \geq 2^{2-k}/\delta$ by Fact (i). Therefore, (6) and (8) imply that for $k \in \{0, 1\}$, the probability that $A_{q+k}2^{q+k}$ is outside $(1 \pm \delta)|S|$ is at most

$$2\exp\left(-2t\left(\frac{2^{2-k} - 1/2}{b}\right)^2\right) < 2\exp(-9t/(2b^2)) .$$

To establish proposition (c) note that if $A_q \geq (1-\delta)\mu_q$, then $A_q \geq (1-\delta)(4/\delta)$ and, thus, $j \geq q$. Finally, to establish proposition (d) observe that, by Fact (iii), the random variables Z_i are non-increasing in i, so that $j \geq q+2$ implies $A_{q+2}2^{q+2} < (1-\delta)(4/\delta)$. To bound the probability of this event we note that $\mu_{q+2} < 2/\delta$. Thus, $\mu_{q+2} + w \geq (1-\delta)(4/\delta)$, implies $w > 2(1-2\delta)/\delta$, which, since $\delta \leq 1/3$, implies $w > 2$. Therefore, (5) implies $\Pr[j \geq q+2] \leq e^{-8t/b^2}$.

7   Homogeneous Distributions

Our goal in Sects. 7–9 is to derive an upper bound for B when the random matrix A corresponds to the parity check matrix of an LDPC code. To that end, in this section we derive an expression for B valid for any random set distribution that satisfies certain symmetry properties. In Sect. 8 we relate the sets R_i corresponding to codewords of LDPC codes to these properties. Finally, in Sect. 9 we discuss how to deal with miscellaneous technical issues arising from the need to be able to work with formulas with an arbitrary number of variables and clauses, while retaining mathematical rigor in our bounding of B.

The analysis in this section is identical to the one in [1] except for requiring that $f(n) = 0$ in the definition of tractability. This has the effect of changing the lower index of summation in the definition of B in Theorem 4 from 0 to 1 which, in turn, makes a significant difference in practice.

Definition 3. *An i-uniform distribution, $\mathcal{D}_i$ is homogeneous if there exists a function f, called the density of $\mathcal{D}_i$, such that for all $\sigma, \tau \in \{0,1\}^n$, if $R \sim \mathcal{D}_i$, then $\Pr[\tau \in R \mid \sigma \in R] = f(\text{Hamming}(\sigma, \tau))$.*

Definition 4. *A homogenous distribution is tractable if its density f satisfies: $f(j) \geq f(j+1)$ for $j < n/2$, $f(j) \leq f(n-j)$ for $j \geq n/2$, and $f(n) = 0$.*

For any $S \subset \{0,1\}^n$ and $\sigma \in S$, let $H_\sigma^S(d)$ denote the number of elements of S at Hamming distance d from σ. In [1] it was shown that for any homogenous distribution $\mathscr{D}_i$, and any $M \geq 1$,

$$\text{Boost}(\mathscr{D}_i, M) \leq \max_{\substack{S \subseteq \{0,1\}^n \\ |S| \geq M \\ \sigma \in S}} \frac{2^i}{|S|-1} \sum_{d=1}^{n} H_\sigma^S(d) f(d). \tag{9}$$

To bound (9), we assume that $|S| \geq 2n+1$ so that there exists $2 \leq z \leq n/2$ such that $(|S|-1)/2 = \binom{n}{1} + \binom{n}{2} + \cdots + \binom{n}{z-1} + \alpha\binom{n}{z}$, for some $\alpha \in [0,1)$. (If $|S| < 2n+1$, then we can estimate $|S$ by using a handful of long parity constraints.) Fact $f(j) \leq f(n-j)$ for $j \geq n/2$ implies (10). Facts $f(j) \geq f(j+1)$ for $j < n/2$ and $f(n) = 0$ imply (11). Finally, the fact $f(z-1) \geq f(z)$ implies (13).

$$\frac{\sum_{d=1}^{n} H_\sigma^S(d) f(d)}{|S|-1} \leq \frac{\sum_{d=1}^{n/2} H_\sigma^S(d) f(d) + \sum_{d>n/2} H_\sigma^S(d) f(n-d)}{|S|-1} \tag{10}$$

$$\leq \frac{2\left(\sum_{d=1}^{z-1} \binom{n}{d} f(d) + \alpha\binom{n}{z} f(z)\right)}{|S|-1} \tag{11}$$

$$= \frac{\sum_{d=1}^{z-1} \binom{n}{d} f(d) + \alpha\binom{n}{z} f(z)}{\sum_{d=1}^{z-1} \binom{n}{d} + \alpha\binom{n}{z}} \tag{12}$$

$$\leq \frac{\sum_{d=1}^{z-1} \binom{n}{d} f(d)}{\sum_{d=1}^{z-1} \binom{n}{d}} \tag{13}$$

$$:= B(z). \tag{14}$$

To bound $B(z)$ observe that since $f(j) \geq f(j+1)$ for $j < n/2$ it follows that $B(j) \geq B(j+1)$ for $j < n/2$. Thus, to bound $B(z)$ from above it suffices to bound z from below. Let $h : x \mapsto -x \log_2 x - (1-x) \log_2 x$ be the binary entropy function and let $h^{-1} : [0,1] \mapsto [0,1]$ map y to the smallest number x such that $h(x) = y$. It is well-known that $\sum_{d=1}^{z} \binom{n}{d} \leq 2^{nh(z/n)}$, for every integer $1 \leq z \leq n/2$. Therefore, $z \geq \lceil nh^{-1}(\log_2(|S|/2)/n) \rceil$, which combined with (9) and (14) implies the following.

Theorem 4. *If $\mathscr{D}_i$ is a tractable distribution with density f, then*

$$\text{Boost}(\mathscr{D}_i, M) \leq 2^i B\left(\left\lceil nh^{-1}\left(\frac{\log_2 M - 1}{n}\right)\right\rceil\right), \tag{15}$$

where $B(z) = \sum_{d=1}^{z-1} \binom{n}{d} f(d) / \sum_{d=1}^{z-1} \binom{n}{d}$ and $h^{-1} : [0,1] \mapsto [0,1]$ maps y to the smallest number x such that $h(x) = y$, where h is the binary entropy function.

8   Low Density Parity Check Codes

We will consider the set of all matrices $\{0,1\}^{i \times n}$ where:

(i) Every column (variable) has exactly $1 \geq 3$ non-zero elements.
(ii) Every row (equation) has $\lfloor r \rfloor$ or $\lceil r \rceil$ non-zero elements, where $r = 1n/i$.

Given n, i, and 1, let i_0 denote the number of equations with $\lfloor r \rfloor$ variables and let $i_1 = i - i_0$. Let A be selected uniformly at random[2] among all matrices satisfying (i)–(ii). Let $R = \{\sigma \in \{0,1\}^n : A\sigma = b\}$, where $b \in \{0,1\}^i$ is uniformly random. Lemma 3.157 of [12] implies that for every $\sigma \in \{0,1\}^n$, if $\sigma \in R$, then the expected number of codewords at distance d from σ, denoted by $\mathrm{codewords}(d)$, is independent of σ (due to the row- and column-symmetry in the distribution of A) and equals the coefficient of x^{d1} in the polynomial

$$\binom{n}{d} \frac{\left(\sum_j \binom{r}{2j} x^{2j}\right)^{i_0} \left(\sum_j \binom{r+1}{2j} x^{2j}\right)^{i_1}}{\binom{n1}{d1}}.$$

If $\mathscr{D}_i$ denotes the distribution of R, the uniformity in the choice of b implies that $\mathscr{D}_i$ is i-uniform. The fact that for every $\sigma \in \{0,1\}^n$, conditional on $\sigma \in R$, the expected number of codewords at distance d from σ is independent of σ implies that for any fixed $\tau \neq \sigma$, $\Pr[\text{both } \sigma, \tau \in R] = 2^{-i} f(d)$, where $f(d) = \mathrm{codewords}(d)/\binom{n}{d}$, making $\mathscr{D}_i$ homogeneous with density f.

Regarding tractability, we begin by noting that if any equation has an odd number of variables, then the complement of a codeword can not be a codeword, implying $\mathrm{codewords}(n) = 0$. When r is an ever integer we achieve $i_1 > 0$ by adding a single dummy Boolean variable to the formula (and reducing all our estimates of $|S|$ by 2). To simplify exposition in the following we assume $i_1 > 0$.

It is also well-known [12] that $\mathrm{codewords}(j) \geq \mathrm{codewords}(j+1)$ for $j < n/2$, so that we are left to establish $f(j) \geq f(j+1)$ for all $0 \leq j < n/2$. Unfortunately, this is not strictly true for a trivial reason: in the vicinity of $n/2$ the function f is non-monotone, exhibiting minuscule fluctuations (due to finite-scale-effects) around its globally minimum value at $n/2$. While this prevents us from applying Theorem 4 immediately, it is easy to overcome. Specifically, for the proof of Theorem 4 to go through it is enough that $f(j) \geq f(j+1)$ for all $1 \leq j < z$ (instead of all $1 \leq j < n/2$), something which for most sets of interest holds, as $z \ll n/2$. Thus, to provide a rigorous upper bound on Boost, it is enough to verify the monotonicity of f up to z while evaluating $B(z)$.

9   Bounding B in Practice

In defining our systems of parity equations based on LDPC codes in the previous sections, we made sure that every variable participates in an even number of equations, we used equations whose lengths are successive integers, and we insisted on always having at least one equation of odd length. These seemingly

---

[2] This can be done by selecting a uniformly random permutation of size $[1n]$ and using it to map each of the $1n$ non-zeros to equations; when $1, r \in O(1)$, the variables in each equation will be distinct with probability $\Omega(1)$, so that a handful of trials suffice to generate a matrix as desired.

minor tricks make a very big difference in the bound of Boost in Theorem 4. Unfortunately, the number of iterations, t, needed by our (δ, ϵ)-approximation algorithm of Sect. 6 has a very large leading constant factor, in order to simplify the mathematical analysis. (This is *not* the case for our upper-bounding algorithm of Sect. 5.) For example, if the approximation factor $\delta = 1/3$ and the error probability $\theta = 1/5$, even in the ideal case where $B = 1$, i.e., the case of pairwise independence, $t = 3,709$. In reality, when $B = 1$, a dozen repetitions are more than enough to get an approximation with this δ, θ. Far worse, when $B = 2$, the number of repetitions t explodes to over 1 million, making the derivation of rigorous (δ, ϵ)-approximations via Theorem 4 unrealistic. That said, we believe that further sharpening of Theorem 4 is within grasp.

Luckily, our algorithms for deriving rigorous upper and lower bounds have much better constant-factor behavior. Moreover, as we will see experimentally, the heuristic estimate for $|S|$ that can be surmised from their (ultra-fast) execution appears to be *excellent* in practice. Below we describe a set of experiments we performed showing that one can get rigorous results in realistic times using our tools for formulas that are largely outside the reach of all known other model counters.

10   Experiments

We compare Algorithms 1, 2, i.e., our lower and upper bounding algorithms, with the deterministic, exact model counter sharpSAT [16] and the probabilistic, approximate model counter ApproxMC2 (AMC2) [4]. We consider the same 387 formulas as [4] except for 2 unsatisfiable formulas and 10 formulas whose number of solutions (and, thus, equations) is so small that our parity equations devolve into long XOR equations. Of the remaining 375 formulas, sharpSAT solves 245 in under 2 s, in every case significantly faster than all other methods. At the other extreme, 40 formulas are not solved by any method within the given time limit of 8 h. We report on the remaining 90, most interesting, formulas. All experiments were run on a modern cluster of 13 nodes, each with 16 cores and 128 GB RAM.

We use an improved implementation of CryptoMiniSat [14] tuned for hashing-based algorithms by Mate Soos and Kuldeep Meel, which is pending publication. This also allows to deal with the fact that 10 of the 90 formulas come with a *sampling set*, i.e., a subset of variables V such that the goal is to count the size of the projection of the set of all models on V. Since sharpSAT does not provide such constrained counting functionality, we do not run it on these formulas.

To provide a sense of the tradeoff between the length of the parity constraints and B, we note that when every variable appears in 6 parity constraints, then $B < 30$ for all but 3 formulas, while for all but 1 formula all equations have length at most 16. When every variable appears in 12 parity constraints, then $B < 3$ for all but 3 formulas, while for all but 6 formulas all equations have length at most 28.

Our Algorithms 1, 2 terminated within the allotted time for 87 of the 90 formulas, providing a rigorous lower bound and a rigorous upper bound. By

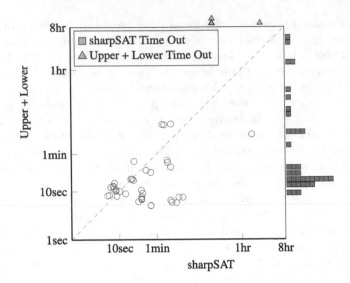

Fig. 1. The sum of the running times of the lower and upper bounding algorithms in F2 vs. the running time of sharpSAT.

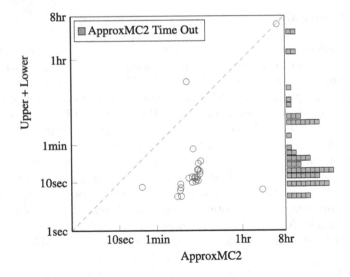

Fig. 2. The sum of the running times of the lower and upper bounding algorithms in F2 vs. the running time of ApproxMC2.

comparison, sharpSAT terminated on 45 formulas (out of $90 - 10 = 80$), while ApproxMC2 on 25 of 90.

For most formulas the ratio between our two rigorous bounds is between 8 and 16 and for none more than 64. For the 48 formulas for which the model count is known, either exactly via sharpSAT or approximately via ApproxMC2, the ratio between our upper bound and the known count was *typically less than 2 and never more than 3*. This is in spite of the fact that the time to derive it is often just a handful of *seconds* for formulas for which ApproxMC2 and/or `sharpSAT` time out given 8 h.

In Figs. 1 and 2, we plot the sum of the running time of our two algorithms, against the running time of sharpSat and ApproxMC2, respectively. (Marks outside the 8hr × 8hr box, indicate a time-out and only one of their two coordinates is meaningful.)

11 Conclusions

We have shown that by using systems off parity constraints corresponding to LDPC matrices, one can get rigorous lower bounds *and* rigorous upper bounds. While these bounds do not come with a priori guarantees about how close they will be to one another, in practice they are typically within a small multiplicative factor, e.g., 2–3. We believe that for many practical applications such bounds will be quite useful, as they are both rigorous and fast to derive. In particular, when $(\log_2 |S|)/n$ is not too small, the constraint lengths can remain bounded, for arbitrarily large n. As a result, our tool F2 can deliver rigorous results for formulas that appear outside the reach of tools based on long parity equations, such as ApproxMC2.

Acknowledgements. We are grateful to Kuldeep Meel and Moshe Vardi for sharing their code and formulas and for several valuable conversations. We thank Ben Sherman and Kostas Zampetakis for comments on earlier versions. Finally, we are grateful to the anonymous reviewers for several suggestions that improved the presentation.

References

1. Achlioptas, D., Theodoropoulos, P.: Probabilistic model counting with short XORs. In: Gaspers, S., Walsh, T. (eds.) SAT 2017. LNCS, vol. 10491, pp. 3–19. Springer, Cham (2017). https://doi.org/10.1007/978-3-319-66263-3_1
2. Chakraborty, S., Fremont, D.J., Meel, K.S., Seshia, S.A., Vardi, M.Y.: Distribution-aware sampling and weighted model counting for SAT. In: Brodley, C.E., Stone, P. (eds.) Proceedings of the Twenty-Eighth AAAI Conference on Artificial Intelligence, 27–31 July 2014, Québec City, Canada, pp. 1722–1730. AAAI Press (2014)
3. Chakraborty, S., Meel, K.S., Vardi, M.Y.: A scalable approximate model counter. In: Schulte, C. (ed.) CP 2013. LNCS, vol. 8124, pp. 200–216. Springer, Heidelberg (2013). https://doi.org/10.1007/978-3-642-40627-0_18
4. Chakraborty, S., Meel, K.S., Vardi, M.Y.: Algorithmic improvements in approximate counting for probabilistic inference: from linear to logarithmic SAT calls. In: Kambhampati, S. (ed.) Proceedings of the Twenty-Fifth International Joint Conference on Artificial Intelligence, IJCAI 2016, New York, NY, USA, 9–15 July 2016, pp. 3569–3576. IJCAI/AAAI Press (2016)
5. Ermon, S., Gomes, C.P., Sabharwal, A., Selman, B.: Taming the curse of dimensionality: discrete integration by hashing and optimization. In: Proceedings of the 30th International Conference on Machine Learning (ICML) (2013)

6. Ermon, S., Gomes, C.P., Sabharwal, A., Selman, B.: Low-density parity constraints for hashing-based discrete integration. In: Proceedings of the 31st International Conference on Machine Learning (ICML), pp. 271–279 (2014)
7. Gomes, C.P., Hoffmann, J., Sabharwal, A., Selman, B.: Short XORs for model counting: from theory to practice. In: Theory and Applications of Satisfiability Testing (SAT), pp. 100–106 (2007)
8. Gomes, C.P., Sabharwal, A., Selman, B.: Model counting: a new strategy for obtaining good bounds. In: Proceedings of the 21st National Conference on Artificial Intelligence (AAAI), pp. 54–61 (2006)
9. Ivrii, A., Malik, S., Meel, K.S., Vardi, M.Y.: On computing minimal independent support and its applications to sampling and counting. Constraints **21**(1), 41–58 (2016)
10. Maurer, A.: A bound on the deviation probability for sums of non-negative random variables. JIPAM J. Inequal. Pure Appl. Math. **4**(1), 6 (2003). Article 15
11. Moskewicz, M.W., Madigan, C.F., Zhao, Y., Zhang, L., Malik, S.: Chaff: engineering an efficient SAT solver. In: Proceedings of the 38th Annual Design Automation Conference, DAC 2001, pp. 530–535. ACM, New York (2001)
12. Richardson, T., Urbanke, R.: Modern Coding Theory. Cambridge University Press, Cambridge (2008)
13. Sipser, M.: A complexity theoretic approach to randomness. In: Proceedings of the 15th ACM Symposium on Theory of Computing (STOC), pp. 330–335 (1983)
14. Soos, M.: CryptoMiniSat - a SAT solver for cryptographic problems (2009). https://www.msoos.org/cryptominisat5/
15. Stockmeyer, L.: On approximation algorithms for #P. SIAM J. Comput. **14**(4), 849–861 (1985)
16. Thurley, M.: sharpSAT – counting models with advanced component caching and implicit BCP. In: Biere, A., Gomes, C.P. (eds.) SAT 2006. LNCS, vol. 4121, pp. 424–429. Springer, Heidelberg (2006). https://doi.org/10.1007/11814948_38
17. Valiant, L.G., Vazirani, V.V.: NP is as easy as detecting unique solutions. Theor. Comput. Sci. **47**, 85–93 (1986)
18. Zhao, S., Chaturapruek, S., Sabharwal, A., Ermon, S.: Closing the gap between short and long XORs for model counting. In: Schuurmans, D., Wellman, M.P. (eds.) Proceedings of the Thirtieth AAAI Conference on Artificial Intelligence, 12–17 February 2016, Phoenix, Arizona, USA, pp. 3322–3329. AAAI Press (2016)

Exploiting Treewidth for Projected Model Counting and Its Limits

Johannes K. Fichte[✉], Markus Hecher[✉], Michael Morak[✉],
and Stefan Woltran[✉]

Institute of Logic and Computation, TU Wien, Vienna, Austria
{fichte,hecher,morak,woltran}@dbai.tuwien.ac.at

Abstract. In this paper, we introduce a novel algorithm to solve *projected model counting* (PMC). PMC asks to count solutions of a Boolean formula with respect to a given set of *projected variables*, where multiple solutions that are identical when restricted to the projected variables count as only one solution. Our algorithm exploits small treewidth of the primal graph of the input instance. It runs in time $\mathcal{O}(2^{2^{k+4}} n^2)$ where k is the treewidth and n is the input size of the instance. In other words, we obtain that the problem PMC is fixed-parameter tractable when parameterized by treewidth. Further, we take the exponential time hypothesis (ETH) into consideration and establish lower bounds of bounded treewidth algorithms for PMC, yielding asymptotically tight runtime bounds of our algorithm.

Keywords: Parameterized algorithms · Tree decompositions
Multi-pass dynamic programming · Projected model counting
Propositional logic

1 Introduction

A problem that has been used to solve a large variety of real-world questions is the *model counting problem* (#SAT) [2,11,14,16,33,37,40,42,45]. It asks to compute the number of solutions of a Boolean formula [24] and is theoretically of high worst-case complexity (#·P-complete [38,43]). Lately, both #SAT and its approximate version have received renewed attention in theory and practice [9,16,31,39]. A concept that allows very natural abstractions of data and query results is projection. Projection has wide applications in databases [1] and declarative problem modeling. The problem *projected model counting* (PMC) asks to count solutions of a Boolean formula with respect to a given set of *projected variables*, where multiple solutions that are identical when restricted to the projected variables count as only one solution. If all variables of the formula

---

The work has been supported by the Austrian Science Fund (FWF), Grants Y698 and P26696, and the German Science Fund (DFG), Grant HO 1294/11-1. The first two authors are also affiliated with the University of Potsdam, Germany.

© Springer International Publishing AG, part of Springer Nature 2018
O. Beyersdorff and C. M. Wintersteiger (Eds.): SAT 2018, LNCS 10929, pp. 165–184, 2018.
https://doi.org/10.1007/978-3-319-94144-8_11

are projected variables, then PMC is the #SAT problem and if there are no projected variables then it is simply the SAT problem. Projected variables allow for solving problems where one needs to introduce auxiliary variables, in particular, if these variables are functionally independent of the variables of interest, in the problem encoding, e.g., [21,23].

When we consider the computational complexity of PMC it turns out that under standard assumptions the problem is even harder than #SAT, more precisely, complete for the class #· NP [17]. Even though there is a PMC solver [3] and an ASP solver that implements projected enumeration [22], PMC has received very little attention in parameterized algorithmics so far. Parameterized algorithms [12,15,20,34] tackle computationally hard problems by directly exploiting certain structural properties (parameter) of the input instance to solve the problem faster, preferably in polynomial-time for a fixed parameter value. In this paper, we consider the treewidth of graphs associated with the given input formula as parameter, namely the primal graph [41]. Roughly speaking, small *treewidth* of a graph measures its tree-likeness and sparsity. Treewidth is defined in terms of *tree decompositions (TDs)*, which are arrangements of graphs into trees. When we take advantage of small treewidth, we usually take a TD and evaluate the considered problem in parts, via *dynamic programming (DP)* on the TD.

New Contributions.

1. We introduce a novel algorithm to *solve projected model counting (PMC)* in time $\mathcal{O}(2^{2^{k+4}} n^2)$ where k is the treewidth of the primal graph of the instance and n is the size of the input instance. Similar to recent DP algorithms for problems on the second level of the polynomial hierarchy [19], our algorithm traverses the given tree decomposition multiple times (multi-pass). In the first traversal, we run a dynamic programming algorithm on tree decompositions to solve SAT [41]. In a second traversal, we construct equivalence classes on top of the previous computation to obtain model counts with respect to the projected variables by exploiting combinatorial properties of intersections.
2. We establish that our *runtime bounds are asymptotically tight under the exponential time hypothesis (ETH)* [28] using a recent result by Lampis and Mitsou [32], who established lower bounds for the problem ∃∀-SAT assuming ETH. Intuitively, ETH states a complexity theoretical lower bound on how fast satisfiability problems can be solved. More precisely, one *cannot* solve 3-SAT in time $2^{s \cdot n} \cdot n^{\mathcal{O}(1)}$ for some $s > 0$ and number n of variables.

2   Preliminaries

For a set X, let 2^X be the *power set of X* consisting of all subsets Y with $\emptyset \subseteq Y \subseteq X$. Recall the well-known combinatorial inclusion-exclusion principle [25], which states that for two finite sets A and B it is true that $|A \cup B| = |A| + |B| - |A \cap B|$. Later, we need a generalized version for arbitrary many sets. Given for some integer n a family of finite sets $X_1, X_2, \ldots, X_n$, the number of elements in the union over all sets is $|\bigcup_{j=1}^n X_j| = \sum_{I \subseteq \{1,\ldots,n\}, I \neq \emptyset} (-1)^{|I|-1} |\bigcap_{i \in I} X_i|$.

Satisfiability. A literal is a (Boolean) variable x or its negation $\neg x$. A *clause* is a finite set of literals, interpreted as the disjunction of these literals. A *(CNF) formula* is a finite set of clauses, interpreted as the conjunction of its clauses. A 3-CNF has clauses of length at most 3. Let F be a formula. A *sub-formula* S of F is a subset $S \subseteq F$ of F. For a clause $c \in F$, we let var(c) consist of all variables that occur in c and var$(F) := \bigcup_{c \in F}$ var(c). A (partial) *assignment* is a mapping $\alpha : \text{var}(F) \rightarrow \{0, 1\}$. For $x \in \text{var}(F)$, we define $\alpha(\neg x) := 1 - \alpha(x)$. The formula F *under the assignment* $\alpha \in 2^{\text{var}(F)}$ is the formula $F_{|\alpha}$ obtained from F by removing all clauses c containing a literal set to 1 by α and removing from the remaining clauses all literals set to 0 by α. An assignment α is *satisfying* if $F_{|\alpha} = \emptyset$ and F is *satisfiable* if there is a satisfying assignment α. Let V be a set of variables. An *interpretation* is a set $J \subseteq V$ and its induced assignment $\alpha_{J,V}$ of J with respect to V is defined as follows $\alpha_{J,V} := \{v \mapsto 1 \mid v \in J \cap V\} \cup \{v \mapsto 0 \mid v \in V \setminus J\}$. We simply write α_J for $\alpha_{J,V}$ if $V = \text{var}(F)$. An interpretation J is a *model* of F, denoted by $J \models F$, if its induced assignment α_J is satisfying. Given a formula F; the problem SAT asks whether F is satisfiable and the problem #SAT asks to output the number of models of F, i.e., $|S|$ where S is the set of all models of F.

Projected Model Counting. An instance of the projected model counting problem is a pair (F, P) where F is a (CNF) formula and P is a set of Boolean variables such that $P \subseteq \text{var}(F)$. We call the set P *projection variables* of the instance. The *projected model count* of a formula F with respect to P is the number of total assignments α to variables in P such that the formula $F_{|\alpha}$ under α is satisfiable. The *projected model counting problem (PMC)* [3] asks to output the projected model count of F, i.e., $|\{M \cap P \mid M \in S\}|$ where S is the set of all models of F.

Example 1. Consider formula $F := \{\overbrace{\neg a \vee b \vee p_1}^{c_1}, \overbrace{a \vee \neg b \vee \neg p_1}^{c_2}, \overbrace{a \vee p_2}^{c_3}, \overbrace{a \vee \neg p_2}^{c_4}\}$ and set $P := \{p_1, p_2\}$ of projection variables. The models of formula F are $\{a, b\}$, $\{a, p_1\}$, $\{a, b, p_1\}$, $\{a, b, p_2\}$, $\{a, p_1, p_2\}$, and $\{a, b, p_1, p_2\}$. However, projected to the set P, we only have models $\emptyset$, $\{p_1\}$, $\{p_2\}$, and $\{p_1, p_2\}$. Hence, the model count of F is 6 whereas the projected model count of instance (F, P) is 4. ∎

Computational Complexity. We assume familiarity with standard notions in computational complexity [35] and use counting complexity classes as defined by Hemaspaandra and Vollmer [27]. For parameterized complexity, we refer to standard texts [12,15,20,34]. Let Σ and Σ' be some finite alphabets. We call $I \in \Sigma^*$ an *instance* and $\|I\|$ denotes the size of I. Let $L \subseteq \Sigma^* \times \mathbb{N}$ and $L' \subseteq \Sigma'^* \times \mathbb{N}$ be two parameterized problems. An *fpt-reduction* r from L to L' is a many-to-one reduction from $\Sigma^* \times \mathbb{N}$ to $\Sigma'^* \times \mathbb{N}$ such that for all $I \in \Sigma^*$ we have $(I, k) \in L$ if and only if $r(I, k) = (I', k') \in L'$ such that $k' \leq g(k)$ for a fixed computable function $g : \mathbb{N} \rightarrow \mathbb{N}$, and there is a computable function f and a constant c such that r is computable in time $O(f(k)\|I\|^c)$ [20]. A *witness function* is a function $\mathcal{W}: \Sigma^* \rightarrow 2^{\Sigma'^*}$ that maps an instance $I \in \Sigma^*$ to a finite subset of Σ'^*. We call the set $\mathcal{W}(I)$ the *witnesses*. A *parameterized counting problem* $L : \Sigma^* \times \mathbb{N}_0 \rightarrow \mathbb{N}_0$ is a function that maps a given instance $I \in \Sigma^*$ and an

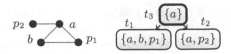

Fig. 1. Primal graph P_F of F from Example 2 (left) with a TD $\mathcal{T}$ of graph P_F (right).

integer $k \in \mathbb{N}$ to the cardinality of its witnesses $|\mathcal{W}(I)|$. We call k the *parameter*. The *exponential time hypothesis* (ETH) states that the (decision) problem SAT on 3-CNF formulas *cannot* be solved in time $2^{s \cdot n} \cdot n^{\mathcal{O}(1)}$ for some $s > 0$ where n is the number of variables [28].

Tree Decompositions and Treewidth. For basic terminology on graphs, we refer to standard texts [8,13]. For a tree $T = (N, A, n)$ with root n and a node $t \in N$, we let children(t, T) be the sequence of all nodes t' in arbitrarily but fixed order, which have an edge $(t, t') \in A$. Let $G = (V, E)$ be a graph. A *tree decomposition (TD)* of graph G is a pair $\mathcal{T} = (T, \chi)$ where $T = (N, A, n)$ is a rooted tree, $n \in N$ the root, and χ a mapping that assigns to each node $t \in N$ a set $\chi(t) \subseteq V$, called a *bag*, such that the following conditions hold: (i) $V = \bigcup_{t \in N} \chi(t)$ and $E \subseteq \bigcup_{t \in N} \{uv \mid u, v \in \chi(t)\}$; (ii) for each $r, s, t \in N$ such that s lies on the path from r to t, we have $\chi(r) \cap \chi(t) \subseteq \chi(s)$. Then, width$(\mathcal{T}) := \max_{t \in N} |\chi(t)| - 1$. The *treewidth* $tw(G)$ of G is the minimum width$(\mathcal{T})$ over all tree decompositions $\mathcal{T}$ of G. For arbitrary but fixed $w \geq 1$, it is feasible in linear time to decide if a graph has treewidth at most w and, if so, to compute a tree decomposition of width w [5]. In order to simplify case distinctions in the algorithms, we always use so-called nice tree decompositions, which can be computed in linear time without increasing the width [7] and are defined as follows. For a node $t \in N$, we say that type(t) is *leaf* if children$(t, T) = \langle \rangle$; *join* if children$(t, T) = \langle t', t'' \rangle$ where $\chi(t) = \chi(t') = \chi(t'') \neq \emptyset$; *int* ("introduce") if children$(t, T) = \langle t' \rangle$, $\chi(t') \subseteq \chi(t)$ and $|\chi(t)| = |\chi(t')| + 1$; *rem* ("removal") if children$(t, T) = \langle t' \rangle$, $\chi(t') \supseteq \chi(t)$ and $|\chi(t')| = |\chi(t)| + 1$. If for every node $t \in N$, type$(t) \in \{leaf, join, int, rem\}$ and bags of leaf nodes and the root are empty, then the TD is called *nice*.

3   Dynamic Programming on TDs for SAT

Before we introduce our algorithm, we need some notations for dynamic programming on tree decompositions and recall how to solve the decision problem SAT by exploiting small treewidth.

Graph Representation of SAT Formulas. In order to use tree decompositions for satisfiability problems, we need a dedicated graph representation of the given formula F. The *primal graph* P_F of F has as vertices the variables of F and two variables are joined by an edge if they occur together in a clause of F. Further, we define some auxiliary notation. For a given node t of a tree decomposition (T, χ) of the primal graph, we let $F_t := \{c \mid c \in F, \text{var}(c) \subseteq \chi(t)\}$, i.e., clauses entirely covered by $\chi(t)$. The set $F_{\leq t}$ denotes the union over F_s for all descendant nodes $s \in N$ of t. In the following, we sometimes simply write *tree* *decomposition of formula F* or *treewidth of F* and omit the actual graph representation of F.

Example 2. Consider formula F from Example 1. The primal graph P_F of formula F and a tree decomposition T of P_F are depicted in Fig. 1. Intuitively, T allows to evaluate formula F in parts. When evaluating $F_{\leq t_3}$, we split into $F_{\leq t_1} = \{c_1, c_2\}$ and $F_{\leq t_2} = \{c_3, c_4\}$, respectively. ∎

Dynamic Programming on TDs. Algorithms that solve SAT or #SAT [41] in linear time for input formulas of bounded treewidth proceed by dynamic programming along the tree decomposition (in post-order) where at each node t of the tree information is gathered [6] in a table τ_t. A *table* τ is a set of rows, where a *row* $\boldsymbol{u} \in \tau$ is a sequence of fixed length. Tables are derived by an algorithm, which we therefore call *table algorithm* $\mathbb{A}$. The actual length, content, and meaning of the rows depend on the algorithm $\mathbb{A}$ that derives tables. Therefore, we often explicitly state $\mathbb{A}$-*row* if rows of this *type* are syntactically used for table algorithm $\mathbb{A}$ and similar $\mathbb{A}$-*table* for tables. For sake of comprehension, we specify the rows before presenting the actual table algorithm for manipulating tables. The rows used by a table algorithm SAT have in common that the first position of these rows manipulated by SAT consists of an interpretation. The remaining positions of the row depend on the considered table algorithm. For each sequence $\boldsymbol{u} \in \tau$, we write $I(\boldsymbol{u})$ to address the interpretation (first) part of the sequence $\boldsymbol{u}$. Further, for a given positive integer i, we denote by $\boldsymbol{u}_{(i)}$ the i-th element of row $\boldsymbol{u}$ and define $\tau_{(i)}$ as $\tau_{(i)} := \{\boldsymbol{u}_{(i)} \mid \boldsymbol{u} \in \tau\}$.

Then, the dynamic programming approach for propositional satisfiability performs the following steps:

1. Construct the primal graph P_F of F.
2. Compute a tree decomposition (T, χ) of P_F, where $T = (N, \cdot, n)$.
3. Run $\mathrm{DP}_{\mathrm{SAT}}$ (see Listing 1), which executes a table algorithm SAT for every node t in post-order of the nodes in N, and returns SAT-Comp mapping every node t to its table. SAT takes as input[1] bag $\chi(t)$, sub-formula F_t, and tables Child-Tabs previously computed at children of t and outputs a table τ_t.
4. Print the result by interpreting the table for root n of T.

Listing 2 presents table algorithm SAT that uses the primal graph representation. We provide only brief intuition, for details we refer to the original source [41]. The main idea is to store in table τ_t only interpretations that are a model of sub-formula $F_{\leq t}$ when restricted to bag $\chi(t)$. Table algorithm SAT transforms at node t certain row combinations of the tables (Child-Tabs) of child nodes of t into rows of table τ_t. The transformation depends on a case where variable a is added or not added to an interpretation (*int*), removed from an interpretation

---

[1] Actually, SAT takes in addition as input *PP*-Tabs, which contains a mapping of nodes of the tree decomposition to tables, i.e., tables of the previous pass. Later, we use this for a second traversal to pass results (SAT-Comp) from the first traversal to the table algorithm PROJ for projected model counting in the second traversal.

Listing 1. Algorithm $\text{DP}_{\mathbb{A}}((F,P),\mathcal{T},PP\text{-}Tabs)$ for DP on TD $\mathcal{T}$ [18].

In: Table algorithm $\mathbb{A}$, TD $\mathcal{T} = (T,\chi)$ of F s.t. $T = (N,\cdot,n)$, tables PP-Tabs.
Out: Table $\mathbb{A}$-Comp, which maps each TD node $t \in N$ to some computed
    table τ_t.

1 **for** iterate t *in* post-order(T,n) **do**
2   Child-Tabs $:= \langle \mathbb{A}\text{-Comp}[t_1],\ldots,\mathbb{A}\text{-Comp}[t_\ell]\rangle$ where children$(t,T) = \langle t_1,\ldots,t_\ell\rangle$
3     $\mathbb{A}\text{-Comp}[t] \leftarrow \mathbb{A}(t,\chi(t),F_t,P \cap \chi(t),\text{Child-Tabs},PP\text{-}Tabs)$
4 **return** $\mathbb{A}\text{-}Comp$

Listing 2. Table algorithm $\mathbb{SAT}(t,\chi_t,F_t,\cdot,\text{Child-Tabs},\cdot)$ [41].

In: Node t, bag χ_t, clauses F_t, sequence Child-Tabs of tables. **Out:** Table τ_t.

1 **if** type$(t) = $ *leaf* **then** $\tau_t \leftarrow \{\langle\emptyset\rangle\}$
2 **else if** type$(t) = $ *int*, $a \in \chi_t$ is introduced, and Child-Tabs $= \langle\tau'\rangle$ **then**
3     $\tau_t \leftarrow \{\langle K\rangle$ $\mid \langle J\rangle \in \tau', K \in \{J, J \cup \{a\}\}, K \vDash F_t\}$
4 **else if** type$(t) = $ *rem*, $a \notin \chi_t$ is removed, and Child-Tabs $= \langle\tau'\rangle$ **then**
5     $\tau_t \leftarrow \{\langle J \setminus \{a\}\rangle$ $\mid \langle J\rangle \in \tau'\}$
6 **else if** type$(t) = $ *join*, and Child-Tabs $= \langle\tau',\tau''\rangle$ **then**
7     $\tau_t \leftarrow \{\langle J\rangle$ $\mid \langle J\rangle \in \tau', \langle J\rangle \in \tau''\}$
8 **return** τ_t

(*rem*), or where coinciding interpretations are required (*join*). In the end, an interpretation $I(\boldsymbol{u})$ from a row $\boldsymbol{u}$ of the table τ_n at the root n proves that there is a supset $J \supseteq I(\boldsymbol{u})$ that is a model of $F = F_{\leq n}$, and hence that the formula is satisfiable. Example 3 lists selected tables when running algorithm DP_{SAT}.

Example 3. Consider formula F from Example 2. Figure 2 illustrates a tree decomposition $\mathcal{T}' = (\cdot,\chi)$ of the primal graph of F and tables $\tau_1, \ldots, \tau_{12}$ that are obtained during the execution of $\text{DP}_{\text{SAT}}((F,\cdot),\mathcal{T}',\cdot)$. We assume that each row in a table τ_t is identified by a number, i.e., row i corresponds to $\boldsymbol{u}_{t.i} = \langle J_{t.i}\rangle$.

Table $\tau_1 = \{\langle\emptyset\rangle\}$ as type$(t_1) = $ *leaf*. Since type$(t_2) = $ *int*, we construct table τ_2 from τ_1 by taking $J_{1.i}$ and $J_{1.i} \cup \{a\}$ for each $\langle J_{1.i}\rangle \in \tau_1$. Then, t_3 introduces p_1 and t_4 introduces b. $F_{t_1} = F_{t_2} = F_{t_3} = \emptyset$, but since $\chi(t_4) \subseteq \text{var}(c_1)$ we have $F_{t_4} = \{c_1,c_2\}$ for t_4. In consequence, for each $J_{4.i}$ of table τ_4, we have $\{c_1,c_2\} \vDash J_{4.i}$ since $\mathbb{SAT}$ enforces satisfiability of F_t in node t. Since type$(t_5) = $ *rem*, we remove variable p_1 from all elements in τ_4 to construct τ_5. Note that we have already seen all rules where p_1 occurs and hence p_1 can no longer affect interpretations during the remaining traversal. We similarly create $\tau_6 = \{\langle\emptyset\rangle,\langle a\rangle\}$ and $\tau_{10} = \{\langle a\rangle\}$. Since type$(t_{11}) = $ *join*, we build table τ_{11} by taking the intersection of τ_6 and τ_{10}. Intuitively, this combines interpretations agreeing on a. By definition (primal graph and TDs), for every $c \in F$, variables var(c) occur together in at least one common bag. Hence, $F = F_{\leq t_{12}}$ and since $\tau_{12} = \{\langle\emptyset\rangle\}$, we can reconstruct for example model $\{a,b,p_2\} = J_{11.1} \cup J_{5.4} \cup J_{9.2}$ of F using highlighted (yellow) rows in Fig. 2. On the other hand, if F was unsatisfiable, τ_{12} would be empty ($\emptyset$). ∎

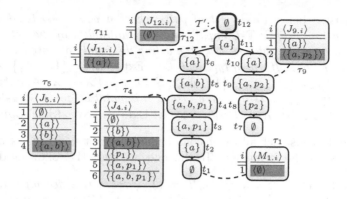

Fig. 2. Selected tables obtained by algorithm DP$_{\mathrm{PRIM}}$ on tree decomposition $\mathcal{T}'$.

The following definition simplifies the presentation. At a node t and for a row $\boldsymbol{u}$ of the table $\mathbb{SAT}$-Comp$[t]$, it yields the rows in the tables of the children of t that were involved in computing row $\boldsymbol{u}$ by algorithm $\mathbb{SAT}$.

Definition 1 (c.f., [19]). *Let F be a formula, $\mathcal{T} = (T, \chi)$ be a tree decomposition of F, t be a node of T that has ℓ children, and $\tau_1, \ldots, \tau_\ell$ be the $\mathbb{SAT}$-tables computed by* DP$_{\mathbb{SAT}}((F, \cdot), \mathcal{T}, \cdot)$ *where* children$(t, T) = \langle t_1, \ldots, t_\ell \rangle$. *Given a sequence $\boldsymbol{s} = \langle s_1, \ldots, s_\ell \rangle$, we let $\langle\!\langle \boldsymbol{s} \rangle\!\rangle := \langle \{s_1\}, \ldots, \{s_\ell\} \rangle$, for technical reasons.*

For a given $\mathbb{SAT}$-row $\boldsymbol{u}$, we define the originating $\mathbb{SAT}$-rows of $\boldsymbol{u}$ in node t by
$\mathbb{SAT}$-origins$(t, \boldsymbol{u}) := \{ \boldsymbol{s} \mid \boldsymbol{s} \in \tau_1 \times \cdots \times \tau_\ell, \tau = \mathbb{SAT}(t, \chi(t), F_t, \cdot, \langle\!\langle \boldsymbol{s} \rangle\!\rangle, \cdot), \boldsymbol{u} \in \tau \}$.
We extend this to a $\mathbb{SAT}$-table σ by $\mathbb{SAT}$-origins$(t, \sigma) := \bigcup_{\boldsymbol{u} \in \sigma} \mathbb{SAT}$-origins$(t, \boldsymbol{u})$.

Remark 1. An actual implementation would not compute origins, but store and reuse them without side-effects to worst-case complexity during tree traversal.

Example 4. Consider formula F, tree decomposition $\mathcal{T}' = (T, \chi)$, and tables $\tau_1, \ldots, \tau_{12}$ from Example 3. We focus on $\boldsymbol{u}_{1.1} = \langle J_{1.1} \rangle = \langle \emptyset \rangle$ of table τ_1 of the leaf t_1. The row $\boldsymbol{u}_{1.1}$ has no preceding row, since type$(t_1) = leaf$. Hence, we have $\mathbb{SAT}$-origins$(t_1, \boldsymbol{u}_{1.1}) = \{\langle\rangle\}$. The origins of row $\boldsymbol{u}_{5.1}$ of table τ_5 are given by $\mathbb{SAT}$-origins$(t_5, \boldsymbol{u}_{5.1})$, which correspond to the preceding rows in table t_4 that lead to row $\boldsymbol{u}_{5.1}$ of table τ_5 when running algorithm $\mathbb{SAT}$, i.e., $\mathbb{SAT}$-origins$(t_5, \boldsymbol{u}_{5.1}) = \{\langle \boldsymbol{u}_{4.1} \rangle, \langle \boldsymbol{u}_{4.4} \rangle\}$. Observe that $\mathbb{SAT}$-origins$(t_i, \boldsymbol{u}) = \emptyset$ for any row $\boldsymbol{u} \notin \tau_i$. For node t_{11} of type *join* and row $\boldsymbol{u}_{11.1}$, we obtain $\mathbb{SAT}$-origins$(t_{11}, \boldsymbol{u}_{11.1}) = \{\langle \boldsymbol{u}_{6.2}, \boldsymbol{u}_{10.1} \rangle\}$ (see Example 3). More general, when using algorithm $\mathbb{SAT}$, at a node t of type *join* with table τ we have $\mathbb{SAT}$-origins$(t, \boldsymbol{u}) = \{\langle \boldsymbol{u}, \boldsymbol{u} \rangle\}$ for row $\boldsymbol{u} \in \tau$. ∎

Definition 1 talked about a top-down direction for rows and their origins. In addition, we need definitions to talk about a recursive version of these origins from a node t down to the leafs, mainly to state properties for our algorithms.

Definition 2. *Let F be a formula, $\mathcal{T} = (T, \chi)$ be a tree decomposition with $T = (N, \cdot, n)$, $t \in N$, $\mathbb{SAT}$-Comp$[t']$ be obtained by* DP$_{\mathbb{SAT}}((F, \cdot), \mathcal{T}, \cdot)$ *for each node t' of the* induced *sub-tree $T[t]$ rooted at t, and $\boldsymbol{u}$ be a row of $\mathbb{SAT}$-Comp$[t]$.*

An extension below t *is a set of pairs where a pair consists of a node* t' *of* $T[t]$ *and a row* $\boldsymbol{v}$ *of* $\mathbb{SAT}$-*Comp*$[t']$ *and the cardinality of the set equals the number of nodes in the sub-tree* $T[t]$. *We define the family of* extensions below t *recursively as follows. If* t *is of type leaf, then* $\text{Ext}_{\leq t}(\boldsymbol{u}) := \{\{\langle t, \boldsymbol{u}\rangle\}\}$; *otherwise* $\text{Ext}_{\leq t}(\boldsymbol{u}) := \bigcup_{v \in \mathbb{SAT}\text{-origins}(t, \boldsymbol{u})} \{\{\langle t, \boldsymbol{u}\rangle\} \cup X_1 \cup \ldots \cup X_\ell \mid X_i \in \text{Ext}_{\leq t_i}(\boldsymbol{v}_{(i)})\}$ *for the* ℓ *children* $t_1, \ldots, t_\ell$ *of* t. *We extend this notation for a* $\mathbb{SAT}$-*table* σ *by* $\text{Ext}_{\leq t}(\sigma) := \bigcup_{\boldsymbol{u} \in \sigma} \text{Ext}_{\leq t}(\boldsymbol{u})$. *Further, we let* $\text{Exts} := \text{Ext}_{\leq n}(\mathbb{SAT}\text{-}Comp[n])$.

If we would construct all extensions below the root n, it allows us to also obtain all models of a formula F. To this end, we state the following definition.

Definition 3. *Let* F *be a formula,* $\mathcal{T} = (T, \chi)$ *be a tree decomposition of* F, t *be a node of* T, *and* $\sigma \subseteq \mathbb{SAT}$-*Comp*$[t]$ *be a set of* $\mathbb{SAT}$-*rows that have been computed by* $\text{DP}_{\mathbb{SAT}}((F, \cdot), \mathcal{T}, \cdot)$ *at* t. *We define the* satisfiable extensions below t *for* σ *by* $\text{SatExt}_{\leq t}(\sigma) := \bigcup_{\boldsymbol{u} \in \sigma}\{X \mid X \in \text{Ext}_{\leq t}(\boldsymbol{u}), X \subseteq Y, Y \in \text{Exts}\}$.

Observation 1. *Let* F *be a formula,* $\mathcal{T}$ *be a tree decomposition with root* n *of* F. *Then,* $\text{SatExt}_{\leq n}(\mathbb{SAT}\text{-}Comp[t]) = \text{Exts}$.

Next, we define an auxiliary notation that gives us a way to reconstruct interpretations from families of extensions.

Definition 4. *Let* (F, P) *be an instance of* PMC, $\mathcal{T} = (T, \chi)$ *be a tree decomposition of* F, t *be a node of* T. *Further, let* E *be a family of extensions below* t, *and* P *be a set of projection variables. We define the set* $I(E)$ *of* interpretations *of* E *by* $I(E) := \{\bigcup_{\langle \cdot, \boldsymbol{u}\rangle \in X} I(\boldsymbol{u}) \mid X \in E\}$ *and the set* $I_P(E)$ *of* projected interpretations *by* $I_P(E) := \{\bigcup_{\langle \cdot, \boldsymbol{u}\rangle \in X} I(\boldsymbol{u}) \cap P \mid X \in E\}$.

Example 5. Consider again formula F and tree decomposition $\mathcal{T}'$ with root n of F from Example 3. Let $X = \{\langle t_{12}, \langle\emptyset\rangle\rangle, \langle t_{11}, \langle\{a\}\rangle\rangle, \langle t_6, \langle\{a\}\rangle\rangle, \langle t_5, \langle\{a, b\}\rangle\rangle, \langle t_4, \langle\{a, b\}\rangle\rangle, \langle t_3, \langle\{a\}\rangle\rangle, \langle t_2, \langle\{a\}\rangle\rangle, \langle t_1, \langle\emptyset\rangle\rangle, \langle t_{10}, \langle\{a\}\rangle\rangle, \langle t_9, \langle\{a, p_2\}\rangle\rangle, \langle t_8, \langle\{p_2\}\rangle\rangle, \langle t_7, \langle\emptyset\rangle\rangle\}$ be an extension below n. Observe that $X \in \text{Exts}$ and that Fig. 2 highlights those rows of tables for nodes $t_{12}, t_{11}, t_9, t_5, t_4$ and t_1 that also occur in X (in yellow). Further, $I(\{X\}) = \{a, b, p_2\}$ computes the corresponding model of X, and $I_P(\{X\}) = \{p_2\}$ derives the projected model of X. $I(\text{Exts})$ refers to the set of models of F, whereas $I_P(\text{Exts})$ is the set of projected models of F. ∎

4   Counting Projected Models by Dynamic Programming

In this section, we introduce the dynamic programming algorithm $\text{PCNT}_{\mathbb{SAT}}$ to solve the projected model counting problem (PMC) for Boolean formulas. Our algorithm traverses the tree decomposition twice following a multi-pass dynamic programming paradigm [19]. Similar to the previous section, we construct a graph representation and heuristically compute a tree decomposition of this graph. Then, we run $\text{DP}_{\mathbb{SAT}}$ (see Listing 1) in Step 3a as first traversal. Step 3a can also be seen as a preprocessing step for projected model counting, from which we immediately know whether the problem has a solution. Afterwards we remove all rows from the $\mathbb{S}$AT-tables which cannot be extended to a solution for the SAT problem (*"Purge non-solutions"*). In other words, we keep only rows $\boldsymbol{u}$ in table $\mathbb{S}$AT-Comp[t] at node t if its interpretation $I(\boldsymbol{u})$ can be extended to a model of F, more formally, $(t, \boldsymbol{u}) \in X$ for some $X \in \mathrm{SatExt}_{\leq t}(\mathbb{S}\mathrm{AT\text{-}Comp}[t])$. Thereby, we avoid redundancies and can simplify the description of our next step, since we then only have to consider (parts of) models. In Step 3b (DP$_{\mathbb{PROJ}}$), we traverse the tree decomposition a second time to count projections of interpretations of rows in $\mathbb{S}$AT-tables. In the following, we only describe the table algorithm $\mathbb{PROJ}$, since the traversal in DP$_{\mathbb{PROJ}}$ is the same as before. For $\mathbb{PROJ}$, a row at a node t is a pair $\langle \sigma, c \rangle$ where σ is a $\mathbb{S}$AT-table, in particular, a subset of $\mathbb{S}$AT-Comp[t] computed by DP$_{\mathbb{SAT}}$, and c is a non-negative integer. In fact, we store in integer c a count that expresses the number of "all-overlapping" solutions (ipmc), whereas in the end we aim for the projected model count (pmc), clarified in the following.

Definition 5. *Let F be a formula, $\mathcal{T} = (T, \chi)$ be a tree decomposition of F, t be a node of T, $\sigma \subseteq \mathbb{S}\mathrm{AT\text{-}Comp}[t]$ be a set of $\mathbb{S}$AT-rows that have been computed by* DP$_{\mathbb{SAT}}((F, \cdot), \mathcal{T}, \cdot)$ *at node t in T. Then, the* projected model count pmc$_{\leq t}(\sigma)$ *of σ below t is the size of the union over projected interpretations of the satisfiable extensions of σ below t, formally,* pmc$_{\leq t}(\sigma) := |\bigcup_{\boldsymbol{u} \in \sigma} I_P(\mathrm{SatExt}_{\leq t}(\{\boldsymbol{u}\}))|$.

The intersection projected model count ipmc$_{\leq t}(\sigma)$ *of σ below t is the size of the intersection over projected interpretations of the satisfiable extensions of σ below t, i.e.,* ipmc$_{\leq t}(\sigma) := |\bigcap_{\boldsymbol{u} \in \sigma} I_P(\mathrm{SatExt}_{\leq t}(\{\boldsymbol{u}\}))|$.

The next definitions provide central notions for grouping rows of tables according to the given projection of variables.

Definition 6. *Let (F, P) be an instance of* PMC *and σ be a $\mathbb{S}$AT-table. We define the relation $=_{\mathrm{P}} \subseteq \sigma \times \sigma$ to consider equivalent rows with respect to the projection of its interpretations by $=_{\mathrm{P}} := \{(\boldsymbol{u}, \boldsymbol{v}) \mid \boldsymbol{u}, \boldsymbol{v} \in \sigma, I(\boldsymbol{u}) \cap P = I(\boldsymbol{v}) \cap P\}$.*

Observation 2. *The relation $=_{\mathrm{P}}$ is an equivalence relation.*

Definition 7. *Let τ be a $\mathbb{S}$AT-table and $\boldsymbol{u}$ be a row of τ. The relation $=_{\mathrm{P}}$ induces equivalence classes $[\boldsymbol{u}]_P$ on the $\mathbb{S}$AT-table τ in the usual way, i.e., $[\boldsymbol{u}]_P = \{\boldsymbol{v} \mid \boldsymbol{v} =_{\mathrm{P}} \boldsymbol{u}, \boldsymbol{v} \in \tau\}$ [44]. We denote by* buckets$_P(\tau)$ *the set of equivalence classes of τ, i.e.,* buckets$_P(\tau) := (\tau / =_{\mathrm{P}}) = \{[\boldsymbol{u}]_P \mid \boldsymbol{u} \in \tau\}$. *Further, we define the set* sub-buckets$_P(\tau)$ *of all sub-equivalence classes of τ by* sub-buckets$_P(\tau) := \{S \mid \emptyset \subsetneq S \subseteq B, B \in \mathrm{buckets}_P(\tau)\}$.

Example 6. Consider again formula F and set P of projection variables from Example 1 and tree decomposition $\mathcal{T}' = (T, \chi)$ and $\mathbb{S}$AT-table τ_4 from Fig. 2. We have $\boldsymbol{u}_{4.1} =_P \boldsymbol{u}_{4.2}$ and $\boldsymbol{u}_{4.4} =_P \boldsymbol{u}_{4.5}$. We obtain the set $\tau_4 / =_P$ of equivalence classes of τ_4 by buckets$_P(\sigma) = \{\{\boldsymbol{u}_{4.1}, \boldsymbol{u}_{4.2}, \boldsymbol{u}_{4.3}\}, \{\boldsymbol{u}_{4.4}, \boldsymbol{u}_{4.5}, \boldsymbol{u}_{4.6}\}\}$. ∎

Since $\mathbb{PROJ}$ stores a counter in $\mathbb{PROJ}$-tables together with a $\mathbb{S}$AT-table, we need an auxiliary definition that given $\mathbb{S}$AT-table allows us to select the respective counts from a $\mathbb{PROJ}$-table. Later, we use the definition in the context of looking up the already computed projected counts for tables of *children* of a given node.

Definition 8. *Given a $\mathbb{PROJ}$-table ι and a $\mathbb{SAT}$-table σ we define the stored ipmc for all rows of σ in ι by* s-ipmc$(\iota,\sigma) := \sum_{\langle \sigma,c\rangle \in \iota} c$. *Later, we apply this to rows from several origins. Therefore, for a sequence s of $\mathbb{PROJ}$-tables of length ℓ and a set O of sequences of $\mathbb{SAT}$-rows where each sequence is of length ℓ, we let* s-ipmc$(s, O) = \prod_{i \in \{1,\dots,\ell\}}$ s-ipmc$(s_{(i)}, O_{(i)})$.

When computing s-ipmc in Definition 8, we select the i-th position of the sequence together with sets of the i-th position from the set of sequences. We need this somewhat technical construction, since later at node t we apply this definition to $\mathbb{PROJ}$-tables of children of t and origins of subsets of $\mathbb{SAT}$-tables. There, we may simply have several children if the node is of type *join* and hence we need to select from the right children.

Now, we are in position to give a core definition for our algorithm that solves PMC. Intuitively, when we are at a node t in the Algorithm $\mathrm{DP}_{\mathbb{PROJ}}$ we already computed all tables $\mathbb{SAT}$-Comp by $\mathrm{DP}_{\mathbb{SAT}}$ according to Step 3a, purged non-solutions, and computed $\mathbb{PROJ}$-Comp$[t']$ for all nodes t' below t and in particular the $\mathbb{PROJ}$-tables Child-Tabs of the children of t. Then, we compute the projected model count of a subset σ of the $\mathbb{SAT}$-rows in $\mathbb{SAT}$-Comp$[t]$, which we formalize in the following definition, by applying the generalized inclusion-exclusion principle to the stored projected model count of origins.

Definition 9. *Let (F, P) be an instance of PMC, $\mathcal{T} = (T, \chi)$ be a tree decomposition of F, $\mathbb{SAT}$-Comp$[s]$ be the $\mathbb{SAT}$-tables computed by $\mathrm{DP}_{\mathbb{SAT}}((F, \cdot), \mathcal{T}, \cdot)$ for every node s of T. Further, let t be a node of T with ℓ children, Child-Tabs $= \langle \mathbb{PROJ}$-Comp$[t_1], \dots, \mathbb{PROJ}$-Comp$[t_\ell]\rangle$ be the sequence of $\mathbb{PROJ}$-tables computed by $\mathrm{DP}_{\mathbb{PROJ}}((F, P), \mathcal{T}, \mathbb{SAT}$-Comp$)$ where children$(t, T) = \langle t_1, \dots, t_\ell \rangle$, and $\sigma \subseteq \mathbb{SAT}$-Comp$[t]$ be a table. We define the* (inductive) projected model count of σ:

$$\mathrm{pmc}(t, \sigma, \textit{Child-Tabs}) := \sum_{\emptyset \subsetneq O \subseteq \mathbb{SAT}\text{-origins}(t,\sigma)} (-1)^{(|O|-1)} \cdot \text{s-ipmc}(\textit{Child-Tabs}, O).$$

Vaguely speaking, pmc determines the $\mathbb{SAT}$-origins of the set σ of rows, goes over all subsets of these origins and looks up the stored counts (s-ipmc) in the $\mathbb{PROJ}$-tables of the children of t. Example 7 provides an idea on how to compute the projected model count of tables of our running example using pmc.

Example 7. The function defined in Definition 9 allows us to compute the projected count for a given $\mathbb{SAT}$-table. Therefore, consider again formula F and tree decomposition $\mathcal{T}'$ from Example 2 and Fig. 2. Say we want to compute the projected count pmc$(t_5, \{u_{5.4}\}, \textit{Child-Tabs})$ where Child-Tabs $:= \{\langle\{u_{4.3}\}, 1\rangle, \langle\{u_{4.6}\}, 1\rangle\}$ for row $u_{5.4}$ of table τ_5. Note that t_5 has $\ell = 1$ child nodes $\langle t_4 \rangle$ and therefore the product of Definition 8 consists of only one factor. Observe that $\mathbb{SAT}$-origins$(t_5, u_{5.4}) = \{\langle u_{4.3}\rangle, \langle u_{4.6}\rangle\}$. Since the rows $u_{4.3}$ and $u_{4.6}$ do not

---

Listing 3. Table algorithm $\mathbb{PROJ}(t, \cdot, \cdot, P, \text{Child-Tabs}, \mathbb{SAT}\text{-Comp})$.

In: Node t, set P of projection variables, Child-Tabs, and $\mathbb{SAT}$-Comp.
Out: Table ι_t consisting of pairs $\langle \sigma, c \rangle$, where $\sigma \subseteq \mathbb{SAT}\text{-Comp}[t]$ and $c \in \mathbb{N}$.
1 $\iota_t \leftarrow \left\{ \langle \sigma, \text{ipmc}(t, \sigma, \text{Child-Tabs}) \rangle \mid \sigma \in \text{sub-buckets}_P(\mathbb{SAT}\text{-Comp}[t]) \right\}$
2 **return** ι_t

---

occur in the same $\mathbb{SAT}$-table of Child-Tabs, only the value of s-ipmc for the two singleton origin sets $\{\langle u_{4.3} \rangle\}$ and $\{\langle u_{4.6} \rangle\}$ is non-zero; for the remaining set of origins we have zero. Hence, we obtain $\text{pmc}(t_5, \{u_{5.4}\}, \text{Child-Tabs}) = 2$.   ∎

Before we present algorithm $\mathbb{PROJ}$ (Listing 3), we give a definition that allows us at a certain node t to compute the intersection pmc for a given $\mathbb{SAT}$-table σ by computing the pmc (using stored ipmc values from $\mathbb{PROJ}$-tables for children of t), and subtracting and adding ipmc values for subsets $\emptyset \subsetneq \rho \subsetneq \sigma$ accordingly.

Definition 10. *Let $\mathcal{T} = (T, \cdot)$ be a tree decomposition, t be a node of T, ρ be $\mathbb{SAT}$-table, and Child-Tabs be a sequence of tables. Then, we define the (recursive) ipmc of σ as follows:*

$$\text{ipmc}(t, \sigma, \text{Child-Tabs}) := \begin{cases} 1, & \text{if } \text{type}(t) = \text{leaf}, \\ \Big| \text{pmc}(t, \sigma, \text{Child-Tabs}) + \\ \sum_{\emptyset \subsetneq \rho \subsetneq \sigma} (-1)^{|\rho|} \cdot \text{ipmc}(t, \rho, \text{Child-Tabs}) \Big|, & \text{otherwise.} \end{cases}$$

In other words, if a node is of type *leaf* the ipmc is one, since by definition of a tree decomposition the bags of nodes of type *leaf* contain only one projected interpretation (the empty set). Otherwise, using Definition 9, we are able to compute the ipmc for a given $\mathbb{SAT}$-table σ, which is by construction the same as $\text{ipmc}_{\leq t}(\sigma)$ (c.f. proof of Theorem 3 later). In more detail, we want to compute for a $\mathbb{SAT}$-table σ its ipmc that represents "all-overlapping" counts of σ with respect to set P of projection variables, that is, $\text{ipmc}_{\leq t}(\sigma)$. Therefore, for ipmc, we rearrange the inclusion-exclusion principle. To this end, we take pmc, which computes the "non-overlapping" count of σ with respect to P, by once more exploiting the inclusion-exclusion principle on $\mathbb{SAT}$-origins of σ (as already discussed) such that we count every projected model only once. Then we have to alternately subtract and add ipmc values for strict subsets ρ of σ, accordingly.

Finally, Listing 3 presents table algorithm $\mathbb{PROJ}$, which stores for given node t a $\mathbb{PROJ}$-table consisting of every sub-bucket of the given table $\mathbb{SAT}$-Comp$[t]$ together with its ipmc (as presented above).

Example 8. Recall instance (F, P), tree decomposition $\mathcal{T}'$, and tables $\tau_1, \ldots,$ τ_{12} from Examples 1, 2, and Fig. 2. Figure 3 depicts selected tables of $\iota_1, \ldots \iota_{12}$ obtained after running DP$_{\mathbb{PROJ}}$ for counting projected interpretations. We assume numbered rows, i.e., row i in table ι_t corresponds to $v_{t.i} = \langle \sigma_{t.i}, c_{t.i} \rangle$. Note that for some nodes t, there are rows among different $\mathbb{SAT}$-tables that occur in $\text{Ext}_{\leq t}$, but not in $\text{SatExt}_{\leq t}$. These rows are removed during purging. In fact, rows $u_{4.1}, u_{4.2}$, and $u_{4.4}$ do not occur in table ι_4. Observe that purging is

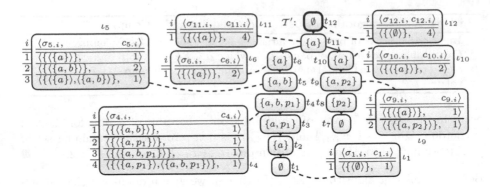

Fig. 3. Selected tables obtained by DP_PROJ on TD $\mathcal{T}'$ using DP_SAT (c.f., Fig. 2).

a crucial trick here that avoids to correct stored counters c by backtracking whenever a certain row of a table has no succeeding row in the parent table.

Next, we discuss selected rows obtained by DP_PROJ$((F, P), \mathcal{T}', \text{SAT-Comp})$. Tables $\iota_1, \ldots, \iota_{12}$ that are computed at the respective nodes of the tree decomposition are shown in Fig. 3. Since type$(t_1) = leaf$, we have $\iota_1 = \langle\{\langle\emptyset\rangle\}, 1\rangle$. Intuitively, up to node t_1 the SAT-row $\langle\emptyset\rangle$ belongs to 1 bucket. Node t_2 introduces variable a, which results in table $\iota_2 := \{\langle\{\langle\{a\}\rangle\}, 1\rangle\}$. Note that the SAT-row $\langle\emptyset\rangle$ is subject to purging. Node t_3 introduces p_1 and node t_4 introduces b. Node t_5 removes projected variable p_1. The row $\mathbf{v_{5.2}}$ of PROJ-table ι_5 has already been discussed in Example 7 and row $\mathbf{v_{5.1}}$ works similar. For row $\mathbf{v_{5.3}}$ we compute the count ipmc$(t_5, \{\mathbf{u_{5.2}}, \mathbf{u_{5.4}}\}, \langle\iota_4\rangle)$ by means of pmc. Therefore, take for ρ the sets $\{\mathbf{u_{5.2}}\}$, $\{\mathbf{u_{5.4}}\}$, and $\{\mathbf{u_{5.2}}, \mathbf{u_{5.4}}\}$. For the singleton sets, we simply have pmc$(t_5, \{\mathbf{u_{5.2}}\}, \langle\iota_4\rangle) = $ ipmc$(t_5, \{\mathbf{u_{5.2}}\}, \langle\iota_4\rangle) = c_{5.1} = 1$ and pmc$(t_5, \{\mathbf{u_{5.4}}\}, \langle\iota_4\rangle) = $ ipmc$(t_5, \{\mathbf{u_{5.4}}\}, \langle\iota_4\rangle) = c_{5.2} = 2$. To compute pmc$(t_5, \{\mathbf{u_{5.2}}, \mathbf{u_{5.4}}\}, \langle\iota_4\rangle)$ following Definition 9, take for O the sets $\{\mathbf{u_{4.5}}\}$, $\{\mathbf{u_{4.3}}\}$, and $\{\mathbf{u_{4.6}}\}$ into account, since all other nonempty subsets of origins of $\mathbf{u_{5.2}}$ and $\mathbf{u_{5.4}}$ in ι_4 do not occur in ι_4. Then, we take the sum over the values s-ipmc$(\langle t_4\rangle, \{\langle\mathbf{u_{4.5}}\rangle\}) = 1$, s-ipmc$(\langle t_4\rangle, \{\langle\mathbf{u_{4.3}}\rangle\}) = 1$, and s-ipmc$(\langle t_4\rangle, \{\langle\mathbf{u_{4.6}}\rangle\}) = 1$; and subtract s-ipmc$(\langle t_4\rangle, \{\langle\mathbf{u_{4.5}}\rangle, \langle\mathbf{u_{4.6}}\rangle\}) = 1$. Hence, pmc$(t_5, \{\mathbf{u_{5.2}}, \mathbf{u_{5.4}}\}, \langle\iota_4\rangle) = 2$. In order to compute ipmc$(t_5, \{\mathbf{u_{5.2}}, \mathbf{u_{5.4}}\}, \langle\iota_4\rangle) = |pmc(t_5, \{\mathbf{u_{5.2}}, \mathbf{u_{5.4}}\}, \langle\iota_4\rangle) - $ipmc$(t_5, \{\mathbf{u_{5.2}}\}, \langle\iota_4\rangle) - $ipmc$(t_5, \{\mathbf{u_{5.4}}\}, \langle\iota_4\rangle)| = |2 - 1 - 2| = |-1| = 1$. Hence, $c_{5.3} = 1$ represents the number of projected models, both rows $\mathbf{u_{5.2}}$ and $\mathbf{u_{5.4}}$ have in common. We then use it for table t_6.

For node t_{11} of type *join* one simply in addition multiplies stored s-ipmc values for SAT-rows in the two children of t_{11} accordingly (see Definition 8). In the end, the projected model count of F corresponds to s-ipmc$(\iota_{12}, \cdot) = 4$. ∎

5   Runtime (Upper and Lower Bounds)

In this section, we first present asymptotic upper bounds on the runtime of our Algorithm DP_PROJ. For the analysis, we assume $\gamma(n)$ to be the costs for multiplying two n-bit integers, which can be achieved in time $n \cdot \log n \cdot \log \log n$ [26,30].

Then, our main result is a lower bound that establishes that there cannot be an algorithm that solves PMC in time that is only single exponential in the treewidth and polynomial in the size of the formula unless the exponential time hypothesis (ETH) fails. This result establishes that there *cannot* be an algorithm exploiting treewidth that is *asymptotically better* than our presented algorithm, although one can likely improve on the analysis and give a better algorithm.

Theorem 1. *Given a PMC instance (F, P) and a tree decomposition $\mathcal{T} = (T, \chi)$ of F of width k with g nodes. Algorithm $\mathrm{DP_{PROJ}}$ runs in time $\mathcal{O}(2^{2^{k+4}} \cdot \gamma(\|F\|) \cdot g)$.*

Proof. Let $d = k + 1$ be maximum bag size of $\mathcal{T}$. For each node t of T, we consider table $\tau = \mathbb{SAT}\text{-Comp}[t]$ which has been computed by $\mathrm{DP_{SAT}}$ [41]. The table τ has at most 2^d rows. In the worst case we store in $\iota = \mathbb{PROJ}\text{-Comp}[t]$ each subset $\sigma \subseteq \tau$ together with exactly one counter. Hence, we have 2^{2^d} many rows in ι. In order to compute ipmc for σ, we consider every subset $\rho \subseteq \sigma$ and compute pmc. Since $|\sigma| \leq 2^d$, we have at most 2^{2^d} many subsets ρ of σ. For computing pmc, there could be each subset of the origins of ρ for each child table, which are less than $2^{2^{d+1}} \cdot 2^{2^{d+1}}$ (join and remove case). In total, we obtain a runtime bound of $\mathcal{O}(2^{2^d} \cdot 2^{2^d} \cdot 2^{2^{d+1}} \cdot 2^{2^{d+1}} \cdot \gamma(\|F\|)) \subseteq \mathcal{O}(2^{2^{d+3}} \cdot \gamma(\|F\|))$ since we also need multiplication of counters. Then, we apply this to every node t of the tree decomposition, which results in running time $\mathcal{O}(2^{2^{d+3}} \cdot \gamma(\|F\|) \cdot g)$. $\square$

Corollary 1. *Given an instance (F, P) of PMC where F has treewidth k. Algorithm $\mathrm{PCNT_{SAT}}$ runs in time $\mathcal{O}(2^{2^{k+4}} \cdot \gamma(\|F\|) \cdot \|F\|)$.*

Proof. We compute in time $2^{\mathcal{O}(k^3)} \cdot |V|$ a tree decomposition $\mathcal{T}'$ of width at most k [5] of primal graph P_F. Then, we run a decision version of the algorithm $\mathrm{DP_{SAT}}$ by Samer and Szeider [41] in time $\mathcal{O}(2^k \cdot \gamma(\|F\|) \cdot \|F\|)$. Then, we again traverse the decomposition, thereby keeping rows that have a satisfying extension ("purging"), in time $\mathcal{O}(2^k \cdot \|F\|)$. Finally, we run $\mathrm{DP_{PROJ}}$ and obtain the claim by Theorem 1 and since $\mathcal{T}'$ has linearly many nodes [5]. $\square$

The next results also establish the lower bounds for our worst-cases.

Theorem 2. *Unless ETH fails, PMC cannot be solved in time $2^{2^{o(k)}} \cdot \|F\|^{o(k)}$ for a given instance (F, P) where k is the treewidth of the primal graph of F.*

Proof. Assume for proof by contradiction that there is such an algorithm. We show that this contradicts a recent result [32], which states that one cannot decide the validity of a QBF [4,29] $Q = \exists V_1 . \forall V_2 . E$ in time $2^{2^{o(k)}} \cdot \|E\|^{o(k)}$ under ETH. Given an instance (Q, k) of $\exists\forall\text{-SAT}$ when parameterized by the treewidth k of E, we provide a reduction to an instance $(\forall V_1 . \exists V_2 . E', k)$ of $\forall\exists\text{-}$ SAT where $E' \equiv \neg E$ and E' is in CNF. Observe that the primal graphs of E and E' are isomorphic and therefore have the same treewidth k [32]. Then, given an instance $(\forall V_1.\exists V_2.E', k)$ of $\forall\exists$-SAT when parameterized by the treewidth k, we provide a reduction to an instance $((F, P, n), k)$ of decision version PMC-exactly-n of PMC such that $F = E$, $P = V_1$, and the number n of solutions is exactly $2^{|V_1|}$. The reduction gives a yes instance $((F, P, n), k)$ of PMC-exactly-n if and only if $(\forall V_1.\exists V_2.E', k)$ is a yes instance of $\forall\exists$-SAT. The reduction is also an fpt-reduction, since the treewidth of F is exactly k.   □

Corollary 2. *Given an instance* (F, P) *of* PMC *where* F *has treewidth* k. *Then, Algorithm* $\mathrm{PCNT_{SAT}}$ *runs in time* $2^{2^{\Theta(k)}} \cdot \gamma(\|F\|) \cdot \|F\|$.

6   Correctness of the Algorithm

In the following, we state definitions required for the correctness proofs of our algorithm $\mathbb{PROJ}$. In the end, we only store rows that are restricted to the bag content to maintain runtime bounds. Similar to related work [18,41], we proceed in two steps. First, we define properties of so-called $\mathbb{PROJ}$-*solutions up to* t, and then restrict these to $\mathbb{PROJ}$-*row solutions* at t.

For the following statements, we assume that we have given an arbitrary instance (F, P) of PMC and a tree decomposition $\mathcal{T} = (T, \chi)$ of formula F, where $T = (N, A, n)$, node $n \in N$ is the root and $\mathcal{T}$ is of width k. Moreover, for every $t \in N$ of tree decomposition $\mathcal{T}$, we let $\mathbb{SAT}$-Comp$[t]$ be the tables that have been computed by running algorithm $\mathrm{DP_{SAT}}$ for the dedicated input. Analogously, let $\mathbb{PROJ}$-Comp$[t]$ be the tables computed by running $\mathrm{DP_{PROJ}}$ for the input.

Definition 11. *Let* $\emptyset \subsetneq \sigma \subseteq \mathbb{SAT}$-*Comp*$[t]$ *be a table with* $\sigma \in$ sub-buckets$_P$ ($\mathbb{SAT}$-*Comp*$[t]$). *We define a* $\mathbb{PROJ}$-*solution up to* t *to be the sequence* $\langle \hat{\sigma} \rangle = \langle \mathrm{SatExt}_{\leq t}(\sigma) \rangle$.

Next, we recall that we can reconstruct all models from the tables.

Proposition 1. $I(\mathrm{SatExt}_{\leq n}(\mathbb{SAT}$-*Comp*$[n])) = I(\mathrm{Exts}) = \{J \in 2^{\mathrm{var}(F)} | J \models F\}$.

Proof (Idea). We use a construction similar to Samer and Szeider [41] and Pichler et al. [36, Fig. 1], where we simply collect preceding rows.   □

Before we present equivalence results between ipmc$_{\leq t}(\ldots)$ and the recursive version ipmc$(t, \ldots)$ (Definition 10) used during the computation of $\mathrm{DP_{PROJ}}$, recall that ipmc$_{\leq t}$ and pmc$_{\leq t}$ (Definition 5) are key to compute the projected model count. The following corollary states that computing ipmc$_{\leq n}$ at the root n actually suffices to compute the projected model count pmc$_{\leq n}$ of the formula.

Corollary 3. ipmc$_{\leq n}(\mathbb{SAT}$-*Comp*$[n]) = $ pmc$_{\leq n}(\mathbb{SAT}$-*Comp*$[n]) = |I_P(\mathrm{SatExt}_{\leq n}$ ($\mathbb{SAT}$-*Comp*$[n]))| = |I_P(\mathrm{Exts})| = |\{J \cap P \mid J \in 2^{\mathrm{var}(F)}, J \models F\}|$.

Proof. The corollary immediately follows from Proposition 1 and the observation that $|\text{SAT-Comp}[n]| \leq 1$ by properties of algorithm SAT and since $\chi(n) = \emptyset$. $\square$

The following lemma establishes that the PROJ-solutions up to root n of a given tree decomposition solve the PMC problem.

Lemma 1. *The value* $c = \sum_{\langle\hat\sigma\rangle \text{ is a PROJ-solution up to } n} |I_P(\hat\sigma)|$ *if and only if* c *is the projected model count of* F *with respect to the set* P *of projection variables.*

Proof. Assume that $c = \sum_{\langle\hat\sigma\rangle \text{ is a PROJ-solution up to } n} |I_P(\hat\sigma)|$. Observe that there can be at most one projected solution up to n, since $\chi(n) = \emptyset$. If $c = 0$, then $\text{SAT-Comp}[n]$ contains no rows. Hence, F has no models, c.f., Proposition 1, and obviously also no models projected to P. Consequently, c is the projected model count of F. If $c > 0$ we have by Corollary 3 that c is equivalent to the projected model count of F with respect to P. We proceed similar in the if direction. $\square$

In the following, we provide for a given node t and a given PROJ-solution up to t, the definition of a PROJ-row solution at t.

Definition 12. *Let* $t, t' \in N$ *be nodes of a given tree decomposition* $\mathcal{T}$, *and* $\hat\sigma$ *be a* PROJ-*solution up to* t. *Then, we define the local table for* t' *as* $\text{local}(t', \hat\sigma) := \{\langle u\rangle | \langle t', u\rangle \in \hat\sigma\}$, *and if* $t = t'$, *the* PROJ-*row solution at* t *by* $\langle \text{local}(t, \hat\sigma), |I_P(\hat\sigma)|\rangle$.

Observation 3. *Let* $\langle\hat\sigma\rangle$ *be a* PROJ-*solution up to a node* $t \in N$. *There is exactly one corresponding* PROJ-*row solution* $\langle\text{local}(t, \hat\sigma), |I_P(\hat\sigma)|\rangle$ *at* t.

Vice versa, let $\langle\sigma, c\rangle$ *be a* PROJ-*row solution at* t *for some integer* c. *Then, there is exactly one corresponding* PROJ-*solution* $\langle\text{SatExt}_{\leq t}(\sigma)\rangle$ *up to* t.

We need to ensure that storing PROJ-row solutions at a node suffices to solve the PMC problem, which is necessary to obtain runtime bounds (c.f. Corollary 1).

Lemma 2. *Let* $t \in N$ *be a node of the tree decomposition* $\mathcal{T}$. *There is a* PROJ-*row solution at root* n *if and only if the projected model count of* F *is larger than 0.*

Proof. ("$\Longrightarrow$"): Let $\langle\sigma, c\rangle$ be a PROJ-row solution at root n where σ is a SAT-table and c is a positive integer. Then, by Definition 12, there also exists a corresponding PROJ-solution $\langle\hat\sigma\rangle$ up to n such that $\sigma = \text{local}(n, \hat\sigma)$ and $c = |I_P(\hat\sigma)|$. Moreover, since $\chi(n) = \emptyset$, we have $|\text{SAT-Comp}[n]| = 1$. Then, by Definition 11, $\hat\sigma = \text{SAT-Comp}[n]$. By Corollary 3, we have $c = |I_P(\text{SAT-Comp}[n])|$. Finally, the claim follows. ("$\Longleftarrow$"): Similar to the only-if direction. $\square$

Observation 4. *Let* $X_1, \ldots, X_n$ *be finite sets. The number* $|\bigcap_{i \in X} X_i|$ *is given by* $|\bigcap_{i \in X} X_i| = ||\bigcup_{j=1}^{n} X_j| + \sum_{\emptyset \subsetneq I \subsetneq X} (-1)^{|I|} |\bigcap_{i \in I} X_i||$.

Lemma 3. *Let* $t \in N$ *be a node of the tree decomposition* $\mathcal{T}$ *with children* $(t, T) = \langle t_1, \ldots, t_\ell\rangle$ *and let* $\langle\sigma, \cdot\rangle$ *be a* PROJ-*row solution at* t. *Then,*

1. $\mathrm{ipmc}(t, \sigma, \langle \mathrm{PROJ\text{-}Comp}[t_1], ..., \mathrm{PROJ\text{-}Comp}[t_\ell]\rangle) = \mathrm{ipmc}_{\leq t}(\sigma)$

2. If type$(t) \neq$ leaf: $\mathrm{pmc}(t, \sigma, \langle \mathrm{PROJ\text{-}Comp}[t_1], ..., \mathrm{PROJ\text{-}Comp}[t_\ell]\rangle) = \mathrm{pmc}_{\leq t}(\sigma)$.

Proof (Sketch). We prove the statement by simultaneous induction. ("Induction Hypothesis"): Lemma 3 holds for the nodes in children(t, T) and also for node t, but on strict subsets $\rho \subsetneq \sigma$. ("Base Cases"): Let type$(t) =$ *leaf.* By definition, $\mathrm{ipmc}(t, \emptyset, \langle\rangle) = \mathrm{ipmc}_{\leq t}(\emptyset) = 1$. Recall that for pmc the equivalence does not hold for leaves, but we use a node t that has a node $t' \in N$ with type$(t') =$ *leaf* as child for the base case. Observe that by definition t has exactly one child. Then, we have $\mathrm{pmc}(t, \sigma, \langle \mathrm{PROJ\text{-}Comp}[t']\rangle) = \sum_{\emptyset \subsetneq O \subseteq \mathrm{SAT\text{-}origins}(t,\sigma)} (-1)^{(|O|-1)} \cdot$ s-ipmc$(\langle \mathrm{SAT\text{-}Comp}[t']\rangle, O) = |\bigcup_{u \in \sigma} I_P(\mathrm{SatExt}_{\leq t}(\{u\}))| = \mathrm{pmc}_{\leq t}(\sigma) = 1$ for PROJ-row solution $\langle \sigma, \cdot \rangle$ at t. ("Induction Step"): We proceed by case distinction. Assume that type$(t) =$ *int.* Let $a \in (\chi(t) \setminus \chi(t'))$ be the introduced variable. We have two cases. Assume Case (i): a also belongs to $(\mathrm{var}(F) \setminus P)$, i.e., a is not a projection variable. Let $\langle \sigma, c \rangle$ be a PROJ-row solution at t for some integer c. By construction of algorithm SAT there are many rows in the table $\mathrm{SAT\text{-}Comp}[t]$ for one row in the table $\mathrm{SAT\text{-}Comp}[t']$, more precisely, $|\mathrm{buckets}_P(\sigma)| = 1$. As a result, $\mathrm{pmc}_{\leq t}(\sigma) = \mathrm{pmc}_{\leq t'}(\mathrm{SAT\text{-}origins}(t, \sigma))$ by applying Observation 3. We apply the inclusion-exclusion principle on every subset ρ of the origins of σ in the definition of pmc and by induction hypothesis we know that $\mathrm{ipmc}(t', \rho, \langle \mathrm{PROJ\text{-}Comp}[t']\rangle) = \mathrm{ipmc}_{\leq t'}(\rho)$, therefore, s-ipmc$(\mathrm{PROJ\text{-}Comp}[t'], \rho) = \mathrm{ipmc}_{\leq t'}(\rho)$. This concludes Case (i) for pmc. The induction step for ipmc works similar by applying Observation 4 and comparing corresponding PROJ-solutions up to t or t', respectively. Further, for showing the lemma for ipmc, one has to additionally apply the hypothesis for node t, but on strict subsets $\emptyset \subsetneq \rho \subsetneq \sigma$ of σ. Assume that we have Case (ii): a also belongs to P, i.e., a is a projection variable. This is a special case of Case (i) since $|\mathrm{buckets}_P(\sigma)| = 1$. Similarly, for join and remove nodes. $\square$

Lemma 4 (Soundness). *Let $t \in N$ be a node of the tree decomposition T with* children$(t, T) = \langle t_1, \ldots, t_\ell \rangle$. *Then, each row $\langle \tau, c \rangle$ at node t obtained by PROJ is a PROJ-row solution for t.*

Proof (Idea). Observe that Listing 3 computes a row for each sub-bucket $\sigma \in$ sub-buckets$_P(\mathrm{SAT\text{-}Comp}[t])$. The resulting row $\langle \sigma, c \rangle$ obtained by ipmc is indeed a PROJ-row solution for t according to Lemma 3. $\square$

Lemma 5 (Completeness). *Let $t \in N$ be a node of tree decomposition T where* children$(t, T) = \langle t_1, \ldots, t_\ell \rangle$ *and* type$(t) \neq$ *leaf. Given a PROJ-row solution $\langle \sigma, c \rangle$ at t. Then, there is $\langle C_1, \ldots, C_\ell \rangle$ where each C_i is a set of PROJ-row solutions at t_i with $\sigma = \mathrm{PROJ}(t, \cdot, \cdot, P, \langle C_1, \ldots, C_\ell \rangle, \mathrm{SAT\text{-}Comp})$.*

Proof (Idea). Since $\langle \sigma, c \rangle$ is a PROJ-row solution for t, there is by Definition 12 a corresponding PROJ-solution $\langle \hat{\sigma} \rangle$ up to t such that local$(t, \hat{\sigma}) = \sigma$. We proceed again by case distinction. Assume type$(t) =$ *int* and $t' = t_1$. Then we define $\hat{\sigma}' := \{(t', \hat{\rho}) \mid (t', \hat{\rho}) \in \sigma, t \neq t'\}$. Then, for each subset $\emptyset \subsetneq \rho \subseteq$ local$(t', \hat{\sigma}')$, we define $\langle \rho, |I_P(\mathrm{SatExt}_{\leq t}(\rho))| \rangle$ in accordance with

Definition 12. By Observation 3, we have that $\langle \rho, |I_P(\text{SatExt}_{\leq t}(\rho))|\rangle$ is a SAT-row solution at t'. Since we defined PROJ-row solutions for t' for all respective PROJ-solutions up to t', we encountered every PROJ-row solution for t' required for deriving $\langle \sigma, c \rangle$ via PROJ (c.f. Definitions 9 and 10). Similarly, for remove and join nodes.   □

Theorem 3. *The algorithm* DP$_{\text{PROJ}}$ *is correct. More precisely,* DP$_{\text{PROJ}}((F, P), \mathcal{T},$ *SAT-Comp) returns tables* PROJ-*Comp such that* $c = $ s-ipmc(SAT-*Comp*$[n], \cdot)$ *is the projected model count of F with respect to the set P of projection variables.*

Proof. By Lemma 4 we have soundness for every node $t \in N$ and hence only valid rows as output of table algorithm PROJ when traversing the tree decomposition in post-order up to the root n. By Lemma 2 we know that the projected model count c of F is larger than zero if and only if there exists a certain PROJ-row solution for n. This PROJ-row solution at node n is of the form $\langle \{\langle \emptyset, \ldots \rangle\}, c \rangle$. If there is no PROJ-row solution at node n, then SAT-Comp$[n] = \emptyset$ since the table algorithm SAT is correct (c.f. Proposition 1). Consequently, we have $c = 0$. Therefore, $c = $ s-ipmc(SAT-Comp$[n], \cdot$) is the pmc of F w.r.t. P in both cases.

Next, we establish completeness by induction starting from root n. Let therefore, $\langle \hat{\sigma} \rangle$ be the PROJ-solution up to n, where for each row in $\boldsymbol{u} \in \hat{\sigma}$, $I(\boldsymbol{u})$ corresponds to a model of F. By Definition 12, we know that for n we can construct a PROJ-row solution at n of the form $\langle \{\langle \emptyset, \ldots \rangle\}, c \rangle$ for $\hat{\sigma}$. We already established the induction step in Lemma 5. Finally, we stop at the leaves.   □

Corollary 4. *The algorithm* PCNT$_{\text{SAT}}$ *is correct, i.e.,* PCNT$_{\text{SAT}}$ *solves* PMC.

Proof. The result follows, since PCNT$_{\text{SAT}}$ consists of pass DP$_{\text{SAT}}$, a purging step and DP$_{\text{PROJ}}$. For correctness of DP$_{\text{SAT}}$ we refer to other sources [18,41]. By Proposition 1, "purging" neither destroys soundness nor completeness of DP$_{\text{SAT}}$.   □

7   Conclusions

We introduced a dynamic programming algorithm to solve projected model counting (PMC) by exploiting the structural parameter treewidth. Our algorithm is asymptotically optimal under the exponential time hypothesis (ETH). Its runtime is double exponential in the treewidth of the primal graph of the instance and polynomial in the size of the input instance. We believe that our results can also be extended to another graph representation, namely the incidence graph. Our approach is very general and might be applicable to a wide range of other hard combinatorial problems, such as projection for ASP [18] and QBF [10].

References

1. Abiteboul, S., Hull, R., Vianu, V.: Foundations of Databases: The Logical Level, 1st edn. Addison-Wesley, Boston (1995)
2. Abramson, B., Brown, J., Edwards, W., Murphy, A., Winkler, R.L.: Hailfinder: a Bayesian system for forecasting severe weather. Int. J. Forecast. **12**(1), 57–71 (1996)
3. Aziz, R.A., Chu, G., Muise, C., Stuckey, P.: #∃SAT: projected model counting. In: Heule, M., Weaver, S. (eds.) SAT 2015. LNCS, vol. 9340, pp. 121–137. Springer, Cham (2015). https://doi.org/10.1007/978-3-319-24318-4_10
4. Biere, A., Heule, M., van Maaren, H., Walsh, T. (eds.): Handbook of Satisfiability, Frontiers in Artificial Intelligence and Applications, vol. 185. IOS Press, Amsterdam (2009)
5. Bodlaender, H.L.: A linear-time algorithm for finding tree-decompositions of small treewidth. SIAM J. Comput. **25**(6), 1305–1317 (1996)
6. Bodlaender, H.L., Kloks, T.: Efficient and constructive algorithms for the pathwidth and treewidth of graphs. J. Algorithms **21**(2), 358–402 (1996)
7. Bodlaender, H.L., Koster, A.M.C.A.: Combinatorial optimization on graphs of bounded treewidth. Comput. J. **51**(3), 255–269 (2008)
8. Bondy, J.A., Murty, U.S.R.: Graph theory, Graduate Texts in Mathematics, vol. 244. Springer Verlag, New York (2008)
9. Chakraborty, S., Meel, K.S., Vardi, M.Y.: Improving approximate counting for probabilistic inference: from linear to logarithmic SAT solver calls. In: Kambhampati, S. (ed.) Proceedings of 25th International Joint Conference on Artificial Intelligence (IJCAI 2016), pp. 3569–3576. The AAAI Press, New York City, July 2016
10. Charwat, G., Woltran, S.: Dynamic programming-based QBF solving. In: Lonsing, F., Seidl, M. (eds.) Proceedings of the 4th International Workshop on Quantified Boolean Formulas (QBF 2016), vol. 1719, pp. 27–40. CEUR Workshop Proceedings (CEUR-WS.org) (2016). Co-located with 19th International Conference on Theory and Applications of Satisfiability Testing (SAT 2016)
11. Choi, A., Van den Broeck, G., Darwiche, A.: Tractable learning for structured probability spaces: a case study in learning preference distributions. In: Yang, Q. (ed.) Proceedings of 24th International Joint Conference on Artificial Intelligence (IJCAI 2015). The AAAI Press (2015)
12. Cygan, M., Fomin, F.V., Kowalik, Ł., Lokshtanov, D., Marx, D., Pilipczuk, M., Pilipczuk, M., Saurabh, S.: Parameterized Algorithms. Springer, Cham (2015). https://doi.org/10.1007/978-3-319-21275-3
13. Diestel, R.: Graph Theory. Graduate Texts in Mathematics, vol. 173, 4th edn. Springer, Heidelberg (2012)
14. Domshlak, C., Hoffmann, J.: Probabilistic planning via heuristic forward search and weighted model counting. J. Artif. Intell. Res. **30**, 565–620 (2007)
15. Downey, R.G., Fellows, M.R.: Fundamentals of Parameterized Complexity. TCS. Springer, London (2013). https://doi.org/10.1007/978-1-4471-5559-1
16. Dueñas-Osorio, L., Meel, K.S., Paredes, R., Vardi, M.Y.: Counting-based reliability estimation for power-transmission grids. In: Singh, S.P., Markovitch, S. (eds.) Proceedings of the Thirty-First AAAI Conference on Artificial Intelligence (AAAI 2017), pp. 4488–4494. The AAAI Press, February 2017
17. Durand, A., Hermann, M., Kolaitis, P.G.: Subtractive reductions and complete problems for counting complexity classes. Theor. Comput. Sci. **340**(3), 496–513 (2005)

18. Fichte, J.K., Hecher, M., Morak, M., Woltran, S.: Answer set solving with bounded treewidth revisited. In: Balduccini, M., Janhunen, T. (eds.) LPNMR 2017. LNCS (LNAI), vol. 10377, pp. 132–145. Springer, Cham (2017). https://doi.org/10.1007/978-3-319-61660-5_13

19. Fichte, J.K., Hecher, M., Morak, M., Woltran, S.: DynASP2.5: dynamic programming on tree decompositions in action. In: Lokshtanov, D., Nishimura, N. (eds.) Proceedings of the 12th International Symposium on Parameterized and Exact Computation (IPEC 2017). Dagstuhl Publishing (2017)

20. Flum, J., Grohe, M.: Parameterized Complexity Theory. TTCS, vol. XIV. Springer, Berlin (2006). https://doi.org/10.1007/3-540-29953-X

21. Gebser, M., Schaub, T., Thiele, S., Veber, P.: Detecting inconsistencies in large biological networks with answer set programming. Theory Pract. Log. Program. **11**(2–3), 323–360 (2011)

22. Gebser, M., Kaufmann, B., Schaub, T.: Solution enumeration for projected boolean search problems. In: van Hoeve, W.-J., Hooker, J.N. (eds.) CPAIOR 2009. LNCS, vol. 5547, pp. 71–86. Springer, Heidelberg (2009). https://doi.org/10.1007/978-3-642-01929-6_7

23. Ginsberg, M.L., Parkes, A.J., Roy, A.: Supermodels and robustness. In: Rich, C., Mostow, J. (eds.) Proceedings of the 15th National Conference on Artificial Intelligence and 10th Innovative Applications of Artificial Intelligence Conference (AAAI/IAAI 1998), pp. 334–339. The AAAI Press, Madison, July 1998

24. Gomes, C.P., Sabharwal, A., Selman, B.: Chapter 20: model counting. In: Biere, A., Heule, M., van Maaren, H., Walsh, T. (eds.) Handbook of Satisfiability, Frontiers in Artificial Intelligence and Applications, vol. 185, pp. 633–654. IOS Press, Amsterdam (2009)

25. Graham, R.L., Grötschel, M., Lovász, L.: Handbook of Combinatorics, vol. I. Elsevier Science Publishers, North-Holland (1995)

26. Harvey, D., van der Hoeven, J., Lecerf, G.: Even faster integer multiplication. J. Complex. **36**, 1–30 (2016)

27. Hemaspaandra, L.A., Vollmer, H.: The satanic notations: Counting classes beyond #P and other definitional adventures. SIGACT News **26**(1), 2–13 (1995)

28. Impagliazzo, R., Paturi, R., Zane, F.: Which problems have strongly exponential complexity? J. Comput. Syst. Sci. **63**(4), 512–530 (2001)

29. Kleine Büning, H., Lettman, T.: Propositional Logic: Deduction and Algorithms. Cambridge University Press, Cambridge (1999)

30. Knuth, D.E.: How fast can we multiply? In: The Art of Computer Programming, Seminumerical Algorithms, 3 edn., vol. 2, chap. 4.3.3, pp. 294–318. Addison-Wesley (1998)

31. Lagniez, J.M., Marquis, P.: An improved decision-DNNF compiler. In: Sierra, C. (ed.) Proceedings of the Twenty-Sixth International Joint Conference on Artificial Intelligence (IJCAI 2017). The AAAI Press (2017)

32. Lampis, M., Mitsou, V.: Treewidth with a quantifier alternation revisited. In: Lokshtanov, D., Nishimura, N. (eds.) Proceedings of the 12th International Symposium on Parameterized and Exact Computation (IPEC 2017). Dagstuhl Publishing (2017)

33. Manning, C.D., Raghavan, P., Schütze, H.: Introduction to Information Retrieval. Cambridge University Press, Cambridge (2008)

34. Niedermeier, R.: Invitation to Fixed-Parameter Algorithms. Oxford Lecture Series in Mathematics and its Applications, vol. 31. Oxford University Press, New York (2006)

35. Papadimitriou, C.H.: Computational Complexity. Addison-Wesley, Boston (1994)
36. Pichler, R., Rümmele, S., Woltran, S.: Counting and enumeration problems with bounded treewidth. In: Clarke, E.M., Voronkov, A. (eds.) LPAR 2010. LNCS (LNAI), vol. 6355, pp. 387–404. Springer, Heidelberg (2010). https://doi.org/10.1007/978-3-642-17511-4_22
37. Pourret, O., Naim, P., Bruce, M.: Bayesian Networks - A Practical Guide to Applications. Wiley, Hoboken (2008)
38. Roth, D.: On the hardness of approximate reasoning. Artif. Intell. $82(1-2)$, 273–302 (1996)
39. Sæther, S.H., Telle, J.A., Vatshelle, M.: Solving #SAT and MAXSAT by dynamic programming. J. Artif. Intell. Res. 54, 59–82 (2015)
40. Sahami, M., Dumais, S., Heckerman, D., Horvitz, E.: A Bayesian approach to filtering junk e-mail. In: Joachims, T. (ed.) Proceedings of the AAAI-98 Workshop on Learning for Text Categorization, vol. 62, pp. 98–105 (1998)
41. Samer, M., Szeider, S.: Algorithms for propositional model counting. J. Discret. Algorithms $8(1)$, 50–64 (2010)
42. Sang, T., Beame, P., Kautz, H.: Performing Bayesian inference by weighted model counting. In: Veloso, M.M., Kambhampati, S. (eds.) Proceedings of the 29th National Conference on Artificial Intelligence (AAAI 2005). The AAAI Press (2005)
43. Valiant, L.: The complexity of enumeration and reliability problems. SIAM J. Comput. $8(3)$, 410–421 (1979)
44. Wilder, R.L.: Introduction to the Foundations of Mathematics, 2nd edn. Wiley, New York (1965)
45. Xue, Y., Choi, A., Darwiche, A.: Basing decisions on sentences in decision diagrams. In: Hoffmann, J., Selman, B. (eds.) Proceedings of the 26th AAAI Conference on Artificial Intelligence (AAAI 2012). The AAAI Press (2012)

Quantified Boolean Formulae

Circuit-Based Search Space Pruning in QBF

Mikoláš Janota[⊠]

IST/INESC-ID, University of Lisbon, Lisbon, Portugal
mikolas.janota@gmail.com

Abstract. This paper describes the algorithm implemented in the QBF solver CQESTO, which has placed second in the non-CNF track of the last year's QBF competition. The algorithm is inspired by the CNF-based solver QESTO. Just as QESTO, CQESTO invokes a SAT solver in a black-box fashion. However, it directly operates on the circuit representation of the formula. The paper analyzes the individual operations that the solver performs.

1 Introduction

Since the indisputable success of SAT and SMT, research has been trying to push the frontiers of automated logic-based solving. Reasoning with quantifiers represents a nontrivial challenge. Indeed, even in the Boolean case adding quantifiers bumps the complexity class from NP-complete to PSPACE-complete. Yet, quantifiers enable modeling a number of interesting problems [3].

This paper aims at the advancement of solving with quantifiers in the Boolean domain (QBF). Using CNF as input causes intrinsic issues in QBF solving [1,15,28]. Consequently, there have been efforts towards solvers operating directly on a non-clausal representation [2,8,12,18,25,28]. This line of research is supported by the circuit-like QBF format QCIR [16].

This paper presents the solver CQESTO, which reads in a circuit-like representation of the problem and keeps on solving *directly* on this representation. For each quantification level, the solver creates a propositional formula that determines the possible assignments to the variables of that particular level. If one of these formulas becomes unsatisfiable, a formula in one of the preceding levels is to be strengthened. *The main focus of this paper is the analysis of the operations that take place during this strengthening.*

CQESTO extends the family of solvers that repeatedly call a SAT solver (exponentially many times in the worst case). The solver RAReQS [10,12,13] delegates to the SAT solver partial expansions of the QBF. QESTO [14] and CAQE [23] are CNF siblings of CQESTO. QuAbS [25] is similar to CQESTO but it operates on a *literal abstraction*, while CQESTO operates *directly* on the circuit. The workings of CQESTO is also similar to algorithms developed for theories in SMT [4,7,24].

© Springer International Publishing AG, part of Springer Nature 2018
O. Beyersdorff and C. M. Wintersteiger (Eds.): SAT 2018, LNCS 10929, pp. 187–198, 2018.
https://doi.org/10.1007/978-3-319-94144-8_12

The principle contributions of the paper are: (1) Description of the algorithm of the solver CQESTO. (2) Analysis of the operations used in the circuit. (3) Linking these operations to related solvers.

The rest of the paper is organized as follows. Section 2 introduces concepts and notation used throughout the paper; Sect. 3 describes the CQESTO algorithm; Sect. 4 reports on experimental evaluation. Finally, Sect. 5 summarizes the paper and outlines directions for future work.

2   Preliminaries

Standard concepts from propositional logic are assumed. Propositional formulas are built from variables, negation ($\neg$), and conjunction ($\wedge$). For convenience we also consider the constants $0, 1$ representing false and true, respectively. The results immediately extend to other connectives, e.g., $(\phi \Rightarrow \psi) = \neg(\phi \wedge \neg\psi)$, $(\phi \vee \psi) = \neg(\neg\phi \wedge \neg\psi)$. A *literal* is either a variable or its negation. An *assignment* is a mapping from variables to $\{0, 1\}$. Assignments are represented as sets of literals, i.e., $\{x, \neg y\}$ corresponds to $\{x \mapsto 1, y \mapsto 0\}$. For a formula ϕ and an assignment σ, the expression $\phi|_\sigma$ denotes *substitution*, i.e., the simultaneous replacement of variables with their corresponding value. With some abuse of notation, we treat a set of formulas $\mathcal{I}$ and $\bigwedge_{\phi \in \mathcal{I}} \phi$ interchangeably. The paper makes use of the well-established notion of *subformula polarity*. Intuitively, the polarity of a subformula is determined by whether the number of negations above it is odd or even. Formally, we defined priority as follows.

Definition 1 (polarity [19]). *The following rules annotate each occurrence of a subformula of a formula α with its polarity $\in \{+, -\}$.*

$$\alpha^+ \qquad\qquad\qquad \textit{top rule, } \alpha \textit{ is positive}$$
$$(\neg\phi)^\pi \quad \rightsquigarrow \neg\phi^{-\pi} \qquad \textit{negation flips polarity}$$
$$(\phi \wedge \psi)^\pi \rightsquigarrow (\phi^\pi \wedge \psi^\pi) \quad \textit{conjunction maintains polarity}$$

Quantified Boolean Formulas (QBF). QBFs [17] extend propositional logic by quantifiers over Boolean variables. Any propositional formula ϕ is also a QBF with all variables *free*. If Φ is a QBF with a free variable x, the formulas $\exists x.\Phi$ and $\forall x.\Phi$ are QBFs with x *bound*, i.e. not free. Note that we disallow expressions such as $\exists x.\exists x.\, x$. Whenever possible, we write $\exists x_1 \ldots x_k$ instead of $\exists x_1 \ldots \exists x_k$; analogously for $\forall$. For a QBF $\Phi = \forall x.\Psi$ we say that x is *universal* in Φ and is *existential* in $\exists x.\Psi$. Analogously, a literal l is universal (resp. existential) if $\mathsf{var}(l)$ is universal (resp. existential). Semantically a QBF corresponds to a compact representation of a propositional formula. In particular, the formula $\forall x.\Psi$ is satisfied by the same truth assignments as $\Psi|_{\{\neg x\}} \wedge \Psi|_{\{x\}}$ and $\exists x.\Psi$ by $\Psi|_{\{\neg x\}} \vee \Psi|_{\{x\}}$. Since $\forall x \forall y.\Phi$ and $\forall y \forall x.\Phi$ are semantically equivalent, we allow writing $\forall X$ for a set of variables X; analogously for $\exists$. A QBF with no free variables is *false* (resp. *true*), iff it is semantically equivalent to the constant 0 (resp. 1).

A QBF is *closed* if it does not contain any free variables. A QBF is in *prenex form* if it is of the form $Q_1 X_1 \ldots Q_k X_k.\phi$, where $Q_i \in \{\exists, \forall\}$, $Q_i \neq Q_{i+1}$, and

---

Algorithm 1. QBF solving with circuit-based pruning

input : $Q_1 X_1 \ldots \ldots Q_n X_n Q_{n+1} X_{n+1}.\phi$, where X_{n+1} empty, $Q_i \neq Q_{i+1}$
output : truth value

```
1   αᵢ ← 1, for i ∈ 1..n − 1                          // minimalistic initialization
2   αₙ ← (Qₙ = ∃) ? φ : ¬φ
3   αₙ₊₁ ← (Qₙ = ∃) ? ¬φ : φ
4   i ← 1
5   while true do                                    // invariant 1 ≤ i ≤ n + 1
6   │   I ← proj(αᵢ, ⋃ⱼ∈₁..ᵢ₋₁ σⱼ)
7   │   (σᵢ, C) ← SAT(I, αᵢ)
8   │   if σᵢ = ⊥ then
9   │   │   if i ≤ 2 then return Qᵢ = ∀               // nowhere to backtrack
10  │   │   ξ_f ← forget(Xᵢ, ¬C)                      // eliminate Xᵢ
11  │   │   ξ_s ← ξ_f|σᵢ₋₁                            // eliminate Xᵢ₋₁ by substitution
12  │   │   αᵢ₋₂ ← αᵢ₋₂ ∧ ξ_s                         // strengthen
13  │   │   i ← i − 2                                 // backtrack
14  │   else
15  │   │   i ← i + 1                                 // move on
```

ϕ is propositional. The propositional part ϕ is called the *matrix* and the rest the *prefix*. For a variable $x \in X_i$ we say that x is at *level i*. Unless specified otherwise, QBFs are assumed to be closed and in prenex form.

QBF as Games. A closed and prenex QBF $Q_1 X_1 \ldots Q_k X_k.\phi$, represents a two-player game, c.f. [9,18]. The existential player tries to make the matrix true and conversely the universal player tries to make it false. Each player assigns a value only to a variable that belongs to the player and can only assign a variable once all preceding variables have already been assigned. Hence the two players assign values to variables following the order of the prefix alternating on a quantifier. A *play* is a sequence of assignments $\sigma_1, \ldots, \sigma_n$ where σ_i is an assignment to X_i. Within a play, the ith assignment is referred to as the ith *move*. The ith move belongs to player Q_i. A QBF Φ is true iff there exists a winning strategy for $\exists$; it is false iff there exists a winning strategy for $\forall$. The game semantics enables treating $\exists$ and $\forall$ symmetrically, i.e. we are concerned with deciding which player has a winning strategy.

3   CQESTO Algorithm

The algorithm decides a closed, prenex QBF of n quantification levels. For the sake of uniformity we add a quantification level $n+1$ with no variables belonging to the player Q_{n-1}. So the formula being solved is $Q_1 X_1 \ldots Q_n X_n Q_{n+1} X_{n+1}.\phi$, where X_{n+1} is empty and $Q_n \neq Q_{n+1}$. Like so it is guaranteed that eventually either of the player must lose as the play progresses, i.e. there's no need for handling especially a play where all blocks up till Q_n have value—if this happens, Q_{n+1} loses.

The algorithm's pseudocode is presented as Algorithm 1 and its overall intuition is as follows. For each quantification level i there is a propositional formula α_i, which constrains the moves of player Q_i. The algorithm builds assignments $\sigma_1, \ldots, \sigma_k$ so that each σ_i represents the ith move of player Q_i. A SAT solver is used to calculate a new σ_i from α_i. Backtracking occurs when α_i becomes unsatisfiable under the current assignments $\sigma_1, \ldots, \sigma_{i-1}$. Upon backtracking, player Q_i needs to change some moves that he had made earlier. Hence, the algorithm strengthens α_{i-2} and continues from there.[1] Note that if $i = 1$, the union $\bigcup_{j \in 1..i-1} \sigma_j$ is empty and the SAT call is on α_1 with empty assumptions.

The algorithm observes the following invariants regarding the constraints α_i.

Invariant 1 (syntactic). *Each α_i only contains variables $X_1 \cup \cdots \cup X_i$.*

Invariant 2 (semantic). *If player Q_i violates α_i, the player is bound to lose. More formally, if for a partial play $\sigma = \sigma_1, \ldots, \sigma_i$ it holds that $\sigma \vDash \neg\alpha_i$ then there is a winning strategy for the opponent from that position.*

The invariants are established upon initialization by setting all α_i to true except for α_n and α_{n+1}. The constraint α_n is set to the matrix or its negation depending on whether Q_i is existential or universal. The constraint α_{n+1} is set analogously. Note that since $\alpha_{n+1} = \neg\alpha_n$, once α_n is satisfied by $\sigma_1 \cup \cdots \cup \sigma_n$, the constraint α_{i+1} is immediately unsatisfiable since there are no further variables.

The algorithm uses several auxiliary functions. The function SAT models a SAT call on propositional formulas. The function proj is used to propagate information into the current α_i while the solver is moving forward. The function forget enables the solver to strengthen previous restrictions upon backtracking. Let us look at these mechanisms in turn.

3.1   Projection (proj)

The function $\text{proj}(\alpha_i, \sigma)$ produces a set of formulas $\mathcal{I}$ implied by the assignment $\sigma = \sigma_1 \cup \ldots \cup \sigma_{i-1}$ in α_i. The motivation for $\mathcal{I}$ is akin to the one for 1-UIP [20]. The set $\mathcal{I}$ may be envisioned as a cut in the circuit representing the formula α_i. In the context of the algorithm, proj enables generalizing the concrete variable assignment to subformulas. Rather than finding the move σ_i by satisfying $\alpha_i \wedge \bigwedge_{j \in 1..i-1} \sigma_j$, it must satisfy $\alpha_i \wedge \mathcal{I}$. Upon backtracking, $\mathcal{I}$ is used to strengthen α_{i-2}. This gives a better strengthening than a particular assignment.

As a motivational example, consider the formula $\exists xy \forall u \exists z.(x \vee y) \Rightarrow (z \wedge \neg z)$ and the assignment $\{x, \neg y\}$. In this case, the function proj returns $(x \vee y)$ because it is implied by the assignment and keeps forcing a contradiction. This yields the SAT call on $((x \vee y) \Rightarrow (z \wedge \neg z)) \wedge (x \vee y)$. The formula is unsatisfiable

---

[1] The implementation enables jumping across multiple levels by backtracking to the maximum level of variables in the core belonging to Q_i.

and the reason is that $(x \vee y)$ is true. This lets us conclude that at the first level, $\neg(x \vee y)$ must be true—the concrete assignment to x and y is not important.

The function `proj` operates in two phases, first it *propagates* the assignment σ in α_i and then it collects the most general sub-formulas of α_i propagated by σ. To formalize the definition we introduce an auxiliary concept of propagation $\sigma \vdash_p \phi$ meaning that ϕ follows from σ by propagation.

Definition 2 ($\vdash_p$). *For an assignment σ and formula ϕ, the relation $\sigma \vdash_p \phi$ is defined according to the following rules.*

$$\sigma \vdash_p 1 \qquad \sigma \vdash_p \psi \wedge \phi, \textit{if } \sigma \vdash_p \psi \textit{ and } \sigma \vdash_p \phi$$
$$\sigma \vdash_p l, \textit{if } l \in \sigma \qquad \sigma \vdash_p \neg(\psi \wedge \phi), \textit{if } \sigma \vdash_p \neg\psi \textit{ or } \sigma \vdash_p \neg\phi$$

The function `proj` operates recursively on subformulas. It first checks if a subformula or its negation is inferred by propagation. If so, it immediately returns the given subformula or its negation, respectively. Otherwise, it dives into the subformula's structure. Subformulas unaffected by the assignment are ignored. The definition follows (see also Examples 1–3).

Definition 3 (proj). *For a formula ϕ and assignment σ, $\mathtt{proj}(\phi, \sigma)$ is defined by the following equalities.*

$$\mathtt{proj}(\phi, \sigma) = \{\phi\} \qquad\qquad\qquad \textit{if } \sigma \vdash_p \phi$$
$$\mathtt{proj}(\phi, \sigma) = \{\neg\phi\} \qquad\qquad\qquad \textit{if } \sigma \vdash_p \neg\phi$$
$$\mathtt{proj}(\psi \wedge \phi, \sigma) = \mathtt{proj}(\psi, \sigma) \cup \mathtt{proj}(\phi, \sigma) \textit{ if above does not apply}$$
$$\mathtt{proj}(\neg\psi, \sigma) = \mathtt{proj}(\psi, \sigma) \qquad\quad \textit{if above does not apply}$$
$$\mathtt{proj}(l, \sigma) = \emptyset \qquad\qquad\qquad\quad \textit{if above does not apply}$$

Note that `proj` is well defined also for an empty $\sigma = \emptyset$. Since we only have $\emptyset \vdash_p 1$ the projection $\mathtt{proj}(\phi, \emptyset)$ will give the empty set, except for the special cases $\mathtt{proj}(1, \emptyset) = \mathtt{proj}(0, \emptyset) = \{1\}$.

3.2 SAT Calls

SAT calls are used to obtain a move σ_i at position i in a straightforward fashion (line 7). If the SAT calls deem the query unsatisfiable, it provides a core that is used to inform backtracking. In a call $\mathtt{SAT}(\mathcal{I}, \alpha_i)$, $\mathcal{I}$ is a set of propositional formulas modeling assumptions. The function returns a pair $(\sigma_i, \mathcal{C})$, where σ_i is a satisfying assignment to $\alpha_i \wedge \mathcal{I}$ if such exists and it is $\bot$ otherwise. If there is no satisfying assignment, $\mathcal{C} \subseteq \mathcal{I}$ is a *core*, i.e. $\phi \wedge \mathcal{C}$ is also unsatisfiable. Since modern SAT solvers typically only accept formulas in CNF, standard translation from formulas to CNF may be used via fresh variables [26].

3.3 Backtracking

The backtracking mechanism is triggered once the SAT call deems $\alpha_i \wedge \mathcal{I}$ unsatisfiable. Only a core $\mathcal{C} \subseteq \mathcal{I}$ is used, which gives a stronger constraint than using the whole of $\mathcal{I}$. The SAT call guarantees that $\alpha_i \wedge \mathcal{C}$ is unsatisfiable—and therefore losing for player Q_i. The objective is to derive a strengthening for α_{i-2}. *To that end we remove the sets of variables X_i and X_{i-1} from the core $\mathcal{C}$.*

Variables X_{i-1} are removed by substituting the opponent's move σ_{i-1}. The intuition is that the opponent can always play that same move σ_{i-1} in the future. In another words, player Q_i must account for *any* move of the opponent and in this case Q_i prepares for σ_{i-1}. This is best illustrated by an example that already does not contain any of the variables X_i—so we only need to worry about X_{i-1}.

Example 1. Consider $\exists x_1 x_2 \forall y \exists z. (x_1 \wedge z) \wedge (x_2 \vee y)$ with $\sigma_1 = \{x_1, \neg x_2\}$, $\sigma_2 = \{\neg y\}$. Propagation gives $\sigma_1 \cup \sigma_2 \vdash_p x_1$ and $\sigma_1 \cup \sigma_2 \vdash_p \neg(x_2 \vee y)$. SAT gives the core $\mathcal{C} = \neg(x_2 \vee y)$. Negating the core and substituting σ_2 gives $\xi_f = (x_2 \vee y)|_{\{\neg y\}} = x_2$ leading to the strengthening $\alpha_1 \leftarrow \alpha_1 \wedge x_2$.

The removal of the variables X_i relies on their polarity (see Definition 1). Each positive occurrence of a variable is replaced by 1 and each negative occurrence by 0. This operation guarantees that the resulting formula is weaker than the derived core (see Lemma 2 for justification).

Definition 4 (forget). *For a set of variables X and a formula ϕ, the transformation $\texttt{forget}(X, \phi)$ is defined as follows. The definition uses an auxiliary function $\texttt{pol}(\psi, X, c)$ where ψ is a formula and $c \in \{0, 1\}$ a constant. The constant c is determined by the polarity of ψ within ϕ (see Definition 1). If ψ is annotated positively (ψ^+), c is 1; if ψ is annotated negatively (ψ^-), c is 0.*

$$\texttt{forget}(X, \phi) = \texttt{pol}(\phi, X, 1)$$
$$\texttt{pol}(x, X, c) = c, \text{ if } x \in X \quad \texttt{pol}(\phi \wedge \psi, X, c) = \texttt{pol}(\phi, X, c) \wedge \texttt{pol}(\psi, X, c)$$
$$\texttt{pol}(x, X, c) = x, \text{ if } x \notin X \quad \texttt{pol}(\neg \phi, X, c) = \neg \texttt{pol}(\phi, X, \neg c)$$

Example 2. Consider $\exists x_1 x_2 \forall y \exists z. (x_1 \wedge z) \wedge (x_2 \vee y)$ and $\sigma_1 = \{\neg x_1, x_2\}$, $\sigma_2 = \{\neg y\}$. By propagation obtain $\sigma_1 \cup \sigma_2 \vdash_p \neg(x_1 \wedge z)$ and $\sigma_1 \cup \sigma_2 \vdash_p (x_2 \vee y)$. Yielding core $\mathcal{C} = \neg(x_1 \wedge z)$. Negating the core gives $(x_1 \wedge z)$, applying $\texttt{forget}(\{z\}, (x_1 \wedge z)) = x_1$. Hence we obtain the strengthening step $\alpha_1 \leftarrow \alpha_1 \wedge x_1$.

We conclude by an example where both X_i, X_{i-1} are removed at the same time.

Example 3. $\exists x_1 x_2 x_3 \forall y \exists zw. ((x_1 \wedge x_2 \vee w) \Rightarrow \neg z) \wedge ((x_3 \wedge y \vee \neg w) \Rightarrow z)$ for $\sigma_1 = \{x_1, x_2, x_3\}$, $\sigma_2 = \{y\}$ obtain $\sigma_1 \cup \sigma_2 \vdash_p (x_1 \wedge x_2 \vee w)$ and $\sigma_1 \cup \sigma_2 \vdash_p (x_3 \wedge y \vee \neg w)$. SAT giving the core $\mathcal{C} = \{(x_1 \wedge x_2 \vee w), (x_3 \wedge y \vee \neg w)\}$. Negate: $\neg(x_1 \wedge x_2 \vee w) \vee \neg(x_3 \wedge y \vee \neg w)$; apply $\texttt{forget}$: $\neg(x_1 \wedge x_2) \vee \neg(x_3 \wedge y)$; substitute $\{y\}$: $\neg(x_1 \wedge x_2) \vee \neg x_3$.

3.4 Discussion

CQESTO hinges on two operations: projection and forgetting. Arguably, any backtracking QBF solver must perform these operations in some form, while observing the properties outlined in the following section (Sect. 3.5). Here I remark that the implementation of CQESTO enables deviations from the current presentation. In particular, Invariant 1 may not be strictly observed: upon initialization all α_i are initialized with the original (negated) matrix. Like so, downstream variables also appear in α_i.

Several QBF solvers are characterized by repeated SAT calls. RAReQS [10, 12, 13] performs a heavy-handed expansion of quantifiers requiring recursive calls: this may turn unwieldy in formulas with many quantification levels. The operation `forget` in RAReQS corresponds to the creation of new copies of the variables in the refinement. QELL [27] can be seen as a variant of RAReQS where variables are removed by greedy elimination.

Both QESTO [14] and CAQE [23] can be seen as specializations of CQESTO for CNF input. A similar approach has also been used in SMT [4].

QuAbS [25] is similar to CQESTO but with some important differences. Conceptually, CQESTO *directly* works on the given circuit. In contrast, QuAbS encodes the circuit into clauses and then it operates on the literals representing subformulas. The representing literals are effectively Tseitin variables (accommodating for semantics of quantification).[2] In this sense, CQESTO is more flexible because QuAbS loses the information about the circuit upon translation to clauses. Observe that for instance that Examples 1 and 2 operate on the same formula but `proj` gives different sub-formulas.

One important consequence of this flexibility is that CQESTO calls the SAT solver with fewer assumptions. As an example consider a sub-circuit ϕ that is set to 1 by propagation of values on previous quantification levels. Further, there are some sub-circuits of ϕ, also set to 1 by previous levels, let's say $\psi_1, \ldots, \psi_k$. To communicate to the current level that these are already set to true, QuAbS invokes the SAT solver with the assumptions $t_\phi, t_{\psi_1}, \ldots, t_{\psi_k}$, where t_γ is the representing literal. Consequently, in QuAbS, any of these assumptions may appear in the core. However, t_ϕ is the more desirable core because it "covers" its sub-circuits. In CQESTO, it is guaranteed to obtain such core because only ϕ will be using the assumptions (thanks to the function `proj`).

3.5 Correctness and Termination

This section shows correctness and termination by showing specific properties of the used operations. We begin by a lemma that intuitively shows that $(\alpha_i \wedge \sigma)$ is roughly the same as $(\alpha_i \wedge \mathtt{proj}(\alpha_i, \sigma))$. This is relevant to the SAT call on line 7.

Lemma 1. *For $\sigma = \bigcup_{j \in 1..i-1} \sigma_j$ and $\mathcal{I} = \mathtt{proj}(\alpha_i, \sigma)$ it holds that*

1. $\sigma \Rightarrow \mathcal{I}$
2. Models restricted to X_i of $\alpha_i \wedge \sigma$ and $\alpha_i \wedge \mathcal{I}$ are the same.

---

[2] This was also done similarly in Z3 when implementing [4].

Proof (sketch). (1) By induction on expression depth. If for a literal l, it holds that $\sigma \vdash_p l$, then by Definition 2 also $l \in \sigma$ and therefore $\sigma \Rightarrow l$. For composite expressions, the implication holds by standard semantics of $\wedge$ and $\neg$.

(2) The models restricted to X_i of $\alpha_i \wedge \sigma$ are the same as of $\alpha_i|_\sigma$. Hence, instead of conjoining σ to α_i we imagine we substitute it into α_i and then apply standard simplification, e.g. $0 \wedge \phi = 0$. This results into the same formula as if we substituted directly 1 for $\phi \in \mathcal{I}$ with and 0 for $\neg\phi \in \mathcal{I}$. E.g. for $\sigma = \{x, y\}$ and $\alpha_i = (x \vee z) \wedge (y \vee q) \wedge o$, we obtain $\sigma \vdash_p (x \vee z)$ and $\sigma \vdash_p (y \vee q)$, and $\alpha_i|_\sigma$ gives o, which is equivalent replacing $(x \vee z)$ and $(x \vee z)$ with 1.

We continue by inspecting the operation $\texttt{forget}(X_i, \psi)$, important in abstraction strengthening. We show that the operation is a weakening of ψ, i.e. it does not rule out permissible moves. At the same time, however, we need to show that the result is not too weak. In particular, that the performed strengthening on lines 10–12 does not allow repeating a play that was once already lost.

Lemma 2. *Let ψ be a formula, $\sigma = \bigcup_{j \in 1..i-1} \sigma_j$ and $\mathcal{C} \subseteq \texttt{proj}(\alpha_i, \sigma)$ s.t. $\alpha_i \wedge \mathcal{C}$ is unsatisfiable.*

1. $\psi \Rightarrow \texttt{forget}(X_i, \psi)$
2. $\sigma \wedge \texttt{forget}(X_i, \neg\mathcal{C})$ is unsatisfiable
3. $\big(\sigma_1 \wedge \cdots \wedge \sigma_{i-2} \wedge \texttt{forget}(X_i, \neg\mathcal{C})\big)|_{\sigma_{i-1}}$ is unsatisfiable

Proof (sketch). (1) A positive occurrence of a formula with a weaker one, or replacing a negative occurrence of a formula with a stronger one leads to a weaker formula [19, The Polarity Proposition]. The operation $\texttt{forget}$ is a special case of this because a positive occurrence of a variable is replaced by 1 (trivially weaker) and a negative occurrence by 0 (trivially stronger).

(2) Since the elements of the core $\mathcal{C}$ must have been obtained by $\texttt{proj}$ (see ln. 7), we have $\sigma \vdash_p \phi$ for $\phi \in \mathcal{C}$ where all variables X_i are unassigned in σ. Hence, replacing the X_i variables with arbitrary expressions preserves the $\vdash_p$ relation, e.g. $\{x\} \vdash_p (x \vee z)$ but also $\{x\} \vdash_p (x \vee 0)$. Bullet (3) is a immediate consequence of (2).

Lemma 3. *Algorithm 1 preserves Invariants 1 and 2.*

Proof (sketch). Invariant 1 is trivially satisfied upon initialization and is preserved by backtracking. Invariant 2 is trivially satisfied upon initialization.

Since $\mathcal{C}$ is a core, $\alpha_i \wedge \mathcal{C}$ is unsatisfiable and therefore $\alpha_i \Rightarrow \neg\mathcal{C}$. This means that player Q_i must satisfy $\neg\mathcal{C}$ because he must satisfy α_i. The operation $\texttt{forget}$ is a weakening (Lemma 2(1)) and therefore the player also must satisfy $\texttt{forget}(\neg\mathcal{C})$. Since the opponent can always decide to play σ_{i-1}, player Q_i must also satisfy $\big(\sigma_1 \wedge \cdots \wedge \sigma_{i-2} \wedge \texttt{forget}(X_i, \neg\mathcal{C})\big)|_{\sigma_{i-1}}$ at level $i-2$. Therefore the backtracking operation preserves the invariant.

Theorem 1. *The algorithm is correct, i.e. it returns the validity of the formula.*

Table 1. Result summary.

Solver	Solved (320)
QFUN	**118**
CQESTO	112
QuAbS	103
GQ	87
Qute	83

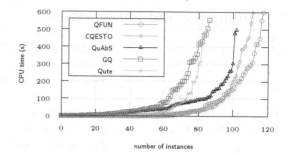

Fig. 1. Cactus plot. A point at (x, y) means that the solver solved x instances each within y sec.

Proof (sketch). The algorithm terminates only if α_i becomes unsatisfiable for $i \in 1..2$. From Invariant 2, the opponent of Q_i has a winning strategy for whatever move Q_i plays. Since Q_i has no previous moves to alter, there's a winning strategy for the opponent for the whole game. The algorithm returns true iff the losing player is $\forall$, i.e. iff there is a winning strategy for $\exists$.

Theorem 2. *The algorithm is terminating.*

Proof (sketch). We show that if the solver backtracks upon an assignment $\sigma = \bigcup_{j \in 1..i-1} \sigma_j$, the same assignment will not appear again. For contradiction let us assume that σ appears in a future run. This means that σ_{i-2} was obtained from $\text{SAT}(\mathcal{I}, \alpha_{i-2})$ with $\mathcal{I} = \text{proj}(\alpha_{i-2}, \bigcup_{j \in 1..i-3} \sigma_j)$. From Lemma 2(3) we have that $\bigcup_{j \in 1..i-2} \sigma_j$ is not a model of α_{i-2}. From Lemma 1(2) $\alpha_{i-2} \wedge \mathcal{I}$ have the same models as $\alpha_{i-2} \wedge \bigcup_{j \in 1..i-3} \sigma_j$, which gives a contradiction.

4   Experimental Evaluation

The prototype CQESTO was implemented in C++ where logical gates are hash-consed as to avoid redundant sub-circuits. SAT calls are delegated to minisat 2.2 [6]. It differs from the Algorithm 1 by starting with stronger α_i: An α_i is initialized by ϕ where opponent's moves are fixed to a constant value.

CQESTO is compared to the solvers QFUN [11], QuAbS [25], Qute [21], and GhostQ [18] on the QBF Eval '17 instances [22] on Intel Xeon E5-2630 2.60 GHz with 64 GB memory; the limits were set to 32 GB and 600 s.

The results are summarized in Table 1 and Fig. 1; detailed results can be found online [5]. There's a clear division between SAT-based solvers (QFUN, CQESTO, QuAbS) and resolution-based solvers (GhostQ, Qute). QFUN is in the lead closely followed by CQESTO. QuAbS is only 9 instances behind CQESTO but the cactus plot shows a notable slow-down early on. I remark that detailed inspection reveals that Qute fares much better on instances with high number of quantification levels. It is a subject of future work to better understand if the difference between QuAbS and CQESTO is due to implementation or the different calculation of strengthening.

5   Summary and Future Work

This paper contributes to the understanding of QBF solving by studying the algorithm CQESTO, which is characterized by maintaining propositional restrictions, in a circuit form, on the possible moves of the corresponding player at each quantification level. *Projection* is used to propagate the current assignment into the circuit. Once the SAT solver provides a contradiction at the current level, this needs to be transferred to the level to which we backtrack. Upon backtracking, CQESTO performs two operations: *substitution* of the opponent's move, *forgetting* of variables belonging to the player. Identifying these operations helps us making a link to other solvers, such as RAReQS.

The presented operations open several avenues for future work. They may enable connecting CNF-based learning [29] and the circuit-based approach by extending propagation (Definition 2). The discussed connection between CQESTO and RAReQS also opens the possibility of combining the two methods, which would in particularly be beneficial in formulas with high number of quantifiers, where RAReQS may be too heavy-handed. The recently proposed use of machine learning for RAReQS implemented in QFUN [11] could also be used in CQESTO as a look-ahead for future moves of the opponent.

Acknowledgments. This work was supported by national funds through Fundação para a Ciência e a Tecnologia (FCT) with reference UID/CEC/50021/2013. The author would like to thank Nikolaj Bjørner and João Marques-Silva for the helpful discussions on the topic.

References

1. Ansótegui, C., Gomes, C.P., Selman, B.: The Achilles' heel of QBF. In: National Conference on Artificial Intelligence and the Seventeenth Innovative Applications of Artificial Intelligence Conference (AAAI), pp. 275–281 (2005)
2. Balabanov, V., Jiang, J.-H.R., Scholl, C., Mishchenko, A., Brayton, R.K.: 2QBF: challenges and solutions. In: Creignou, N., Le Berre, D. (eds.) SAT 2016. LNCS, vol. 9710, pp. 453–469. Springer, Cham (2016). https://doi.org/10.1007/978-3-319-40970-2_28
3. Benedetti, M., Mangassarian, H.: QBF-based formal verification: experience and perspectives. J. Satisfiability Bool. Model. Comput. (JSAT) **5**(1–4), 133–191 (2008)
4. Bjørner, N., Janota, M.: Playing with quantified satisfaction. In: International Conferences on Logic for Programming LPAR-20, Short Presentations, vol. 35, pp. 15–27. EasyChair (2015)
5. CQESTO website. http://sat.inesc-id.pt/~mikolas/sw/cqesto/res.html
6. Eén, N., Sörensson, N.: An extensible SAT-solver. In: Giunchiglia, E., Tacchella, A. (eds.) SAT 2003. LNCS, vol. 2919, pp. 502–518. Springer, Heidelberg (2004). https://doi.org/10.1007/978-3-540-24605-3_37
7. Farzan, A., Kincaid, Z.: Strategy synthesis for linear arithmetic games. In: Proceedings of the ACM on Programming Languages 2 (POPL), pp. 61:1–61:30, December 2017. https://doi.org/10.1145/3158149

8. Goultiaeva, A., Seidl, M., Biere, A.: Bridging the gap between dual propagation and CNF-based QBF solving. In: Design, Automation & Test in Europe (DATE), pp. 811–814 (2013)
9. Goultiaeva, A., Van Gelder, A., Bacchus, F.: A uniform approach for generating proofs and strategies for both true and false QBF formulas. In: International Joint Conference on Artificial Intelligence (IJCAI), pp. 546–553 (2011)
10. Janota, M., Klieber, W., Marques-Silva, J., Clarke, E.: Solving QBF with counterexample guided refinement. In: Cimatti, A., Sebastiani, R. (eds.) SAT 2012. LNCS, vol. 7317, pp. 114–128. Springer, Heidelberg (2012). https://doi.org/10.1007/978-3-642-31612-8_10
11. Janota, M.: Towards generalization in QBF solving via machine learning. In: AAAI Conference on Artificial Intelligence (2018)
12. Janota, M., Klieber, W., Marques-Silva, J., Clarke, E.: Solving QBF with counterexample guided refinement. Artif. Intell. **234**, 1–25 (2016)
13. Janota, M., Marques-Silva, J.: Abstraction-based algorithm for 2QBF. In: Sakallah, K.A., Simon, L. (eds.) SAT 2011. LNCS, vol. 6695, pp. 230–244. Springer, Heidelberg (2011). https://doi.org/10.1007/978-3-642-21581-0_19
14. Janota, M., Marques-Silva, J.: Solving QBF by clause selection. In: International Joint Conference on Artificial Intelligence (IJCAI) (2015)
15. Janota, M., Marques-Silva, J.: An Achilles' heel of term-resolution. In: Conference on Artificial Intelligence (EPIA), pp. 670–680 (2017)
16. Jordan, C., Klieber, W., Seidl, M.: Non-CNF QBF solving with QCIR. In: Proceedings of BNP (Workshop) (2016)
17. Kleine Büning, H., Bubeck, U.: Theory of quantified Boolean formulas. In: Biere, A., Heule, M., van Maaren, H., Walsh, T. (eds.) Handbook of Satisfiability. Frontiers in Artificial Intelligence and Applications, vol. 185, pp. 735–760. IOS Press (2009)
18. Klieber, W., Sapra, S., Gao, S., Clarke, E.: A non-prenex, non-clausal QBF solver with game-state learning. In: Strichman, O., Szeider, S. (eds.) SAT 2010. LNCS, vol. 6175, pp. 128–142. Springer, Heidelberg (2010). https://doi.org/10.1007/978-3-642-14186-7_12
19. Manna, Z., Waldinger, R.: The Logical Basis for Computer Programming, vol. 2. Addison-Wesley, Reading (1985)
20. Marques Silva, J.P., Sakallah, K.A.: GRASP: a search algorithm for propositional satisfiability. IEEE Trans. Comput. **48**(5), 506–521 (1999)
21. Peitl, T., Slivovsky, F., Szeider, S.: Dependency learning for QBF. In: Gaspers, S., Walsh, T. (eds.) SAT 2017. LNCS, vol. 10491, pp. 298–313. Springer, Cham (2017). https://doi.org/10.1007/978-3-319-66263-3_19
22. QBF Eval 2017 (2017). http://www.qbflib.org/event_page.php?year=2017
23. Rabe, M.N., Tentrup, L.: CAQE: a certifying QBF solver. In: Formal Methods in Computer-Aided Design, FMCAD, pp. 136–143 (2015)
24. Reynolds, A., King, T., Kuncak, V.: Solving quantified linear arithmetic by counterexample-guided instantiation. Formal Methods Syst. Des. **51**(3), 500–532 (2017). https://doi.org/10.1007/s10703-017-0290-y
25. Tentrup, L.: Non-prenex QBF solving using abstraction. In: Creignou, N., Le Berre, D. (eds.) SAT 2016. LNCS, vol. 9710, pp. 393–401. Springer, Cham (2016). https://doi.org/10.1007/978-3-319-40970-2_24
26. Tseitin, G.S.: On the complexity of derivations in the propositional calculus. In: Slisenko, A.O. (ed.) Studies in Constructive Mathematics and Mathematical Logic Part II (1968)

27. Tu, K.-H., Hsu, T.-C., Jiang, J.-H.R.: QELL: QBF reasoning with extended clause learning and levelized SAT solving. In: Heule, M., Weaver, S. (eds.) SAT 2015. LNCS, vol. 9340, pp. 343–359. Springer, Cham (2015). https://doi.org/10.1007/978-3-319-24318-4_25
28. Zhang, L.: Solving QBF by combining conjunctive and disjunctive normal forms. In: National Conference on Artificial Intelligence and the Eighteenth Innovative Applications of Artificial Intelligence Conference (AAAI) (2006)
29. Zhang, L., Malik, S.: Conflict driven learning in a quantified Boolean satisfiability solver. In: International Conference On Computer Aided Design (ICCAD), pp. 442–449 (2002)

Symmetries of Quantified Boolean Formulas

Manuel Kauers[1]([✉]) and Martina Seidl[2]

[1] Institute for Algebra, JKU Linz, Linz, Austria
manuel.kauers@jku.at
[2] Institute for Formal Models and Verification, JKU Linz, Linz, Austria
martina.seidl@jku.at

Abstract. While symmetries are well understood for Boolean formulas and successfully exploited in practical SAT solving, less is known about symmetries in quantified Boolean formulas (QBF). There are some works introducing adaptions of propositional symmetry breaking techniques, with a theory covering only very specific parts of QBF symmetries. We present a general framework that gives a concise characterization of symmetries of QBF. Our framework naturally incorporates the duality of universal and existential symmetries resulting in a general basis for QBF symmetry breaking.

1 Introduction

Mathematicians are generally advised [1] to exploit the symmetry in a given problem for solving it. In automated reasoning, however, symmetries are often exploited by destroying them. In this context, to destroy a symmetry means to enrich the given problem by additional constraints which tell the solver that certain parts of the search space are equivalent, so that it investigates only one of them. Such symmetry breaking techniques have been studied for a long time. They are particularly well developed in SAT [2] and CSP [3]. In CSP [4] it has been observed that it is appropriate to distinguish two kinds of symmetries: those of the problem itself and those of the solution set. In the present paper, we apply this idea to Quantified Boolean Formulas (QBF) [5].

Symmetry breaking for QBF has already been studied more than ten years ago [6–9], and it can have a dramatic effect on the performance of QBF solvers. As an extreme example, the instances of the KBKF benchmark set [10] are highly symmetric. For some problem sizes n, we applied the two configurations QRes (standard Q-resolution) and LD (long-distance resolution) of the solver DepQBF [11] to this benchmark set. For LD it is known that it performs exponentially better than QRes on the KBKF formulas [12]. The table on the previous page shows the runtimes of DepQBF without and with symmetry breaking (SB). In particular, we enriched the formulas with symmetry breaking formulas over

Parts of this work were supported by the Austrian Science Fund (FWF) under grant numbers NFN S11408-N23 (RiSE), Y464-N18, and SFB F5004.

O. Beyersdorff and C. M. Wintersteiger (Eds.): SAT 2018, LNCS 10929, pp. 199–216, 2018.
https://doi.org/10.1007/978-3-319-94144-8_13

the existential variables. While QRes-DepQBF only solves two formulas without symmetry breaking, with symmetry breaking it even outperforms LD-DepQBF. Also for the LD configuration, the symmetry breaking formulas are beneficial. While this is an extreme example, symmetries appear not only in crafted formulas. In fact, we found that about 60% of the benchmarks used in the PCNF track of QBFEval [13] have nontrivial symmetries that could be exploited.

n	Solving times (in sec)			
	W/o SB		With SB	
	QRes	LD	QRes	LD
10	0.3	0.5	0.4	0.4
20	160	0.5	0.4	0.4
40	>3600	0.5	0.4	0.4
80	>3600	0.7	0.4	0.4
160	>3600	2.2	0.5	0.4
320	>3600	12.3	0.6	0.5
640	>3600	36.8	1.0	0.8
1280	>3600	241.1	22.6	19.7
2560	>3600	>3600	215.7	155.2
5120	>3600	>3600	1873.2	1042.6

In this paper, we develop an explicit, uniform, and general theory for symmetries of QBFs. The theory is developed from scratch, and we include detailed proofs of all theorems. The pioneering work on QBF symmetries [6–9] largely consisted in translating symmetry breaking techniques well-known from SAT to QBF. This is not trivial, as universal quantifiers require special treatment. Since then, however, research on QBF symmetry breaking almost stagnated. We believe that more work is necessary. For example, we have observed that universal symmetry breakers as introduced in [8] fail to work correctly in modern clause-and-cube-learning QBF solvers when compactly provided as cubes. Although the encoding of the symmetry breaker for universal variables is provably correct in theory, it turns out to be incompatible with pruning techniques like pure literal elimination for which already the compatibility with learning is not obvious [14]. The cubes obtained from symmetry breaking are conceptually different than the learned cubes, because they do not encode a (partial) satisfying assignment of the clauses. As the pruning techniques usually only consider the clausal part of the formula, it can happen that they are wrongly applied in the presence of cubes stemming from a symmetry breaking formula over universal variables, affecting the correctness of the solving result.

We hope that the theory developed in this paper will help to resuscitate the interest in symmetries for QBF, lead to a better understanding of the interplay between symmetry breaking and modern optimization techniques, provide a starting point for translating recent progress made in SAT and CSP to the QBF world, and produce special symmetry breaking formulas that better exploit the unique features of QBF. Potential applications of our framework are the development of novel symmetry breaking formulas based on different orderings then the currently considered lexicographic ordering, the transfer of recent improvements in static symmetry breaking for SAT to QBF, as well as the establishment of dynamic symmetry breaking.

2   Quantified Boolean Formulas

Let $X = \{x_1, \ldots, x_n\}$ be a finite set of propositional variables and $\mathrm{BF}(X)$ be a set of *Boolean formulas* over X. The elements of $\mathrm{BF}(X)$ are well-formed formulas built from the variables of X, truth constants $\top$ (true) and $\bot$ (false), as well as logical connectives according to a certain grammar. For most of the paper, we will not need to be specific about the structure of the elements of $\mathrm{BF}(X)$. We assume a well-defined semantics for the logical connectives, i.e., for every $\phi \in \mathrm{BF}(X)$ and every assignment $\sigma \colon X \to \{\top, \bot\}$ there is a designated value $[\phi]_\sigma \in \{\top, \bot\}$ associated to ϕ and σ. In particular, we use $\wedge$ (conjunction), $\vee$ (disjunction), $\leftrightarrow$ (equivalence), $\to$ (implication), $\oplus$ (xor), and $\neg$ (negation) with their standard semantics for combining and negating formulas. Two formulas $\phi, \psi \in \mathrm{BF}(X)$ are *equivalent* if for every assignment $\sigma \colon X \to \{\top, \bot\}$ we have $[\phi]_\sigma = [\psi]_\sigma$. We use lowercase Greek letters for Boolean formulas and assignments.

If $f \colon \mathrm{BF}(X) \to \mathrm{BF}(X)$ is a function and $\sigma \colon X \to \{\top, \bot\}$ is an assignment, the assignment $f(\sigma) \colon X \to \{\top, \bot\}$ is defined through $f(\sigma)(x) = [f(x)]_\sigma$ ($x \in X$). A partial assignment is a function $\sigma \colon Y \to \{\top, \bot\}$ with $Y \subseteq X$. If σ is such a partial assignment and $\phi \in \mathrm{BF}(X)$, then $[\phi]_\sigma$ shall refer to an element of $\mathrm{BF}(X \setminus Y)$ such that for every assignment $\tau \colon X \to \{\top, \bot\}$ with $\tau|_Y = \sigma$ we have $[[\phi]_\sigma]_\tau = [\phi]_\tau$. For example, $[\phi]_\sigma$ could be the formula obtained from ϕ by replacing every variable $y \in Y$ by the truth value $\sigma(y)$ and then simplifying.

We use uppercase Greek letters to denote *quantified Boolean formulas* (QBFs). A QBF has the form $\varPhi = P.\phi$ where $\phi \in \mathrm{BF}(X)$ is a Boolean formula and P is a quantifier prefix for X, i.e., $P = Q_1 x_1 Q_2 x_2 \ldots Q_n x_n$ for $Q_1, \ldots, Q_n \in \{\forall, \exists\}$. We only consider closed formulas, i.e., each element of X appears in the prefix. For a fixed prefix $P = Q_1 x_1 Q_2 x_2 \ldots Q_n x_n$, two variables x_i, x_j are said to belong to the same *quantifier block* if $Q_{\min(i,j)} = \cdots = Q_{\max(i,j)}$.

Every QBF is either true or false. The truth value is defined recursively as follows: $\forall x P.\phi$ is true iff both $P.[\phi]_{\{x=\top\}}$ and $P.[\phi]_{\{x=\bot\}}$ are true, and $\exists x P.\phi$ is true iff $P.[\phi]_{\{x=\top\}}$ or $P.[\phi]_{\{x=\bot\}}$ is true. For example, $\forall x_1 \exists x_2.(x_1 \leftrightarrow x_2)$ is true and $\exists x_1 \forall x_2.(x_1 \leftrightarrow x_2)$ is false. The semantics of a QBF $P.\phi$ can also be described as a game for two players [15]: In the ith move, the truth value of x_i is chosen by the existential player if $Q_i = \exists$ and by the universal player if $Q_i = \forall$. The existential player wins if the resulting formula is true and the universal player wins if the resulting formula is false. In this interpretation, a QBF is true if there is a winning strategy for the existential player and it is false if there is a winning strategy for the universal player.

Strategies can be described as trees. Let $P = Q_1 x_1 Q_2 x_2 \ldots Q_n x_n$ be a prefix. An *existential strategy* for P is a tree of height $n + 1$ where every node at level $k \in \{1, \ldots, n\}$ has one child if $Q_k = \exists$ and two children if $Q_k = \forall$. In the case $Q_k = \forall$, the two edges to the children are labeled by $\top$ and $\bot$, respectively. In the case $Q_k = \exists$, the edge to the only child is labeled by either $\top$ or $\bot$. *Universal strategies* are defined analogously, the only difference being that the roles of the quantifiers are exchanged, i.e., nodes at level k have two successors if $Q_k = \exists$ (one labeled $\bot$ and one labeled $\top$) and one successor if $Q_k = \forall$ (labeled either $\bot$ or $\top$). Here are the four existential strategies and the two universal strategies for the prefix $\forall x_1 \exists x_2$:

We write $\mathbb{S}_\exists(P)$ for the set of all existential strategies and $\mathbb{S}_\forall(P)$ for the set of all universal strategies. As shown in the following lemma, the set of paths of a given existential strategy for prefix P is never disjoint from the set of paths of a given universal strategy. Unless otherwise stated, by a path, we mean a complete path starting at the root and ending at a leaf, together with the corresponding truth value labels.

Lemma 1. *If P is a prefix and $s \in \mathbb{S}_\exists(P)$, $t \in \mathbb{S}_\forall(P)$, then s and t have a path in common.*

Proof. A common path can be constructed by induction on the length of the prefix. There is nothing to show for prefixes of length 0. Suppose the claim holds for all prefixes of length n and consider a prefix $P' = P\,Q_{n+1}x_{n+1}$ of length $n+1$. Let $s \in \mathbb{S}_\exists(P')$, $t \in \mathbb{S}_\forall(P')$ be arbitrary. By chopping off the leafs of s and t, we obtain elements of $\mathbb{S}_\exists(P)$ and $\mathbb{S}_\forall(P)$, respectively, and these share a common path σ_0 by induction hypothesis. If $Q_{n+1} = \exists$, then σ_0 has a unique continuation in s, with an edge labeled either $\top$ or $\bot$, and σ_0 has two continuations in t, one labeled $\top$ and one labeled $\bot$, so the continuation of σ_0 in s must also appear in t. If $Q_{n+1} = \forall$, the argumentation is analogous. $\square$

Every path in a strategy for a prefix P corresponds to an assignment $\sigma \colon X \to \{\top, \bot\}$. An existential strategy for QBF $P.\phi$ is a *winning strategy* (for the existential player) if all its paths are assignments for which ϕ is true. A universal strategy is a *winning strategy* (for the universal player) if all its paths are assignments for which ϕ is false. For a QBF $P.\phi$ and an existential strategy $s \in \mathbb{S}_\exists(P)$, we define $[P.\phi]_s = \bigwedge_\sigma [\phi]_\sigma$, where σ ranges over all the assignments corresponding to a path of s. (Recall that our assignments are total unless otherwise stated, and our paths go from the root to a leaf unless otherwise stated.) Then we have $[P.\phi]_s = \top$ if and only if s is an existential winning strategy. For a universal strategy $t \in \mathbb{S}_\forall(P)$, we define $[P.\phi]_t = \bigvee_\tau [\phi]_\tau$, where τ ranges over all the assignments corresponding to a path of t. Then $[P.\phi]_s = \bot$ if and only if t is a universal winning strategy.

The definitions made in the previous paragraph are consistent with the interpretation of QBFs introduced earlier: a QBF is true if and only if there is an existential winning strategy, and it is false if and only if there is a universal winning strategy. Lemma 1 ensures that a QBF is either true or false. As another consequence of Lemma 1, observe that for every QBF $P.\phi$ we have

$$\left(\exists\, s \in \mathbb{S}_\exists(P) : [P.\phi]_s = \top \right) \iff \left(\forall\, t \in \mathbb{S}_\forall(P) : [P.\phi]_t = \top \right)$$

$$\text{and} \quad \left(\forall\, s \in \mathbb{S}_\exists(P) : [P.\phi]_s = \bot \right) \iff \left(\exists\, t \in \mathbb{S}_\forall(P) : [P.\phi]_t = \bot \right).$$

We will also need the following property, the proof of which is straightforward.

Lemma 2. *Let P be a prefix for X, and let $\phi, \psi \in \mathrm{BF}(X)$. Then for all $s \in \mathbb{S}_\exists(P)$ we have $[P.(\phi \wedge \psi)]_s = [P.\phi]_s \wedge [P.\psi]_s$, and for all $t \in \mathbb{S}_\forall(P)$ we have $[P.(\phi \vee \psi)]_t = [P.\phi]_t \vee [P.\psi]_t$.*

3   Groups and Group Actions

Symmetries can be described using groups and group actions [16]. Recall that a group is a set G together with an associative binary operation $G \times G \to G$, $(g, h) \mapsto gh$. A group has a neutral element and every element has an inverse in G. A typical example for a group is the set $\mathbb{Z}$ of integers together with addition. Another example is the group of permutations. For any fixed $n \in \mathbb{N}$, a permutation is a bijective function $\pi \colon \{1, \ldots, n\} \to \{1, \ldots, n\}$. The set of all such functions together with composition forms a group, called the symmetric group and denoted by S_n.

A (nonempty) subset H of a group G is called a subgroup of G if it is closed under the group operation and taking inverses. For example, the set $2\mathbb{Z}$ of all even integers is a subgroup of $\mathbb{Z}$, and the set $\{\mathrm{id}, \left(\begin{smallmatrix} 1 & 2 & 3 \\ 1 & 3 & 2 \end{smallmatrix}\right)\}$ is a subgroup of S_3. In general, a subset E of G is not a subgroup. However, for every subset E we can consider the intersection of all subgroups of G containing E. This is a subgroup and it is denoted by $\langle E \rangle$. The elements of E are called *generators* of the subgroup. For example, we have $2\mathbb{Z} = \langle 2 \rangle$, but also $2\mathbb{Z} = \langle 4, 6 \rangle$. A set of generators for S_3 is $\{\left(\begin{smallmatrix} 1 & 2 & 3 \\ 2 & 3 & 1 \end{smallmatrix}\right), \left(\begin{smallmatrix} 1 & 2 & 3 \\ 2 & 1 & 3 \end{smallmatrix}\right)\}$.

If G is a group and S is a set then a *group action* is a map $G \times S \to S$, $(g, x) \mapsto g(x)$ which is compatible with the group operation, i.e., for all $g, h \in G$ and $x \in S$ we have $(gh)(x) = g(h(x))$ and $e(x) = x$, where e is the neutral element of G. Note that when we have a group action, every element $g \in G$ can be interpreted as a bijective function $g \colon S \to S$.

For example, for $G = S_n$ and $S = \{1, \ldots, n\}$ we have a group action by the definition of the elements of S_n. Alternatively, we can let S_n act on a set of tuples of length n, say on $S = \{\square, \bigcirc, \triangle\}^n$, via permutation of the indices, i.e., $\pi(x_1, \ldots, x_n) = (x_{\pi(1)}, \ldots, x_{\pi(n)})$. For example, for $g = \left(\begin{smallmatrix} 1 & 2 & 3 \\ 1 & 3 & 2 \end{smallmatrix}\right)$ we would have $g(\square, \bigcirc, \square) = (\square, \square, \bigcirc)$, $g(\triangle, \triangle, \square) = g(\triangle, \square, \triangle)$, $g(\bigcirc, \triangle, \triangle) = (\bigcirc, \triangle, \triangle)$, etc. As one more example, we can consider the group $G = S_n \times S_m$ consisting of all pairs of permutations. The operation for this group is defined component-wise, i.e., $(\pi, \sigma)(\pi', \sigma') = (\pi\pi', \sigma\sigma')$. We can let G act on a set of two dimensional arrays with shape $n \times m$, say on $S = \{\square, \bigcirc, \triangle\}^{n \times m}$, by letting the first component of a group element permute the row index and the second component permute the column index. For example, for $g = (\left(\begin{smallmatrix} 1 & 2 & 3 \\ 1 & 3 & 2 \end{smallmatrix}\right), \left(\begin{smallmatrix} 1 & 2 & 3 \\ 2 & 3 & 1 \end{smallmatrix}\right))$ we then have

$$g\left(\begin{array}{|c|c|c|} \hline \square & \bigcirc & \triangle \\ \hline \square & \triangle & \square \\ \hline \bigcirc & \bigcirc & \square \\ \hline \end{array}\right) = \begin{array}{|c|c|c|} \hline \bigcirc & \triangle & \square \\ \hline \bigcirc & \square & \bigcirc \\ \hline \triangle & \square & \square \\ \hline \end{array}.$$

If we have a group action $G \times S \to S$, we can define an equivalence relation on S via $x \sim y \iff \exists\, g \in G : x = g(y)$. The axioms of groups and group actions ensure that $\sim$ is indeed an equivalence relation. The equivalence classes are called the *orbits* of the group action. For example, for the action of S_3 on $\{\square, \bigcirc, \triangle\}^3$ discussed above, there are some orbits of size 1 (e.g., $\{(\bigcirc, \bigcirc, \bigcirc)\}$), some orbits of size 3 (e.g., $\{(\square, \square, \triangle), (\square, \triangle, \square), (\triangle, \square, \square)\}$), and there is one orbit of size 6 ($\{(\square, \bigcirc, \triangle), (\square, \triangle, \bigcirc), (\bigcirc, \triangle, \square), (\bigcirc, \square, \triangle), (\triangle, \bigcirc, \square), (\triangle, \square, \bigcirc)\}$).

4   Syntactic Symmetries

In previous work [9], symmetries are characterized as permutations of literals with certain properties like being closed under negation, taking into account the order of the quantifiers, and, when extended to full formulas, always mapping a QBF to itself. As we will argue in the following, this point of view on QBF symmetries covers only a part of the full theory. We use group actions to describe symmetries of QBFs. Two kinds of group actions are of interest. On the one hand, we consider transformations that map formulas to formulas, i.e., a group action $G \times \mathrm{BF}(X) \to \mathrm{BF}(X)$. On the other hand, we consider transformations that map strategies to strategies, i.e., a group action $G \times \mathbb{S}_\exists(P) \to \mathbb{S}_\exists(P)$ or $G \times \mathbb{S}_\forall(P) \to \mathbb{S}_\forall(P)$. In both cases, we consider groups G which preserve the set of winning strategies for a given QBF $P.\phi$.

Let us first consider group actions $G \times \mathrm{BF}(X) \to \mathrm{BF}(X)$. In this case, we need to impose a technical restriction introduced in the following definition.

Definition 3. *Let P be a prefix for X. A bijective function $f\colon \mathrm{BF}(X) \to \mathrm{BF}(X)$ is called* admissible *(w.r.t. P) if*

1. *for every assignment $\sigma\colon X \to \{\top, \bot\}$ and every formula $\phi \in \mathrm{BF}(X)$ we have $[\phi]_{f(\sigma)} = [f(\phi)]_\sigma$;*
2. *for every variable $x \in X$ the formula $f(x)$ only contains variables that belong to the same quantifier block of P as x.*

The first condition ensures that an admissible function f preserves propositional satisfiability. In particular, it implies that for any $\phi, \psi \in \mathrm{BF}(X)$, the formulas $f(\neg\phi)$ and $\neg f(\phi)$ are equivalent, as are $f(\phi \circ \psi)$ and $f(\phi) \circ f(\psi)$ for every binary connective $\circ$. It follows that the inverse of an admissible function is again admissible. It also follows that an admissible function f is essentially determined by its values for the variables. Note that according to Definition 3 variables can be mapped to arbitrary formulas. The second condition may be replaced by a less restricted version, but for simplicity we use the conservative version of above.

Example 4. *Let $X = \{x, y, a, b\}$ and $P = \forall x \forall y \exists a \exists b$. There is an admissible function f with $f(x) = \neg x, f(y) = y, f(a) = b, f(b) = a$. For such a function, we may have $f(x \vee (a \to y)) = \neg x \vee (b \to y)$. A function g with $g(x) = b$ cannot be admissible, because of the second condition. By the first condition, a function h with $h(x) = x$ and $h(y) = \neg x$ cannot be admissible.*

Next we show that admissible functions not only preserve satisfiability of Boolean formulas, but also the truth of QBFs.

Theorem 5. *Let P be a prefix for X and $f\colon \mathrm{BF}(X) \to \mathrm{BF}(X)$ be admissible for P. For any $\phi \in \mathrm{BF}(X)$ the formula $P.\phi$ is true if and only if $P.f(\phi)$ is true.*

Proof. Since the inverse of an admissible function is admissible, it suffices to show "$\Rightarrow$". To do so, we proceed by induction on the number of quantifier blocks in P.

There is nothing to show when P is empty. Suppose the claim is true for all prefixes with k quantifier blocks, and consider a prefix $P = Qx_1 Qx_2 \cdots Qx_i P'$

for some $i \in \{1,\ldots,n\}$, $Q \in \{\forall, \exists\}$, and a prefix P' for $x_{i+1},\ldots,x_n$ with at most k quantifier blocks whose top quantifier is not Q. By the admissibility, we may view f as a pair of functions $f_1\colon \mathrm{BF}(\{x_1,\ldots,x_i\}) \to \mathrm{BF}(\{x_1,\ldots,x_i\})$ and $f_2\colon \mathrm{BF}(\{x_{i+1},\ldots,x_n\}) \to \mathrm{BF}(\{x_{i+1},\ldots,x_n\})$, where f_2 is admissible for P'. Let $s \in \mathbb{S}_\exists(P)$ be a winning strategy for $P.\phi$. We construct a winning strategy $t \in \mathbb{S}_\exists(P)$ for $P.f(\phi)$.

Case 1: $Q = \exists$. In this case, the upper i levels of s and t consist of single paths. Let $\sigma\colon \{x_1,\ldots,x_i\} \to \{\top,\bot\}$ be the assignment corresponding to the upper i levels of s. The subtree s_σ of s rooted at the end of σ (level $i+1$) is a winning strategy for $P'.[\phi]_\sigma$. By induction hypothesis, $P'.f_2([\phi]_\sigma)$ has a winning strategy. Let t have an initial path corresponding to the assignment $\tau = f_1^{-1}(\sigma)$ followed by a winning strategy of $P'.f_2([\phi]_\sigma)$. (Since f_1 is invertible and independent of $x_{i+1},\ldots,x_n$, the assignment τ is well-defined.) Then t is a winning strategy of $P.f(\phi)$. To see this, let ρ be an arbitrary path of t. We show that $[f(\phi)]_\rho = \top$. Indeed,

$$
[f(\phi)]_\rho \overset{\downarrow}{=} [[f(\phi)]_\tau]_\rho \overset{\downarrow}{=} [[f(\phi)]_{f_1^{-1}(\sigma)}]_\rho \overset{\downarrow}{=} [[f_1(f_2(\phi))]_{f_1^{-1}(\sigma)}]_\rho
$$

with labels: *t starts with τ*, *Def. of τ*, *Def. of f_1, f_2*

$$
\overset{\uparrow}{=} [[f_1^{-1}(f_1(f_2(\phi)))]_\sigma]_\rho = [[f_2(\phi)]_\sigma]_\rho \overset{\uparrow}{=} [f_2([\phi]_\sigma)]_\rho \overset{\uparrow}{=} \top .
$$

with labels: *f_1 admissible*, *f_2 admissible*, *choice of t*

Case 2: $Q = \forall$. In this case, the upper i levels of both s and t form complete binary trees in which every path corresponds to an assignment for the variables $x_1,\ldots,x_i$. Let $\tau\colon \{x_1,\ldots,x_i\} \to \{\top,\bot\}$ be such an assignment, and let $\sigma = f_1(\tau)$. Let s_σ be the subtree of s rooted at σ. This is a winning strategy for the formula $P'.[\phi]_\sigma$ obtained from $P.\phi$ by instantiating the variables $x_1,\ldots,x_i$ according to σ and dropping the corresponding part of the prefix. By induction hypothesis, $P'.f_2([\phi]_\sigma)$ has a winning strategy. Pick one and use it as the subtree of t rooted at τ. The same calculation as in Case 1 shows that t is a winning strategy for $P.f(\phi)$. $\qquad\square$

Example 6. *Consider the true QBF $\Phi = P.\phi = \forall x \forall y \exists a \exists b.((x \leftrightarrow a) \wedge (y \leftrightarrow b))$. If f is an admissible function with $f(x) = y$, $f(y) = x$, $f(a) = b$, $f(b) = a$, then obviously, $P.f(\phi)$ is true as well. If g is a non-admissible function with $g(x) = b$, $g(b) = x$, then $P.g(\phi)$ is false.*

Next we introduce the concept of a *syntactic symmetry group*. The attribute 'syntactic' shall emphasize that this group acts on formulas, in contrast to the 'semantic' symmetry group introduced later, which acts on strategies. Our distinction between syntactic and semantic symmetries corresponds to the distinction between the problem and solution symmetries made in CSP [4].

Definition 7. *Let $P.\phi$ be a QBF and let $G \times \mathrm{BF}(X) \to \mathrm{BF}(X)$ be a group action such that every $g \in G$ is admissible w.r.t. P. We call G a syntactic symmetry group for $P.\phi$ if ϕ and $g(\phi)$ are equivalent (i.e. $\phi \leftrightarrow g(\phi)$ is a tautology) for all $g \in G$.*

It should be noticed that being a 'symmetry group' is strictly speaking not a property of the group itself but rather a property of the action of G on $BF(X)$. The elements of a symmetry group are called symmetries. In general, we call a group action admissible if every $g \in G$ is admissible. Definition 7 implies that when G is a syntactic symmetry group for $P.\phi$, then for every element $g \in G$ the QBF $P.g(\phi)$ has the same set of winning strategies as $P.\phi$. Note that this is not already a consequence of Theorem 5, which only said that $P.g(\phi)$ is true if and only if $P.\phi$ is true, which does not imply that they have the same winning strategies.

Example 8. *Consider the QBF $\Phi = P.\phi = \forall x \forall y \exists a \exists b.((x \leftrightarrow a) \land (y \leftrightarrow b))$. A syntactic symmetry group for Φ is $G = \{\mathrm{id}, f\}$, where f is an admissible function with $f(x) = y$, $f(y) = x$, $f(a) = b$, $f(b) = a$.*

Symmetries are often restricted to functions which map variables to literals [9]. But this restriction is not necessary. Also the admissible function g defined by $g(x) = x$, $g(y) = x \oplus y$, $g(a) = a$, $g(b) = a \oplus b$ is a syntactic symmetry for Φ.

5   Semantic Symmetries

In SAT, considering syntactic symmetries is enough, because the solutions of Boolean formulas are variable assignments. As introduced in Sect. 2, the solutions of QBFs are tree-shaped strategies. In order to be able to permute certain subtrees of a strategy while keeping others untouched, we introduce semantic symmetry groups. For the definition of semantic symmetry groups, no technical requirement like the admissibility is needed. Every permutation of strategies that maps winning strategies to winning strategies is fine.

Definition 9. *Let $\Phi = P.\phi$ be a QBF and let G be a group acting on $\mathbb{S}_\exists(P)$ (or on $\mathbb{S}_\forall(P)$). We call G a semantic symmetry group for Φ if for all $g \in G$ and all $s \in \mathbb{S}_\exists(P)$ (or all $s \in \mathbb{S}_\forall(P)$) we have $[\Phi]_s = [\Phi]_{g(s)}$.*

A single syntactic symmetry can give rise to several distinct semantic symmetries, as shown in the following example.

Example 10. *Consider again $\Phi = P.\phi = \forall x \forall y \exists a \exists b.((x \leftrightarrow a) \land (y \leftrightarrow b))$. The function f of the previous example, which exchanges x with y and a with b in the formula, can be translated to a semantic symmetry $\tilde{f}$:*

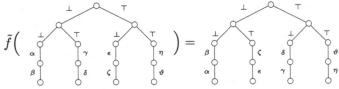

This symmetry exchanges the labels of level 3 and level 4 and swaps the existential parts of the two paths in the middle. Regardless of the choice of $\alpha, \ldots, \eta \in$

$\{\bot, \top\}$, *the strategy on the left is winning if and only if the strategy on the right is winning, so $\tilde{f}$ maps winning strategies to winning strategies.*

Some further semantic symmetries can be constructed from f. For example, in order to be a winning strategy, it is necessary that $\alpha = \beta = \bot$. So we can take a function that just flips α and β but does not touch the rest of the tree. For the same reason, also a function that just flips η and ϑ but does not affect the rest of the tree is a semantic symmetry. The composition of these two functions and the function $\tilde{f}$ described before (in an arbitrary order) yields a symmetry that exchanges γ with ζ and δ with ϵ but keeps $\alpha, \beta, \eta, \vartheta$ fixed. Also this function is a semantic symmetry.

The construction described in the example above works in general. Recall that for an assignment $\sigma \colon X \to \{\top, \bot\}$ and a function $f \colon \mathrm{BF}(X) \to \mathrm{BF}(X)$, the assignment $f(\sigma) \colon X \to \{\top, \bot\}$ is defined by $f(\sigma)(x) = [f(x)]_\sigma$ for $x \in X$.

Lemma 11. *Let P be a prefix for X and g be an element of a group acting admissibly on $\mathrm{BF}(X)$. Then there is a function $f \colon \mathbb{S}_\exists(P) \to \mathbb{S}_\exists(P)$ such that for all $s \in \mathbb{S}_\exists(P)$ we have that σ is a path of $f(s)$ if and only if $g(\sigma)$ is a path of s.*

Proof. Since g is an admissible function, it acts independently on variables belonging to different quantifier blocks. Therefore it suffices to consider the case where P consists of a single quantifier block. If all quantifiers are existential, then s consists of a single path, so the claim is obvious. If there are only universal quantifiers, then s consists of a complete binary tree containing all possible paths, so the claim is obvious as well. □

Starting from a syntactic symmetry group G_{syn}, we can consider all the semantic symmetries that can be obtained from it like in the example above. All these semantic symmetries form a semantic symmetry group, which we call the semantic symmetry group associated to G_{syn}.

Definition 12. *Let P be a prefix for X and let $G_{\mathrm{syn}} \times \mathrm{BF}(X) \to \mathrm{BF}(X)$ be an admissible group action. Let G_{sem} be the set of all bijective functions $f \colon \mathbb{S}_\exists(P) \to \mathbb{S}_\exists(P)$ such that for all $s \in \mathbb{S}_\exists(P)$ and every path σ of $f(s)$ there exists a $g \in G_{\mathrm{syn}}$ such that $g(\sigma)$ is a path of s. This G_{sem} is called the* associated group *of G_{syn}.*

Again, it would be formally more accurate but less convenient to say that the action of G_{sem} on $\mathbb{S}_\exists(P)$ is associated to the action of G_{syn} on $\mathrm{BF}(X)$.

Theorem 13. *If G_{syn} is a syntactic symmetry group for a QBF Φ, then the associated group G_{sem} of G_{syn} is a semantic symmetry group for Φ.*

Proof. Let $\Phi = P.\phi$. Obviously, G_{sem} is a group. To show that it is a symmetry group, let $s \in \mathbb{S}_\exists(P)$ be a winning strategy for Φ, and let $g_{\mathrm{sem}} \in G_{\mathrm{sem}}$. We show that $g_{\mathrm{sem}}(s)$ is again a winning strategy. Let σ be a path of $g_{\mathrm{sem}}(s)$. By Definition 12, there exists a $g_{\mathrm{syn}} \in G_{\mathrm{syn}}$ such that $g_{\mathrm{syn}}(\sigma)$ is a path of s. Since s is a winning strategy, $[\phi]_{g_{\mathrm{syn}}(\sigma)} = \top$, and since g_{syn} is admissible, $[\phi]_{g_{\mathrm{syn}}(\sigma)} = [g_{\mathrm{syn}}(\phi)]_\sigma$. Since g_{syn} is a symmetry, $[g_{\mathrm{syn}}(\phi)]_\sigma = [\phi]_\sigma$, so reading backwards we have $[\phi]_\sigma = [g_{\mathrm{syn}}(\phi)]_\sigma = [g_{\mathrm{syn}}(\phi)]_\sigma = [\phi]_{g_{\mathrm{syn}}(\sigma)} = \top$. Hence every path of $g_{\mathrm{sem}}(s)$ is a satisfying assignment for ϕ, so $g_{\mathrm{sem}}(s)$ is a winning strategy. □

The distinction between a syntactic and a semantic symmetry group is immaterial when the prefix consists of a single quantifier block. In particular, SAT problems can be viewed as QBFs in which all quantifiers are $\exists$. For such formulas, each tree in $\mathbb{S}_\exists(P)$ consists of a single path, so in this case the requirement $\forall\, s \in \mathbb{S}_\exists(P) : [\Phi]_s = [\Phi]_{g(s)}$ from Definition 9 boils down to the requirement that $[\phi]_\sigma = [\phi]_{f(\sigma)}$ should hold for all assignments $\sigma \colon X \to \{\top, \bot\}$. This reflects the condition of Definition 7 that ϕ and $f(\phi)$ are equivalent.

As we have seen in Example 10, there is more diversity for prefixes with several quantifier blocks. In such cases, a single element of a syntactic symmetry group can give rise to a lot of elements of the associated semantic symmetry group. In fact, the associated semantic symmetry group is very versatile. For example, when there are two strategies $s, s' \in \mathbb{S}_\exists(P)$ and some element f of an associated semantic symmetry group G_{sem} such that $f(s) = s'$, then there is also an element $h \in G_{\mathrm{sem}}$ with $h(s) = s'$, $h(s') = s$ and $h(r) = r$ for all $r \in \mathbb{S}_\exists(P) \setminus \{s, s'\}$. The next lemma is a generalization of this observation which indicates that G_{sem} contains elements that exchange subtrees across strategies.

Lemma 14. *Let $P = Q_1 x_1 \ldots Q_n x_n$ be a prefix and $G_{\mathrm{syn}} \times \mathrm{BF}(X) \to \mathrm{BF}(X)$ be an admissible group action. Let G_{sem} be the associated group of G_{syn}. Let $s \in \mathbb{S}_\exists(P)$ and let σ be a path of s. Let $i \in \{1, \ldots, n\}$ be such that $[x_j]_\sigma = [g(x_j)]_\sigma$ for all $g \in G_{\mathrm{syn}}$ and all $j < i$.*

Further, let $f \in G_{\mathrm{sem}}$ and $s' = f(s)$. Let σ' be a path of s' such that the first $i - 1$ edges of σ' agree with the first $i - 1$ edges of σ. By the choice of i such a σ' exists. Let $t, t' \in \mathbb{S}_\exists(Q_i x_i \ldots Q_n x_n)$ be the subtrees of s, s' rooted at the ith node of σ, σ', respectively, and let $s'' \in \mathbb{S}_\exists(P)$ be the strategy obtained from s by replacing t by t', as illustrated in the picture below. Then there exists $h \in G_{\mathrm{sem}}$ with $h(s) = s''$.

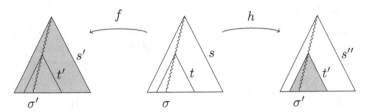

Proof. Define $h \colon \mathbb{S}_\exists(P) \to \mathbb{S}_\exists(P)$ by $h(s) = s''$, $h(s'') = s$, and $h(r) = r$ for all $r \in \mathbb{S}_\exists(P) \setminus \{s, s''\}$. Obviously, h is a bijective function from $\mathbb{S}_\exists(P)$ to $\mathbb{S}_\exists(P)$. To show that h belongs to G_{sem}, we must show that for every $r \in \mathbb{S}_\exists(P)$ and every path ρ of $h(r)$ there exists $g \in G_{\mathrm{syn}}$ such that $g(\rho)$ is a path of r. For $r \in \mathbb{S}_\exists(P) \setminus \{s, s''\}$ we have $h(r) = r$, so there is nothing to show.

Consider the case $r = s$. Let ρ be a path of $h(r) = s''$. If ρ does not end in the subtree t', then the same path ρ also appears in r and we can take $g = \mathrm{id}$. Now suppose that ρ does end in the subtree t'. Then ρ is also a path of $s' = f(s)$, because all paths of s and s' ending in t or t' agree above the ith node. Since $f \in G_{\mathrm{sem}}$, there exists $g \in G_{\mathrm{syn}}$ such that $g(\rho)$ is a path of s.

Finally, consider the case $r = s''$. Let ρ be a path of $h(r) = s$. If ρ does not end in the subtree t, then the same path ρ also appears in r and we can take $g = \mathrm{id}$. Now suppose that ρ does end in the subtree t. Then the first $i - 1$ edges of ρ agree with those of σ. Since $s = f^{-1}(s')$, there exists $g \in G_{\mathrm{syn}}$ such that $g(\rho)$ is a path of s'. By assumption on G_{syn}, the element g fixes first $i - 1$ edges of ρ, so $g(\rho)$ ends in t' and is therefore a path of s'', as required. $\square$

6   Existential Symmetry Breakers

The action of a syntactic symmetry group of a QBF $P.\phi$ splits $\mathrm{BF}(X)$ into orbits. For all the formulas ψ in the orbit of ϕ, the QBF $P.\psi$ has exactly the same winning strategies as $P.\phi$. For finding a winning strategy, we therefore have the freedom of exchanging ϕ with any other formula in its orbit.

The action of a semantic symmetry group on $\mathbb{S}_\exists(P)$ splits $\mathbb{S}_\exists(P)$ into orbits. In this case, every orbit either contains only winning strategies for $P.\phi$ or no winning strategies for $P.\phi$ at all:

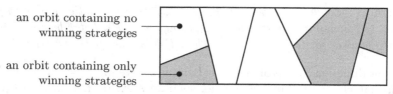

Instead of checking all elements of $\mathbb{S}_\exists(P)$, it is sufficient to check one element per orbit. If a winning strategy exists, then any such sample contains one.

To avoid inspecting strategies that belong to the same orbit symmetry breaking introduces a formula $\psi \in \mathrm{BF}(X)$ which is such that $P.\psi$ has at least one winning strategy in every orbit. Such a formula is called a *symmetry breaker*. The key observation is that instead of solving $P.\phi$, we can solve $P.(\phi \wedge \psi)$. Every winning strategy for the latter will be a winning strategy for the former, and if the former has at least one winning strategy, then so does the latter. By furthermore allowing transformations of ϕ via a syntactic symmetry group, we get the following definition.

Definition 15. *Let P be a prefix for X, let G_{syn} be a group acting admissibly on $\mathrm{BF}(X)$ and let G_{sem} be a group action on $\mathbb{S}_\exists(P)$. A formula $\psi \in \mathrm{BF}(X)$ is called an* existential symmetry breaker *for P (w.r.t. the actions of G_{syn} and G_{sem}) if for every $s \in \mathbb{S}_\exists(P)$ there exist $g_{\mathrm{syn}} \in G_{\mathrm{syn}}$ and $g_{\mathrm{sem}} \in G_{\mathrm{sem}}$ such that $[P.g_{\mathrm{syn}}(\psi)]_{g_{\mathrm{sem}}(s)} = \top$.*

Example 16. *Consider the formula $\Phi = P.\phi = \forall x \exists y \exists z.(y \leftrightarrow z)$. All the elements of $\mathbb{S}_\exists(P)$ have the form depicted on the right. As syntactic symmetries, we have the admissible functions $f, g \colon \mathrm{BF}(X) \to \mathrm{BF}(X)$ defined by $f(x) = x$, $f(y) = z$, $f(z) = y$, and $g(x) = x$, $g(y) = \neg y$, $g(z) = \neg z$, respectively, so we can take $G_{\mathrm{syn}} = \langle f, g \rangle$ as a syntactic symmetry group.*

According to [8,9] the formula $\neg y$ is a symmetry breaker for $P.\phi$. When considering G_{syn} together with $G_{\mathrm{sem}} = \{\mathrm{id}\}$ (what would be sufficient for SAT), *the complications for QBF become obvious. The orbit of $\neg y$ is $O = \{y, z, \neg y, \neg z\}$. Now consider the strategy with $\alpha = \top, \beta = \bot, \gamma = \bot, \delta = \top$. For any $\psi \in O$, this strategy does not satisfy $P.\psi$, because ψ is true on one branch, but false on the other. Using semantic symmetries can overcome this problem.*

Semantic symmetries can act differently on different paths. Let $f_1 \colon \mathbb{S}_\exists(P) \to \mathbb{S}_\exists(P)$ be the function which exchanges α, β and leaves γ, δ fixed, let $g_1 \colon \mathbb{S}_\exists(P) \to \mathbb{S}_\exists(P)$ be the function which replaces α, β by $\neg\alpha, \neg\beta$ and leaves γ, δ fixed, and let $f_2, g_2 \colon \mathbb{S}_\exists(P) \to \mathbb{S}_\exists(P)$ be defined like f_1, g_1 but with the roles of α, β and γ, δ exchanged. The group $G_{\mathrm{sem}} = \langle f_1, g_1, f_2, g_2 \rangle$ is a semantic symmetry group for Φ. This group splits $\mathbb{S}_\exists(P)$ into four orbits: one orbit consists of all strategies with $\alpha = \beta$, $\gamma = \delta$, one consists of those with $\alpha = \beta$, $\gamma \neq \delta$, one consists of those with $\alpha \neq \beta$, $\gamma = \delta$, and on consists of those with $\alpha \neq \beta$, $\gamma \neq \delta$.

Taking $G_{\mathrm{syn}} = \{\mathrm{id}\}$ together with this group G_{sem}, the formula $\neg y$ is a symmetry breaker, because each orbit contains one element with $\alpha = \gamma = \bot$.

The following theorem is the main property of symmetry breakers.

Theorem 17. *Let $\Phi = P.\phi$ be a QBF. Let G_{syn} be a syntactic symmetry group and G_{sem} be a semantic symmetry group acting on $\mathbb{S}_\exists(P)$. Let ψ be an existential symmetry breaker for G_{syn} and G_{sem}. Then $P.\phi$ is true iff $P.(\phi \wedge \psi)$ is true.*

Proof. The direction "$\Leftarrow$" is obvious (by Lemma 2). We show "$\Rightarrow$". Let $s \in \mathbb{S}_\exists(P)$ be such that $[\Phi]_s = \top$. Since Φ is true, such an s exists. Let $g_{\mathrm{syn}} \in G_{\mathrm{syn}}$ and $g_{\mathrm{sem}} \in G_{\mathrm{sem}}$ be such that $[P.g_{\mathrm{syn}}(\psi)]_{g_{\mathrm{sem}}(s)} = \top$. Since ψ is an existential symmetry breaker, such elements exist. Since G_{syn} and G_{sem} are symmetry groups, $[P.g_{\mathrm{syn}}(\phi)]_{g_{\mathrm{sem}}(s)} = [P.\phi]_s = \top$. Lemma 2 implies $[P.(g_{\mathrm{syn}}(\phi) \wedge g_{\mathrm{syn}}(\psi))]_{g_{\mathrm{sem}}(s)} = \top$. By the compatibility with logical operations (admissibility),

$$[P.g_{\mathrm{syn}}(\phi \wedge \psi)]_{g_{\mathrm{sem}}(s)} = [P.(g_{\mathrm{syn}}(\phi) \wedge g_{\mathrm{syn}}(\psi))]_{g_{\mathrm{sem}}(s)} = \top.$$

Now by Theorem 5 applied with g_{syn}^{-1} to $P.g_{\mathrm{syn}}(\phi \wedge \psi)$, it follows that there exists s' such that $[P.(\phi \wedge \psi)]_{s'} = \top$, as claimed.   □

As a corollary, we may remark that for an existential symmetry breaker ψ for the prefix P the formula $P.\psi$ is always true. To see this, choose $\phi = \top$ and observe that any groups G_{syn} and G_{sem} are symmetry groups for ϕ. By the theorem, $P.(\phi \wedge \psi)$ is true, so $P.\psi$ is true.

7   Universal Symmetry Breakers

An inherent property of reasoning about QBFs is the duality between "existential" and "universal" reasoning [17], i.e., the duality between proving and refuting a QBF. For showing that a QBF is true, an existential strategy has to be found that is an existential winning strategy. An existential symmetry breaker tightens the pool of existential strategies among which the existential winning strategy can be found (in case there is one).

If the given QBF is false, then a universal strategy has to be found that is a universal winning strategy. In this case, an existential symmetry breaker is not useful. Recall that a universal winning strategy is a tree in which all paths are falsifying assignments. Using an existential symmetry breaker as in Theorem 17 tends to increase the number of such paths and thus increases the number of potential candidates. To aid the search for a universal winning strategy, it would be better to increase the number of paths corresponding to satisfying assignments, because this reduces the search space for universal winning strategies. For getting symmetry breakers serving this purpose, we can use a theory that is analogous to the theory of the previous section.

Definition 18. *Let P be a prefix for X, let G_{syn} be a group acting admissibly on $BF(X)$ and let G_{sem} be a group action on $\mathbb{S}_\forall(P)$. A formula $\psi \in BF(X)$ is called a* universal symmetry breaker *for P (w.r.t. the actions of G_{syn} and G_{sem}) if for every $t \in \mathbb{S}_\forall(P)$ there exist $g_{syn} \in G_{syn}$ and $g_{sem} \in G_{sem}$ such that $[P.g_{syn}(\psi)]_{g_{sem}(t)} = \bot$.*

No change is needed for the definition of syntactic symmetry groups. A semantic symmetry group for $\Phi = P.\phi$ is now a group acting on $\mathbb{S}_\forall(P)$ in such a way that $[P.\phi]_t = [P.\phi]_{g(t)}$ for all $g \in G$ and all $t \in \mathbb{S}_\forall(P)$. With these adaptions, we have the following analog of Theorem 17.

Theorem 19. *Let $\Phi = P.\phi$ be a QBF. Let G_{syn} be a syntactic symmetry group and G_{sem} be a semantic symmetry group acting on $\mathbb{S}_\forall(P)$. Let ψ be a universal symmetry breaker for G_{syn} and G_{sem}. Then $P.\phi$ is false iff $P.(\phi \vee \psi)$ is false.*

The proof is obtained from the proof of Theorem 17 by replacing $\mathbb{S}_\exists(P)$ by $\mathbb{S}_\forall(P)$, every $\wedge$ by $\vee$, every $\top$ by $\bot$, and "existential" by "universal".

We have seen before that for an existential symmetry breaker $\psi_\exists$ the QBF $P.\psi_\exists$ is necessarily true. Likewise, for a universal symmetry breaker $\psi_\forall$, the QBF $P.\psi_\forall$ is necessarily false. This has the important consequence that existential and universal symmetry breakers can be used in combination, even if they are not defined with respect to the same group actions.

Theorem 20. *Let $\Phi = P.\phi$ be a QBF. Let $G_{syn}^\exists$ and $G_{syn}^\forall$ be syntactic symmetry groups of Φ, let $G_{sem}^\exists$ be a semantic symmetry group of Φ acting on $\mathbb{S}_\exists(P)$ and let $G_{sem}^\forall$ be a semantic symmetry group of Φ acting on $\mathbb{S}_\forall(P)$. Let $\psi_\exists$ be an existential symmetry breaker for $G_{syn}^\exists$ and $G_{sem}^\exists$, and let $\psi_\forall$ be a universal symmetry breaker for $G_{syn}^\forall$ and $G_{sem}^\forall$. Then $P.\phi$ is true iff $P.((\phi \vee \psi_\forall) \wedge \psi_\exists)$ is true iff $P.((\phi \wedge \psi_\exists) \vee \psi_\forall)$ is true.*

Proof. For the first equivalence, we have

$$P.\phi \text{ is true} \overset{\text{Thm. 19}}{\Longleftrightarrow} P.(\phi \vee \psi_\forall) \text{ is true}$$

$$\overset{\text{Def.}}{\Longleftrightarrow} \exists\, s \in \mathbb{S}_\exists(P) : [P.(\phi \vee \psi_\forall)]_s = \top$$

$$\Longleftrightarrow \exists\, s \in \mathbb{S}_\exists(P) : [P.(\phi \vee \psi_\forall)]_s \wedge \underbrace{[P.\psi_\exists]_s}_{=\top} = \top$$

$$\overset{\text{Lem. 2}}{\Longleftrightarrow} \exists\, s \in \mathbb{S}_\exists(P) : [P.((\phi \vee \psi_\forall) \wedge \psi_\exists)]_s = \top$$

$$\overset{\text{Def.}}{\Longleftrightarrow} P.((\phi \vee \psi_\forall) \wedge \psi_\exists) \text{ is true.}$$

The proof of the second equivalence is analogous.   □

Next we relate existential symmetry breakers to universal symmetry breakers. Observe that when P is a prefix and $\tilde{P}$ is the prefix obtained from P by changing all quantifiers, i.e., replacing each $\exists$ by $\forall$ and each $\forall$ by $\exists$, then $\mathbb{S}_\exists(P) = \mathbb{S}_\forall(\tilde{P})$. For any formula $\phi \in \mathrm{BF}(X)$ and any $s \in \mathbb{S}_\exists(P) = \mathbb{S}_\forall(\tilde{P})$ we have $\neg[P.\phi]_s = [\tilde{P}.\neg\phi]_s$. Therefore, if G_syn is a group acting admissibly on $\mathrm{BF}(X)$ and G_sem is a group acting on $\mathbb{S}_\exists(P) = \mathbb{S}_\forall(\tilde{P})$, we have

ψ is an existential symmetry breaker for G_syn and G_sem

$$\Longleftrightarrow \forall\, s \in \mathbb{S}_\exists(P) \; \exists\, g_\text{syn} \in G_\text{syn}, g_\text{sem} \in G_\text{sem} : [P.g_\text{syn}(\psi)]_{g_\text{sem}(s)} = \top$$

$$\Longleftrightarrow \forall\, s \in \mathbb{S}_\forall(\tilde{P}) \; \exists\, g_\text{syn} \in G_\text{syn}, g_\text{sem} \in G_\text{sem} : [\tilde{P}.\neg g_\text{syn}(\psi)]_{g_\text{sem}(s)} = \bot$$

$$\Longleftrightarrow \forall\, s \in \mathbb{S}_\forall(\tilde{P}) \; \exists\, g_\text{syn} \in G_\text{syn}, g_\text{sem} \in G_\text{sem} : [\tilde{P}.g_\text{syn}(\neg\psi)]_{g_\text{sem}(s)} = \bot$$

$$\Longleftrightarrow \neg\psi \text{ is a universal symmetry breaker for } G_\text{syn} \text{ and } G_\text{sem},$$

where admissibility of g_syn is used in the third step. We have thus proven the following theorem, which captures Property 2 of the symmetry breaker introduced in [8] by relating existential and universal symmetry breakers.

Theorem 21. *Let P be a prefix for X and let $\tilde{P}$ be the prefix obtained from P by flipping all the quantifiers. Let G_syn be a group acting admissibly on $\mathrm{BF}(X)$ and let G_sem be a group acting on $\mathbb{S}_\exists(P) = \mathbb{S}_\forall(\tilde{P})$. Then $\psi \in \mathrm{BF}(X)$ is an existential symmetry breaker for G_syn and G_sem if and only if $\neg\psi$ is a universal symmetry breaker for G_syn and G_sem.*

8   Construction of Symmetry Breakers

Because of Theorem 21, it suffices to discuss the construction of existential symmetry breakers. A universal symmetry breaker is obtained in a dual manner. Given a symmetry group, the basic idea is similar as for SAT (see also the French thesis of Jabbour [9] for a detailed discussion on lifting SAT symmetry breaking techniques to QBF). First an order on $\mathbb{S}_\exists(P)$ is imposed, so that every orbit contains an element which is minimal with respect to the order. Then we construct a formula $\psi_\exists$ for which (at least) the minimal elements of the orbits are winning strategies. Any such formula is an existential symmetry breaker. One way of constructing an existential symmetry breaker is given in the following theorem, which generalizes the symmetry breaking technique by Crawford et al. [18]. We give a formal proof that we obtain indeed a QBF symmetry breaker and conclude with lifting a CNF encoding used in recent SAT solving technology [19] to QBF.

Theorem 22. *Let* $P = Q_1 x_1 \ldots Q_n x_n$ *be a prefix for* X, *let* G_{syn} *be a group acting admissibly on* $\mathrm{BF}(X)$, *and let* G_{sem} *be the associated group of* G_{syn}. *Then*

$$\psi = \bigwedge_{\substack{i\,=\,1 \\ Q_i\,=\,\exists}}^{n} \bigwedge_{g \in G_{\mathrm{syn}}} \left(\left(\bigwedge_{j<i} (x_j \leftrightarrow g(x_j)) \right) \to \left(x_i \to g(x_i) \right) \right)$$

is an existential symmetry breaker for G_{syn} *and* G_{sem}.

Proof. All elements of $\mathbb{S}_\exists(P)$ are trees with the same shape. Fix a numbering of the edge positions in these trees which is such that whenever two edges are connected by a path, the edge closer to the root has the smaller index. (One possibility is breadth first search order.) For any two distinct strategies $s_1, s_2 \in \mathbb{S}_\exists(P)$, there is then a minimal k such that the labels of the kth edges of s_1, s_2 differ. Define $s_1 < s_2$ if the label is $\bot$ for s_1 and $\top$ for s_2, and $s_1 > s_2$ otherwise.

Let $s \in \mathbb{S}_\exists(P)$. We need to show that there are $g_{\mathrm{syn}} \in G_{\mathrm{syn}}$ and $g_{\mathrm{sem}} \in G_{\mathrm{sem}}$ such that $[g_{\mathrm{syn}}(\psi)]_{g_{\mathrm{sem}}(s)} = \top$. Let $g_{\mathrm{syn}} = \mathrm{id}$ and let g_{sem} be such that $\tilde{s} := g_{\mathrm{sem}}(s)$ is as small as possible in the order defined above. We show that $[\psi]_{\tilde{s}} = \top$. Assume otherwise. Then there exists $i \in \{1, \ldots, n\}$ with $Q_i = \exists$ and $g \in G_{\mathrm{syn}}$ and a path σ in $\tilde{s}$ with $[x_j]_\sigma = [g(x_j)]_\sigma$ for all $j < i$ and $[x_i]_\sigma = \top$ and $[g(x_i)]_\sigma = \bot$. By Lemma 11, the element $g \in G_{\mathrm{syn}}$ can be translated into an element $f \in G_{\mathrm{sem}}$ which maps $\tilde{s}$ to a strategy $f(\tilde{s})$ which contains a path that agrees with σ on the upper $i-1$ edges but not on the ith. By Lemma 14, applied to the subgroup $H \subseteq G_{\mathrm{syn}}$ consisting of all $h \in G_{\mathrm{syn}}$ with $[x_j]_\sigma = [h(x_j)]_\sigma$ for all $j < i$, we may assume that $f(\tilde{s})$ and $\tilde{s}$ only differ in edges that belong to the subtree rooted at the ith node of σ. As all these edges have higher indices, we have $\tilde{s} < s$, in contradiction to the minimality assumption on s. $\square$

Note that we do not need to know the group G_{sem} explicitly. It is only used implicitly in the proof.

In nontrivial applications, G_{syn} will have a lot of elements. It is not necessary (and not advisable) to use them all, although Theorem 22 would allow us to do so. In general, if a formula $\psi_1 \wedge \psi_2$ is an existential symmetry breaker, then so are ψ_1 and ψ_2, so we are free to use only parts of the large conjunctions. A reasonable choice is to pick a set E of generators for G_{syn} and let the inner conjunction run over (some of) the elements of E.

The formula ψ of Theorem 22 can be efficiently encoded as conjunctive normal form (CNF), adopting the propositional encoding of [2,19]: let $g \in G_{\mathrm{syn}}$ and let

$\{y_0^g, \ldots, y_{n-1}^g\}$ be a set of fresh variables. First, we define a set I^g of clauses that represent all implications $x_i \rightarrow g(x_i)$ of ψ from Theorem 17,

$$I^g = \{(\neg y_{i-1}^g \vee \neg x_i \vee g(x_i)) \mid 1 \leq i \leq n, Q_i = \exists\}.$$

When x_i is existentially quantified, by using Tseitin variables y_{i-1}^g we can recycle the implications $x_i \rightarrow g(x_i)$ in the encoding of the equivalences $x_j \leftrightarrow g(x_j)$ that appear in the outer implication:

$$E^g = \{(y_j^g \vee \neg y_{j-1}^g \vee \neg x_j) \wedge (y_j \vee \neg y_{j-1} \vee g(x_j)) \mid 1 \leq j < n, Q_j = \exists\}.$$

If variable x_j is universally quantified, the recycling is not possible, so we use

$$U^g = \{(y_j^g \vee \neg y_{j-1}^g \vee \neg x_j \vee \neg g(x_j)) \wedge (y_j^g \vee \neg y_{j-1}^g \vee x_j \vee g(x_j)) \mid 1 \leq j < n, Q_j = \forall\}$$

instead. The CNF encoding of ψ is then the conjunction of y_0^g and all the clauses in I^g, E^g, and U^g, for all desired $g \in G_{\text{syn}}$. The prefix P has to be extended by additional quantifiers which bind the Tseitin variables y_i^g. As explained in [20], the position of such a new variable in the prefix has to be behind the quantifiers of the variables occurring in its definition. The encoding of universal symmetry breakers works similarly and results in a formula in disjunctive normal form (DNF), i.e., a disjunction of cubes (where a cube is a conjunction of literals). In this case the auxiliary variables are universally quantified. The obtained cubes could be used by solvers that simultaneously reason on the CNF and DNF representation of a formula (e.g., [21,22]) or by solvers that operate on formulas of arbitrary structure (e.g., [22–24]). The practical evaluation of this approach is a separate topic which we leave to future work.

Besides the practical evaluation of the discussed symmetry breakers in connection with recent QBF solving technologies there are many more promising directions for future work. Also different orderings than the lexicographic order applied in Theorem 22 could be used [25] for the construction of novel symmetry breakers. Recent improvements of static symmetry breaking [19] for SAT could be lifted to QBF and applied in combination with recent preprocessing techniques. Also dynamic symmetry breaking during the solving could be beneficial, for example in the form of symmetric explanation learning [26].

An other interesting direction would be the relaxation of the quantifier ordering. Our symmetry framework assumes a fixed quantifier prefix with a strict ordering. In recent works it has been shown that relaxing this order by the means of dependency schemes is beneficial for QBF solving both in theory and in practice [27,28]. In a similar way as proof systems have been parameterized with dependency schemes, our symmetry framework can also be parameterized with dependency schemes. It can be expected that a more relaxed notion of quantifier dependencies induces more symmetries resulting in more powerful symmetry breakers.

References

1. Polya, G.: How to Solve It: A New Aspect of Mathematical Method. Princeton University Press, Princeton (1945)
2. Sakallah, K.A.: Symmetry and satisfiability. In: Biere, A., Heule, M., Van Maaren, H., Walsh, T. (eds.) Handbook of Satisfiability. Frontiers in Artificial Intelligence and Applications, vol. 185, pp. 289–338. IOS Press, Amsterdam (2009)
3. Gent, I.P., Petrie, K.E., Puget, J.: Symmetry in constraint programming. In: Rossi, F., Walsh, T., van Beek, P. (eds.) Handbook of Constraint Programming. Foundations of Artificial Intelligence, vol. 2, pp. 329–376. Elsevier, Amsterdam (2006)
4. Cohen, D.A., Jeavons, P., Jefferson, C., Petrie, K.E., Smith, B.M.: Constraint symmetry and solution symmetry. In: Proceedings of the 21st National Conference on Artificial Intelligence and the 18th Innovative Applications of Artificial Intelligence Conference (AAAI/IAAI 2006), pp. 1589–1592. AAAI Press (2006)
5. Kleine Büning, H., Bubeck, U.: Theory of quantified Boolean formulas. In: Biere, A., Heule, M., Van Maaren, H., Walsh, T. (eds.) Handbook of Satisfiability. Frontiers in Artificial Intelligence and Applications, vol. 185, pp. 735–760. IOS Press, Amsterdam (2009)
6. Audemard, G., Mazure, B., Sais, L.: Dealing with symmetries in quantified Boolean formulas. In: Proceedings of the 7th International Conference on Theory and Applications of Satisfiability Testing (SAT 2004), Online Proceedings (2004)
7. Audemard, G., Jabbour, S., Sais, L.: Symmetry breaking in quantified Boolean formulae. In: Proceedings of the 20th International Joint Conference on Artificial Intelligence (IJCAI 2007), pp. 2262–2267 (2007)
8. Audemard, G., Jabbour, S., Sais, L.: Efficient symmetry breaking predicates for quantified Boolean formulae. In: Proceedings of Workshop on Symmetry and Constraint Satisfaction Problems (SymCon 2007) (2007). 7 pages
9. Jabbour, S.: De la satisfiabilité propositionnelle aux formules booléennes quantifiées. Ph.D. thesis, CRIL, Lens, France (2008)
10. Kleine Büning, H., Karpinski, M., Flögel, A.: Resolution for quantified Boolean formulas. Inf. Comput. **117**(1), 12–18 (1995)
11. Lonsing, F., Egly, U.: DepQBF 6.0: a search-based QBF solver beyond traditional QCDCL. In: de Moura, L. (ed.) CADE 2017. LNCS (LNAI), vol. 10395, pp. 371–384. Springer, Cham (2017). https://doi.org/10.1007/978-3-319-63046-5_23
12. Egly, U., Lonsing, F., Widl, M.: Long-distance resolution: proof generation and strategy extraction in search-based QBF solving. In: McMillan, K., Middeldorp, A., Voronkov, A. (eds.) LPAR 2013. LNCS, vol. 8312, pp. 291–308. Springer, Heidelberg (2013). https://doi.org/10.1007/978-3-642-45221-5_21
13. Pulina, L., Seidl, M.: The QBFEval 2017. http://www.qbflib.org/qbfeval17
14. Giunchiglia, E., Narizzano, M., Tacchella, A.: Monotone literals and learning in QBF reasoning. In: Wallace, M. (ed.) CP 2004. LNCS, vol. 3258, pp. 260–273. Springer, Heidelberg (2004). https://doi.org/10.1007/978-3-540-30201-8_21
15. Papadimitriou, C.H.: Computational Complexity. Addison-Wesley, Reading (1994)
16. Artin, M.: Algebra. Pearson Prentice Hall, Upper Saddle River (2011)
17. Sabharwal, A., Ansotegui, C., Gomes, C.P., Hart, J.W., Selman, B.: QBF modeling: exploiting player symmetry for simplicity and efficiency. In: Biere, A., Gomes, C.P. (eds.) SAT 2006. LNCS, vol. 4121, pp. 382–395. Springer, Heidelberg (2006). https://doi.org/10.1007/11814948_35
18. Crawford, J.M., Ginsberg, M.L., Luks, E.M., Roy, A.: Symmetry-breaking predicates for search problems. In: Proceedings of the 5th International Conference on

Principles of Knowledge Representation and Reasoning (KR 1996), pp. 148–159. Morgan Kaufmann (1996)

19. Devriendt, J., Bogaerts, B., Bruynooghe, M., Denecker, M.: Improved static symmetry breaking for SAT. In: Creignou, N., Le Berre, D. (eds.) SAT 2016. LNCS, vol. 9710, pp. 104–122. Springer, Cham (2016). https://doi.org/10.1007/978-3-319-40970-2_8

20. Egly, U., Seidl, M., Tompits, H., Woltran, S., Zolda, M.: Comparing different prenexing strategies for quantified Boolean formulas. In: Giunchiglia, E., Tacchella, A. (eds.) SAT 2003. LNCS, vol. 2919, pp. 214–228. Springer, Heidelberg (2004). https://doi.org/10.1007/978-3-540-24605-3_17

21. Goultiaeva, A., Seidl, M., Biere, A.: Bridging the gap between dual propagation and CNF-based QBF solving. In: Proceedings of the International Conference on Design, Automation and Test in Europe (DATE 2013), EDA Consortium, San Jose, CA, USA, pp. 811–814. ACM DL (2013)

22. Janota, M., Klieber, W., Marques-Silva, J., Clarke, E.M.: Solving QBF with counterexample guided refinement. Artif. Intell. **234**, 1–25 (2016)

23. Janota, M.: QFUN: towards machine learning in QBF. CoRR abs/1710.02198 (2017)

24. Tentrup, L.: Non-prenex QBF solving using abstraction. In: Creignou, N., Le Berre, D. (eds.) SAT 2016. LNCS, vol. 9710, pp. 393–401. Springer, Cham (2016). https://doi.org/10.1007/978-3-319-40970-2_24

25. Narodytska, N., Walsh, T.: Breaking symmetry with different orderings. In: Schulte, C. (ed.) CP 2013. LNCS, vol. 8124, pp. 545–561. Springer, Heidelberg (2013). https://doi.org/10.1007/978-3-642-40627-0_41

26. Devriendt, J., Bogaerts, B., Bruynooghe, M.: Symmetric explanation learning: effective dynamic symmetry handling for SAT. In: Gaspers, S., Walsh, T. (eds.) SAT 2017. LNCS, vol. 10491, pp. 83–100. Springer, Cham (2017). https://doi.org/10.1007/978-3-319-66263-3_6

27. Blinkhorn, J., Beyersdorff, O.: Shortening QBF proofs with dependency schemes. In: Gaspers, S., Walsh, T. (eds.) SAT 2017. LNCS, vol. 10491, pp. 263–280. Springer, Cham (2017). https://doi.org/10.1007/978-3-319-66263-3_17

28. Peitl, T., Slivovsky, F., Szeider, S.: Dependency learning for QBF. In: Gaspers, S., Walsh, T. (eds.) SAT 2017. LNCS, vol. 10491, pp. 298–313. Springer, Cham (2017). https://doi.org/10.1007/978-3-319-66263-3_19

Local Soundness for QBF Calculi

Martin Suda and Bernhard Gleiss[(✉)]

TU Wien, Vienna, Austria
{msuda,bgleiss}@forsyte.at

Abstract. We develop new semantics for resolution-based calculi for Quantified Boolean Formulas, covering both the CDCL-derived calculi and the expansion-derived ones. The semantics is centred around the notion of a partial strategy for the universal player and allows us to show in a local, inference-by-inference manner that these calculi are sound. It also helps us understand some less intuitive concepts, such as the role of tautologies in long-distance resolution or the meaning of the "star" in the annotations of IRM-calc. Furthermore, we show that a clause of any of these calculi can be, in the spirit of Curry-Howard correspondence, interpreted as a specification of the corresponding partial strategy. The strategy is total, i.e. winning, when specified by the empty clause.

1 Introduction

The ongoing interest in the problem of Quantified Boolean Formulas (QBF) has resulted in numerous solving techniques, e.g. [10,11,19,22,23], as well as various resolution-based, clausal calculi [2,5,20,21,28] which advance our understanding of the techniques and formalise the involved reasoning.

While a substantial progress in terms of understanding these calculi has already been made on the front of proof complexity [2,4–8,13,17,18,20,26], the question of semantics of the involved intermediate clauses has until now received comparatively less attention. In many cases, the semantics is left only implicit, determined by the way in which the clauses are allowed to interact via inferences. This is in stark contrast with propositional or first-order logic, in which a clause can always be identified with the set of its models.

In this paper, we propose to use strategies, more specifically, the partial strategies for the universal player, as the central objects manipulated within a refutation. We show how strategies arise from the formula matrix and identify operations for obtaining new strategies by combining old ones. We then provide the missing meaning to the intermediate clauses of the existing calculi by seeing them as abstractions of these strategies. This way, we obtain soundness of the calculi in a purely local, modular way, in contrast to the global arguments known from the literature, which need to manipulate the whole refutation, c.f. [5,15,16]. While the advantage of having a *general model theory* could be (as in other logics)

This work was supported by ERC Starting Grant 2014 SYMCAR 639270 and the Austrian research projects FWF S11403-N23 and S11409-N23.

O. Beyersdorff and C. M. Wintersteiger (Eds.): SAT 2018, LNCS 10929, pp. 217–234, 2018.
https://doi.org/10.1007/978-3-319-94144-8_14

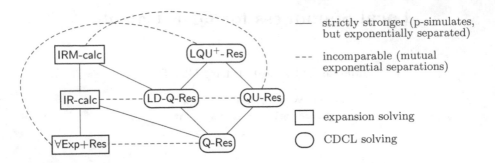

Fig. 1. QBF resolution calculi [6] and their simulation order.

immense, modularity in itself is already a very useful property as it enables the notion of a *sound inference*, an inference which can be added to a calculus without the need to reprove soundness of the whole calculus.

Semantical arguments of soundness have already appeared in the literature, but so far they only targeted simpler calculi (see "the lower part" of Fig. 1) and each with a different method. Semantical soundness is straightforward for Q-Res [24,27] and can be extended to LD-Q-Res via the notion of a *shadow clause* [3] introduced for the purpose of strategy extraction [1]. On the front of expansion-derived calculi, a translation from QBF to first-order logic [25] suggests how to interpret derivations of (up to) IR-calc with the help of first-order model theory [9,14]. Strategies introduced in this paper provide a single semantic concept for proving soundness of all the calculi in Fig. 1, including the expansion-derived calculus IRM-calc and the CDCL-derived calculus LQU$^+$-Res, covering the remaining weaker calculi via simulations.

We are able to view the above mentioned abstraction as providing a specification for a strategy when understood as a program. This relates our approach to the Curry-Howard correspondence: We can see the specification clause as a type and the derivation which lead to it and for which a strategy is the semantical denotation as the implementing program. The specification of the empty clause can then be read as "my strategy is total and therefore winning."

Contributions. The main contributions of this paper are as follows.

- We introduce winning strategies for the universal player as the central notion of a new semantics for QBF calculi (Sect. 3). Subsequently, we identify operations to manipulate and combine strategies and prove them sound in a semantical and local way (Sect. 4).
- We argue that the inference rules in both CDCL-derived calculi such as LQU$^+$-Res and the expansion-derived ones including IRM-calc can be seen as abstractions of operations on strategies (Sects. 5 and 6).
  - A strategy abstracting to the empty clause can be readily used to certify that the input formula is false. We show that there are small IRM-calc refutations

which only have exponential winning strategies for the universal player in our formalism (Sect. 7). This opens the question whether there are more compact representations of strategies that could be manipulated as easily.

2   Preliminaries

A Quantified Boolean Formula (QBF) in the prenex form $\Phi = \Pi.\varphi$ consists of a *quantifier prefix* Π and a *matrix* φ. The prefix Π is a sequence of distinct quantified variables $Q_1 v_1 \ldots Q_k v_k$, where each Q_i is either the existential quantifier $\exists$, in which case v_i is called an existential variable, or the universal quantifier $\forall$, in which case v_i is called a universal variable. Each variable is assigned an *index* $\mathrm{ind}(v_i) = i$. We denote the set of all the existential variables $\mathcal{X}$ and the set of all the universal variables $\mathcal{U}$. The matrix φ is a propositional formula. We say that a QBF Φ is closed if the variables of the matrix $\mathrm{var}(\varphi)$ are amongst $\mathcal{V} = \{v_1, \ldots, v_k\} = \mathcal{X} \,\dot{\cup}\, \mathcal{U}$. We will only consider closed QBFs here.

A literal l is either a variable v, in which case it has *polarity* $\mathrm{pol}(v) = 1$, or a negation $\bar{v}$, which has polarity $\mathrm{pol}(\bar{v}) = 0$. We define the variable of a literal $\mathrm{var}(l) = v$ in both cases. We also extend index to literals via $\mathrm{ind}(l) = \mathrm{ind}(\mathrm{var}(l))$. By $\bar{l}$ we denote the complement of a literal l, i.e. $\bar{l} = v$ if $l = \bar{v}$ and $\bar{l} = \bar{v}$ if $l = v$. Accordingly, $\mathrm{pol}(\bar{l}) = 1 - \mathrm{pol}(l)$.

We will be dealing with QBFs with the matrix in Conjunctive Normal Form (CNF). A clause is a disjunction of literals. A clause is called a tautology if it contains a complementary pair of literals. A propositional formula φ is in CNF if it is a conjunction of clauses. It is customary to treat a clause as the set of its literals and to treat a formula in CNF as the set of its clauses.

An assignment $\alpha : \mathcal{S} \to \{0, 1\}$ is a mapping from a set of variables $\mathcal{S}$ to the Boolean domain $\{0, 1\}$. Whenever $\mathcal{S} \supseteq \mathrm{var}(\varphi)$, the assignment α can be used to evaluate a propositional formula φ in the usual sense. We say that two assignments are *compatible*, if they agree on the intersection of their respective domains. We denote by $\sigma \parallel \tau$ that σ and τ are not compatible, i.e. that there is $v \in \mathrm{dom}(\sigma) \cap \mathrm{dom}(\tau)$ such that $\sigma(v) \neq \tau(v)$.

In the context of a fixed QBF $\Phi = \Pi.\varphi$, we represent assignments as strings of literals strictly ordered by the variable index. For example, given a QBF with prefix $\Pi = \forall x \exists y \forall u$ the assignment $\alpha = \{0/x, 1/u\}$ can be written simply as $\bar{x}u$. We introduce the prefix order relation on strings $\preceq$, where $\sigma \preceq \tau$ denotes that there is a string ξ such that $\sigma\xi = \tau$. An assignment α is called *full* if $\mathrm{dom}(\alpha) = \mathcal{V}$.

3   Policies and Strategies

A QBF is often seen as specifying a game of the existential player against the universal player who alternate at assigning values to their respective variables trying to make the formula true (resp. false) under the obtained assignment. In such a game it is natural to represent the individual moves by literals.

The central notion of our semantics is a strategy, which we obtain as a special case of a policy. Policies are best understood as (non-complete) *binary trees* with nodes labeled by variables (in an order respecting the index) and edges labeled by the Boolean values. However, to streamline the later exposition we adopt an equivalent set-theoretical approach for representing trees in the form of prefix-closed sets of strings. The correspondence will be demonstrated on examples.

A *policy* P is a set of assignments such that for every assignment σ and for every literal l and k

(1) $\sigma l \in P$ implies $\sigma \in P$ (P is prefix-closed),
(2) $\sigma l, \sigma k \in P$ implies $\mathrm{var}(l) = \mathrm{var}(k)$ (P is consistently branching).

The *trivial* policy $P_\epsilon = \{\epsilon\}$ where ϵ is the empty string (which stands for the empty assignment $\epsilon : \emptyset \to \{0,1\}$), will be in figures denoted by $\bot$.

An assignment σ is *maximal* in P, if $\sigma \in P$ and for every $\tau \succeq \sigma$ if $\tau \in P$ then $\tau = \sigma$. A full assignment $\alpha : \mathcal{V} \to \{0,1\}$ is *according to* a policy P, also written

$$P \models \alpha,$$

if it is compatible with some σ maximal in P. We say that a policy P *suggests* a move l in the context σ if $\sigma l \in P$, but $\sigma \bar{l} \notin P$. We say that a policy P *branches* on a variable x in the context σ if both $\sigma x \in P$ and $\sigma \bar{x} \in P$.

Example 1. Any full assignment α is according to P_ϵ. On the other hand, there is no full assignment α according to the empty policy $P_\emptyset = \emptyset$.

For the given prefix $\exists x \exists y \forall z$ consider the policy $P = \{\epsilon, x, xz, \bar{x}, \bar{x}\bar{z}\}$. It suggests the move z in the context x and the move $\bar{z}$ in the context $\bar{x}$. It does not suggest a move for the variable x, but it branches on x, and neither suggests a move for nor branches on y.

Policy P is rendered as a tree in Fig. 2. Each node of the tree corresponds to a string in P, the root to the empty string ϵ, and each Boolean value labelling an edge marks the polarity of the "last" literal in a corresponding string.

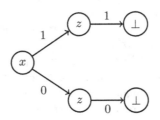

Fig. 2. A tree representation of the policy P from Example 1.

The following central definition captures the notion a strategy. A policy P is a *strategy for the universal player* if, when both players play according to P, the universal player wins by making the matrix false. Moreover, a strategy is *winning* if the existential player cannot "escape her fate" by ignoring some moves suggested to her and thus playing out the game in a way for which the policy does not provide any guarantees to the universal player.

Definition 1. *Let us fix a QBF $\Phi = \Pi.\varphi$. A policy P is a* partial strategy *for the universal player, or simply a* strategy, *if for every full assignment α*

$$P \models \alpha \quad \Rightarrow \quad \alpha \not\models \varphi.$$

A strategy P is total *or* winning, *if it is non-empty and does not suggest any move for the existential player, i.e. whenever it suggests a move l then $\mathrm{var}(l) \in \mathcal{U}$.*

Example 2. Let us consider the false QBF $\Phi = \exists x \exists y \forall z.(x \vee z) \wedge (\bar{x} \vee \bar{z})$. The policy P from Example 1 is a strategy for the universal player, because $xyz, x\bar{y}z, \bar{x}y\bar{z}$ and $\bar{x}\bar{y}\bar{z}$, i.e. all the maximal assignments according to P, each make the formula's matrix false. P is actually a winning strategy, as it is non-empty and does not suggest a move for either x or y.

Lemma 1. *A closed QBF $\Phi = \Pi.\varphi$ is false if and only if there is a policy P which is a winning strategy for the universal player.*

A winning strategy for the universal player is essentially the same object[1] as a Q-counter-model as defined, e.g., by Samulowitz and Bacchus [24]. Thus, since every false QBF has a Q-counter-model, it also has a winning strategy in the sense of Definition 1. Complementarily, if there is a winning strategy for the universal player, the corresponding QBF must be false.

4   Operations on Strategies

Our aim is to give meaning to the clauses manipulated by the various resolution-based calculi for QBF in terms of partial strategies. Before we can do that, we equip ourselves with a set of operations which introduce partial strategies and create new strategies from old ones. Notice that the property of preserving strategies constitutes the core of a local soundness argument: if a sequence of operations turns a set of policies that are partial strategies into a total strategy, we have certified that an input formula must be false.

Axiom: To turn a non-tautologous clause C from the matrix φ into a partial strategy P^C, we just form the prefix closure of the assignment $\bar{C}$ falsifying C:

$$P^C = \{\sigma \mid \sigma \preceq \bar{C}\}.$$

P^C is obviously a non-empty policy. To check that P^C is indeed a partial strategy we notice it suggests exactly the moves which make C false.

Specialisation: Specialisation is an operation which takes a policy P and adds an extra obligation for one of the players by suggesting a move. At the same time the sub-strategy that follows is specialised for the new, more specific context.

---

[1] For technical reasons, we allow branching on universal variables.

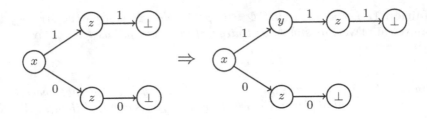

Fig. 3. Specialising a policy at x with y.

Definition 2 (Specialisation). *Let P be a policy, $\sigma \in P$ an assignment and k a literal. We can* specialise P at σ with k, *provided*

(1) if $\sigma = \sigma_0 l_0$ for some assignment σ_0 and a literal l_0 then $\mathrm{ind}(l_0) < \mathrm{ind}(k)$,
(2) if there is a literal l_1 such that $\sigma l_1 \in P$ then $\mathrm{ind}(k) < \mathrm{ind}(l_1)$.[2]

Under such conditions the specialisation *of P at σ with k is defined as*

$$P^{\sigma,k} = \{\xi \mid \xi \in P, \xi \preceq \sigma\} \cup \{\xi \mid \xi \in P, \xi \parallel \sigma\} \cup \{\sigma k \tau \mid \sigma \tau \in P\}.$$

Conditions (1) and (2) ensure that $P^{\sigma,k}$ is a set of assignments. Checking that $P^{\sigma,k}$ is a policy is a tedious exercise. Finally, to see that $P^{\sigma,k}$ is a partial strategy whenever P is, let us consider a full assignment α such that $P^{\sigma,k} \models \alpha$. This means that α is compatible with some ξ maximal in $P^{\sigma,k}$. Now it is easy to see that ξ is either also maximal in P or it is of the form $\sigma k \tau$ and $\sigma \tau$ is maximal in P. In the latter case, since α is compatible with $\sigma k \tau$ it is also compatible with $\sigma \tau$. Thus we learn that $\alpha \not\models \varphi$ as we assumed P to be a partial strategy.

Example 3. When viewing a strategy as a tree, specialisation becomes simply an insertion of a node. In Fig. 3, we specialise the policy P from our running example at the assignment x (i.e. the upper branch) with the move y. The resulting strategy $P^{x,y} = \{\epsilon, x, xy, xyz, \bar{x}, \bar{x}\bar{z}\}$ is visualized in the right tree in Fig. 3. Note that we are able to insert y at that position, since $x < y < z$.

Combining: Policies P and Q can be combined if they, at respective contexts $\sigma \in P$ and $\tau \in R$, suggest a move over the same variable v but of opposite polarity. The combined policy R extends both P and Q in a specific way and creates a new branching on v at the point where the contexts σ and τ "meet". In full generality, there can be more than one such context $\sigma_i \in P$ and $\tau_j \in R$ and the combined policy caters for every pair (σ_i, τ_j) in the described way.

Before we formally define Combining, we need to introduce some auxiliary notation: We make use of the fact that for any non-empty non-trivial policy P, all non-empty assignments which are according to P start with the same variable v (either positive or negated). We can therefore decompose P into the

---

[2] Note that l_1 may not be unique, but its index is (because of consistent branching).

set containing the empty assignment, the set containing all the assignments of P which start with v and all the assignments of P which start with $\bar{v}$.[3]

Lemma 2 (Decomposition). *For every non-empty, non-trivial policy P there is a unique variable v such that P can be decomposed as*

$$P = P_\epsilon \,\dot{\cup}\, v(P^v) \,\dot{\cup}\, \bar{v}(P^{\bar{v}}),$$

where $P_\epsilon = \{\epsilon\}$ is the trivial policy, and for any set of assignments R and a literal l we define $R^l = \{\sigma \mid l\sigma \in R\}$ and $lR = \{l\sigma \mid \sigma \in R\}$.

The sets P^v and $P^{\bar{v}}$ are actually policies and at least one of them is non-empty. We call the variable v the principal variable *of P.*

Proof. A non-empty, non-trivial policy P contains an assignment l of length one (P is prefix-closed) and if it contains another assignment of length one $k \neq l$ then $k = \bar{l}$ (P is consistently branching). The decomposition then follows. □

We now formally introduce Combining. The definition is recursive and proceeds by case distinction.

Definition 3 (Combining). *Let P suggest a move l at every context $\sigma \in S \subseteq P$ and Q suggest a move $\bar{l}$ at every context $\tau \in T \subseteq Q$. The combined policy $P\,[S/T]\,Q$ (the literal l being left implicit) is defined recursively as follows:*

- *The base case: $P\,[\{\epsilon\}/\{\epsilon\}]\,Q = P \cup Q$.*
- *The corner cases: $P\,[\emptyset/T]\,Q = P$, $P\,[S/\emptyset]\,Q = Q$, and $P\,[\emptyset/\emptyset]\,Q = P$.[4]*
- *For the recursive cases, let $P = P_\epsilon \cup vP^v \cup \bar{v}P^{\bar{v}}$ and $Q = Q_\epsilon \cup wQ^w \cup \bar{w}Q^{\bar{w}}$ be the decompositions of P and Q. We compare the indices of v and w:*
  - *If $\mathrm{ind}(v) < \mathrm{ind}(w)$, we set*

$$P\,[S/T]\,Q = P_\epsilon \cup v\left(P^v\,[S^v/T]\,Q\right) \cup \bar{v}\left(P^{\bar{v}}\,[S^{\bar{v}}/T]\,Q\right),$$

  - *if $\mathrm{ind}(v) > \mathrm{ind}(w)$, we set*

$$P\,[S/T]\,Q = P_\epsilon \cup w\left(P\,[S/T^w]\,Q^w\right) \cup \bar{w}\left(P\,[S/T^{\bar{w}}]\,Q^{\bar{w}}\right),$$

  - *and, finally, if $v = w$, we set:*

$$P\,[S/T]\,Q = P_\epsilon \cup v\left(P^v\,[S^v/T^v]\,Q^v\right) \cup \bar{v}\left(P^{\bar{v}}\,[S^{\bar{v}}/T^{\bar{v}}]\,Q^{\bar{v}}\right).$$

Let us comment on the individual cases and how they relate to each other. First, because a policy cannot suggest the same move at two distinct but compatible contexts, we observe that the contexts in S (and also in T) must be pairwise incompatible. Thus if $\epsilon \in S$ then, in fact, $S = \{\epsilon\}$. This justifies why the base case only focuses on the singletons. Second, the corner cases are special in that we do not intend to combine policies for an empty set of contexts

---

[3] In the tree perspective, decomposition basically just says that every non-empty tree has a root node labeled by some variable v and a left and right sub-tree.

[4] The last is an arbitrary choice.

S or T, but they are useful as they make the recursive cases simpler. Finally, to justify that for the recursive cases we can always assume that the argument policies are non-empty, non-trivial (and therefore have a decomposition), we notice that neither the empty nor the trivial policy suggest any move at any context. Therefore, their presence as arguments is covered by the corner cases.

Example 4. In Fig. 4, we combine a strategy P_1 at position x and a strategy P_2 at position $\bar{y}$ into strategy P_3. Note that P_1 and P_2 are implicitly getting specialised using $\bar{y}$ resp. x so that they share a common prefix, i.e. $x\bar{y}$.

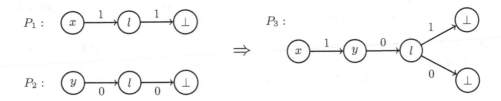

Fig. 4. An example which *combines* strategies P_1 and P_2 into strategy P_3.

It should be clear that combining two policies gives a policy. Furthermore, one can check that whenever P and Q are non-empty, then so is $P\,[S/T]\,Q$. This observation will be used in the soundness proof below, but is also important in its own right. We never want to end up with the empty strategy as the result of performing an operation as the empty strategy can never be a winning one.

Soundness of the Combining operation can be proven under the condition that a pair of involved contexts $\sigma \in S$ and $\tau \in T$ never disagree on suggesting a move "along the way" to l. We formalise this intuition by setting for any σ in P

$$\sigma/P = \{k \mid \tau \preceq \sigma, \tau \neq \sigma, P \text{ suggest } k \text{ in } \tau\},$$

and defining that P and Q are *combinable along* S and T if σ/P is compatible with τ/Q for every $\sigma \in S$ and $\tau \in T$.

Lemma 3 (Soundness of Combining). *Let P and Q be non-trivial strategies with $S \subseteq P$ and $T \subseteq Q$ as in Definition 3. Furthermore, let $S \neq \emptyset \neq T$ and P and Q be combinable along S and T. Then for every full assignment α*

$$P\,[S/T]\,Q \models \alpha \quad \Rightarrow \quad P \models \alpha \text{ or } Q \models \alpha.$$

In other words, the Combining operation is sound under the stated conditions.[5]

The statement of soundness in Lemma 3 may appear counter-intuitive at first sight in that it, rather than providing an implication with a conjunction on the left-hand side, shows an implication with a disjunction on the right-hand side. This form, caused by our focus on the universal player, is, however, what we need here. Intuitively, we ultimately obtain a winning strategy, which can for each play provide a clause from the input matrix that has been made false.

---

[5] The proof of Lemma 3 is omitted due to lack of space.

$$\frac{}{\left\{l^{\tau_C^x} \mid l \in C, x = \mathrm{var}(l), x \in \mathcal{X}\right\}} \quad \text{(Axiom)}$$

$C \in \varphi$ is a non-tautological clause and $\tau_C^x = \{\bar{k} \mid k \in C, \mathrm{var}(k) \in \mathcal{U}_x^<\}$.

$$\frac{\{x^{\tau \cup \xi}\} \cup C_1 \qquad \{\bar{x}^{\tau \cup \sigma}\} \cup C_2}{\mathsf{inst}_\sigma(C_1) \cup \mathsf{inst}_\xi(C_2)} \quad \text{(Resolution)}$$

$\mathrm{dom}(\tau), \mathrm{dom}(\xi), \mathrm{dom}(\sigma)$ are mutually disjoint and $\mathrm{rng}(\tau) \subseteq \{0,1\}$.

$$\frac{C \cup \{l^\mu\} \cup \{l^\sigma\}}{C \cup \{l^\xi\}} \quad \text{(Merging)}$$

$\mathrm{dom}(\mu) = \mathrm{dom}(\sigma)$ and $\xi = \{\mu(u)/u \mid \mu(u) = \sigma(u)\} \cup \{*/u \mid \mu(u) \neq \sigma(u)\}$.

$$\frac{C}{\mathsf{inst}_\tau(C)} \quad \text{(Instantiation)}$$

τ is an assignment to universal variables with $\mathrm{rng}(\tau) \subseteq \{0,1\}$.

Fig. 5. The rules of the expansion-derived calculus IRM-calc.

5 Local Soundness of Expansion-Derived Calculi

Let us recall the expansion-derived (also called instantiation-based) calculi for QBF [5]. These operate on *annotated clauses*, clauses consisting of literals with annotations. An annotation can be described as a partial mapping from variables to $\{0, 1, *\}$. We will treat them analogously to assignments.

An annotated literal l^σ consists of a literal l over an existential variable $\mathrm{var}(l) = x$ and, as an annotation, carries an assignment σ with $\mathrm{rng}(\sigma) \subseteq \{0,1\}$, resp. $\{0, 1, *\}$ in the case of IRM-calc, and with $\mathrm{dom}(\sigma) \subseteq \mathcal{U}_x^<$, where

$$\mathcal{U}_x^< = \{u \in \mathcal{U} \mid \mathrm{ind}(u) < \mathrm{ind}(x)\}$$

denotes the set of the universal dependencies of $x \in \mathcal{X}$. An annotated clause is a set of annotated literals. An auxiliary instantiation function $\mathsf{inst}_\tau(C)$ "applies" an assignment τ to all the literals in C maintaining the above domain restriction:

$$\mathsf{inst}_\tau(C) = \left\{l^{(\sigma\tau) \restriction \mathcal{U}_x^<} \mid l^\sigma \in C \text{ and } \mathrm{var}(l) = x\right\}.$$

Figure 5 describes the rules of the most complex expansion-derived calculus IRM-calc. One obtains IR-calc by dropping the Merging rule, which is the only rule introducing the value $*$ into annotations.[6] Moreover, ∀Exp+Res combines Axiom

---

[6] There is also a simpler way of describing the Resolution rule for IR-calc, which does not rely on inst. However, the presentation in Fig. 5 is equivalent to it.

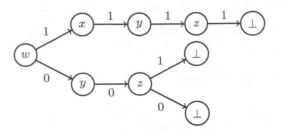

Fig. 6. A strategy P for the prefix $\exists w \forall x \exists y \exists z$.

with Instantiation to obtain "ground" annotated axioms in the first step. In other words, for any conclusion C of the Axiom rule as stated in Fig. 5 and any substitution τ with $\text{dom}(\tau) = \mathcal{U}$ and $\text{rng}(\tau) \subseteq \{0,1\}$, $\text{inst}_\tau(C)$ is an Axiom in $\forall\text{Exp+Res}$. Standalone Instantiation is then not needed in $\forall\text{Exp+Res}$.

5.1   Local Soundness for IR-calc

We start by providing semantics to the clauses of IR-calc and proving local soundness of this calculus. This, while not being the most general result, allows us to explain the key concepts in the cleanest way.

Our plan is to equip ourselves with an abstraction mapping which turns a partial strategy into an IR-calc clause and, in particular, any winning strategy into the empty clause. We then show that IR-calc is sound by considering its inferences one by one and observing that whenever there are strategies which abstract to the premises of an inference, there is a sound operation on the strategies (in the sense of Sect. 4) the result of which abstracts to its conclusion.

Definition 4 (IR-calc abstraction). *The IR-calc abstraction of a policy P is*

$$\mathcal{A}_{\text{IR}}(P) = \left\{ l^{(\sigma \restriction \mathcal{U})} \mid P \text{ suggests a move } \bar{l} \text{ in the context } \sigma, \text{var}(l) \in \mathcal{X} \right\}.$$

We can see that $\mathcal{A}_{\text{IR}}(P)$ records the moves suggested for the existential player as literals and the presence of universal variables in the corresponding contexts as annotations. $\mathcal{A}_{\text{IR}}(P)$ is understood as a clause, i.e. as a formal disjunction.

Example 5. Consider the strategy P visualized in Fig. 6. We have $\mathcal{A}_{\text{IR}}(P) = \bar{y}^{1/x} \vee \bar{z}^{1/x} \vee y$. Note that the first two literals of the clause correspond to the upper branch of P, while the third literal corresponds to the lower branch. Also notice how the branching on w is abstracted away in $\mathcal{A}_{\text{IR}}(P)$.

Axiom: It is easy to see that the IR-calc Axiom corresponding to C is actually $\mathcal{A}_{\text{IR}}(P^C)$, where P^C is the axiom strategy corresponding to C as defined in Sect. 4. Notice that P^C does not forget the universal literals past the last existential one, which cannot be restored from the corresponding IR-calc axiom.

Example 6. Consider a formula $\exists x \forall u \exists y \forall v . \varphi$, where φ contains a clause $C = x \vee u \vee \bar{y} \vee v$. The Axiom strategy corresponding to C is $P^C = \{\epsilon, \bar{x}, \bar{x}\bar{u}, \bar{x}\bar{u}y, \bar{x}\bar{u}y\bar{v}\}$. Furthermore, we have

$$\mathcal{A}_{\mathsf{IR\text{-}calc}}(P^C) = x \vee \bar{y}^{0/u},$$

which is exactly the Axiom IR-calc introduces for C.

Instantiation: The Instantiation inference in IR-calc takes a clause C and τ, an assignment to some universal variables with $\mathsf{rng}(\tau) \subseteq \{0,1\}$, and derives

$$\mathsf{inst}_\tau(C) = \left\{ l^{(\sigma\tau) \restriction \mathcal{U}_x^\le} \mid l^\sigma \in C \text{ and } \mathsf{var}(l) = x \right\}.$$

We show that Instantiation of clauses corresponds to Specialisation of strategies.

Lemma 4. *Let P be a partial strategy and τ an assignment with $\mathsf{dom}(\tau) \subseteq \mathcal{U}$ and $\mathsf{rng}(\tau) \subseteq \{0,1\}$ as above. Then there is a partial strategy P_τ which can be derived from P by a sequence of Specialisation operations such that*

$$\mathsf{inst}_\tau(\mathcal{A}_{\mathsf{IR}}(P)) = \mathcal{A}_{\mathsf{IR}}(P_\tau).$$

Proof (Sketch). We start working with P and modify it in several steps, denoting the intermediate strategy P' (as if it was a variable in an imperative programming language). We take the bindings l from τ one by one and for each modify P' by consecutively specialising it with l at every context $\sigma \in P'$ for which it is allowed (in the sense of Definition 2). This, in particular, means we skip those contexts at which P' already suggests a move for $\mathsf{var}(l)$. ☐

Resolution: The Resolution inference in IR-calc can be defined as:

$$\frac{C_0 \cup \{l^\tau\} \qquad D_0 \cup \{\bar{l}^\tau\}}{C_0 \cup D_0}.$$

Our aim is to simulate resolution of clauses as combining of strategies. We will succeed provided IR-calc does not derive a tautology and, in some cases, our new strategy will be actually stronger than what IR-calc is allowed to believe.

Lemma 5. *Let $C = C_0 \cup \{l^\tau\}$ and $D = D_0 \cup \{\bar{l}^\tau\}$ be IR-calc clauses. For every partial strategy P_C and P_D such that $C = \mathcal{A}_{\mathsf{IR}}(P_C)$ and $D = \mathcal{A}_{\mathsf{IR}}(P_D)$ if $C_0 \cup D_0$ does not contain a complementary pair of literals then there exists a partial strategy P obtained as a combination of P_C and P_D over the literal l such that*

$$\mathcal{A}_{\mathsf{IR}}(P) \subseteq C_0 \cup D_0.$$

Proof (Sketch). Let us define

$$
\begin{aligned}
S &= \{\sigma_C \mid P_C \text{ suggests } \bar{l} \text{ at } \sigma_C \text{ and } (\sigma_C \restriction \mathcal{U}) = \tau\}, \\
T &= \{\sigma_D \mid P_D \text{ suggests } l \text{ at } \sigma_D \text{ and } (\sigma_D \restriction \mathcal{U}) = \tau\}.
\end{aligned}
\tag{1}
$$

and set $P = P_C \, [S/T] \, P_D$.

To see that P is indeed a partial strategy we appeal to Lemma 3. Since $l^\tau \in \mathcal{A}_{IR}(P_C)$ we obtain $S \neq \emptyset$ and similarly for $\bar{l}^\tau \in \mathcal{A}_{IR}(P_D)$ and $T \neq \emptyset$. Furthermore, to see that P_C and P_D are combinable along S and T, let us, for the sake of contradiction, assume that there is a $\sigma_C \in S$ and $\sigma_D \in T$ such that σ_C/P_C and σ_D/P_D are not compatible. This means that P_C suggests a move k at some context $\tau_C \prec \sigma_C$ and P_D suggests a move $\bar{k}$ at some context $\tau_D \prec \sigma_D$, with $\mathrm{var}(k) \in \mathcal{X}$. However, this contradicts our assumption that $C_0 \cup D_0$ does not contain a complementary pair of literals, because it implies that $k^{\tau_0} \in \mathcal{A}_{IR}(P_C) = C_0$ and $\bar{k}^{\tau_0} \in \mathcal{A}_{IR}(P_D) = D_0$ for the unique $\tau_0 = (\tau_C \restriction \mathcal{U}) = (\tau_D \restriction \mathcal{U})$. This verifies the assumptions of Lemma 3.

The second part of our claim, i.e. $\mathcal{A}_{IR}(P) \subseteq C_0 \cup D_0$, is, similarly to the proof of Lemma 3, shown by induction along the computation of $P_C \, [S/T] \, P_D$. Formally, we check there that

$$\mathcal{A}_{IR}(P_C) \cup \mathcal{A}_{IR}(P_D) \supseteq \mathcal{A}_{IR}(P_C \, [S/T] \, P_D),$$

and, moreover, that whenever S and T are defined by the comprehensions (1) then $\mathcal{A}_{IR}(P_C \, [S/T] \, P_D) \cap \{l^\tau, \bar{l}^\tau\} = \emptyset$, i.e. all the occurrences of the pivot get eliminated from the abstraction of the combined strategy.   □

Lemma 5 reveals that it is not always the case that $\mathcal{A}_{IR}(P) = C_0 \cup D_0$, as our abstraction can sometimes become stronger than what the calculus realises. To formally capture this discrepancy, we extend our exposition by one additional "twist", which we will bring to much greater use below when providing analogous semantics for IRM-calc and LQU$^+$-Res. Namely, we will use our abstraction mapping to provide a simulation *relation* between the clauses of a calculus and partial strategies. In the case of IR-calc here, we define

$$C \sim_{IR} P \quad \equiv \quad C \supseteq \mathcal{A}_{IR}(P).$$

Now we just need to reprove Lemma 5 under the assumptions $C \sim_{IR} P_C$ instead of $C = \mathcal{A}_{IR}(P_C)$ (and similarly for D and P_D). This is straightforward if we recall the corner cases of the combining operation on strategies. Here, we can resolve over a pivot "which is not there" by simply reusing as P the strategy corresponding to such vacuous premise and calling it the result. It can be seen that this way we obtain an $\mathcal{A}_{IR}(P)$ that is a subset of $C_0 \cup D_0$ as required.

5.2   What Needs to Be Done Differently for IRM-calc?

The IRM-calc extends IR-calc by allowing for the $*$ value in annotations that is obtained by Merging together literals l^μ and l^σ which do not fully agree in their respective annotations, i.e. $\mu \parallel \sigma$.[7] This is complemented by a more general version of Resolution, which behaves as "unifying" the annotations of the pivots while treating opposing $*$ as non-unifiable (recall Fig. 5).

While we do not show it here in full detail due to lack of space, we claim that the $*$ of IRM-calc does not, per se, carry any logical meaning, but simply

---

[7] We actually do not need the usually stated assumption $\mathsf{dom}(\mu) = \mathsf{dom}(\sigma)$.

$$\frac{D \cup \{u\}}{D} \ (\forall\text{-Red})$$

Literal u is universal and $\text{ind}(u) \geq \text{ind}(l)$ for all $l \in D$.

$$\frac{C_0 \cup \{v\} \qquad D_0 \cup \{\bar{v}\}}{C_0 \cup D_0} \ (\text{Res})$$

Whenever $l \in C_0$ and $\bar{l} \in D_0$ for a literal l then $\text{var}(l) \in \mathcal{U}$ and $\text{ind}(l) > \text{ind}(v)$.

Fig. 7. The rules of the most general CDCL-derived calculus LQU$^+$-Res.

provides a commitment of the calculus to resolve away the involved literals in a specific way. In other words, it is always sound to set a binding to $*$ in an annotation (even for a previously "unbound" universal variable).

We say that an annotation σ^* is a $*$-*specialisation* of an annotation σ if for any $u \in \text{dom}(\sigma^*)$ whenever $\sigma^*(u) \neq *$ then $\sigma^*(u) = \sigma(u)$.

Definition 5 (IRM-calc Simulation Relation). *We say that an IRM-calc clause C is simulated by a strategy P_C, written $C \sim_{IRM} P_C$, if*

$$C \supseteq \left\{ l^{\sigma^*} \mid l^\sigma \in \mathcal{A}_{IR}(P_C), \sigma^* \text{ is a } *\text{-specialisation of } \sigma \right\}.$$

Analogously to Lemma 5, we can simulate IRM-calc Resolution via the Combining operation on strategies. The l moves of the pivot literals in the premise strategies are not in general suggested at "universally identical contexts" (c.f. (1) from the proof of Lemma 5), but at compatible contexts nevertheless, because of unifiability of the corresponding IRM-calc pivots.

6   Local Soundness for CDCL-Derived Calculi

Figure 7 presents the rules of LQU$^+$-Res, the strongest CDCL-derived calculus we study in this paper. It combines the $\forall$-Red rule common to all CDCL-derived calculi with a particular resolution rule Res, the pivot of which can be any variable $v \in \mathcal{V}$. Notice that LQU$^+$-Res is allowed to create a tautology, provided the new complementary pair is universal and has an index greater than the pivot. We will learn that these tautologies are never logically vacuous – in the corresponding strategy the complementary pair is "separated" by the pivot.

The $\forall$-Red rule is extra-logical from the perspective of our semantics. It does not correspond to any operation on the side of the interpreting strategy, which stays the same. We resolve this nuance by providing an abstraction which exposes a strategy as a fully $\forall$-reduced clause, but we allow for non-reduced clauses in derivations via our simulation relation. We start with an auxiliary definition.

We say that a context σ is *universally trailing* in a policy P, if for every $\tau \succeq \sigma$ if P suggests a move l in τ then $\text{var}(l) \in \mathcal{U}$.

Definition 6 (LQU$^+$-Res Abstraction and Simulation). *The LQU$^+$-Res abstraction $\mathcal{A}_{\text{LQU}^+}$ of a policy P and the simulation relation $\sim_{\text{LQU}^+}$ between a LQU$^+$-Res clause and a policy are defined, respectively, as follows:*

$$\mathcal{A}_{\text{LQU}^+}(P) = \left\{ l \mid P \text{ suggests } \bar{l} \text{ in } \sigma \text{ and } \sigma \text{ is not universally trailing in } P \right\},$$

$$C \sim_{\text{LQU}^+} P \quad \equiv \quad C \supseteq \mathcal{A}_{\text{LQU}^+}(P).$$

Let us now show that $\sim_{\text{LQU}^+}$ is indeed a simulation of LQU$^+$-Res derivations in terms of operations on partial strategies.

Axiom: Let P^C be the axiom strategy corresponding to $C \in \varphi$ as defined in Sect. 4. One can check that $\mathcal{A}_{\text{LQU}^+}(P^C)$ is the $\forall$-reduct of C and we thus have $C \sim_{\text{LQU}^+} P^C$ because a reduct only possibly *removes* literals.
$\forall - Red$: As discussed above, the $\forall$-Red is simulated by the identity mapping on the side of strategies. To see this is always possible we just realise the following.

Lemma 6. *Let a policy P suggest a move $\bar{l}$ in context σ which is not universally trailing in P. Then there is a literal $k \in \mathcal{A}_{\text{LQU}^+}(P)$ such that $\text{ind}(k) > \text{ind}(l)$.*

Example 7. Let us work in the context of $\Pi = \exists x \forall u \exists y$. LQU$^+$-Res can derive the clause $C = u \vee y$ by resolving the axioms $\bar{x} \vee u$ and $x \vee y$ over the pivot x. Notice that C cannot be $\forall$-reduced. At the same time, the corresponding strategy $P = \{\epsilon, x, x\bar{u}, \bar{x}, \bar{x}\bar{y}\}$ still records that x is universally trailing and $\mathcal{A}_{\text{LQU}^+}(P) = \{y\}$.

Resolution: Both the possibility of a universal pivot and the creation of tautologies can be uniformly handled on the side of strategies.

Lemma 7. *Let $C = C_0 \cup \{v\}$ and $D = D_0 \cup \{\bar{v}\}$ be the premises of a LQU$^+$-Res Resolution inference. Furthermore, let P_C and P_D be partial strategies such that $C \sim_{\text{LQU}^+} P_C$ and $D \sim_{\text{LQU}^+} P_D$. Then there exists a partial strategy P obtained as a combination of P_C and P_D over the literal $\bar{v}$ such that*

$$(C_0 \cup D_0) \sim_{\text{LQU}^+} P.$$

Proof (Sketch). Analogously to the proof of Lemma 5 we define

$$S = \{\sigma \mid P_C \text{ suggests } \bar{v} \text{ in } \sigma \text{ and } \sigma \text{ is not universally trailing in } P_C\},$$
$$T = \{\tau \mid P_D \text{ suggests } v \text{ in } \tau \text{ and } \tau \text{ is not universally trailing in } P_D\}.$$

and set $P = P_C \, [S/T] \, P_D$. $\qquad \qquad \square$

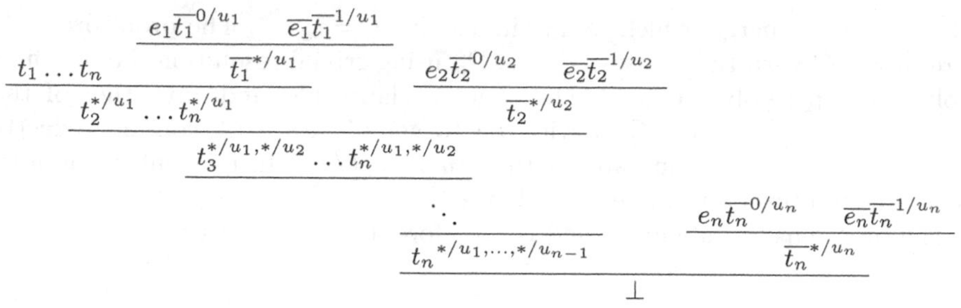

$$\dfrac{t_1 \ldots t_n \quad \dfrac{e_1\overline{t_1}^{0/u_1} \quad \overline{e_1}\overline{t_1}^{1/u_1}}{\overline{t_1}^{*/u_1}}}{t_2^{*/u_1} \ldots t_n^{*/u_1}}$$

$$\dfrac{e_2\overline{t_2}^{0/u_2} \quad \overline{e_2}\overline{t_2}^{1/u_2}}{\overline{t_2}^{*/u_2}}$$

$$\dfrac{t_3^{*/u_1,*/u_2} \ldots t_n^{*/u_1,*/u_2}}{}$$

$$\dfrac{\ddots \quad \dfrac{e_n\overline{t_n}^{0/u_n} \quad \overline{e_n}\overline{t_n}^{1/u_n}}{\overline{t_n}^{*/u_n}}}{t_n^{*/u_1,\ldots,*/u_{n-1}}}$$

$$\bot$$

Fig. 8. A refutation of $\mathcal{C}(F_n)$ from Example 8.

7 Winning Strategies Are Worst-Case Exponential for IRM-calc Proofs

There is a family of QBFs which do not have polynomial winning strategies in the sense of Definition 1, but do have polynomial IRM-calc refutations. This has two main consequences: (1) It is not possible to design an algorithm which generates winning strategies from IRM-calc refutations such that the strategies are polynomial in the size of the refutation. (2) We cannot use partial strategies as a calculus for polynomially simulating IRM-calc.

Example 8. For every natural n consider the false formula

$$F_n := \quad \exists e_1 \ldots e_n \forall u_1 \ldots u_n. \bigvee_i (e_i \leftrightarrow u_i).$$

If P is a winning strategy for the universal player on F_n, it needs to assign u_i to 1 if and only if the existential player assigns e_i to 1. In order words, P needs to branch on every e_i. Therefore, each e_i doubles the number of branches of P from which we conclude that the size of P is exponential in n.

We clausify F_n using Tseitin-variables $t_1, \ldots, t_n$ for the disjuncts and use De Morgan's laws for the negated equivalences. This gives the following formula:

$$\mathcal{C}(F_n) := \quad \exists e_1 \ldots e_n \forall u_1 \ldots u_n \exists t_1 \ldots t_n. \quad (t_1 \vee \cdots \vee t_n)$$
$$\wedge \quad (e_1 \vee u_1 \vee \overline{t_1}) \quad \wedge \quad (\overline{e_1} \vee \overline{u_1} \vee \overline{t_1})$$
$$\vdots$$
$$\wedge \quad (e_n \vee u_n \vee \overline{t_n}) \quad \wedge \quad (\overline{e_n} \vee \overline{u_n} \vee \overline{t_n})$$

Now consider the IRM-calc refutation of $\mathcal{C}(F_n)$ shown in Fig. 8. The proof starts from the clause $C := t_1 \vee \cdots \vee t_n$ and contains n auxiliary sub-proofs where the i-th sub-proof resolves the axiom clauses $e_i \vee \overline{t_i}^{0/u_i}$ and $\overline{e_i} \vee \overline{t_i}^{1/u_i}$ over the pivot

e_i followed by a merge, which results in a unit $D_i = \overline{t_i}^{*/u_i}$. The proof proceeds by resolving C with the clauses $D_1, \ldots, D_n$ using trivial resolution, i.e. the first resolution step resolves C with D_1 to get a clause C_1, and any other of the i resolution steps resolves C_{i-1} with D_i to get C_i. Each C_i contains exactly the literals $t_{i+1}, \ldots, t_n$ annotated with $(*/u_1, \ldots, */u_i)$. In particular, the n-th resolution step results in the empty clause.

The proof has $2n$ inferences and is therefore linear in the size of n.

8   Conclusion and Future Work

We showed how partial strategies can be used as the central semantic objects in QBF. We identified operations which manipulate and combine strategies and proved their soundness in a local, modular way. Furthermore, we described how existing state-of-the-art calculi can be seen to operate on abstractions of these strategies and clarified the local semantics behind their inferences.

While a general model theory does not need to be computationally effective to be useful, in the case of QBF the computational aspects pertaining to strategies seem of great practical importance. Our paper opens several streams of future work along these lines: (1) We intend to combine the operations on strategies presented in this work with the solving-algorithm from [10], which uses strategies directly in the solving process. (2) We would like to use the obtained insights to derive a uniform calculus which polynomially simulates both IRM and LQU+-Res. (3) Continuing the direction of Sect. 7, we would like to clarify whether the exponential separation between strategies and refutations can be extended from IRM-calc to IR-calc or even to ∀Exp+Res. (4) We want to generalise our strategies by using more expressive data structures. In particular, we would like to see whether the operations we identified can be extended to BDDs, i.e. to a representation in which strategies are fully reduced and merged. We envision that doing so could yield a polynomial strategy extraction algorithm for IRM-calc which produces much simpler strategies than existing algorithms.

Acknowledgements. We thank Olaf Beyersdorff, Leroy Chew, Uwe Egly, Mikoláš Janota, Adrián Rebola-Pardo, and Martina Seidl for interesting comments and inspiring discussions on the semantics of QBF.

References

1. Balabanov, V., Jiang, J.R., Janota, M., Widl, M.: Efficient extraction of QBF (counter)models from long-distance resolution proofs. In: Bonet, B., Koenig, S. (eds.) Proceedings of the Twenty-Ninth AAAI Conference on Artificial Intelligence, Austin, Texas, USA, 25–30 January 2015, pp. 3694–3701. AAAI Press (2015)
2. Balabanov, V., Widl, M., Jiang, J.-H.R.: QBF resolution systems and their proof complexities. In: Sinz, C., Egly, U. (eds.) SAT 2014. LNCS, vol. 8561, pp. 154–169. Springer, Cham (2014). https://doi.org/10.1007/978-3-319-09284-3_12

3. Beyersdorff, O., Blinkhorn, J.: Dependency schemes in QBF calculi: semantics and soundness. In: Rueher, M. (ed.) CP 2016. LNCS, vol. 9892, pp. 96–112. Springer, Cham (2016). https://doi.org/10.1007/978-3-319-44953-1_7

4. Beyersdorff, O., Bonacina, I., Chew, L.: Lower bounds: from circuits to QBF proof systems. In: Proceedings of the ACM Conference on Innovations in Theoretical Computer Science (ITCS 2016), pp. 249–260. ACM (2016)

5. Beyersdorff, O., Chew, L., Janota, M.: On unification of QBF resolution-based calculi. In: Csuhaj-Varjú, E., Dietzfelbinger, M., Ésik, Z. (eds.) MFCS 2014. LNCS, vol. 8635, pp. 81–93. Springer, Heidelberg (2014). https://doi.org/10.1007/978-3-662-44465-8_8

6. Beyersdorff, O., Chew, L., Janota, M.: Proof complexity of resolution-based QBF calculi. In: Proceedings of the STACS. LIPIcs, vol. 30, pp. 76–89. Schloss Dagstuhl (2015)

7. Beyersdorff, O., Chew, L., Mahajan, M., Shukla, A.: Feasible interpolation for QBF resolution calculi. In: Halldórsson, M.M., Iwama, K., Kobayashi, N., Speckmann, B. (eds.) ICALP 2015. LNCS, vol. 9134, pp. 180–192. Springer, Heidelberg (2015). https://doi.org/10.1007/978-3-662-47672-7_15

8. Beyersdorff, O., Chew, L., Mahajan, M., Shukla, A.: Are short proofs narrow? QBF resolution is not simple. In: Proceedings of the Symposium on Theoretical Aspects of Computer Science (STACS 2016) (2016)

9. Beyersdorff, O., Chew, L., Schmidt, R.A., Suda, M.: Lifting QBF resolution calculi to DQBF. In: Creignou, N., Le Berre, D. (eds.) SAT 2016. LNCS, vol. 9710, pp. 490–499. Springer, Cham (2016). https://doi.org/10.1007/978-3-319-40970-2_30

10. Bjørner, N., Janota, M., Klieber, W.: On conflicts and strategies in QBF. In: Fehnker, A., McIver, A., Sutcliffe, G., Voronkov, A. (eds.) 20th International Conferences on Logic for Programming, Artificial Intelligence and Reasoning - Short Presentations, LPAR 2015, Suva, Fiji, 24–28 November 2015. EPiC Series in Computing, vol. 35, pp. 28–41. EasyChair (2015). http://www.easychair.org/publications/paper/255082

11. Bloem, R., Braud-Santoni, N., Hadzic, V.: QBF solving by counterexample-guided expansion. CoRR abs/1611.01553 (2016). http://arxiv.org/abs/1611.01553

12. Cimatti, A., Sebastiani, R. (eds.): SAT 2012. LNCS, vol. 7317. Springer, Heidelberg (2012). https://doi.org/10.1007/978-3-642-31612-8

13. Egly, U.: On sequent systems and resolution for QBFs. In: Cimatti and Sebastiani [12], pp. 100–113

14. Egly, U.: On stronger calculi for QBFs. In: Creignou, N., Le Berre, D. (eds.) SAT 2016. LNCS, vol. 9710, pp. 419–434. Springer, Cham (2016). https://doi.org/10.1007/978-3-319-40970-2_26

15. Egly, U., Lonsing, F., Widl, M.: Long-distance resolution: proof generation and strategy extraction in search-based QBF solving. In: McMillan, K., Middeldorp, A., Voronkov, A. (eds.) LPAR 2013. LNCS, vol. 8312, pp. 291–308. Springer, Heidelberg (2013). https://doi.org/10.1007/978-3-642-45221-5_21

16. Goultiaeva, A., Gelder, A.V., Bacchus, F.: A uniform approach for generating proofs and strategies for both true and false QBF formulas. In: Walsh, T. (ed.) Proceedings of the 22nd International Joint Conference on Artificial Intelligence, IJCAI 2011, Barcelona, Catalonia, Spain, 16–22 July 2011, pp. 546–553. IJCAI/AAAI (2011), https://doi.org/10.5591/978-1-57735-516-8/IJCAI11-099

17. Heule, M.J., Seidl, M., Biere, A.: Efficient extraction of Skolem functions from QRAT proofs. In: Formal Methods in Computer-Aided Design (FMCAD), pp. 107–114. IEEE (2014)

18. Heule, M.J.H., Seidl, M., Biere, A.: A unified proof system for QBF preprocessing. In: Demri, S., Kapur, D., Weidenbach, C. (eds.) IJCAR 2014. LNCS (LNAI), vol. 8562, pp. 91–106. Springer, Cham (2014). https://doi.org/10.1007/978-3-319-08587-6_7

19. Janota, M., Klieber, W., Marques-Silva, J., Clarke, E.M.: Solving QBF with counterexample guided refinement. In: Cimatti and Sebastiani [12], pp. 114–128

20. Janota, M., Marques-Silva, J.: Expansion-based QBF solving versus Q-resolution. Theor. Comput. Sci. **577**, 25–42 (2015)

21. Kleine Büning, H., Karpinski, M., Flögel, A.: Resolution for quantified Boolean formulas. Inf. Comput. **117**(1), 12–18 (1995)

22. Lonsing, F., Biere, A.: DepQBF: a dependency-aware QBF solver. JSAT **7**(2–3), 71–76 (2010)

23. Rabe, M.N., Tentrup, L.: CAQE: a certifying QBF solver. In: Kaivola, R., Wahl, T. (eds.) Formal Methods in Computer-Aided Design, FMCAD 2015, Austin, Texas, USA, 27–30 September 2015, pp. 136–143. IEEE (2015)

24. Samulowitz, H., Bacchus, F.: Binary clause reasoning in QBF. In: Biere, A., Gomes, C.P. (eds.) SAT 2006. LNCS, vol. 4121, pp. 353–367. Springer, Heidelberg (2006). https://doi.org/10.1007/11814948_33

25. Seidl, M., Lonsing, F., Biere, A.: qbf2epr: a tool for generating EPR formulas from QBF. In: Proceedings of the PAAR-2012. EPiC, vol. 21, pp. 139–148. EasyChair (2013)

26. Slivovsky, F., Szeider, S.: Variable dependencies and Q-resolution. In: Sinz, C., Egly, U. (eds.) SAT 2014. LNCS, vol. 8561, pp. 269–284. Springer, Cham (2014). https://doi.org/10.1007/978-3-319-09284-3_21

27. Slivovsky, F., Szeider, S.: Soundness of Q-resolution with dependency schemes. Theor. Comput. Sci. **612**, 83–101 (2016). https://doi.org/10.1016/j.tcs.2015.10.020

28. Gelder, A.: Contributions to the theory of practical quantified Boolean formula solving. In: Milano, M. (ed.) CP 2012. LNCS, pp. 647–663. Springer, Heidelberg (2012). https://doi.org/10.1007/978-3-642-33558-7_47

QBF as an Alternative to Courcelle's Theorem

Michael Lampis[1], Stefan Mengel[2], and Valia Mitsou[3(✉)]

[1] Université Paris-Dauphine, PSL Research University, CNRS,
UMR 7243, Paris, France
[2] CNRS, CRIL UMR 8188, Paris, France
[3] Université Paris-Diderot, IRIF, CNRS, UMR 8243, Paris, France
vmitsou@liris.cnrs.fr

Abstract. We propose reductions to quantified Boolean formulas (QBF) as a new approach to showing fixed-parameter linear algorithms for problems parameterized by treewidth. We demonstrate the feasibility of this approach by giving new algorithms for several well-known problems from artificial intelligence that are in general complete for the second level of the polynomial hierarchy. By reduction from QBF we show that all resulting algorithms are essentially optimal in their dependence on the treewidth. Most of the problems that we consider were already known to be fixed-parameter linear by using Courcelle's Theorem or dynamic programming, but we argue that our approach has clear advantages over these techniques: on the one hand, in contrast to Courcelle's Theorem, we get concrete and tight guarantees for the runtime dependence on the treewidth. On the other hand, we avoid tedious dynamic programming and, after showing some normalization results for CNF-formulas, our upper bounds often boil down to a few lines.

1 Introduction

Courcelle's seminal theorem [8] states that every graph property definable in monadic second order logic can be decided in linear time on graphs of constant treewidth. Here treewidth is the famous width measure used to measure intuitively how similar a graph is to a tree. While the statement of Courcelle's Theorem might sound abstract to the unsuspecting reader, the consequences are tremendous. Since a huge number of computational problems can be encoded in monadic second order logic, this gives automatic linear time algorithms for a wealth of problems in such diverse fields as combinatorial algorithms, artificial intelligence and databases; out of the plethora of such papers let us only cite [11,21] that treat problems that will reappear in this paper. This makes Courcelle's Theorem one of the cornerstones of the field of parameterized algorithms.

Unfortunately, its strength comes with a price: while the runtime dependence on the size of the problem instance is linear, the dependence on the treewidth is unclear when using this approach. Moreover, despite recent progress (see e.g.

© Springer International Publishing AG, part of Springer Nature 2018
O. Beyersdorff and C. M. Wintersteiger (Eds.): SAT 2018, LNCS 10929, pp. 235–252, 2018.
https://doi.org/10.1007/978-3-319-94144-8_15

the survey [29]) Courcelle's Theorem is largely considered impractical due to the gigantic constants involved in the construction. Since generally these constants are unavoidable [20], showing linear time algorithms with Courcelle's Theorem can hardly be considered as a satisfying solution.

As a consequence, linear time algorithms conceived with the help of Courcelle's Theorem are sometimes followed up with more concrete algorithms with more explicit runtime guarantees often by dynamic programming or applications of a datalog approach [13,21,23]. Unfortunately, these hand-written algorithms tend to be very technical, in particular for decision problems outside of NP. Furthermore, even this meticulous analysis usually gives algorithms with a dependance on treewidth that is a tower of exponentials.

The purpose of this paper is two-fold. On the one hand we propose reductions to QBF combined with the use of a known QBF-algorithm by Chen [7] as a simple approach to constructing linear-time algorithms for problems beyond NP parameterized by treewidth. In particular, we use the proposed method in order to construct (alternative) algorithms for a variety of problems stemming from artificial intelligence: abduction, circumscription, abstract argumentation and the computation of minimal unsatisfiable sets in unsatisfiable formulas. The advantage of this approach over Courcelle's Theorem or tedious dynamic programming is that the algorithms we provide are almost straightforward to produce, while giving bounds on the treewidth that asymptotically match those of careful dynamic programming. On the other hand, we show that our algorithms are asymptotically best possible, giving matching complexity lower bounds.

Our algorithmic approach might at first sight seem surprising: since QBF with a fixed number of alternations is complete for the different levels of the polynomial hierarchy, there are trivially reductions from all problems in that hierarchy to the corresponding QBF problem. So what is new about this approach? The crucial observation here is that in general reductions to QBF guaranteed by completeness do not maintain the treewidth of the problem. Moreover, while Chen's algorithm runs in linear time, there is no reason for the reduction to QBF to run in linear time which would result in an algorithm with overall non-linear runtime.

The runtime bounds that we give are mostly of the form $2^{2^{O(k)}} n$ where k is the treewidth and n the size of the input. Furthermore, starting from recent lower bounds for QBF [28], we also show that these runtime bounds are essentially tight as there are no algorithms with runtime $2^{2^{o(k)}} 2^{o(n)}$ for the considered problems. Our lower bounds are based on the *Exponential Time Hypothesis (ETH)* which posits that there is no algorithm for 3SAT with runtime $2^{o(n)}$ where n is the number of variables in the input. ETH is by now widely accepted as a standard assumption in the fields of exact and parameterized algorithms for showing tight lower bounds, see e.g., the survey [32]. We remark that our bounds confirm the observation already made in [33] that problems complete for the second level of the polynomial hierarchy parameterized by treewidth tend to have runtime double-exponential in the treewidth.

As a consequence, the main contribution of this paper is to show that reductions to QBF can be used as a simple technique to show algorithms with essentially optimal runtime for a wide range of problems.

Our Contributions. We show upper bounds of the form $2^{2^{O(k)}} n$ for instances of treewidth k and size n for abstract argumentation, abduction, circumscription and the computation of minimal unsatisfiable sets in unsatisfiable formulas. For the former three problems it was already known that there are linear time algorithms for bounded treewidth instances: for abstract argumentation, this was shown in [11] with Courcelle's theorem and a tighter upper bound of the form $2^{2^{O(k)}} n$ was given by dynamic programming in [13]. For abduction, there was a linear time algorithm in [21] for all abduction problems we consider and a $2^{2^{O(k)}} n$ algorithm based on a datalog encoding for some of the problems. The upper bound that we give for so-called *necessity* is new. For circumscription, a linear time algorithm was known [21] but we are the first to give concrete runtime bounds. Finally, we are the first to give upper bounds for minimal unsatisfiable subsets for CNF-formulas of bounded treewidth.

We complement our upper bounds with ETH-based lower bounds for all problems mentioned above, all of which are the first such bounds for these problems.

Finally, we apply our approach to abduction with $\subseteq$-preferences but giving a linear time algorithm with triple exponential dependence on the treewidth, refining upper bounds based on Courcelle's theorem [21] by giving an explicit treewidth dependence.

2   Preliminaries

In this section, we only introduce notation that we will use in all parts of the paper. The background for the problems on which we demonstrate our approach will be given in the individual sections in which these problems are treated.

2.1   Treewidth

Throughout this paper, all graphs will be undirected and simple unless explicitly stated otherwise. A *tree decomposition* $(T, (B_t)_{t \in T})$ of a graph $G = (V, E)$ consists of a tree T and a subset $B_t \subseteq V$ for every node t of T with the following properties:

- every vertex $v \in V$ is contained in at least one set B_t,
- for every edge $uv \in E$, there is a set B_t that contains both u and v, and
- for every $v \in V$, the set $\{t \mid v \in B_t\}$ induces a subtree of T.

The last condition is often called the connectivity condition. The sets B_t are called *bags*. The *width* of a tree decomposition is $\max_{t \in T}(|B_t|) - 1$. The *treewidth* of G is the minimum width of a tree decomposition of G. We will sometimes tacitly use the fact that any tree decomposition can always be assumed to be of size linear in $|V|$ by standard simplifications. Computing the treewidth of a graph is NP-hard [2], but for every fixed k there is a linear time algorithm that decides if a given graph has treewidth at most k and if so computes a tree decomposition witnessing this [4].

A tree decomposition is called *nice* if every node t of T is of one of the following types:

- **leaf node:** t is a leaf of T.
- **introduce node:** t has a single child node t' and $B_t = B_{t'} \cup \{v\}$ for a vertex $v \in V \setminus B_{t'}$.
- **forget node:** t has a single child node t' and $B_t = B_{t'} \setminus \{v\}$ for a vertex $v \in B_{t'}$.
- **join node:** t has exactly two children t_1 and t_2 with $B_t = B_{t_1} = B_{t_2}$.

Nice tree decompositions were introduced in [25] where it was also shown that given a tree decomposition of a graph G, one can in linear time compute a nice tree decomposition of G with the same width.

2.2 CNF Formulas

A *literal* is a propositional variable or the negation of a propositional variable. A *clause* is a disjunction of literals and a CNF-formula is a conjunction of clauses. For technical reasons we assume that there is an injective mapping from the variables in a CNF formula ϕ to $\{0, \ldots, cn\}$ for an arbitrary but fixed constant c where n is the number of variables in ϕ and that we can evaluate this mapping in constant time. This assumption allows us to easily create lists, in linear time in n, which store data assigned to the variables that we can then look up in constant time. Note that formulas in the DIMACS format [9], the standard encoding for CNF formulas, generally have this assumed property. Alternatively, we could use perfect hashing to assign the variables to integers, but this would make some of the algorithms randomized.

Let ϕ and ϕ' be two CNF formulas. We say that ϕ is a projection of ϕ' if and only if $\mathsf{var}(\phi) \subseteq \mathsf{var}(\phi')$ and $a : \mathsf{var}(\phi) \to \{0, 1\}$ is a model of ϕ if and only if a can be extended to a model of ϕ'.

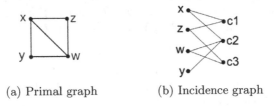

(a) Primal graph   (b) Incidence graph

Fig. 1. Primal and incidence graphs for $\phi = (\neg x \vee z) \wedge (x \vee y \vee \neg w) \wedge (\neg z \vee w)$.

To every CNF formula ϕ we assign a graph called *primal graph* whose vertex set is the set of variables of ϕ. Two vertices are connected by an edge if and only if they appear together in a clause of ϕ (see Fig. 1a). The *primal treewidth* of a CNF formula is the treewidth of its primal graph. We will also be concerned with the following generalization of primal treewidth: the *incidence graph* of a CNF formula has as vertices the variables *and* the clauses of the formula. Two vertices are connected by an edge if and only if one vertex is a variable and the other is a clause such that the variable appears in the clause (see Fig. 1b). The *incidence treewidth* of a formula is then the treewidth of its incidence graph.

It is well-know that the primal treewidth of a CNF-formula can be arbitrarily higher than the incidence treewidth (for example consider a single clause of size n). The other way round, formulas of primal treewidth k can easily be seen to be of incidence treewidth at most $k + 1$ [19].

2.3   From Primal to Incidence Treewidth

While in general primal and incidence treewidth are two different parameters, in this section we argue that when dealing with CNF formulas we don't need to distinguish between the two: first, since incidence treewidth is more general, the lower bounds for primal treewidth transfer automatically to it; second, while the same cannot generally be said for algorithmic results, it is easy to see that the primal treewidth is bounded by the product of the incidence treewidth the arity (clause size), so it suffices to show that we can transform any CNF formula to an equivalent one having bounded arity while roughly maintaining its incidence treewidth. Proposition 1 suggests a linear time transformation achieving this. In the following we can then interchangeably work with incidence treewidth or primal treewidth, whichever is more convenient in the respective situation.

Proposition 1. *There is an algorithm that, given a CNF formula ϕ of incidence treewidth k, computes in time $2^{O(k)}|\phi|$ a 3CNF formula ϕ' of incidence treewidth $O(k)$ with $\mathsf{var}(\phi) \subseteq \mathsf{var}(\phi')$ such that ϕ is a projection of ϕ'.*

Proof (Sketch). We use the classic reduction from SAT to 3SAT that cuts big clauses into smaller clauses by introducing new variables. During this reduction we have to take care that the runtime is in fact linear and that we can bound the treewidth appropriately.   □

It is well-known that if the clauses in a formula ϕ of incidence treewidth k have at most size d, then the primal treewidth of ϕ is at most $(k + 1)d$, see e.g. [19] so the following result follows directly.

Corollary 1. *There is an algorithm that, given a CNF-formula ϕ of incidence treewidth k, computes in time $O(2^k|\phi|)$ a 3CNF-formula ϕ' of primal treewidth $O(k)$ such that ϕ is a projection of ϕ'.*

We will in several places in this paper consider Boolean combinations of functions expressed by CNF formulas of bounded treewidth. The following technical lemma states that we can under certain conditions construct CNF formulas of bounded treewidth for the these Boolean combinations.

Lemma 1. *(a) There is an algorithm that, given a 3CNF-formula ϕ and a tree decomposition $(T, (B_t)_{t \in T})$ of its incidence graph of width $O(k)$, computes in time $\mathrm{poly}(k)n$ a CNF-formula ϕ' and a tree decomposition $(T', (B'_t)_{t \in T})$ of the incidence graph of ϕ' such that $\neg\phi$ is a projection of ϕ', for all $t \in T$ we have $B'_t \cap \mathrm{var}(\phi) = B_t$ and the width of $(T', (B'_t)_{t \in T})$ is $O(k)$.*

(b) There is an algorithm that, given two 3CNF-formulas ϕ_1, ϕ_2 and two tree decompositions $(T, (B^i_t)_{t \in T})$ for $i = 1, 2$ of the incidence graphs of ϕ_i of width $O(k)$ such that for every bag either $B^1_t \cap B^2_t = \emptyset$ or $B^1_t \cap \mathrm{var}(\phi_1) = B^2_t \cap \mathrm{var}(\phi_2)$, computes in time $\mathrm{poly}(k)n$ a tree decomposition $(T', (B'_t)_{t \in T})$ of the incidence graph of $\phi_1 \wedge \phi_2$ such that $\phi' \equiv \phi_1 \wedge \phi_2$, for all $t \in T$ we have $B^1_t \cup B^2_t = B'_t$ and the width of $(T, (B_t)_{t \in T})$ is $O(k)$.

Proof. (a) Because every clause has at most 3 literals, we assume w.l.o.g. that every bag B that contains a clause C contains also all variables of C.

In a first step, we add for every clause $C = \ell_1 \vee \ell_2 \vee \ell_3$ a variable x_C and substitute C by clauses with at most 3-variables encoding the constraint $C = x_C \leftrightarrow \ell_1 \vee l_2 \vee l_3$ introducing some new variables. The result is a CNF-formula ϕ_1 in which every assignment a to $\mathrm{var}(\phi)$ can be extended uniquely to a satisfying assignment a_1 and in a_1 the variable x_C is true if and only if C is satisfied by a. Note that, since every clause has at most 3 variables, the clauses for C can be constructed in constant time. Moreover, we can construct a tree decomposition of width $O(k)$ for ϕ_1 from that of ϕ by adding all new clauses for C and x_C to every bag containing C.

In a next step, we introduce a variable x_t for every $t \in T$ and a constraint $\mathcal{T}$ defining $x_t \leftrightarrow (x_{t_1} \wedge x_{t_2} \wedge \bigwedge_{C \in B_t} x_C)$ where t_1, t_2 are the children of t and the variables are omitted in case they do not appear. The resulting CNF formula ϕ_2 is such that every assignment a to $\mathrm{var}(\phi)$ can be uniquely extended to a satisfying assignment a_2 of ϕ_2 and x_t is true in a_2 if and only if all clauses that appear in the subtree of T rooted in t are satisfied by a. Since every constraint $\mathcal{T}$ has at most k variables, we can construct the 3CNF-formula simulating it in time $O(k)$, e.g. by Tseitin transformation. We again bound the treewidth as before.

The only thing that remains is to add a clause $\neg x_r$ where r is the root of T. This completes the proof of (a).

(b) We simply set $B'_t = B^1_t \cup B^2_t$. It is readily checked that this satisfies all conditions. $\square$

Lemma 2. *There is an algorithm that, given a 3CNF formula ϕ with a tree decomposition $(T, (B_t)_{t \in T})$ of width k of the incidence graph of ϕ and sequences of variables $X := (x_1, \ldots, x_\ell)$, $Y = (y_1, \ldots, y_\ell) \subseteq \mathrm{var}(\phi)^\ell$ such that for every $i \in [\ell]$ there is a bag B_t with $\{x_i, y_i\} \in B_t$, computes in time $\mathrm{poly}(k)|\phi|$ a formula ψ that is a projection of $X \subseteq Y = \bigwedge_{i=1}^{\ell}(x_i \leq y_i)$ and a tree decomposition $(T, (B_t)_{t \in T})$ of ψ of width $O(1)$. The same is true for $\subset$ instead of $\subseteq$.*

Proof. For the case $\subseteq$, ψ is simply $\bigwedge_{i=1}^{\ell}(x_i \leq y_i) = \bigwedge_{i=1}^{\ell} \neg x_i \vee y_i$. ψ satisfies all properties even without projection and with the same tree decomposition.

The case $\subset$ is slightly more complex. We first construct $\bigwedge_{i=1}^{\ell}(x_i = y_i) = \bigwedge_{i=1}^{\ell}(\neg x_i \vee y_i) \wedge (x_i \vee \neg y_i)$. Then we apply Lemma 1 (a) to get a CNF formula that has $X \neq Y$ as a projection. Finally, we use Lemma 1 to get a formula for $X \subset Y = (X \subseteq Y) \wedge (X \neq Y)$. It is easy to check that this formula has the right properties for the tree decomposition.   □

3   2-QBF

Our main tool in this paper will be QBF, the quantified version of CNF. In particular, we will be concerned with the version of QBF which only has two quantifier blocks which is often called 2-QBF. Let us recall some standard definitions. A $\forall\exists$-QBF is a formula of the form $\forall X \exists Y \phi(X, Y)$ where X and Y are disjoint vectors of variables and $\phi(X, Y)$ is a CNF-formula called the *matrix*. We assume the usual semantics for $\forall\exists$-QBF. Moreover, we sometimes consider Boolean combinations of QBF-formulas which we assume to be turned into prenex form again with the help of the usual transformations.

It is well-known that deciding if a given $\forall\exists$-QBF is true is complete for the second level of the polynomial hierarchy, and thus generally considered intractable. Treewidth has been used as an approach for finding tractable fragments of $\forall\exists$-QBF and more generally bounded alternation QBF. Let us define the primal (resp. incidence) treewidth of a $\forall\exists$-QBF to be the primal (resp. incidence) treewidth of the underlying CNF formula. Chen [7] showed the following result.

Theorem 1 [7]. *There is an algorithm that given a $\forall\exists$-QBF of primal treewidth k decides in time $2^{2^{O(k)}} |\phi|$ if ϕ is true.*

We note that the result of [7] is in fact more general than what we state here. In particular, the paper gives a more general algorithm for i-QBF with running time $2^{2^{\cdot^{\cdot^{O(k)}}}} |\phi|$, where the height of the tower of exponentials is i.

In the later parts of this paper, we require a version of Theorem 1 for incidence treewidth which fortunately follows directly from Theorem 1 and Corollary 1.

Corollary 2. *There is an algorithm that given a $\forall\exists$-QBF of incidence treewidth k decides in time $2^{2^{O(k)}} |\phi|$ if ϕ is true.*

We remark that general QBF of bounded treewidth without any restriction on the quantifier prefix is PSPACE-complete [3], and finding tractable fragments by taking into account the structure of the prefix and notions similar to treewidth is quite an active area of research, see e.g. [15,16].

To show tightness of our upper bounds, we use the following theorem from [28].

Theorem 2. *There is no algorithm that, given a $\forall\exists$-QBF ϕ with n variables and primal treewidth k, decides if ϕ is true in time $2^{2^{o(k)}} 2^{o(n)}$, unless ETH is false.*

4 Abstract Argumentation

Abstract argumentation is an area of artificial intelligence which tries to assess the acceptability of arguments within a set of possible arguments based only the relation between them, i.e., which arguments defeat which. Since its creation in [10], abstract argumentation has developed into a major and very active subfield. In this section, we consider the most studied setting introduced in [10].

An argumentation framework is a pair $F = (A, R)$ where A is a finite set and $R \subseteq A \times A$. The elements of A are called *arguments*. The elements of R are called the *attacks* between the arguments and we say for $a, b \in A$ that a attacks b if and only if $ab \in R$. A set $S \subseteq A$ is called *conflict-free* if and only if there are no $a, b \in S$ such that $ab \in R$. We say that a vertex a is *defended* by S if for every b that attacks a, i.e. $ba \in R$, there is an argument $c \in S$ that attacks b. The set S is called *admissible* if and only if it is conflict-free and all elements of S are defended by S. An admissible set S is called *preferred* if and only if it is subset-maximal in the set of all admissible sets.

There are two main notions of acceptance: A set S of arguments is accepted *credulously* if and only if there is a preferred admissible set such that $S \subseteq S'$. The set S is accepted *skeptically* if and only if for all preferred admissible sets S' we have $S \subseteq S'$. Both notions of acceptance have been studied extensively in particular with the following complexity results: it is NP hard to decide, given an argumentation framework $F = (A, R)$ and a set $S \subseteq A$, if S is credulously accepted. For skeptical acceptance, the analogous decision problem is Π_2^p-complete [12]. Credulous acceptance is easier to decide, because when S is contained in any admissible set S' then it is also contained in a preferred admissible set S'': a simple greedy algorithm that adds arguments to S' that are not in any conflicts constructs such an S''.

Concerning treewidth, after some results using Courcelle's Theorem [11], it was shown in [13] by dynamic programming that credulous acceptance can be decided in time $2^{O(k)}n$ while skeptical acceptance can be decided in time $2^{2^{O(k)}}n$ for argument frameworks of size n and treewidth k. Here an argument framework is seen as a directed graph and the treewidth is that of the underlying undirected graph. We reprove these results in our setting. To this end, we first encode conflict free sets in CNF. Given an argumentation framework $F = (A, R)$, construct a CNF formula ϕ_{cf} that has an indicator variable x_a for every $a \in A$ as

$$\phi_{cf} := \bigwedge_{ab \in R} \neg x_a \vee \neg x_b.$$

It is easy to see that the satisfying assignments of ϕ_{cf} encode the conflict-free sets for F. To encode the admissible sets, we add an additional variable P_a for every $a \in A$ and define:

$$\phi_d := \phi_{cf} \wedge \bigwedge_{a \in A} \left((\neg P_a \vee \bigvee_{b:ba \in R} x_b) \wedge \bigwedge_{b:ba \in R} (P_a \vee \neg x_b) \right)$$

The clauses for each P_a are equivalent to $P_a \leftrightarrow \bigvee_{b:ba \in R} x_b$, i.e., P_a is true in a model if and only if a is attacked by the encoded set. Thus by setting

$$\phi_{adm} := \phi_d \wedge \bigwedge_{ba \in R} (\neg P_b \vee \neg x_a)$$

we get a CNF formula whose models restricted to the x_a variables are exactly the admissible sets. We remark that in [27] the authors give a similar SAT-encoding for argumentation problems with slightly different semantics.

Claim 3. *If F has treewidth k, then ϕ_{adm} has incidence treewidth $O(k)$.*

Proof. We start from a tree decomposition $(T, (B_t)_{t \in T})$ of width k of F and construct a tree decomposition of ϕ_{adm}. First note that $(T, (B_t)_{t \in T})$ is also a tree decomposition of the primal graph of ϕ_{cf} up to renaming each a to x_a. For every $ba \in R$ there is thus a bag B that contains both b and a. We connect a new leaf to B containing $\{C_{a,b}, a, b\}$ where $C_{a,b}$ is a clause node for the clause $\neg x_a \vee \neg x_b$ to construct a tree decomposition of the primal graph of ϕ_d.

Now we add P_a to all bags containing x_a, so that for every clause $P_a \vee \neg x_b$ we have a bag containing both variables, and we add new leaves for the corresponding clause nodes as before. Then we add for every clause $C_a := \neg P_a \vee \bigvee_{b:ba \in R} x_b$ the node C_a to every bag containing a. This covers all edges incident to C_a in the incidence graph of ϕ_d and since for every a we only have one such edge, this only increases the width of the decomposition by a constant factor. We obtain a tree decomposition of width $O(k)$ for the incidence graph of ϕ_d.

The additional edges for ϕ_{adm} are treated similarly to above and we get a tree decomposition of width $O(k)$ of ϕ_{adm} of ϕ as desired.   □

Combining Claim 3 with the fact that satisfiability of CNF-formulas of incidence treewidth k can be solved in time $2^{O(k)}$, see e.g. [36], we directly get the first result of [13].

Theorem 4. *There is an algorithm that, given an argumentation framework $F = (A, R)$ of treewidth k and a set $S \subseteq A$, decides in time $2^{O(k)}|A|$ if S is credulously accepted.*

We also give a short reproof of the second result of [13].

Theorem 5. *There is an algorithm that, given an argumentation framework $F = (A, R)$ of treewidth k and a set $S \subseteq A$, decides in time $2^{2^{O(k)}}|A|$ if S is skeptically accepted.*

Proof. Note that the preferred admissible sets of $F = (A, R)$ are exactly the subset maximal assignments to the x_a that can be extended to a satisfying assignment of ϕ_{adm}. Let $X := \{x_a \mid a \in A\}$, then we can express the fact that an assignment is a preferred admissible set by

$$\phi'(X) = \exists P \forall X' \forall P' \left(\phi_{adm}(X, P) \wedge (\neg \phi_{adm}(X', P') \vee \neg (X \subset X')) \right)$$

where the sets P, X' and P' are defined analogously to X. Then S does not appear in all preferred admissible sets if and only if

$$\exists X(\phi'(X) \wedge \bigvee_{a \in S} \neg x_a).$$

After negation we get

$$\forall X \forall P \exists X' \exists P' \left(\neg \phi_{adm}(X, P) \vee (\phi_{adm}(X', P') \wedge (X \subset X')) \vee \bigwedge_{a \in S} x_a \right)$$

and using Lemma 1 afterwards yields a $\forall \exists$-QBF of incidence treewidth $O(k)$ that is true if and only if S appears in all preferred admissible sets. This gives the result with Corollary 2. $\qquad\square$

We remark that QBF encoding of problems in abstract argumentation have been studied in [1,14].

We now show that Theorem 5 is essentially tight.

Theorem 6. *There is no algorithm that, given an argumentation framework $F = (A, R)$ of size n and treewidth k and a set $S \subseteq A$, decides if S is in every preferred admissible set of F in time $2^{2^{o(k)}} 2^{o(n)}$, unless ETH is false.*

Proof. We use a construction from [12,13]: for a given $\forall \exists$-QBF $\forall Y \exists Z \phi$ in variables $Y \cup Z = \{x_1, \ldots, x_n\}$ and clauses $C_1, \ldots, C_m$, define $F_\phi = (A, R)$ with

$$A = \{\phi, C_1, \ldots, C_m\} \vee \{x_i, \bar{x}_i \mid 1 \leq i \leq n\} \cup \{b_1, b_2, b_3\}$$
$$R = \{(C_j, \phi) \mid 1 \leq j \leq m\} \cup \{(x_i, \bar{x}_i), (\bar{x}_i, x_i) \mid 1 \leq i \leq n\}$$
$$\cup \{(x_i, C_j) \mid x_i \text{ in } C_j, 1 \leq j \leq m\} \cup \{(\bar{x}_i, C_j) \mid \neg x_i \text{ in } C_j, 1 \leq j \leq m\}$$
$$\cup \{(\phi, b_1), (\phi, b_2), (\phi, b_3), (b_1, b_2), (b_2, b_3), (b_3, b_1)\} \cup \{(b_1, z), (b_1, \bar{z}) \mid z \in Z\}$$

One can show that ϕ is in every preferred admissible set of F_ϕ if and only if ϕ is true. Moreover, from a tree decomposition of the primal graph of ϕ we get a tree decomposition of F as follows: we add every $\bar{x}_i$ to every bag that contains x_i and we add b_1, b_2, b_3 to all bags. This increases the treewidth from k to $2k+3$ and thus we get the claim with Theorem 2. $\qquad\square$

5   Abduction

In this section, we consider *(propositional) abduction*, a form of non-monotone reasoning that aims to find explanations for observations that are consistent with an underlying theory. A *propositional abduction problem* (short PAP) consists of a tuple $P = (V, H, M, T)$ where T is a propositional formula called the *theory* in variables V, the set $M \subseteq V$ is called the set of *manifestations* and $H \subseteq V$ the set of *hypotheses*. We assume that T is always in CNF. In abduction, one identifies a set $S \subseteq V$ with the formula $\bigwedge_{x \in S} x$. Similarly, given a set $S \subseteq H$, we define $T \cup S := T \wedge \bigwedge_{x \in S} x$. A set $S \subseteq H$ is a *solution* of the PAP, if $T \cup S \models M$, i.e., all models of $T \cup S$ are models of M.

There are three main problems on PAPs that have been previously studied:

- Solvability: Given a PAP P, does it have a solution?
- Relevance: Given a PAP P and $h \in H$, is h contained in *at least one* solution?
- Necessity: Given a PAP P and $h \in H$, is h contained in *all* solutions?

The first two problems are Σ_2^p-complete while necessity is Π_2^p-complete [18]. In [21], it is shown with Courcelle's Theorem that if the theory T of an instance P is of bounded treewidth, then all three above problems can be solved in linear time. Moreover, [21] gives an algorithm based on a Datalog-encoding that solves the solvability and relevance problems in time $2^{2^{O(k)}}|T|$ on instances of treewidth k. Our first result gives a simple reproof of the latter results and gives a similar runtime for necessity.

Theorem 7. *There is a linear time algorithm that, given a PAP $P = (V, H, M, T)$ such that the incidence treewidth of T is k and $h \in H$, decides in time $2^{2^{O(k)}}|T|$ the solvability, relevance and necessity problems.*

Proof. We first consider solvability. We identify the subsets $S \subseteq H$ with assignments to H in the obvious way. Then, for a given choice S, we have that $T \cup S$ is consistent if and only if

$$\psi_1(S) := \exists X T(X) \wedge \bigwedge_{s_i \in H} (s_i \to x_i),$$

is true where X has a variable x_i for every variable $v_i \in V$. Moreover, $T \cup S \models M$ if and only if

$$\psi_2 := \forall X' \left(\bigwedge_{s_i \in H} (s_i \to x_i') \right) \to \left(T(X') \to \bigwedge_{v_i \in M} x_i' \right),$$

where X' similarly to X has a variable x_i for every variable $v_i \in V$. To get a $\forall\exists$-formula, we observe that the PAP has no solution if and only if

$$\forall S \neg(\psi_1(S) \wedge \psi_2(S)) = \forall S \forall X \exists X' \neg (T(X) \wedge S \subseteq X|_H) \vee (S \subseteq X'|_H \wedge T(X') \wedge \neg \bigwedge_{v_i \in M} x_i')$$

is true, where $X|_H$ denotes the restriction of X to the variables of H. Now applying Lemmata 1 and 2 in combination with de Morgan laws to express $\vee$ yields a $\forall\exists$-QBF of incidence treewidth $O(k)$ and the result follows with Corollary 2.

For relevance, we simply add the hypothesis h to T and test for solvability.

For necessity, observe that h is in all solutions if and only if

$$\forall S(\psi_1(S) \wedge \psi_2(S)) \to h,$$

which can easily be brought into $\forall\exists$-QBF slightly extending the construction for the solvability case. $\qquad\square$

Using the Σ_2^p-hardness reduction from [18], it is not hard to show that the above runtime bounds are tight.

Theorem 8. *There is no algorithm that, given a PAP P whose theory has primal treewidth k, decides solvability of P in time $2^{2^{o(k)}} 2^{o(n)}$, unless ETH is false. The same is true for relevance and necessity of a variable h.*

Proof. Let $\phi' = \forall X \exists Y \phi$ be a $\forall\exists$-QBF with $X = \{x_1, \ldots, x_m\}$ and $Y = \{y_1, \ldots, y_\ell\}$. Define a PAP $P = (V, H, M, T)$ as follows

$$V = X \cup Y \cup X' \cup \{s\}$$
$$H = X \cup X'$$
$$M = Y \cup \{s\}$$
$$T = \bigwedge_{i=1}^{m} (x_i \leftrightarrow \neg x_i') \wedge \underbrace{(\phi \rightarrow s \wedge \bigwedge_{j=1}^{\ell} y_j)}_{\psi} \wedge \bigwedge_{j=1}^{\ell} (s \rightarrow y_j)$$

where $X' = \{x_1', \ldots, x_m'\}$ and s are fresh variables. It is shown in [18] that ϕ' is true if and only if P has a solution. We show that T can be rewritten into CNF-formula T' with the help of Lemma 1. The only non-obvious part is the rewriting of ψ. We solve this part by first negating into $(\phi \wedge (\neg s \vee \bigvee_{j=1}^{\ell} \neg y_j)$ and observing that the second conjunct is just a clause, adding it to ϕ only increases the treewidth by 2. Finally, we negate the resulting formula to get a CNF-formula for ψ with the desired properties. The rest of the construction of T' is straightforward. The claim then follows with Theorem 2.

The result is a PAP with theory T' of treewidth $O(k)$ and $O(n)$ variables and the result for solvability follows with Theorem 2. As to the result for relevance and necessity, we point the reader to the proof of Theorem 4.3 in [18]. There for a PAP P a new PAP P' with three additional variables and 5 additional clauses is constructed such that solvability of P reduces to the necessity (resp. relevance) of a variable in P'. Since adding a fixed number of variables and clauses only increases the primal treewidth at most by a constant, the claim follows.   □

5.1   Adding ⊆-Preferences

In abduction there are often preferences for the solution that one wants to consider for a given PAP. One particular interesting case is ⊆-preference where one tries to find (subset-)minimal solutions, i.e. solutions S such that no strict subset $S' \subseteq S$ is a solution. This is a very natural concept as it corresponds to finding minimal explanations for the observed manifestations. We consider two variations of the problems considered above, ⊆-relevance and ⊆-necessity. Surprisingly, complexity-wise, both remain in the second level of the polynomial hierarchy [17]. Below we give a linear-time algorithm for these problems.

Theorem 9. *There is a linear time algorithm that, given a PAP $P = (V, H, M, T)$ such that the incidence treewidth of T is k and $h \in H$, decides in time $2^{2^{2^{O(k)}}} |T|$ the $\subseteq$-relevance and $\subseteq$-necessity problems.*

Proof (sketch). We have seen how to express the property of a set S being a solution as a formula $\psi(S)$ in the proof of Theorem 5. Then expressing that S is a minimal model can written by

$$\psi'(S) := \psi(S) \wedge (\forall S'(S' \subseteq S \rightarrow \neg\psi(S'))).$$

This directly yields QBFs for encoding the $\subseteq$-necessity and $\subseteq$-relevance problems as before which can again be turned into treewidth $O(k)$. The only difference is that we now have three quantifier alternations leading to a triple-exponential dependence on k when applying the algorithm from [7]. □

We remark that [21] already gives a linear time algorithm for $\subseteq$-relevance and $\subseteq$-necessity based on Courcelle's algorithm and thus without any guarantees for the dependence on the runtime. Note that somewhat disturbingly the dependence on the treewidth in Theorem 9 is triple-exponential. We remark that the lower bounds we could get with the techniques from the other sections are only double-exponential. Certainly, having a double-exponential dependency as in our other upper bounds would be preferable and thus we leave this as an open question.

6   Circumscription

In this section, we consider the problem of circumscription. To this end, consider a CNF-formula T encoding a propositional function called the *theory*. Let the variable set X of T be partitioned into three variable sets P, Q, Z. Then a model a of T is called (P, Q, Z)-*minimal* if and only if there is no model a' such that $a'|_P \subset a|_P$ and $a'|_Q = a|_Q$. In words, a is minimal on P for the models that coincide with it on Q. Note that a and a' can take arbitrary values on Z. We denote the (P, Q, Z)-minimal models of T by $\mathrm{MM}(T, P, Q, Z)$. Given a CNF-formula F, we say that $\mathrm{MM}(T, P, Q, Z)$ entails F, in symbols $\mathrm{MM}(T, P, Q, Z) \models F$, if all assignments in $\mathrm{MM}(T, P, Q, Z)$ are models of F. The problem of *circumscription* is, given T, P, Q, Z and F as before, to decide if $\mathrm{MM}(T, P, Q, Z) \models F$.

Circumscription has been studied extensively and is used in many fields, see e.g. [31, 34]. We remark that circumscription can also be seen as a form of closed world reasoning which is equivalent to reasoning under the so-called extended closed world assumption, see e.g. [6] for more context. On general instances circumscription is Π_2^p-complete [17] and for bounded treewidth instances, i.e. if the treewidth of $T \wedge F$ is bounded, there is a linear time algorithm shown by Courcelle's Theorem [21]. There is also a linear time algorithm for the corresponding counting problem based on datalog [23]. We here give a version of the result from [21] more concrete runtime bounds.

Theorem 10. *There is an algorithm that, given an instance T, P, Q, Z and F of incidence treewidth k, decides if $\mathrm{MM}(T, P, Q, Z) \models F$ in time $2^{2^{O(k)}}(|T| + |F|)$.*

Proof. Note that we have $\mathrm{MM}(T, P, Q, Z) \models F$ if and only if for every assignment (a_P, a_Q, a_Z) to P, Q, Z, we have that (a_P, a_Q, a_Z) is not a model of T, or (a_P, a_Q, a_Z) is a model of F or there is a model (a'_P, a'_Q, a'_Z) of T such that $a'_P \subset a_P$ and $a'_Q = a_Q$. This can be written as a $\forall\exists$-formula as follows:

$$\psi := \forall P \forall Q \forall Z \exists P' \exists Z' (\neg T(P, Q, Z) \vee F(P, Q, Z) \vee (T(P', Q, Z') \wedge P' \subset P)).$$

We first compute a tree decomposition of $T \wedge F$ of width $O(k)$ in time $2^{O(k)}(|T|+|F|)$. We can use Lemma 1, Lemma 2 and Proposition 1 to compute in time $\mathrm{poly}(k)(|T| + |F|)$ a CNF-formula ϕ such that the matrix of ψ is a projection of ϕ and ϕ has incidence treewidth $O(k)$. Applying Corollary 2, yields the result. $\qquad\square$

We now show that Theorem 10 is essentially optimal by analyzing the proof in [17].

Theorem 11. *There is no algorithm that, given an instance T, P, Q, Z and F of size n and treewidth k, decides if $\mathrm{MM}(T, P, Q, Z) \models F$ in time $2^{2^{o(k)}} 2^{o(n)}$, unless ETH is false.*

Proof. Let $\psi = \forall X \exists Y \phi$ be a QBF with $X = \{x_1, \ldots, x_m\}$ and $Y = \{y_1, \ldots, y_\ell\}$. We define the theory T as follows:

$$T = \left(\bigwedge_{i=1}^{m} (x_i \neq z_i)\right) \wedge ((u \wedge y_1 \wedge \ldots y_\ell) \vee \phi),$$

where $z_1, \ldots, z_m$ and u are fresh variables. Set $P = \mathrm{var}(T)$ and $Q = \emptyset$ and Z the rest of the variables. In [17], it is shown that $\mathrm{MM}(T, P, Q, Z) \models \neg u$ if and only if ϕ is true. Now using Lemma 1 we turn ψ into a 2-QBF ψ' with the same properties. Note that ψ' has treewidth $O(k)$ and $O(m + \ell)$ variables and thus the claim follows directly with Theorem 2. $\qquad\square$

7 Minimal Unsatisfiable Subsets

Faced with unsatisfiable CNF-formula, it is in many practical settings highly interesting to find the sources of unsatisfiability. One standard way of describing them is by so-called minimal unsatisfiable subsets. A *minimal unsatisfiable set (short MUS)* is an unsatisfiable set C of clauses of a CNF-formula such that every proper subset of C is satisfiable. The computation of MUS has attracted a lot of attention, see e.g. [22,26,38] and the references therein.

In this section, we study the following question: given a CNF-formula ϕ and a clause C, is C contained in a MUS of ϕ? Clauses for which this is the case can in a certain sense be considered as not problematic for the satisfiability of ϕ. As for the other problems studied in this paper, it turns out that the above problem is complete for the second level of the polynomial hierarchy, more specifically for Σ_2^p [30]. Treewidth restrictions seem to not have been considered before, but we show that our approach gives a linear time algorithm in a simple way.

Theorem 12. *There is an algorithm that, given a CNF-formula ϕ incidence treewidth k and a clause C of ϕ, decides C is in a MUS of ϕ in time $2^{2^{O(k)}}|\phi|$.*

Proof. Note that C is in a MUS of ϕ if and only if there is an unsatisfiable clause set $\mathcal{C}$ such that $C \in \mathcal{C}$ and $\mathcal{C} \setminus \{C\}$ is satisfiable. We will encode this in $\forall\exists$-QBF. In a first step, similarly to the proof of Lemma 1, we add a new variable x_C for every clause C of ϕ and substitute ϕ by clauses expressing $C \leftrightarrow x_c$. Call the resulting formula ψ. It is easy to see that the incidence treewidth of ψ is at most double that of ϕ. Moreover, for every assignment a to $\mathsf{var}(\phi)$, there is exactly one extension to a satisfying assignment a' of ψ. Moreover, in a' a clause variable x_C is true if and only if a satisfies the clause C. Let $\mathcal{C}$ be a set of clauses, then $\mathcal{C}$ is unsatisfiable if and only if for every assignment a to $\mathsf{var}(\phi)$, $\mathcal{C}$ is not contained in the set of satisfied clauses. Interpreting sets by assignments as before, we can write this as a formula by

$$\psi'(\mathcal{C}) := \forall X \forall \mathcal{C}' : \psi_C(X, \mathcal{C}') \rightarrow \neg(\mathcal{C} \subseteq \mathcal{C}').$$

Let now $\mathcal{C}$ range over the sets of clauses not containing C. Then we have by the considerations above that C appears in a MUS if and only if

$$\psi^* = \exists \mathcal{C} \psi'(\mathcal{C} \cup \{C\}) \wedge \neg \psi'(\mathcal{C})$$
$$= \exists \mathcal{C} \exists X' \exists \mathcal{C}' \forall X'' \forall \mathcal{C}''(\phi_C(X', \mathcal{C}') \rightarrow \neg(\mathcal{C} \cup \{C\} \subseteq \mathcal{C}')) \wedge \phi_C(X', \mathcal{C}'') \wedge \mathcal{C} \subseteq \mathcal{C}''$$

Negating and rewriting the matrix of the resulting QBF with Lemma 1, we get in linear time a $\forall\exists$-QBF of treewidth $O(k)$ that is true if and only if C does not appear in a MUS of ϕ. Using Theorem 2 completes the proof.   □

We remark that different QBF encodings for MUS membership have also been studied in [24]. We now show that Theorem 12 is essentially tight.

Theorem 13. *There is no algorithm that, given a CNF-formula ϕ with n variables and primal treewidth k and a clause C of ϕ, decides if C is in a MUS of ϕ in time $2^{2^{o(k)}} 2^{o(n)}$, unless ETH is false.*

Proof. Given a $\forall\exists$-QBF $\psi = \forall X \exists Y \phi$ of incidence treewidth k where $C_1, \ldots, C_m$ are the clauses of ϕ, we construct the CNF-formula

$$\phi' = \bigwedge_{x \in X} (x \wedge \neg x) \wedge w \wedge \bigwedge_{i=1}^{m}(\neg w \vee C_i).$$

In [30] it is shown that ψ is true if and only if the clause w appears in a MUS of ϕ'. Note that ϕ' has primal treewidth $k + 1$: in a tree decomposition of the primal graph of ϕ, we can simply add the variable w into all bags to get a tree decomposition of the primal graph of ϕ'. Since clearly $|\phi'| = O(|\phi|)$, any algorithm to check if w is in a MUS of ϕ' in time $2^{2^{o(k)}} 2^{o(n)}$ contradicts ETH with Theorem 2.   □

8 Conclusion

In this paper, we took an alternate approach in the design of optimal algorithms mainly for the second level of the polynomial hierarchy parameterized by treewidth: we used reductions to 2-QBF. We stress that, apart from some technical transformations on CNF-formulas which we reused throughout the paper, our algorithms are straightforward and all complexity proofs very simple. We consider this as a strength of what we propose and not as a lack of depth, since our initial goal was to provide a black-box technique for designing optimal linear-time algorithms with an asymptotically optimal guarantee on the treewidth. We further supplement the vast majority of our algorithms by tight lower-bounds, using ETH reductions again from 2-QBF.

We concentrated on areas of artificial intelligence, investigating a collection of well-studied and diverse problems that are complete for Σ_2^p and Π_2^p. However we conjecture that we could apply our approach to several problems with similar complexity status. Natural candidates are problems complete for classes in the polynomial hierarchy, starting from the second level, see e.g. [37] for an overview (mere NP-complete problems can often be tackled by other successful techniques).

Of course, our approach is no silver bullet that magically makes all other techniques obsolete. On the one hand, for problems whose formulation is more complex than what we consider here, Courcelle's Theorem might offer a richer language to model problems than QBF. This is similar in spirit to some problems being easier to model in declarative languages like ASP than in CNF. On the other hand, handwritten algorithms probably offer better constants than what we get by our approach. For example, the constants in [13] are more concrete and smaller than what we give in Sect. 4. However, one could argue that for double-exponential dependencies, the exact constants probably do not matter too much simply because already for small parameter values the algorithms become infeasible[1]. Despite these issues, in our opinion, QBF encodings offer a great trade-off between expressivity and tightness for the runtime bounds and we consider it as a valuable alternative.

Acknowledgments. Most of the research in this paper was performed during a stay of the first and third authors at CRIL that was financed by the project PEPS INS2I 2017 CODA. The second author is thankful for many valuable discussions with members of CRIL, in particular Jean-Marie Lagniez, Emmanuel Lonca and Pierre Marquis, on the topic of this article.

Moreover, the authors would like to thank the anonymous reviewers whose numerous helpful remarks allowed to improve the presentation of the paper.

---

[1] To give the reader an impression: $2^{2^5} \approx 4.2 \times 10^9$ and already $2^{2^6} \approx 1.8 \times 10^{19}$.

References

1. Arieli, O., Caminada, M.W.A.: A general QBF-based formalization of abstract argumentation theory. In: Verheij, B., Szeider, S., Woltran, S. (eds.) Computational Models of Argument, COMMA 2012, pp. 105–116 (2012)
2. Arnborg, S., Proskurowski, A.: Linear time algorithms for NP-hard problems restricted to partial k-trees. Discrete Appl. Math. **23**(1), 11–24 (1989)
3. Atserias, A., Oliva, S.: Bounded-width QBF is PSPACE-complete. J. Comput. Syst. Sci. **80**(7), 1415–1429 (2014)
4. Bodlaender, H.L.: A linear-time algorithm for finding tree-decompositions of small treewidth. SIAM J. Comput. **25**(6), 1305–1317 (1996)
5. Bodlaender, H.L., Drange, P.G., Dregi, M.S., Fomin, F.V., Lokshtanov, D., Pilipczuk, M.: An o(c^k n) 5-approximation algorithm for treewidth. In: 54th Annual IEEE Symposium on Foundations of Computer Science, FOCS 2013, pp. 499–508 (2013)
6. Cadoli, M., Lenzerini, M.: The complexity of propositional closed world reasoning and circumscription. J. Comput. Syst. Sci. **48**(2), 255–310 (1994)
7. Chen, H.: Quantified constraint satisfaction and bounded treewidth. In: de Mántaras, R.L., Saitta, L. (eds.) Proceedings of the 16th European Conference on Artificial Intelligence, ECAI 2004, pp. 161–165 (2004)
8. Courcelle, B.: The monadic second-order logic of graphs. I. Recognizable sets of finite graphs. Inf. Comput. **85**(1), 12–75 (1990)
9. DIMACS: Satisfiability: Suggested Format. DIMACS Challenge. DIMACS (1993)
10. Dung, P.M.: On the acceptability of arguments and its fundamental role in non-monotonic reasoning, logic programming and n-person games. Artif. Intell. **77**(2), 321–358 (1995)
11. Dunne, P.E.: Computational properties of argument systems satisfying graph-theoretic constraints. Artif. Intell. **171**(10–15), 701–729 (2007)
12. Dunne, P.E., Bench-Capon, T.J.M.: Coherence in finite argument systems. Artif. Intell. **141**(1/2), 187–203 (2002)
13. Dvořák, W., Pichler, R., Woltran, S.: Towards fixed-parameter tractable algorithms for abstract argumentation. Artif. Intell. **186**, 1–37 (2012)
14. Egly, U., Woltran, S.: Reasoning in argumentation frameworks using quantified boolean formulas. In: Dunne, P.E., Bench-Capon, T.J.M. (eds.) Computational Models of Argument, COMMA 2006, pp. 133–144 (2006)
15. Eiben, E., Ganian, R., Ordyniak, S.: Using decomposition-parameters for QBF: mind the prefix! In: Schuurmans, D., Wellman, M.P. (eds.) Proceedings of the Thirtieth AAAI Conference on Artificial Intelligence, pp. 964–970 (2016)
16. Eiben, E., Ganian, R., Ordyniak, S.: Small resolution proofs for QBF using dependency treewidth. CoRR, abs/1711.02120 (2017)
17. Eiter, T., Gottlob, G.: Propositional circumscription and extended closed-world reasoning are Π_2^p-complete. Theor. Comput. Sci. **114**(2), 231–245 (1993)
18. Eiter, T., Gottlob, G.: The complexity of logic-based abduction. J. ACM **42**(1), 3–42 (1995)
19. Fischer, E., Makowsky, J.A., Ravve, E.V.: Counting truth assignments of formulas of bounded tree-width or clique-width. Discrete Appl. Math. **156**(4), 511–529 (2008)
20. Frick, M., Grohe, M.: The complexity of first-order and monadic second-order logic revisited. Ann. Pure Appl. Logic **130**(1–3), 3–31 (2004)

21. Gottlob, G., Pichler, R., Wei, F.: Bounded treewidth as a key to tractability of knowledge representation and reasoning. Artif. Intell. **174**(1), 105–132 (2010)
22. Ignatiev, A., Previti, A., Liffiton, M., Marques-Silva, J.: Smallest MUS extraction with minimal hitting set dualization. In: Pesant, G. (ed.) CP 2015. LNCS, vol. 9255, pp. 173–182. Springer, Cham (2015). https://doi.org/10.1007/978-3-319-23219-5_13
23. Jakl, M., Pichler, R., Rümmele, S., Woltran, S.: Fast counting with bounded treewidth. In: Cervesato, I., Veith, H., Voronkov, A. (eds.) LPAR 2008. LNCS (LNAI), vol. 5330, pp. 436–450. Springer, Heidelberg (2008). https://doi.org/10.1007/978-3-540-89439-1_31
24. Janota, M., Marques-Silva, J.: On deciding MUS membership with QBF. In: Lee, J. (ed.) CP 2011. LNCS, vol. 6876, pp. 414–428. Springer, Heidelberg (2011). https://doi.org/10.1007/978-3-642-23786-7_32
25. Kloks, T. (ed.): Treewidth: Computations and Approximations. LNCS, vol. 842. Springer, Heidelberg (1994). https://doi.org/10.1007/BFb0045375
26. Lagniez, J.-M., Biere, A.: Factoring out assumptions to speed up MUS extraction. In: Järvisalo, M., Van Gelder, A. (eds.) SAT 2013. LNCS, vol. 7962, pp. 276–292. Springer, Heidelberg (2013). https://doi.org/10.1007/978-3-642-39071-5_21
27. Lagniez, J.-M., Lonca, E., Mailly, J.-G.: CoQuiAAS: a constraint-based quick abstract argumentation solver. In: 27th IEEE International Conference on Tools with Artificial Intelligence, ICTAI 2015, pp. 928–935 (2015)
28. Lampis, M., Mitsou, V.: Treewidth with a quantifier alternation revisited (2017)
29. Langer, A., Reidl, F., Rossmanith, P., Sikdar, S.: Practical algorithms for MSO model-checking on tree-decomposable graphs. Comput. Sci. Rev. **13–14**, 39–74 (2014)
30. Liberatore, P.: Redundancy in logic I: CNF propositional formulae. Artif. Intell. **163**(2), 203–232 (2005)
31. Lifschitz, V.: Handbook of Logic in Artificial Intelligence and Logic Programming, vol. 3. Chapter Circumscription, pp. 297–352. Oxford University Press Inc., New York (1994)
32. Lokshtanov, D., Marx, D., Saurabh, S.: Lower bounds based on the exponential time hypothesis. Bull. EATCS **105**, 41–72 (2011)
33. Marx, D., Mitsou, V.: Double-exponential and triple-exponential bounds for choosability problems parameterized by treewidth. In: 43rd International Colloquium on Automata, Languages, and Programming, ICALP 2016, pp. 28:1–28:15 (2016)
34. McCarthy, J.: Applications of circumscription to formalizing common-sense knowledge. Artif. Intell. **28**(1), 89–116 (1986)
35. Pan, G., Vardi, M.Y.: Fixed-parameter hierarchies inside PSPACE. In: 21st IEEE Symposium on Logic in Computer Science, LICS 2006, pp. 27–36 (2006)
36. Samer, M., Szeider, S.: Algorithms for propositional model counting. J. Discrete Algorithms **8**(1), 50–64 (2010)
37. Schaefer, M., Umans, C.: Completeness in the polynomial-time hierarchy: a compendium. SIGACT News **33**(3), 32–49 (2002)
38. Marques-Silva, J., Lynce, I.: On improving MUS extraction algorithms. In: Sakallah, K.A., Simon, L. (eds.) SAT 2011. LNCS, vol. 6695, pp. 159–173. Springer, Heidelberg (2011). https://doi.org/10.1007/978-3-642-21581-0_14

Polynomial-Time Validation of QCDCL Certificates

Tomáš Peitl, Friedrich Slivovsky$^{(\boxtimes)}$, and Stefan Szeider

Algorithms and Complexity Group, TU Wien, Vienna, Austria
{peitl,fslivovsky,sz}@ac.tuwien.ac.at

Abstract. Quantified Boolean Formulas (QBFs) offer compact encodings of problems arising in areas such as verification and synthesis. These applications require that QBF solvers not only decide whether an input formula is true or false but also output a witnessing certificate, i.e. a representation of the winning strategy. State-of-the-art QBF solvers based on Quantified Conflict-Driven Constraint Learning (QCDCL) can emit Q-resolution proofs, from which in turn such certificates can be extracted. The correctness of a certificate generated in this way is validated by substituting it into the matrix of the input QBF and using a SAT solver to check that the resulting propositional formula (the *validation formula*) is unsatisfiable. This final check is often the most time-consuming part of the entire certification workflow. We propose a new validation method that does not require a SAT call and provably runs in polynomial time. It uses the Q-resolution proof from which the given certificate was extracted to directly generate a (propositional) proof of the validation formula in the RUP format, which can be verified by a proof checker such as DRAT-trim. Experiments with a prototype implementation show a robust, albeit modest, increase in the number of successfully validated certificates compared to validation with a SAT solver.

1 Introduction

Quantified Boolean Formulas (QBFs) offer succinct encodings for problems from domains such as formal verification, synthesis, and planning [3,5,7,15,21,22]. Even though SAT-based approaches to these problems are generally still superior, the evolution of QBF solvers in recent years is starting to tip the scales in their favor [9]. In most of these applications, it is required that QBF solvers not only output a simple true/false answer but also produce a *strategy*, or *certificate*, that shows how this answer can be realized. For example, a certificate might encode a counterexample to the soundness of a software system, or a synthesized program.

Most state-of-the-art QBF solvers have the ability to generate such certificates, and some recently developed solvers have been explicitly designed with certification in mind [19,20,23]. Search-based solvers implementing Quantified

---

This research was partially supported by FWF grants P27721 and W1255-N23.

O. Beyersdorff and C. M. Wintersteiger (Eds.): SAT 2018, LNCS 10929, pp. 253–269, 2018.
https://doi.org/10.1007/978-3-319-94144-8_16

Conflict-Driven Constraint Learning (QCDCL) [6,26] can output Q-resolution proofs [4,17,18], from which in turn certificates can be extracted in linear time [1,2].

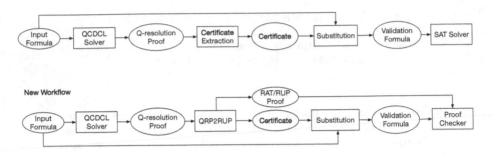

Fig. 1. Certificate extraction and validation for QCDCL solvers.

Since QBF solvers and (to a lesser degree) certificate extraction tools are complex pieces of software that may contain bugs, certificates obtained in this way ought to be independently *validated*. This can be achieved by substituting the certificate back into the matrix of the input QBF and using a SAT solver to check that the resulting propositional formula (which we call the *validation formula*) is unsatisfiable [17]. This certification workflow is illustrated in the top half of Fig. 1. Once a certificate is validated, we can essentially trust its correctness as much as we trust in the correctness of the SAT solver used for validation[1]. However, since certificates tend to be large, the corresponding SAT call frequently amounts to the most time-consuming step in the entire certification workflow and even causes timeouts [17].

In this paper, we propose an alternative validation method for QCDCL that avoids this SAT call. Instead, it uses the Q-resolution proof from which the given certificate was extracted to generate a proof of the validation formula in the RUP format [12], whose correctness can then be verified by a propositional proof checker such as DRAT-trim [25]. This workflow is sketched in the lower half of Fig. 1. Since this RUP proof can be computed from the Q-resolution proof in linear time and checked in polynomial time, we obtain a validation procedure that provably runs in polynomial time.

We implemented this new validation method in a tool named QRP2RUP and tested it on benchmark instances from several recent QBF evaluations. Our experiments show a robust, albeit modest, increase in the number of successfully validated certificates compared to validation with a SAT solver.

---

[1] We still have to make sure that the validation formula is constructed correctly so that it is not trivially unsatisfiable. We discuss this issue in Sect. 8.

2   Preliminaries

A *literal* is a negated or unnegated variable. If x is a variable, we write $\bar{x} = \neg x$ and $\overline{\neg x} = x$, and let $var(x) = var(\neg x) = x$. If X is a set of literals, we write $\overline{X}$ for the set $\{\,\bar{x} : x \in X\,\}$ and let $var(X) = \{\,var(\ell) : \ell \in X\,\}$. An *assignment* to a set X of variables is a mapping $\tau : X \to \{true, false\}$. An assignment σ is an extension of the assignment τ if σ assigns all variables that τ does, and to the same polarity. We extend assignments $\tau : X \to \{true, false\}$ to literals by letting $\tau(\neg x) = \neg \tau(x)$ for $x \in X$.

We consider Boolean *circuits* over $\{\neg, \wedge, \vee, false, true\}$ and write $var(\varphi)$ for the set of variables occurring in a circuit φ. If φ is a circuit and τ an assignment, $\varphi[\tau]$ denotes the circuit obtained by replacing each variable $x \in X \cap var(\varphi)$ by $\tau(x)$ and propagating constants. A circuit φ is *satisfiable* if there is an assignment τ such that $\varphi[\tau] = true$, otherwise it is *unsatisfiable*.

A *clause* (*term*) is a circuit consisting of a disjunction (conjunction) of literals. We write $\bot$ for the empty clause and $\top$ for the empty term. We call a clause *tautological* (and a term *contradictory*) if it contains the same variable negated as well as unnegated. A *CNF formula* (*DNF formula*) is a circuit consisting of a conjunction (disjunction) of non-tautological clauses (non-contradictory terms). Whenever convenient, we treat clauses and terms as sets of literals, and CNF and DNF formulas as sets of sets of literals. Throughout the paper, we make use of the fact that any circuit can be transformed into an equisatisfiable CNF formula of size linear in the size of the circuit [24].

A *unit clause* is a clause containing a single literal. A CNF formula ψ is derived from a CNF formula φ by the *unit clause rule* if (ℓ) is a unit clause of φ and $\psi = \varphi[\{\ell \mapsto true\}]$. *Unit propagation* in a CNF formula consists in repeated applictions of the unit clause rule. Unit propagation is said to *derive* the literal ℓ in a CNF formula φ if a CNF formula ψ with $(\ell) \in \psi$ can be derived from φ by unit propagation. We say that unit propagation causes a *conflict* if $false$ can be derived by unit propagation. If unit propagation does not cause a conflict the set of literals that can be derived by unit propagation induces an assignment. The *closure* of an assignment τ with respect to unit propagation (in φ) is τ combined with the set of literals derivable by unit propagation in $\varphi[\tau]$.

A clause C has the *reverse unit propagation* (RUP) property with respect to a CNF formula φ if unit propagation in $\varphi[\{\,\ell \mapsto false : \ell \in C\,\}]$ causes a conflict. A *RUP proof of unsatisfiability* of a CNF formula φ is a sequence $C_1, \ldots, C_m$ of clauses such that $C_m = \bot$ and each clause C_i has the RUP property with respect to $\varphi \cup \{C_1, \ldots, C_{i-1}\}$, for $1 \le i \le m$.

A (prenex) *Quantified Boolean Formula* $\Phi = \mathcal{Q}.\varphi$ consists of a quantifier prefix $\mathcal{Q}$ and a circuit φ, called the *matrix* of Φ. A *quantifier prefix* is a sequence $\mathcal{Q} = Q_1 x_1 \ldots Q_n x_n$, where the x_i are pairwise distinct variables and $Q_i \in \{\forall, \exists\}$ for $1 \le i \le n$. Relative to Φ, variable x_i and its associated literals are called *existential* (*universal*) if $Q_i = \exists$ ($Q_i = \forall$). We write $E(\Phi)$ and $U(\Phi)$ for the sets of existential and universal variables of Φ, respectively. We assume that the set of variables occurring in φ is precisely $\{x_1, \ldots, x_n\}$ (in particular, we only consider closed QBFs) and let $var(\Phi) = \{x_1, \ldots, x_n\}$. We define a total order $<_\Phi$ on the variables of Φ as $x_i <_\Phi x_j \Leftrightarrow i < j$ and let $D_\Phi(v) = \{w \in var(\Phi) : w <_\Phi v\}$ for $v \in var(\Phi)$. We drop the subscript from $<_\Phi$ and D_Φ whenever Φ is understood.

A *model circuit* of Φ for a variable $v \in var(\Phi)$ is a circuit f_v with $var(f_v) \subseteq D(v)$. A *model* of Φ is an indexed family $\{f_e\}_{e \in E(\Phi)}$ of model circuits such that $\varphi[\tau] = true$ for every assignment $\tau : var(\Phi) \to \{true, false\}$ that satisfies $f_e[\tau] = \tau(e)$ for $e \in E(\Phi)$. A *countermodel* of Φ is an indexed family $\{f_u\}_{u \in U(\Phi)}$ of model circuits such that $\varphi[\tau] = false$ for every assignment $\tau : var(\Phi) \to \{true, false\}$ that satisfies $f_u[\tau] = \tau(u)$ for $u \in U(\Phi)$. A QBF is *true* if it has a model, and *false* if it has a countermodel.

A QBF is a *PCNF (PDNF) formula* if its matrix is a CNF (DNF) formula. *Q-resolution* [14] and *long-distance Q-resolution* [1,27] are proof systems for false PCNF formulas. Let $\Phi = Q.\varphi$ be a PCNF formula. A *Q-resolution refutation* of Φ is a sequence $\mathcal{P} = C_1, \ldots, C_m$ of non-tautological clauses where $C_m = \bot$ and each clause C_i is obtained in one of the following ways:

- $C_i \in \varphi$ is an *input clause*.
- $C_i = (C_j \setminus \{p\}) \cup (C_k \setminus \{\neg p\})$ is the *resolvent* of clauses C_j and C_k on *pivot* variable $p \in E(\Phi)$, where $1 \leq j, k < i$ and $p \in C_j$, $\neg p \in C_k$.
- $C_i = C_j \setminus L$ is obtained from C_j with $1 \leq j < i$ by *universal reduction*. This requires that every literal $\ell \in L$ is universal and that there is no existential variable $e \in var(C_i)$ such that $var(\ell) < e$.

The *size* of $\mathcal{P}$ is defined as $|\mathcal{P}| := \sum_{i=1}^m |C_i|$.

Long-distance Q-resolution [1] is a generalization of Q-resolution that permits the derivation of tautological clauses by modifying the resolution rule in the following way: if $\ell \in C_j$, $\overline{\ell} \in C_k$, and $var(\ell) \neq p$, then ℓ must be universal and $p < var(\ell)$. In this case we say that the literals ℓ and $\overline{\ell}$ are *merged*, and refer to the pair $\ell, \overline{\ell}$ as a *merged literal* of C_i.

Dual proof systems for true PDNF formulas operating on terms are known as *Q-consensus* and *long-distance Q-consensus*. The dual of universal reduction in these proof systems is called *existential reduction*.

3   Validation of Certificates

In this section, we will describe the setting of the problem of QBF certificate validation. Then, in Sects. 4 and 5, we present an algorithm that computes a RUP proof that can be used to replace the final call to the SAT solver by a simple proof check. For the sake of simplicity, we will only focus on false PCNF formulas. The results generalize to true formulas by duality, which will be discussed in Sect. 6.

Let φ be a CNF formula, let C be a boolean circuit. The substitution of C into φ, denoted by $\varphi[C]$, is simply the CNF formula φ in conjunction with a CNF encoding of C (which may contain additional auxiliary variables). Let $\Phi = Q.\varphi$ be a false QBF in PCNF, let C be a boolean circuit whose inputs are existential variables of Φ and whose outputs are universal variables of Φ. The task of verifying that C is a countermodel of Φ is to verify that $\varphi[C]$ is unsatisfiable.

Some QCDCL QBF solvers are capable of outputting a *trace* that contains a (long-distance) Q-resolution refutation of the formula solved. From this refutation, a countermodel circuit can be computed by the Balabanov-Jiang (BJ) algorithm [1], or by the extended Balabanov-Jiang-Janota-Widl (BJJW) algorithm [2] for long-distance Q-resolution. Let $\Phi = \mathcal{Q}.\varphi$ be a QBF, let $\mathcal{P}$ be a (long-distance) Q-resolution refutation of it, let $\mathrm{cc}(\mathcal{P})$ be the countermodel circuit computed by the appropriate version of BJ/BJJW. The CNF formula that results from substitution of $\mathrm{cc}(\mathcal{P})$ into φ as described in the previous paragraph, i.e., $\varphi[\mathrm{cc}(\mathcal{P})]$, is denoted by $\Phi[\mathcal{P}]$, and is called the *validation formula* for the QBF Φ and the proof $\mathcal{P}$. This is the formula that must be checked for unsatisfiability in order to verify the correctness of the certificate $\mathrm{cc}(\mathcal{P})$. We will now present a way how to directly compute a RUP proof for the validation formula out of the proof $\mathcal{P}$, thus obviating the need to use a SAT solver and making validation checks solvable in polynomial time.

4   RUP Proofs from Ordinary Q-Resolution

We will begin by describing a countermodel, and in particular its CNF version obtained by the Tseitin conversion, computed by BJ. For a full explanation of the algorithm we refer to the original paper [1]. We illustrate the certificate extraction process on this example formula

$$\exists x_1, x_2 \,\forall y \,\exists z \quad (x_1 \lor x_2 \lor y \lor z) \land (x_1 \lor x_2 \lor \overline{z}) \land (x_1 \lor \overline{x_2})$$
$$\land (\overline{x_1} \lor \overline{x_2} \lor \overline{y} \lor \overline{z}) \land (\overline{x_1} \lor \overline{x_2} \lor z) \land (\overline{x_1} \lor x_2)$$

along with its Q-resolution refutation:

(1)	$x_1 \lor x_2 \lor y \lor z$	(input)		(7)	$x_1 \lor x_2 \lor y$	$(1,3)$
(2)	$\overline{x_1} \lor \overline{x_2} \lor \overline{y} \lor \overline{z}$	(input)		(8)	$x_1 \lor x_2$	(7)
(3)	$x_1 \lor x_2 \lor \overline{z}$	(input)		(9)	$\overline{x_1} \lor \overline{x_2} \lor \overline{y}$	$(2,4)$
(4)	$\overline{x_1} \lor \overline{x_2} \lor z$	(input)		(10)	$\overline{x_1} \lor \overline{x_2}$	(9)
(5)	$x_1 \lor \overline{x_2}$	(input)		(11)	x_1	$(5,8)$
(6)	$\overline{x_1} \lor x_2$	(input)		(12)	$\overline{x_1}$	$(6,10)$
				(13)	$\bot$	$(11,12)$

Let $\mathcal{P}$ be a Q-resolution refutation of a formula $\Phi = \mathcal{Q}.\varphi$. BJ processes the clauses of $\mathcal{P}$ forward, and everytime a conclusion R of a reduction step $R = R' - L$ (read the set of literals L is reduced from the clause R' to obtain the clause R) is encountered, for every literal ℓ from L either the clause R (if ℓ is positive) or the term $\overline{R}$ (if ℓ is negative) is pushed to what is called the *countermodel array* of $var(\ell)$ (cf. [1]). At the end, the arrays represent the countermodel functions for their respective variables, in the following way:

Let u be a universal variable, and let its countermodel array have the entries $X_1, \ldots, X_n$. This array is interpreted by constructing a set of partial circuits. Let $f_n^u = X_n$. Then we define

$$f_k^u = \begin{cases} X_k \wedge f_{k+1}^u & \text{if } X_k \text{ is a clause,} \\ X_k \vee f_{k+1}^u & \text{if } X_k \text{ is a term,} \end{cases}$$

and finally $f^u = f_1^u$. The circuit f^u represents the countermodel function for the variable u. Intuitively, these circuits find the first reduction step whose conclusion is falsified, and set all of the reduced literals in the premise so that they are falsified too, which ensures that the falsified clause is implied by the conjunction of input clauses and hence at least one of those is falsified too (Fig. 2).

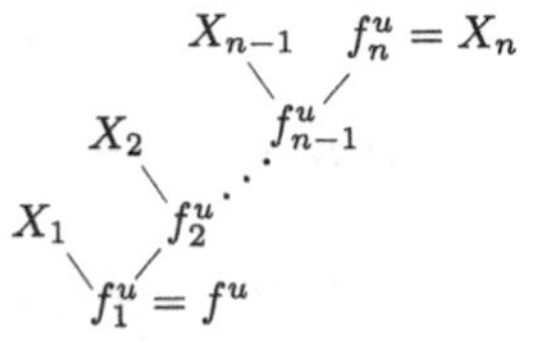

Fig. 2. Schematic depiction of a countermodel circuit extracted by BJ. Each f_i is either an "and" or an "or" gate, depending on the context.

Let us see what this means on the example formula and proof. There is only one universal variable, so we will only build one countermodel array. Processing the clauses forward, the first conclusion of a reduction step that we encounter is (8), y is reduced in positive polarity, so we push the clause $(x_1 \vee x_2)$ to the countermodel array. Next, we encounter the conclusion (10), here y is reduced in negative polarity, so we push the negation of the conclusion $(\overline{x_1} \vee \overline{x_2})$, the term $(x_1 \wedge x_2)$. There are no more reduction steps, so the final countermodel array for y is $[(x_1 \vee x_2), (x_1 \wedge x_2)]$. According to the interpretation above, this results in the circuit $y = ((x_1 \vee x_2) \wedge (x_1 \wedge x_2)) = (x_1 \wedge x_2)$. It can be easily verified that this is indeed a countermodel for the formula.

Let us now examine how the circuit f^u can be translated into CNF for substitution into Φ. We can observe that the circuit f^u has a nested structure, in which first the values of all of the X_k are evaluated, which are then further processed by the circuit to obtain the value for u. Every X_k is either a clause or a term corresponding to a conclusion of a reduction step in $\mathcal{P}$. Let $R_1, \ldots, R_N$ be all conclusions of reduction steps in $\mathcal{P}$, in the same order as they appear in the proof. Then for every X_k there is i_k such that $X_k = R_{i_k}$ or $X_k = \overline{R_{i_k}}$. Let us define variables $g_i = R_i$ for $1 \leq i \leq N$ using the set of clauses

$$G = \{\{(\overline{g_i} \vee R_i)\} \cup \{(g_i \vee \overline{\ell}) \mid \ell \in R_i\} \mid 1 \leq i \leq N\}.$$

Rather than encoding each countermodel circuit using its X_k members, we will leverage the fact that X_k is either equivalent to g_{i_k} or to $\overline{g_{i_k}}$ and replace it by the suitable polarity. This way, the recursive definitions of f_k^u boil down to

$$f_n^u = \begin{cases} g_{i_n} & \text{if } X_n \text{ is a clause,} \\ \overline{g_{i_n}} & \text{if } X_n \text{ is a term,} \end{cases}$$

and for $1 \leq k < n$

$$f_k^u = \begin{cases} g_{i_k} \wedge f_{k+1}^u & \text{if } X_k \text{ is a clause,} \\ \overline{g_{i_k}} \vee f_{k+1}^u & \text{if } X_k \text{ is a term.} \end{cases}$$

At this point, since the countermodel arrays are populated in the order of the proof, we can observe the following:

Observation 1. *Whenever g_{i_k} and $g_{i_{k'}}$ appear in the same circuit and $k < k'$, i.e., g_{i_k} comes before $g_{i_{k'}}$ in the corresponding countermodel array, then also $i_k < i_{k'}$, i.e., the reduction step corresponding to g_{i_k} also comes before the one corresponding to $g_{i_{k'}}$.*

Using the simplified circuits with the variables g_i, we can finally produce an encoding into CNF. By using the Tseitin conversion, we get the clauses

$$F_n^u = \begin{cases} \underbrace{(f_n^u \vee \overline{g_{i_n}}) \wedge (\overline{f_n^u} \vee g_{i_n})}_{F_{n,1}^u \qquad\qquad F_{n,2}^u} & \text{if } X_n \text{ is a clause,} \\ (f_n^u \vee g_{i_n}) \wedge (\overline{f_n^u} \vee \overline{g_{i_n}}) & \text{if } X_n \text{ is a term,} \end{cases}$$

and for $1 \leq k < n$

$$F_k^u = \begin{cases} \underbrace{(f_k^u \vee \overline{g_{i_k}} \vee \overline{f_{k+1}^u})}_{F_{k,1}^u} \wedge \underbrace{(\overline{f_k^u} \vee g_{i_k})}_{F_{k,2}^u} \wedge \underbrace{(\overline{f_k^u} \vee f_{k+1}^u)}_{F_{k,3}^u} & \text{if } X_k \text{ is a clause,} \\ (\overline{f_k^u} \vee \overline{g_{i_k}} \vee f_{k+1}^u) \wedge (f_k^u \vee g_{i_k}) \wedge (f_k^u \vee \overline{f_{k+1}^u}) & \text{if } X_k \text{ is a term.} \end{cases}$$

In our running example, we have two reduction steps, there are therefore two definitions of g-variables, namely $g_1 = (x_1 \vee x_2)$ and $g_2 = (\overline{x_1} \vee \overline{x_2})$. If we replace the actual entries in the countermodel array by the g-variables, we get the array $[g_1, \overline{g_2}]$ and the corresponding circuit $y = g_1 \wedge \overline{g_2}$. Its CNF encoding is

$$(y \vee \overline{g_1} \vee g_2) \wedge (\overline{y} \vee g_1) \wedge (\overline{y} \vee \overline{g_2}).$$

Starting from a formula $\Phi = \mathcal{Q}.\varphi$ and its Q-resolution refutation $\mathcal{P}$, G will denote the set of clauses defining the g_i and F will denote the set of clauses F_k^u (for all universals u and appropriate k) defining the countermodel. The validation formula $\Phi[\mathcal{P}]$ is then $\varphi \wedge G \wedge F$ and we will now present a RUP proof for it.

We will need the following notation. Let x, y be variables of a propositional formula φ, let τ be an assignment to variables of φ. We write $x \cong_\tau^\varphi y$ if, for every extension σ of τ that defines x or y, either unit propagation in $\varphi[\sigma]$ causes a conflict or $\sigma'(x) = \sigma'(y)$, where σ' is the closure of σ with respect to unit propagation. If φ is understood from the context, we may drop the superscript, likewise, if τ is the empty assignment, we may drop the subscript.

Lemma 1. *Let u be a universal variable of Φ whose countermodel array has n entries and the corresponding g-variables are $g_{i_1}, \ldots, g_{i_n}$. For $1 \le k \le n$ let τ_k be a partial assignment (to variables of $\Phi[\mathcal{P}]$) which sets $g_{i_1}, \ldots, g_{i_{k-1}}$ to true. Then $f^u \cong_{\tau_k} f_k^u$.*

Proof. We can see that the clauses $F_{j,2}^u[\tau_k]$ are satisfied by g_{i_k} and $\overline{g_{i_k}}$ disappears from $F_{j,1}^u[\tau_k]$ for $1 \le j < k$. The clauses $F_{j,1}^u[\tau_k]$ and $F_{j,3}^u[\tau_k]$ we are left with encode precisely $f_j^u \cong f_{j+1}^u$. Together, we have that under the assignment τ_k, $f^u = f_1^u \cong f_k^u$, or in other words $f^u \cong_{\tau_k} f_k^u$. $\square$

The following lemma asserts that the intuition about how countermodel circuits find the first falsified conclusion and set the variable accordingly is indeed true.

Lemma 2. *For $1 \le i \le N$ let τ_i be a partial assignment (to variables of $\Phi[\mathcal{P}]$) which sets $g_1, \ldots, g_{i-1}$ to true and g_i to false. Let ℓ be a universal literal that is reduced in the reduction step leading to R_i. Under the assignment τ_i unit propagation (in $\Phi[\mathcal{P}]$) causes a conflict or derives $\overline{\ell}$.*

Proof. Let us assume unit propagation does not cause a conflict. Let $u = var(\ell)$, g_i occurs in the countermodel array of u as some g_{i_k}. If ℓ is positive, $F_{k,2}^u$ together with $\overline{g_{i_k}}$ propagate $\overline{f_k^u}$. If ℓ is negative, $F_{k,2}^u$ together with $\overline{g_{i_k}}$ propagate f_k^u. We can use Observation 1 to see that all $g_{i_{k'}}$ with $k' < k$ are set to true and Lemma 1 applies, so that $f^u \cong f_k^u$ and the value for u propagated is false if ℓ is negative and true if ℓ is positive. Either way, this means that $\overline{\ell}$ is propagated. $\square$

With Lemma 2, we can describe how to construct a RUP proof for $\Phi[\mathcal{P}]$ from $\mathcal{P}$.

Theorem 1. *Let $\mathcal{P}$ be a Q-resolution refutation of the formula $\Phi = \mathcal{Q}.\varphi$. Then there exists a RUP proof of unsatisfiability of the validation formula $\Phi[\mathcal{P}]$ of size $O(|\mathcal{P}|)$, and this proof can be computed in $O(|\mathcal{P}|)$ time.*

Proof. Let $\mathcal{P}'$ be $\mathcal{P}$ with each conclusion R_i replaced by the unit clause (g_i), and with the input clauses omitted. We claim that $\mathcal{P}'$ is a RUP proof of unsatisfiability of $\Phi[\mathcal{P}]$. Since resolvents are always RUP with respect to their premises we only need to verify that all (g_i) are RUP too. Let $R_i = R_i' - L$ be a reduction step, let $\ell \in L$ be one of the universal literals reduced to obtain R_i, let $u = var(l)$. We need to prove that setting (g_i) to false causes a conflict by unit propagation. At the time when (g_i) is inserted into the proof, all (g_j) with $j < i$ have already been inserted and since they are unit clauses, all g_j with $j < i$ are set to true by unit propagation. Adding to that the assignment $\overline{g_i}$, the conditions of Lemma 2 are satisfied and so either unit propagation causes a conflict (in which case we are done), or $\overline{\ell}$ is propagated. Since ℓ was chosen without loss of generality, all literals in L are propagated to false, and since $\overline{g_i}$ trivially propagates all literals of R_i to false, R_i' is falsified and a conflict is reached as required. Clearly, the size of $\mathcal{P}'$ is bounded by the size of $\mathcal{P}$, and it can be computed in time $O(|\mathcal{P}|)$ as the amount of work per each clause of $\mathcal{P}$ is proportional to its size.   □

For example, the RUP proof constructed according to Theorem 1 from the example Q-resolution proof would consist in the following sequence of clauses:

$$(x_1 \vee x_2 \vee y),\ (g_1),\ (\overline{x_1} \vee \overline{x_2} \vee \overline{y}),\ (g_2),\ (x_1),\ (\overline{x_1}),\ \bot$$

5 RUP Proofs from Long-Distance Q-Resolution

With long-distance Q-resolution, we cannot directly use the clauses of the refutation in the RUP proof as we did in the proof of Theorem 1, because these clauses may be tautological. Instead, we adopt the approach that was used in the paper of Balabanov et al. [2] in order to generalize BJ to long-distance Q-resolution proofs. The following definition is taken from the paper of Balabanov et al. [2], with a slight change of notation.

Definition 1. *Let* $\mathcal{P}$ *be a long-distance Q-resolution refutation of the QBF* $\Phi = Q.\varphi$. *Let* $C \in \mathcal{P}$ *be a clause,* $\ell \in C$ *a literal and* $u = var(\ell)$. *The* phase function *of the variable* u *in the clause* C, *denoted by* $u^\phi(C)$, *is a boolean function defined recursively as follows:*

- *if* C *is an input clause, then* $u^\phi(C) = 1$ *if* $\ell = u$, *otherwise* $u^\phi(C) = 0$.
- *if* C *is the result of application of universal reduction on the clause* C', $u^\phi(C) = u^\phi(C')$.
- *if* C *is the resolvent of* C_1 *and* C_2 *on the pivot literal* p, $p \in C_1$, $\overline{p} \in C_2$, *then if* $u \notin var(C_1)$, *then* $u^\phi(C) = u^\phi(C_2)$, *if* $u \notin var(C_2)$ *or* $u^\phi(C_1) = u^\phi(C_2)$, *then* $u^\phi(C) = u^\phi(C_1)$, *otherwise* $u^\phi(C) = (p \wedge u^\phi(C_2)) \vee (\overline{p} \wedge u^\phi(C_1))$.

The effective literal *of* ℓ *in* C, *denoted by* $\ell^\epsilon(C)$, *is a literal that satisfies* $\ell^\epsilon(C) \Leftrightarrow (u \Leftrightarrow u^\phi(C))$. *The* shadow clause *of* C *is the clause* $C^\sigma = \bigvee_{\ell \in C} \ell^\epsilon(C)$.

The phase function intuitively tells us, under a given assignment to previous variables in the quantifier prefix, what is the phase in which a given universal variable would have appeared in a given clause, had we restricted the proof using that assignment. The effective literal is a literal which, based on an assignment to previous existential variables, is equivalent to the polarity of its variable indicated by the phase function. Note that in the case when the phase function is constant, i.e. 0 or 1, the effective literal of any literal is simply the literal itself. In such cases we say that the literal is *unmerged*. Literals that are not unmerged are *merged*.

We will now present a description of the countermodel computed by BJJW from a long-distance Q-resolution refutation. In order to do that, we adapt the notation from Sect. 4. Let $\mathcal{P}$ be a long-distance Q-resolution refutation of a formula $\Phi = Q.\varphi$. The conclusions of reduction steps in $\mathcal{P}$, in the same order as they appear, are denoted by $R_1, \ldots, R_N$. The variables g_i, $1 \le i \le N$, are now equivalent to the shadow clauses $R_i{}^\sigma$ instead of R_i themselves. Since BJJW keeps track of the phase function of every universal variable in every clause, we will use a variable $u^\phi(C)$ to denote the output of the phase function. We will also have variables $\ell^\epsilon(C)$ for the effective literals. In the case of unmerged literals, this will simply be ℓ. By H we will denote the conjunction of all clauses that encode the circuits which define phase variables and effective literals.

The partial countermodel circuits f_k^u from the previous section are slightly more complicated now. Let $R_i = R_i' - L$ be a reduction step, let $\ell \in L$ be a literal that is being reduced, let $u = var(\ell)$. If ℓ is unmerged, $R_i{}^\sigma$ is pushed into the countermodel array of u, similarly as in the case of ordinary Q-resolution. However, if ℓ is merged, we first require that both ℓ and $\bar\ell$ be reduced at the same time (merged literals arise from merges, so they are always in both polarities in a clause), and as such two entries are pushed into the countermodel array of u, namely $R_i{}^\sigma \vee \overline{u^\phi(R_i')}$ and right afterwards $\overline{R_i{}^\sigma} \wedge \overline{u^\phi(R_i')}$. The intuition for why these entries are added is the following: if the phase $u^\phi(R_i')$ of ℓ in R_i' is positive, and the (shadow clause of the) conclusion is falsified, set u to false, otherwise if the phase is negative and the conclusion is falsified, set u to true, each time falsifying the effective literal $\ell^\epsilon(R_i')$. This is analogous to the ordinary case, where when the conclusion is falsified, the reduced literal is set so that it is falsified, only in this case we falsify the effective literal.

Now, for the sake of simplicity of presentation, we will treat unmerged literals the same way as merged ones. This means that even for unmerged reduced literals we push two entries into the countermodel array, $R_i \vee \overline{u^\phi(R_i')}$ and $\overline{R_i} \wedge \overline{u^\phi(R_i')}$. It is easy to see that if $u^\phi(R_i') = 1$, the term becomes falsified and the clause reduces to just R_i, while if $u^\phi(R_i') = 0$, the clause becomes satisfied and the term reduces to just $\overline{R_i}$. In each case, the circuit is equivalent to what we would have produced by pushing just the one entry as previously.

Let $X_1, \ldots, X_{2n}$ be the entries in the countermodel array of a universal variable u. Each X_{2k-1} is $R_{i_k}{}^\sigma \vee \overline{u^\phi(R_{i_k}')}$ and X_{2k} is $\overline{R_{i_k}{}^\sigma} \wedge \overline{u^\phi(R_{i_k}')}$. We have already defined $g_i = R_i{}^\sigma$, but since each entry in the countermodel array still contains two variables even after replacing $R_{i_k}{}^\sigma$ with g_{i_k}, we will define the auxiliary variables $f_{2k-1}'^u = g_{i_k} \vee \overline{u^\phi(R_{i_k}')}$ and $f_{2k}'^u = \overline{g_{i_k}} \wedge \overline{u^\phi(R_{i_k}')}$ using the following sets of clauses (for $1 \le k \le n$):

$$F_{2k-1}'^u = \overbrace{\left(\overline{f_{2k-1}'^u} \vee g_{i_k} \vee \overline{u^\phi(R_{i_k}')}\right)}^{F_{2k-1,1}'^u} \wedge \overbrace{\left(f_{2k-1}'^u \vee \overline{g_{i_k}}\right)}^{F_{2k-1,2}'^u} \wedge \overbrace{\left(f_{2k-1}'^u \vee u^\phi(R_{i_k}')\right)}^{F_{2k-1,3}'^u}$$

$$F_{2k}'^u = \underbrace{\left(f_{2k}'^u \vee g_{i_k} \vee u^\phi(R_{i_k}')\right)}_{F_{2k,1}'^u} \wedge \underbrace{\left(\overline{f_{2k}'^u} \vee \overline{g_{i_k}}\right)}_{F_{2k,2}'^u} \wedge \underbrace{\left(\overline{f_{2k}'^u} \vee \overline{u^\phi(R_{i_k}')}\right)}_{F_{2k,3}'^u}$$

Let F' be the conjunction of all $F_k'^u$ for all universal variables u and all appropriate k. The following is immediate from the clauses F'.

Observation 2. *Setting g_{i_k} to true causes unit propagation to set $f_{2k-1}'^u$ and $\overline{f_{2k}'^u}$.*

Finally, we are ready to present the set F of clauses which encode the counter-model circuit:

$$F_{2n,1}^u = (f_{2n}^u \vee \overline{f_{2n}'^u}), \quad F_{2n,2}^u = (\overline{f_{2n}^u}, f_{2n}'^u),$$

and for $1 \le k < 2n$

$$F_k^u = \begin{cases} \underbrace{(f_k^u \vee \overline{f_k'^u} \vee \overline{f_{k+1}^u})}_{F_{k,1}^u} \wedge \underbrace{(\overline{f_k^u}, f_k'^u)}_{F_{k,2}^u} \wedge \underbrace{(\overline{f_k^u}, f_{k+1}^u)}_{F_{k,3}^u} & \text{if } k \text{ is odd,} \\ \underbrace{(\overline{f_k^u} \vee f_k'^u \vee f_{k+1}^u)}_{} \wedge \underbrace{(f_k^u, \overline{f_k'^u})}_{} \wedge \underbrace{(f_k^u, \overline{f_{k+1}^u})}_{} & \text{if } k \text{ is even.} \end{cases}$$

Similarly as before, let F be the conjunction of all F_k^u for all appropriate u and k, and let G be the conjunction of the clauses defining the equivalences $g_i \Leftrightarrow R_i^\sigma$. Then, the validation formula for Φ and $\mathcal{P}$ is $\Phi[\mathcal{P}] = \varphi \wedge F \wedge F' \wedge G \wedge H$.

The following are analogues of Lemmas 1 and 2.

Lemma 3. *Let u be a universal variable of Φ whose countermodel array has $2n$ entries and the corresponding g-variables are $g_{i_1}, \ldots, g_{i_{2n}}$. For $1 \le k \le 2n$ let τ_k be a partial assignment (to variables of $\Phi[\mathcal{P}]$) which sets $g_{i_1}, \ldots, g_{i_{k-1}}$ to true. Then $f^u \cong_{\tau_k} f_{2k-1}^u$.*

Proof. Let $1 \le j < k$. Applying Observation 2, we see that $f_{2j-1}'^u$ and $\overline{f_{2j}'^u}$ are propagated, in each case, inspecting the restricted clauses that remain, we see that $f_{2j-1}^u \cong_{\tau_k} f_{2j}^u$ and $f_{2j}^u \cong_{\tau_k} f_{2j+1}^u$. Altogether, we get $f^u \cong_{\tau_k} f_k^u$. □

Lemma 4. *For $1 \le i \le N$ let τ_i be a partial assignment (to variables of $\Phi[\mathcal{P}]$) which sets $g_1, \ldots, g_{i-1}$ to true and g_i to false. Let u be a universal variable of Φ in whose countermodel g_i appears as some g_{i_k}. Let R_i be the corresponding reduction step, obtained from R_i'. Then, under either of the assignments $\tau_i \cup u^\phi(R_i')$ and $\tau_i \cup \overline{u^\phi(R_i')}$, unit propagation (in $\Phi[\mathcal{P}]$) causes a conflict or derives $\overline{u^\epsilon(R_i')}$.*

Proof. Assume unit propagation not cause a conflict. Let us assume $u^\phi(R_i')$ first. Since we have $\overline{g_{i_k}} \wedge u^\phi(R_i')$, the clause $F_{2k-1,1}'^u$ propagates $\overline{f_{2k-1}'^u}$, which in turn propagates $\overline{f_{2k-1}^u}$. Since $g_1, \ldots, g_{i-1}$ are set to true, Lemma 3 applies and the value of f_{2k-1}^u is propagated for the value of u, meaning $\overline{u}$ is propagated. Together with the assumption $u^\phi(R_i')$, we have that the effective literal $u^\epsilon(R_i')$ is set to false by unit propagation.

If on the other hand we assume $\overline{u^\phi(R_i')}$, $f_{2k-1}'^u$ is propagated from $F_{2k-1,3}'^u$, which means that the restricted clauses F_{2k-1}^u now encode $f_{2k-1}^u \cong f_{2k}^u$. Also, $F_{2k,1}'^u$ propagates $f_{2k}'^u$, which in turn propagates f_{2k}^u. Since $g_1, \ldots, g_{i-1}$ are set to true, Lemma 3 applies and the value of f_{2k}^u is propagated for the value of u, meaning u is propagated. Together with the assumption $\overline{u^\phi(R_i')}$, we have that the effective literal $u^\epsilon(R_i')$ is set to false by unit propagation. □

While in the case of ordinary Q-resolution, the resolvent of two clauses is always RUP with respect to those clauses, this is not true in the case of long-distance

Q-resolution. This is due to the fact that if a merge occurs, a fresh effective literal is introduced in the resolvent, and just falsifying this new fresh literal without the knowledge of the value of the corresponding phase variable does not cause the effective literals in the premises of the resolution step to become falsified. Therefore, we first prove that a set of extra clauses can be derived from the definitions of phase functions and effective literals. These clauses will then empower unit propagation to deal with merged effective literals the same way as with unmerged ones.

Let C be the resolvent of C_1 and C_2 on the pivot literal $p \in C_1$ (and $\bar{p} \in C_2$). Let $\ell \in C_1$, $\bar{\ell} \in C_2$, $u = var(\ell)$ be a universal literal such that $u^\phi(C_1) \neq u^\phi(C_2)$, i.e. u is being merged in this resolution step. Then the clauses E^u_{C,C_1} and E^u_{C,C_2} are defined as follows:

$$E^u_{C,C_1} = (u^\epsilon(C) \vee p \vee \overline{u^\epsilon(C_1)}), \quad E^u_{C,C_2} = (u^\epsilon(C) \vee \bar{p} \vee \overline{u^\epsilon(C_2)}).$$

We will denote by E the set of all $E^u_{C,D}$ for appropriate premise D, resolvent C, and merged literal u. The clauses of E will provide us with a direct relationship between successive effective literals of one variable. They express one direction of the conditional dependence of an effective literal on the previous effective literals—if an effective literal is false, then based on the value of the pivot variable, the corresponding previous effective literal must be false too.

Lemma 5. *All clauses of E are derivable by RUP from H. The combined size of the RUP proofs is $O(|\mathcal{P}|)$ and they are computable in $O(|\mathcal{P}|)$ time.*

Proof. Let $E^u_{C,D} \in E$, let $p \in D$ be the pivot literal. It can be easily verified by unit propagation on the definitions of phase functions and effective literals that the following is the required RUP proof:

$$(u^\epsilon(C) \vee p \vee \overline{u^\epsilon(D)} \vee u^\phi(C)), (E^u_{C,D})$$

Clearly, per each resolution step, these proofs only take up constant space and are computable in constant time, resulting in an overall linear bound. □

We now state the main result of this section (we omit the proof due to space constraints).

Theorem 2. *Let $\mathcal{P}$ be a long-distance Q-resolution refutation of the formula $\Phi = Q.\varphi$. Then there exists a RUP proof of unsatisfiability of the validation formula $\Phi[\mathcal{P}]$ of size $O(|\mathcal{P}|)$, and this proof can be computed in $O(|\mathcal{P}|)$ time.*

Finally, let us point out that even though we presented concrete CNF encodings for many of the circuits, other encodings can work as well. Namely, it is sufficient if the encodings contain the g-variables (because these are present in the RUP proof) and satisfy the unit-propagation properties of the lemmas.

6   True Formulas

In this section we show how to derive analogues of Theorems 1 and 2 for true formulas. Let us start with the case of a (long-distance) Q-consensus proof $\mathcal{P}$ of a true PDNF formula $\Phi = \mathcal{Q}.\varphi$. In this case the validation formula $\Phi[\mathcal{P}]$ for the model $\mathrm{CC}(\mathcal{P})$ is the DNF φ in disjunction with $\mathrm{DNF}(\mathrm{CC}(\mathcal{P}))$. The task of validation of the model $\mathrm{CC}(\mathcal{P})$ is to check that $\Phi[\mathcal{P}]$ is valid, and checking the validity of $\Phi[\mathcal{P}]$ is equivalent to checking that the CNF $\overline{\Phi[\mathcal{P}]}$ is unsatisfiable.

Theorem 3. *Let $\mathcal{P}$ be a long-distance Q-consensus proof of the PDNF formula $\Phi = \mathcal{Q}.\varphi$. Then there exists a RUP proof of unsatisfiability of the negated validation formula $\overline{\Phi[\mathcal{P}]}$ of size $O(|\mathcal{P}|)$, and it can be computed in $O(|\mathcal{P}|)$ time.*

Proof. We observe that the countermodels extracted by BJ/BJJW from $\mathcal{P}$ and from its negation $\overline{\mathcal{P}}$ are in fact the same (we have not discussed the variants of BJ/BJJW for true formulas here, but check the definitions in [1,2] to see that this trivially holds), which means that their CNF and DNF encodings are negations of one another. This means that

$$\overline{\Phi[\mathcal{P}]} = \overline{\varphi \vee \mathrm{DNF}(\mathrm{CC}(\mathcal{P}))} = \overline{\varphi} \wedge \overline{\mathrm{DNF}(\mathrm{CC}(\mathcal{P}))} = \overline{\varphi} \wedge \mathrm{CNF}(\mathrm{CC}(\overline{\mathcal{P}})) = \overline{\Phi}[\overline{\mathcal{P}}],$$

and we can apply Theorem 2 on $\overline{\Phi}$ and $\overline{\mathcal{P}}$. □

For a Q-consensus proof $\mathcal{P}$ of a true PCNF formula $\Phi = \mathcal{Q}.\varphi$ let us first clarify what the validation formula looks like. We would need to check the validity of $\varphi \vee \mathrm{DNF}(\mathrm{CC}(\mathcal{P}))$, but φ is a CNF and $\mathrm{CC}(\mathcal{P})$ must be encoded as a DNF for validity checking. Therefore, we need to first transform φ to DNF using the Tseitin transformation as follows. Suppose $\varphi = C_1 \wedge \cdots \wedge C_n$. We will define the clause variables $c_i = C_i$ and represent $\mathrm{DNF}(\varphi)$ as follows:

$$\mathrm{DNF}(\varphi) = \bigvee_{i=1}^{n} \left[(c_i \wedge \overline{C_i}) \vee \bigvee_{\ell \in C_i} (\overline{c_i} \wedge \ell) \right] \vee (c_1 \wedge \cdots \wedge c_n).$$

The validation formula $\Phi[\mathcal{P}]$ is then $\mathrm{DNF}(\varphi) \vee \mathrm{DNF}(\mathrm{CC}(\mathcal{P}))$. As before, instead of checking the validity of $\Phi[\mathcal{P}]$, we will check the unsatisfiability of $\overline{\Phi[\mathcal{P}]}$.

Theorem 4. *Let $\mathcal{P}$ be a long-distance Q-consensus proof of the PCNF formula $\Phi = \mathcal{Q}.\varphi$ with the set of initial terms μ. If every clause from $\overline{\mu}$ is RUP with respect to $\mathrm{DNF}(\varphi)$, then there exists a RUP proof of unsatisfiability of the negated validation formula $\overline{\Phi[\mathcal{P}]}$ of size $O(|\mathcal{P}|)$, and it can be computed in $O(|\mathcal{P}|)$ time.*

Proof. Let $M = \mathcal{Q}.\mu$ be the PDNF consisting of the initial terms. Using Theorem 3, we obtain a RUP proof for the negated validation formula $\overline{M[\mathcal{P}]} = \overline{M}[\overline{\mathcal{P}}] = \overline{\mu} \wedge \mathrm{CNF}(\mathrm{CC}(\overline{\mathcal{P}}))$. By prepending $\overline{\mu}$ to this proof, we obtain a RUP proof of $\mathrm{DNF}(\varphi) \wedge \mathrm{CNF}(\mathrm{CC}(\overline{\mathcal{P}})) = \overline{\Phi[\mathcal{P}]}$ of size $O(|\mathcal{P}| + |\mu|) = O(|\mathcal{P}|)$. □

There are two common ways of obtaining initial terms. One is to transform the CNF φ to DNF [13], in which case there is nothing to prove, because the negated initial terms are directly members of $\overline{\text{DNF}(\varphi)}$ and therefore RUP. The other way is to produce hitting sets of the clauses of φ. In this case, since every initial term is a hitting set of the clauses $C_1, \ldots, C_n$, we have that for every initial term I and for every clause C_i, there is always a clause of $\text{CNF}(\overline{\varphi})$ of the form $(c_i \vee \overline{\ell})$, such that $\ell \in I$. Therefore, by assuming the negation of a negated initial term, i.e. the term itself, unit propagation will propagate c_i for all i, which in turn causes a conflict with the clause $(\overline{c_1} \vee \cdots \vee \overline{c_n})$. Therefore, every clause in $\neg\mu$ is indeed RUP with respect to $\text{DNF}(\varphi)$ and Theorem 4 applies.

Finally, in the paragraph above we mentioned that initial terms are hitting sets of the clauses of φ (in one of the cases). In fact, this need not always be true, since the hitting sets might have existential reduction applied to them first according to the *model generation* rule [10]. Since it is no problem for the QBF solver to output the original hitting set without applying existential reduction, but very difficult (NP-hard in general) for the proof-checker to recover it, we suggest to strengthen the conditions on the QRP proof format by requiring that the initial terms be full hitting sets. If this condition is not met our algorithm may fail to produce valid RUP proofs for true PCNF formulas. Fortunately DepQBF always generated terms that happened to be full hitting sets in our experiments.

7   Experiments

We implemented the algorithm of Theorem 2, which generalizes Theorem 1, in a tool called qrp2rup (https://www.ac.tuwien.ac.at/research/certificates/) and evaluated the performance compared to various other approaches to certificate validation. In particular, since our tool is also capable of emitting deletion information for DRAT-trim, we evaluated the following six configurations of certificate extractors and validators:

- qrp2rup with *deletion* information and validation by DRAT-trim,
- qrp2rup without deletion information (*plain*) and validation by DRAT-trim,
- qrp2rup and validation by Lingeling (ignoring the RUP proof),
- qrp2rup and validation by Glucose (ignoring the RUP proof),
- QBFcert and validation by Lingeling,
- QBFcert and validation by Glucose.

We also experimented with configurations of DRAT-trim that used forward checking (instead of the default backward checking), but excluded the results due to systematically inferior performance. Note that since QBFcert cannot handle long-distance Q-resolution, only the first four configurations were used for the experiments with long-distance proofs. To produce both ordinary and long-distance Q-resolution proofs, we used DepQBF 6.03 in a configuration that allowed tracing (i.e., with most of the advanced techniques off) with a cut-off time of 900 CPU seconds and a memory limit of 4 GB. The validation process

Table 1. Ordinary Q-resolution proofs: number of true + false formulas validated.

Year	Total	QBFcert + SAT-solver		qrp2rup + SAT-solver		qrp2rup + DRAT-trim	
		Lingeling	Glucose	Lingeling	Glucose	Deletion	Plain
2010	162 + 230	88 + 215	88 + 216	88 + 225	92 + **228**	**99** + 224	**99** + 223
2016	157 + 206	124 + 196	123 + 197	116 + 202	128 + **203**	**136** + 202	**136** + 200
2017	18 + 62	**12** + 58	**12** + 58	11 + 62	**12** + **63**	**12** + **63**	**12** + 62

Table 2. Long-distance Q-resolution proofs: number of true + false formulas validated.

		qrp2rup + SAT-solver		**qrp2rup** + DRAT-trim	
Year	Total	Lingeling	Glucose	Deletion	Plain
2010	149 + 222	93 + 215	95 + **217**	**100** + 215	**100** + 215
2016	160 + 250	120 + 197	131 + **200**	**137** + 196	**137** + 196
2017	17 + 59	12 + **59**	13 + **59**	13 + **59**	13 + **59**

was limited to 1800 CPU seconds and 7 GB of memory. The experiments were run on a cluster of heterogeneous machines running 64-bit Ubuntu 16.04.3 LTS (GNU/Linux 4.10.0-42). We evaluated the tools on the PCNF benchmark sets from the QBF Evaluations 2017, 2016, and 2010. The numbers of true and false validated instances for each configuration and benchmark set are reported in the tables below. The column "total" reports the total number of proofs for true and false formulas produced by DepQBF.

The results indicate that our approach is beneficial mainly on true formulas, but performs well across the board. Interestingly, even though QBFcert tends to produce smaller certificates than qrp2rup, Glucose performs worse on them. QBFcert internally uses AIG-based optimizations to shrink the certificates, and it is conceivable that these optimizations hurt Glucose's performance (Tables 1 and 2).

8   Concluding Remarks

We have presented a way of using (long-distance) Q-resolution/Q-consensus proofs in the process of validating QBF certificates. Our approach does not require a SAT call and comes with a polynomial runtime guarantee. Since it allows us to generate proofs in a format that is routinely used to verify the answers produced by SAT solvers and that has prompted the development of formally verified checkers [8,11,16], we can have a high degree of confidence in the correctness of certificates validated in this manner.

However, one subtle challenge remains. When constructing the validation formula $\Phi[\mathcal{P}]$, we take the matrix of Φ and append a CNF encoding of the countermodel. In principle, if we instead appended a small unsatisfiable CNF formula such as $(x) \wedge (\overline{x})$, we could be led to believe that it represents a countermodel when in reality it is much more restrictive than a countermodel is allowed to be (a formula that does not encode a set of functions). It would be desirable to have a way of checking that what we appended to the original matrix is indeed a set of functions (with the correct dependencies) for universal variables. This may require formal verification of parts of the certificate extraction algorithm.

A potential limitation of our approach is that it is sensitive to certain aspects of the CNF encoding of the countermodel to be validated, and therefore does not necessarily work with certificates extracted by other tools. However, our method ought to be compatible with simple circuit-level simplifications of certificates. Moreover, we hope to improve performance by generating GRAT [16] proofs of validation formulas as part of future work.

References

1. Balabanov, V., Jiang, J.R.: Unified QBF certification and its applications. Formal Methods Syst. Des. **41**(1), 45–65 (2012)
2. Balabanov, V., Jiang, J.R., Janota, M., Widl, M.: Efficient extraction of QBF (counter) models from long-distance resolution proofs. In: Bonet, B., Koenig, S. (eds.) Proceedings of the Twenty-Ninth AAAI Conference on Artificial Intelligence, 25–30 January 2015, Austin, Texas, USA, pp. 3694–3701. AAAI Press (2015)
3. Benedetti, M., Mangassarian, H.: QBF-based formal verification: experience and perspectives. J Satisf. Boolean Model. Comput. **5**(1–4), 133–191 (2008)
4. Lonsing, F., Biere, A.: Integrating dependency schemes in search-based QBF solvers. In: Strichman, O., Szeider, S. (eds.) SAT 2010. LNCS, vol. 6175, pp. 158–171. Springer, Heidelberg (2010). https://doi.org/10.1007/978-3-642-14186-7_14
5. Bloem, R., Könighofer, R., Seidl, M.: SAT-based synthesis methods for safety specs. In: McMillan, K.L., Rival, X. (eds.) VMCAI 2014. LNCS, vol. 8318, pp. 1–20. Springer, Heidelberg (2014). https://doi.org/10.1007/978-3-642-54013-4_1
6. Cadoli, M., Schaerf, M., Giovanardi, A., Giovanardi, M.: An algorithm to evaluate quantified boolean formulae and its experimental evaluation. J. Automat. Reason. **28**(2), 101–142 (2002)
7. Cashmore, M., Fox, M., Giunchiglia, E.: Partially grounded planning as quantified boolean formula. In: Borrajo, D., Kambhampati, S., Oddi, A., Fratini, S. (eds.) 23rd International Conference on Automated Planning and Scheduling, ICAPS 2013. AAAI (2013)
8. Cruz-Filipe, L., Heule, M.J.H., Hunt, W.A., Kaufmann, M., Schneider-Kamp, P.: Efficient certified RAT verification. In: de Moura, L. (ed.) CADE 2017. LNCS (LNAI), vol. 10395, pp. 220–236. Springer, Cham (2017). https://doi.org/10.1007/978-3-319-63046-5_14
9. Faymonville, P., Finkbeiner, B., Rabe, M.N., Tentrup, L.: Encodings of bounded synthesis. In: Legay, A., Margaria, T. (eds.) TACAS 2017. LNCS, vol. 10205, pp. 354–370. Springer, Heidelberg (2017). https://doi.org/10.1007/978-3-662-54577-5_20
10. Giunchiglia, E., Narizzano, M., Tacchella, A.: Clause/term resolution and learning in the evaluation of quantified boolean formulas. J. Artif. Intell. Res. **26**, 371–416 (2006)
11. Heule, M., Hunt, W., Kaufmann, M., Wetzler, N.: Efficient, verified checking of propositional proofs. In: Ayala-Rincón, M., Muñoz, C.A. (eds.) ITP 2017. LNCS, vol. 10499, pp. 269–284. Springer, Cham (2017). https://doi.org/10.1007/978-3-319-66107-0_18

12. Heule, M., Hunt, W.A., Wetzler, N.: Trimming while checking clausal proofs. In: Formal Methods in Computer-Aided Design, FMCAD 2013, pp. 181–188. IEEE Computer Soc. (2013)
13. Janota, M., Marques-Silva, J.: An Achilles' heel of term-resolution. In: Oliveira, E., Gama, J., Vale, Z., Lopes Cardoso, H. (eds.) EPIA 2017. LNCS (LNAI), vol. 10423, pp. 670–680. Springer, Cham (2017). https://doi.org/10.1007/978-3-319-65340-2_55
14. Büning, H.K., Karpinski, M., Flögel, A.: Resolution for quantified boolean formulas. Inf. Comput. **117**(1), 12–18 (1995)
15. Kronegger, M., Pfandler, A., Pichler, R.: Conformant planning as benchmark for QBF-solvers. In: International Workshop on Quantified Boolean Formulas - QBF 2013 (2013). http://fmv.jku.at/qbf2013/
16. Lammich, P.: Efficient verified (UN)SAT certificate checking. In: de Moura, L. (ed.) CADE 2017. LNCS (LNAI), vol. 10395, pp. 237–254. Springer, Cham (2017). https://doi.org/10.1007/978-3-319-63046-5_15
17. Niemetz, A., Preiner, M., Lonsing, F., Seidl, M., Biere, A.: Resolution-based certificate extraction for QBF. In: Cimatti, A., Sebastiani, R. (eds.) SAT 2012. LNCS, vol. 7317, pp. 430–435. Springer, Heidelberg (2012). https://doi.org/10.1007/978-3-642-31612-8_33
18. Peitl, T., Slivovsky, F., Szeider, S.: Dependency learning for QBF. In: Gaspers, S., Walsh, T. (eds.) SAT 2017. LNCS, vol. 10491, pp. 298–313. Springer, Cham (2017). https://doi.org/10.1007/978-3-319-66263-3_19
19. Rabe, M.N., Seshia, S.A.: Incremental determinization. In: Creignou, N., Le Berre, D. (eds.) SAT 2016. LNCS, vol. 9710, pp. 375–392. Springer, Cham (2016). https://doi.org/10.1007/978-3-319-40970-2_23
20. Rabe, M.N., Tentrup, L.: CAQE: a certifying QBF solver. In: Kaivola, R., Wahl, T. (eds.) Formal Methods in Computer-Aided Design - FMCAD 2015, pp. 136–143. IEEE Computer Soc. (2015)
21. Rintanen, J.: Asymptotically optimal encodings of conformant planning in QBF. In: 22nd AAAI Conference on Artificial Intelligence, pp. 1045–1050. AAAI (2007)
22. Staber, S., Bloem, R.: Fault localization and correction with QBF. In: Marques-Silva, J., Sakallah, K.A. (eds.) SAT 2007. LNCS, vol. 4501, pp. 355–368. Springer, Heidelberg (2007). https://doi.org/10.1007/978-3-540-72788-0_34
23. Tentrup, L.: Non-prenex QBF solving using abstraction. In: Creignou, N., Le Berre, D. (eds.) SAT 2016. LNCS, vol. 9710, pp. 393–401. Springer, Cham (2016). https://doi.org/10.1007/978-3-319-40970-2_24
24. Tseitin, G.S.: On the complexity of derivation in propositional calculus. In: Siekmann, J.H., Wrightson, G. (eds.) Automation of Reasoning. Symbolic Computation (Artificial Intelligence), pp. 466–483. Springer, Heidelberg (1983). https://doi.org/10.1007/978-3-642-81955-1_28
25. Wetzler, N., Heule, M.J.H., Hunt, W.A.: DRAT-trim: efficient checking and trimming using expressive clausal proofs. In: Sinz, C., Egly, U. (eds.) SAT 2014. LNCS, vol. 8561, pp. 422–429. Springer, Cham (2014). https://doi.org/10.1007/978-3-319-09284-3_31
26. Zhang, L., Malik, S.: Conflict driven learning in a quantified boolean satisfiability solver. In: Pileggi, L.T., Kuehlmann, A. (eds.) Proceedings of the 2002 IEEE/ACM International Conference on Computer-aided Design, ICCAD 2002, San Jose, California, USA, 10–14 November 2002, pp. 442–449. ACM / IEEE Computer Society (2002)
27. Zhang, L., Malik, S.: The quest for efficient boolean satisfiability solvers. In: Brinksma, E., Larsen, K.G. (eds.) CAV 2002. LNCS, vol. 2404, pp. 17–36. Springer, Heidelberg (2002). https://doi.org/10.1007/3-540-45657-0_2

Theory

Sharpness of the Satisfiability Threshold for Non-uniform Random k-SAT

Tobias Friedrich[ID] and Ralf Rothenberger[✉][ID]

Hasso Plattner Institute, Potsdam, Germany
{tobias.friedrich,ralf.rothenberger}@hpi.de

Abstract. We study non-uniform random k-SAT on n variables with an arbitrary probability distribution $\boldsymbol{p}$ on the variable occurrences. The number $t = t(n)$ of randomly drawn clauses at which random formulas go from *asymptotically almost surely (a. a. s.)* satisfiable to *a. a. s.* unsatisfiable is called the *satisfiability threshold*. Such a threshold is called *sharp* if it approaches a step function as n increases. We show that a threshold $t(n)$ for random k-SAT with an ensemble $(\boldsymbol{p}_n)_{n \in \mathbb{N}}$ of arbitrary probability distributions on the variable occurrences is sharp if $\|\boldsymbol{p}_n\|_2^2 = \mathcal{O}_n\left(t^{-\frac{2}{k}}\right)$ and $\|\boldsymbol{p}_n\|_\infty = o_n\left(t^{-\frac{k}{2k-1}} \cdot \log^{-\frac{k-1}{2k-1}} t\right)$.

This result generalizes Friedgut's sharpness result from uniform to non-uniform random k-SAT and implies sharpness for thresholds of a wide range of random k-SAT models with heterogeneous probability distributions, for example such models where the variable probabilities follow a power-law distribution.

1 Introduction

One of the most thoroughly researched topics in theoretical computer science is Satisfiability of Propositional Formulas (SAT). It was one of the first problems shown to be NP-complete by Cook [16] and, independently, by Levin [33]. Furthermore, SAT stands at the core of many results of modern complexity theory, like NP-completeness proofs [32] or lower bounds on runtime assuming the (Strong) Exponential Time Hypothesis [11,17,29,30].

Additional to its importance for theoretical research, Propositional Satisfiability also has practical applications. Many practical problems can be transformed into SAT formulas, for example hard- and software verification, automated planning, and circuit design. Such SAT formulas arising from practical and industrial problems are commonly referred to as *industrial SAT instances*. Surprisingly, even large industrial SAT instances with millions of variables can often be solved efficiently by state-of-the-art SAT solvers. This suggests that these instances have a structure which makes them easier to solve than the theoretical worst-case.

Uniform Random SAT and the Satisfiability Threshold Conjecture: In order to study the average-case complexity of Satisfiability, one can generate

© Springer International Publishing AG, part of Springer Nature 2018
O. Beyersdorff and C. M. Wintersteiger (Eds.): SAT 2018, LNCS 10929, pp. 273–291, 2018.
https://doi.org/10.1007/978-3-319-94144-8_17

a formula Φ at random in conjunctive normal form (CNF) with n variables and m clauses. To this end, we assume to have a probability distribution over all formulas with those properties. If the probability distribution is uniform, we will also refer to the model as *uniform random k-SAT*.

One of the most prominent questions related to uniform random k-SAT is trying to prove the satisfiability threshold conjecture. The *satisfiability threshold conjecture* states that for a formula Φ drawn uniformly at random from the set of all k-CNFs with n variables and m clauses, there is a real number r_k such that

$$\lim_{n \to \infty} \Pr\{\Phi \text{ is satisfiable}\} = \begin{cases} 1 & m/n < r_k; \\ 0 & m/n > r_k. \end{cases}$$

For $k = 2$, Chvatal and Reed [12] and, independently, Goerdt [27] proved that $r_2 = 1$. For $k \geqslant 3$, explicit upper and lower bounds have been derived, e.g., $3.52 \leqslant r_3 \leqslant 4.4898$ [18,28,31]. Additionally, the cavity method from statistical mechanics [34] was used to suggest a numerical estimate of $r_3 \approx 4.26$. Coja-Oghlan and Panagiotou [13,14] derived a bound (up to lower order terms) for $k \geqslant 3$ with $r_k = 2^k \log 2 - \frac{1}{2}(1 + \log 2) \pm o_k(1)$. Recently, Ding et al. [19] proved the exact position of the threshold for sufficiently large values of k.

One goal of showing the conjecture is to rigorously connect or disconnect threshold behavior to the average hardness of solving instances. For uniform random k-SAT for example, the on average hardest instances are concentrated around the threshold [35]. However, the conjecture and a lot of related work only consider formulas that are drawn uniformly at random. But what happens if the formulas are drawn according to a different probability distribution?

Non-uniform Random SAT: There is a substantial body of work which analyzes the satisfiability threshold in different SAT models, like regular random k-SAT [8,9,15,42], random geometric k-SAT [10] and $2+p$-SAT [1,36–38]. However, these models are not motivated by trying to model or understand the properties of industrial instances.

One property of industrial instances is community structure [7], i.e. some variables have a bias towards appearing together in clauses. It is clear by definition that such a bias does not exists in uniform random k-SAT. The Community Attachment Model by Giráldez-Cru and Levy [25] creates random formulas with clear community structure. Yet, the work of Mull et al. [39] shows that instances generated by this model have exponentially long resolution proofs with high probability, making them hard for CDCL on average.

Another important property of industrial instances is their degree distribution. The degree distribution of a formula Φ is a function $f \colon \mathbb{N} \to \mathbb{N}$, where $f(x)$ denotes the number of different Boolean variables that appear x times in Φ (negated or unnegated). In uniform random k-SAT this distribution is binomial, but it has been found out that the degree distribution of many families of industrial instances follows a power-law [5,9]. This means that $f(x)/n \sim x^{-\beta}$, where β is a constant intrinsic to the instance. To help close the gap between the degree distribution of uniform random and industrial instances, Ansótegui et al. [5] proposed a power-law random SAT model. Empirical studies [3–6] found that SAT solvers that are specialized in industrial instances also perform better on power-law formulas than on uniform random formulas. However, it looks like a power-law degree distribution alone makes instances a bit easier to solve, but not actually "easy": median runtimes around the threshold still look like they scale exponentially for several state-of-the-art solvers [24].

Recently, Giráldez-Cru and Levy [26] also introduced the popularity-similarity model, which incorporates both power-law degree distribution and community structure. Like most other models inspired by industrial instances it lacks theoretical work regarding the satisfiability threshold.

In this work we want to consider a generalization of the power-law random SAT model by Ansótegui et al. [5]. Our model allows instances with *any* given ensemble of variable distributions, instead of just power laws: The variables of each clause are drawn with a probability proportional to the n-th distribution in the ensemble, then they are negated independently with a probability of $1/2$ each. Let $\mathcal{D}(n, k, (\boldsymbol{p}_n)_{n\in\mathbb{N}}, m)$ be such a model with a variable distribution ensemble $(\boldsymbol{p}_n)_{n\in\mathbb{N}}$, where m clauses of length k over n variables are drawn. We call this the *clause-drawing* model. If we draw clauses in such a way, the variable probability distribution also defines a probability distribution over k-clauses. Instead of drawing exactly m k-clauses over n variables, one can now imagine flipping a coin for each possible k-clause and taking the clause into the formula with the clause probability multiplied with a certain scaling factor s. By doing so, the expected number of clauses in the formula will be exactly s. We will denote this model by $\mathcal{F}(n, k, (\boldsymbol{p}_n)_{n\in\mathbb{N}}, s)$ and call it the *clause-flipping* model.

Although $\mathcal{F}(n, k, (\boldsymbol{p}_n)_{n\in\mathbb{N}}, s)$ and $\mathcal{D}(n, k, (\boldsymbol{p}_n)_{n\in\mathbb{N}}, m)$ cannot represent industrial instances accurately, they might still give us some insights into which properties of real-world instances make them easy to solve. The one property our models provide is degree distribution. They allow us to look at the connection between degree distribution and hardness in an average-case scenario. As one of the steps in analyzing this connection, we would like to find out for which ensembles of variable probability distributions an equivalent of the satisfiability threshold conjecture holds in non-uniform random k-SAT. To see which ingredients we need to prove the conjecture and which of these ingredients this work provides, we first have to introduce the concept of threshold functions formally.

Threshold Functions: Formally, due to [22] a threshold for a monotone property P is defined as follows in the classical context of uniform probability distributions: Let $p \in [0, 1]$ and let $V = \{0, 1\}^N$ be endowed with the product measure $\mu_p(\cdot)$: for $x \in V$ define $\mu_p(x) = p^{\sum x_i}(1 - p)^{N - \sum x_i}$, and, for $W \subseteq V$, $\mu_p(W) = \sum_{x \in W} \mu_p(x)$. Now let $P = P(n)$ be the family of properties. $p^* = p^*(n)$ is an *asymptotic threshold function* for $P(n)$ if for every $p = p(n)$

$$\lim_{n \to \infty} \mu_p(P) = \begin{cases} 0, & \text{if } p \ll p^* \\ 1, & \text{if } p \gg p^*. \end{cases}$$

Here $\ll$ and $\gg$ denote "asymptotically smaller" and "asymptotically bigger" respectively.

Intuitively, a *sharp threshold* means that the change in probability around the threshold becomes steeper and steeper as the problem size increases, converging to a step function as n tends to infinity. Formally, we say that $P(n)$ has a *sharp threshold* if there exists a function $p^* = p^*(n)$ such that for every constant $\varepsilon > 0$ and for every $p = p(n)$

$$\lim_{n \to \infty} \mu_p(P) = \begin{cases} 0, & \text{if } p \leqslant (1 - \varepsilon)p^* \\ 1, & \text{if } p \geqslant (1 + \varepsilon)p^*. \end{cases}$$

Otherwise we call a threshold *coarse*. The region of p where the limit of $\mu_p(P)$ is bounded away from zero and one is called the *threshold interval*.

Note, that this definition only holds for satisfiability in the uniform clause-flipping model. In the case of the uniform clause-drawing model, the sharpness of the threshold is defined the same way, but with respect to m (or $r = m/n$) instead of p on the appropriate probability space.

Proving the Satisfiability Threshold Conjecture: In terms of threshold functions, the conjecture states that there is a sharp threshold for satisfiability at $m = r_k \cdot n$ and the constant r_k is the same for a fixed k and all sufficiently large n. For $k = 2$, Chvatal and Reed [12] and Goerdt [27] proved the conjecture and showed that $r_2 = 1$. However, random 2-SAT is easier to analyze than random k-SAT and their techniques do not work for bigger values of k. For uniform random k-SAT the "recipe" for proving the conjecture is as follows:

1. Show the existence of an asymptotic threshold function, i.e. show constant lower and upper bounds on r_k.
2. Prove that the threshold is sharp. In 1999 Friedgut [21] showed that the satisfiability threshold for uniform random k-SAT is sharp, although its location is not known exactly for all values of k. However, his result does not prove that r_k is the same for a fixed k and all sufficiently large values of n. Friedgut's proof relies on knowing the asymptotic threshold function.
3. Derive the actual constant r_k and that the threshold is sharp around it. Ding et al. [19] were the first to prove the exact value of r_k for values of k bigger than 2. Their proof relies on the result of Friedgut.

The goal of this paper will be to show the second ingredient for proving the satisfiability threshold conjecture for non-uniform random k-SAT, sharpness of the satisfiability threshold. In addition to being a stepping stone to proving the conjecture, sharpness of the threshold is of some independent interest, since a coarse threshold implies that there is a local property which approximates satisfiability or unsatisfiability. For random SAT this means that with constant probability instances have a constant-sized unsatisfiable subformula, making a lot of instances very easy to solve even around the threshold. Moreover, some of the techniques we use could also be used to analyze more sophisticated models, e.g. the popularity-similarity model [26], which was used in the 2017 SAT Competition.

Our Results: We study the sharpness of the satisfiability threshold for non-uniform random k-SAT and identify sufficient conditions on the variable probability distribution which imply a sharp threshold. Therefore, this work provides the second ingredient for proving a version of the satisfiability threshold conjecture for the non-uniform models $\mathcal{D}\left(n, k, (\boldsymbol{p}_n)_{n \in \mathbb{N}}, m\right)$ and $\mathcal{F}\left(n, k, (\boldsymbol{p}_n)_{n \in \mathbb{N}}, s\right)$ introduced earlier. In the context of these models, the classical result of Friedgut [21] reads as follows:

Theorem 1.1 (by [21]). *For all $n \in \mathbb{N}$ let $\boldsymbol{p}_n = (1/n, 1/n, \ldots, 1/n)$ be a variable probability distribution on n variables. If there is an asymptotic satisfiability threshold $m_c = t(n)$ on $\mathcal{D}\left(n, k, (\boldsymbol{p}_n)_{n \in \mathbb{N}}, m\right)$, then satisfiability has a sharp threshold on $\mathcal{F}\left(n, k, (\boldsymbol{p}_n)_{n \in \mathbb{N}}, s\right)$ with respect to s, and a sharp threshold on $\mathcal{D}\left(n, k, (\boldsymbol{p}_n)_{n \in \mathbb{N}}, m\right)$ with respect to m.*

Our main theorem extends this to non-uniform random k-SAT:

Theorem 3.2. *Let $k \geqslant 2$, let $(\boldsymbol{p}_n)_{n \in \mathbb{N}}$ be an ensemble of variable probability distributions on n variables each and let $s_c = t(n)$ be an asymptotic satisfiability threshold for $\mathcal{F}\left(n, k, (\boldsymbol{p}_n)_{n \in \mathbb{N}}, s\right)$ with respect to s. If $\|\boldsymbol{p}_n\|_\infty = o\left(t^{-\frac{k}{2k-1}} \cdot \log^{-\frac{k-1}{2k-1}} t\right)$ and $\|\boldsymbol{p}_n\|_2^2 = \mathcal{O}\left(t^{-2/k}\right)$, then satisfiability has a sharp threshold on $\mathcal{F}\left(n, k, (\boldsymbol{p}_n)_{n \in \mathbb{N}}, s\right)$ with respect to s.*

Furthermore, we show that the same also holds for the clause-drawing model of non-uniform random k-SAT if the asymptotic threshold is not constant.

Theorem 3.3. *Let $k \geqslant 2$, let $(\boldsymbol{p}_n)_{n \in \mathbb{N}}$ be an ensemble of variable probability distributions on n variables each and let $m_c = t(n) = \omega(1)$ be the asymptotic satisfiability threshold for $\mathcal{D}\left(n, k, (\boldsymbol{p}_n)_{n \in \mathbb{N}}, m\right)$ with respect to m. If $\|\boldsymbol{p}_n\|_\infty = o\left(t^{-\frac{k}{2k-1}} \cdot \log^{-\frac{k-1}{2k-1}} t\right)$ and $\|\boldsymbol{p}_n\|_2^2 = \mathcal{O}\left(t^{-2/k}\right)$, then satisfiability has a sharp threshold on $\mathcal{D}\left(n, k, (\boldsymbol{p}_n)_{n \in \mathbb{N}}, m\right)$ with respect to m.*

Our results actually state that the threshold is sharp for a certain, fixed value of n in the sense that the probability function for unsatisfiability is $o(1)$ if $s = (1 - \varepsilon) \cdot s_c$ (or $m = (1 - \varepsilon) \cdot m_c$) and $1 - o(1)$ if $s = (1 + \varepsilon) \cdot s_c$ (or $m = (1 + \varepsilon) \cdot m_c$). It is still possible that the function behaves differently for higher n due to the changing number of variables and probabilities. Nevertheless, Friedgut's original result also only asserts sharpness for a certain, fixed value of n. This is also the reason why the sharp threshold result does not automatically prove the satisfiability threshold conjecture: There could be different sharp threshold functions (including leading constant factors) for different values of n. For example, there could be some strange oscillations of the function.

Techniques: The proof of the main theorem uses Bourgain's Sharp Threshold Theorem in the version from O'Donnell's book [41]. In general, it follows the lines of Friedgut's proof of sharpness for the threshold of uniform random k-SAT [21].

However, we have to generalize Friedgut's results, like showing that no short unsatisfiable subformula can exist with sufficiently high probability. Furthermore, his lemma to bound the maximum slope of the probability for a monotone property at the threshold cannot be applied anymore, even in a more general form. Instead, we use the maximum slope that is implied by assuming a coarse threshold. Also, we had to adapt Friedgut's coverability lemma when considering non-uniform random k-SAT. In his work, a quasi-unsatisfiable subformula can spawn a constant number of clauses of length $k-1$. Now a quasi-unsatisfiable subformula can spawn clauses of any length $l \leqslant k$. Furthermore, there can now be more than a constant number of spawned clauses.

Please note that due to space limitations, we only provide proof sketches for our results. The full proofs can be found in the full version of the paper.

2   Preliminaries

We analyze random k-SAT on n variables and m clauses. We denote by $X_1, \ldots, X_n$ the Boolean variables. A clause is a disjunction of k literals $\ell_1 \vee \ldots \vee \ell_k$, where each literal assumes a (possibly negated) variable. For a literal ℓ_i let $|\ell_i|$ denote the variable of the literal. A formula Φ in conjunctive normal form is a conjunction of clauses $c_1 \wedge \ldots \wedge c_m$. We conveniently interpret a clause c both as a Boolean formula and as a set of literals. We say that Φ is satisfiable if there exists an assignment of variables $X_1, \ldots, X_n$ such that the formula evaluates to 1. Now let $(\boldsymbol{p}_n)_{n \in \mathbb{N}}$ be an ensemble of probability distributions, where $\boldsymbol{p}_n = (p_{n,1}, p_{n,2}, \ldots, p_{n,n})$ is a probability distribution over n variables with $\Pr(X = X_i) = p_{n,i} =: p_n(X_i)$.

Definition 2.1 (Clause-Drawing Random k-SAT). *Let m, n, k be given, and consider any ensemble of probability distributions $(\boldsymbol{p}_n)_{n \in \mathbb{N}}$, where $\boldsymbol{p}_n = (p_{n,1}, p_{n,2}, \ldots, p_{n,n})$ is a probability distribution over n variables with $\sum_{i=1}^{n} p_{n,i} = 1$. The clause-drawing random k-SAT model $\mathcal{D}(n, k, (\boldsymbol{p}_n)_{n \in \mathbb{N}}, m)$ constructs a random SAT formula Φ by sampling m clauses independently at random. Each clause is sampled as follows:*

1. *Select k variables independently at random from the distribution $\boldsymbol{p}_n$. Repeat until no variables coincide.*
2. *Negate each of the k variables independently at random with probability $1/2$.*

The clause-drawing random k-SAT model is equivalent to drawing each clause independently at random from the set of all k-clauses which contain no variable more than once. The probability to draw a clause c over n variables is then

$$q_c := \frac{\prod_{\ell \in c} p_n(|\ell|)}{2^k \sum_{J \in \mathcal{P}_k(\{1,2,\ldots,n\})} \prod_{j \in J} p_{n,j}}, \tag{2.1}$$

where $\mathcal{P}_k(\cdot)$ denotes the set of cardinality-k elements of the power set. The factor 2^k in the denominator comes from the different possibilities to negate variables.

Note that $k! \sum_{J \in \mathcal{P}_k(\{1,2,\ldots,n\})} \prod_{j \in J} p_{n,j}$ is the probability of choosing a k-clause that contains no variable more than once. To see that this probability is almost 1 for most distributions, we apply the generalized birthday paradox from [2]. Thereby, the probability that a k-clause sampled on n variables has collisions is at most $\frac{1}{2}k^2 \|\boldsymbol{p}_n\|_2^2$; so for $\|\boldsymbol{p}_n\|_2^2 = o(1)$ and constant k we obtain that the probability to draw a specific clause over n variables consisting of variables $X \in S$ is

$$q_c = C \frac{k!}{2^k} \prod_{X \in S} p_n(X), \qquad (2.2)$$

where we define $C := 1/ \left(\sum_{J \in \mathcal{P}_k(\{1,2,\ldots,n\})} \prod_{j \in J} p_{n,j} \right) = \left(1 + \Theta \left(\|\boldsymbol{p}_n\|_2^2 \right) \right)$. This effectively hides the denominator of Eq. (2.1) in C and makes clause probabilities easier to handle. We will later see that this is always the case in the variable probability distributions we consider.

We can now define the coin-flipping equivalent of non-uniform random k-SAT, which we will label *clause-flipping random k-SAT*.

Definition 2.2 (Clause-Flipping Random k-SAT). *Let s, n, k be given, and consider any ensemble of probability distributions $(\boldsymbol{p}_n)_{n \in \mathbb{N}}$, where $\boldsymbol{p}_n$ is a probability distribution over n variables with $\sum_{i=1}^n p_i = 1$. The clause-flipping random k-SAT model $\mathcal{F}(n, k, (\boldsymbol{p}_n)_{n \in \mathbb{N}}, s)$ constructs a random SAT formula Φ over n variables by independently flipping a coin for each of the $\binom{n}{k} 2^k$ possible k-clauses. The coin flip for a clause c is a success with probability*

$$q_{n,c}(s) := \min \left(s \cdot q_{n,c}, 1 \right) = \min \left(s \cdot \frac{\prod_{\ell \in c} p_n(|\ell|)}{2^k \sum_{J \in \mathcal{P}_k(\{1,2,\ldots,n\})} \prod_{j \in J} p_{n,j}}, 1 \right).$$

If successful, the clause is added to the random formula.

Lemma 2.1 relates the two models to each other and will be used throughout the paper. Note that in the lemma the clause probabilities do not necessarily have to be products of variable probabilities! Its proof is a simple exercise.

Lemma 2.1. *Given a clause-flipping model $\mathcal{F}$ with clause probabilities $\boldsymbol{q} = (q_i)_{i \in [n]}$ and a clause-drawing model $\mathcal{D}$ with clause probabilities $\boldsymbol{q}' = (q_i')_{i \in [n]}$ so that $q_i' = \frac{q_i/(1-q_i)}{\sum_{j \in [n]} q_j/(1-q_j)}$, then for all $l \in \mathbb{N}$ and all events $\mathcal{E}$ it holds that*

$$\Pr_{\mathcal{F}} \left(\mathcal{E} \mid \{l \text{ clauses flipped}\} \right) = \Pr_{\mathcal{D}} \left(\mathcal{E} \mid \{l \text{ different clauses drawn}\} \right).$$

3 Sharpness of the Threshold

In Sect. 3.1 we establish a notion of asymptotic and sharp thresholds in the context of non-uniform probability distributions. In Sect. 3.2 we relate this notion of sharpness to Bourgain's Sharp Threshold Theorem. In Sect. 3.3 we prove the sharpness of the threshold in $\mathcal{F}(n, k, (\boldsymbol{p}_n)_{n \in \mathbb{N}}, s)$ with the help of the Sharp Threshold Theorem. Finally, in Sect. 3.4 we relate $\mathcal{D}(n, k, (\boldsymbol{p}_n)_{n \in \mathbb{N}}, m)$ to $\mathcal{F}(n, k, (\boldsymbol{p}_n)_{n \in \mathbb{N}}, s)$ in such a way that the sharpness of the satisfiability threshold carries over.

3.1   Non-uniform Sharpness

We want to generalize the definitions for uniform probability distributions to non-uniform probability distributions.

For the clause-drawing random k-SAT model, we can use the same concepts of asymptotic and sharp thresholds with respect to m as in the uniform case.

For the clause-flipping random k-SAT model, the first thing we notice is that we cannot define the thresholds with respect to p anymore since the clause probabilities are now non-uniform. Instead, we want to define the thresholds with respect to s, the scaling factor of the probability space. This will allow us to relate the two models in Subsect. 3.4.

Unless stated otherwise, we will concentrate on $\mathcal{F}(n, k, (\boldsymbol{p}_n)_{n \in \mathbb{N}}, s)$ to establish the result in this model first. We now encode formulas as vectors $x \in \{0, 1\}^N$, where $N := \binom{n}{k} 2^k$ is the number of different k-clauses over n variables. If a clause is chosen to be in the formula, we set its variable to -1, otherwise we set it to 1. With this encoding of k-CNFs in mind, we can define a function $f: \{-1, 1\}^N \to \{-1, 1\}$, which returns -1 if the encoded k-CNF is unsatisfiable and 1 otherwise. It is easy to see that f is monotone in the sense that $f(x) \leqslant f(y)$ whenever $x \leqslant y$ coordinate-wise. This is the case, since setting a coordinate from -1 to 1 is equivalent to removing a clause from the encoded formula. By doing so, a satisfiable formula cannot be made unsatisfiable, i.e. the value of f can only change from -1 to 1, but not the other way around. This encoding is from O'Donnell's book [41] and makes the application of Bourgain's Sharp Threshold Theorem later in the paper easier.

We can now formally describe the product probability space of $\mathcal{F}(n, k, (\boldsymbol{p}_n)_{n \in \mathbb{N}}, s)$ with the notation of O'Donnell. Given a variable probability distribution $\boldsymbol{p}_n = (p_{n,i})_{i=1,\ldots,n}$, the derived clause probability distribution $\boldsymbol{q}_n = (q_{n,i})_{i=1,\ldots,N}$, and the scaling factor s, we define our product space to be $(\Omega, \pi) := \left(\{-1, 1\}^N, \pi_1 \times \pi_2 \times \ldots \times \pi_N \right)$ with $\pi_i(-1) = q_{n,i}(s)$ and $\pi_i(1) = 1 - q_{n,i}(s)$ for $i = 1, 2, \ldots, N$. We let $\mu_{\boldsymbol{p}_n, s}$ denote the product probability measure, i.e. for $x \in \Omega$

$$\mu_{\boldsymbol{p}_n, s}(x) = \prod_{i=1}^N \pi_i(x_i) = \prod_{i \in [N]:\, x_i = -1} q_{n,i}(s) \prod_{i \in [N]:\, x_i = 1} (1 - q_{n,i}(s)).$$

For $S \subseteq \Omega$ we define $\mu_{\boldsymbol{p}_n, s}(S) = \sum_{x \in S} \mu_{\boldsymbol{p}_n, s}(x)$. We will use the shorthand notation μ instead of $\mu_{\boldsymbol{p}_n, s}$ if the probability measure is clear from context. Furthermore, for an N-element vector $x = (x_1, x_2, \ldots, x_N)$ and a subset $T \subseteq [N]$ let $x_T = (x_i)_{i \in T}$ denote the *restriction of x to T*.

The following statement shows the relation between coarseness of a property's threshold and the derivative of its probability function. The uniform equivalent of the statement holds due to Friedgut [21], but we can show that it also holds in the non-uniform case. The proof of the statement is a simple application of the mean value theorem.

Lemma 3.1. *If a threshold is coarse, then there is a point s^* in the threshold interval, where $s^* \cdot \frac{d\mu_{p_{n,s}}(f)}{ds}|_{s=s^*} \leqslant K$ for some constant K.*

3.2 Influence and Bourgain's Sharp Threshold Theorem

Bourgain's Sharp Threshold Theorem will make use of the total influence of a Boolean function f. Intuitively, the *influence* $\mathbf{Inf}_i[f]$ of a function f describes the probability that the value of the i-th coordinate influences the function value. The *total influence* $\mathbf{I}[f]$ of a function f is the sum of the influence values for all coordinates. Both, $\mathbf{Inf}_i[f]$ and $\mathbf{I}[f]$ depend on the probability distribution π, but we will omit this dependence if it is clear from context. The following definition from [41] formalizes our intuitive one.

Definition 3.1 *[Influence Function]. Let $f \in L^2(\Omega, \pi)$ be $\{-1, 1\}$-valued with $\Omega = \{-1, 1\}^N$ and $\pi = \pi_1 \times \dots \times \pi_N$. The influence of the i-th coordinate is $\mathbf{Inf}_i[f] = \mathop{\mathbb{E}}_{x \sim \pi} [f(x)(L_i f)(x)]^1$, where $L_i f = f - E_i f$ and $E_i f(y) = \mathop{\mathbb{E}}_{y_i \sim \pi_i} [f(y_1, y_2, \dots, y_{i-1}, y_i, y_{i+1} \dots, y_{N-1}, y_N)]$. The total influence of f is $\mathbf{I}[f] = \sum_{i=1}^{n} \mathbf{Inf}_i[f]$.*

The following corollary relates this notion of influence to the notion of coarseness due to Friedgut, more precisely to $\frac{d\mu_{p_{n,s}}(P)}{ds} s = \frac{d\mu_{p_{n,s}}(\{x \in \Omega | f(x) = -1\})}{ds} s$. Its proof is a relatively simple exercise.

Corollary 3.1. *Let $f \in L^2\left(\Omega = \{-1, 1\}^N, \pi = \pi_1 \times \pi_2 \times \dots \times \pi_N\right)$ be $\{-1, 1\}$-valued, monotone, and non-constant. For $s < \left(\max_{i \in [N]}(q_{n,i})\right)^{-1}$ it holds that*

$$\mathbf{I}[f] \leqslant 4 \cdot \frac{d\mu_{p_{n,s}}(\{x \in \Omega \mid f(x) = -1\})}{ds} s. \tag{3.1}$$

To prove our main theorem, we will use the Sharp Threshold Theorem by Bourgain [21] in O'Donnell's version [41]. The theorem states that, if a monotone property P has a coarse threshold, and therefore small influence, then there are local structures which approximate this property. The following is a formal definition of these structures.

Definition 3.2 *[τ-booster]. Let $f: \Omega \to \{-1, 1\}$. For $T \subseteq [N]$, $y \in \Omega$, and $\tau > 0$, we say that the restriction y_T is a τ-booster if $\mathop{\mathbb{E}}_{x \sim \pi} [f \mid x_T = y_T] \geqslant \mathbb{E}[f] + \tau$. If $\tau < 0$, we say that y_T is a τ-booster if $\mathop{\mathbb{E}}_{x \sim \pi} [f \mid x_T = y_T] \leqslant \mathbb{E}[f] - |\tau|$.*

The Sharp Threshold Theorem is stated as follows:

Theorem 3.1 *[Bourgain's Sharp Threshold Theorem]. Let $f \in L^2(\Omega, \pi)$ be $\{-1, 1\}$-valued and non-constant with $\mathbf{I}[f] \leqslant K$ for a constant K.*

---

[1] In the paper we let $x \sim \pi$ denote that the random variable x is drawn from the probability distribution with density π.

Then there is some τ (either negative or positive) with $|\tau| \geqslant \mathbf{Var}[f] \cdot$ $\exp(-\mathcal{O}(\mathbf{I}[f]^2/\mathbf{Var}[f]^2))$ such that

$$\Pr_{x \sim \pi} \left(\exists \, T \subseteq [n], \, |T| \leqslant \mathcal{O}\left(\frac{\mathbf{I}[f]}{\mathbf{Var}[f]} \right) \text{ such that } x_T \text{ is a } \tau\text{-booster} \right) \geqslant |\tau|.$$

This Theorem seems to be specific to probability spaces with uniform probability distributions. However, O'Donnell states that Theorem 3.1 in the version with arbitrary product probability spaces also holds. We verify this claim in the full version of the paper. Furthermore, by carefully checking the proof of the theorem, one can see that the asymptotic values and the bases for the exponential terms can actually be substituted by appropriately chosen exact expressions. Also, it has to be noted that Müller [40] already showed that a version of Bourgain's original theorem also holds for arbitrary product probability spaces.

3.3   Proof of Sharpness for Non-uniform Random k-SAT

This subsection will be dedicated to proving our main theorem:

Theorem 3.2. *Let $k \geqslant 2$, let $(\boldsymbol{p}_n)_{n \in \mathbb{N}}$ be an ensemble of variable probability distributions on n variables each and let $s_c = t(n)$ be an asymptotic satisfiability threshold for $\mathcal{F}(n, k, (\boldsymbol{p}_n)_{n \in \mathbb{N}}, s)$ with respect to s. If $\|\boldsymbol{p}_n\|_\infty = o\left(t^{-\frac{k}{2k-1}} \cdot \log^{-\frac{k-1}{2k-1}} t \right)$ and $\|\boldsymbol{p}_n\|_2^2 = \mathcal{O}\left(t^{-2/k} \right)$, then satisfiability has a sharp threshold on $\mathcal{F}(n, k, (\boldsymbol{p}_n)_{n \in \mathbb{N}}, s)$ with respect to s.*

The proof closely follows the one by Friedgut for uniform random k-SAT [21]. We assume toward a contradiction that the threshold is coarse. Then the Sharp Threshold Theorem tells us that there have to be so-called "boosters" of constant size that appear with constant probability in the random formula. These boosters have the property that conditioning on their existence *boosts* the probability of the random formula to be unsatisfiable by at least an additive constant.

One kind of booster are unsatisfiable subformulas of constant size. Conditioning on these would boost the probability to be unsatisfiable to one. We rule these out by showing that they do not appear with constant probability.

Then, we consider subformulas, which give the second highest boost: maximally quasi-unsatisfiable subformulas. These are subformulas which have only *one* satisfying assignment for the variables appearing in them and adding any new clause over those variables makes them unsatisfiable. We want to show that these cannot boost the probability of a formula to be unsatisfiable by a constant.

Again toward a contradiction, we assume, that conditioning on a maximally quasi-unsatisfiable subformula T is enough to boost the unsatisfiability probability by a constant. First, we prove that conditioning on T is equivalent to adding a number of clauses of size shorter than k to the random formula over variables not appearing in T. Then, we use a version of Friedgut's coverability lemma to show that, if adding these clauses of size smaller than k makes the random formula unsatisfiable with constant probability, then so does adding $o(t)$ clauses of size k.

We prove that this probability is dominated by the probability to make the original random formula unsatisfiable for a slightly bigger scaling factor. However, due to the assumption of a coarse threshold, the slope of the probability function for unsatisfiability has to be small at one point in the threshold interval. If we consider this point, the probability to make the original random formula unsatisfiable cannot be increased by a constant with our slightly increased scaling factor. This contradicts our assumption that the probability is boosted by a constant in the first place. Therefore, quasi-unsatisfiable subformulas cannot be boosters.

After showing this, every less restrictive subformula cannot be a booster either. That means, the only possible boosters are unsatisfiable subformulas, which we ruled out already. Therefore, the implication of the Sharp Threshold Theorem does not hold, which contradicts the assumption of a coarse threshold.

Now we are ready to prove our main theorem.

Application of the Sharp Threshold Theorem. We know that the asymptotic threshold is at a scaling factor $s = \Theta(t(n))$. A threshold due to our definition always has to be $t = \Omega(1)$. Otherwise the expected number of clauses would be $O(t) = o(1)$, leading to a probability of $1 - o(1)$ of having an empty, and thereby satisfiable, formula due to Markov's inequality. We can thus assume that $C = \left(1 + o\left(t^{-1/k}\right)\right)$ due to Eq. (2.2).

To prove Theorem 3.2 we assume that the threshold is coarse. Due to Lemma 3.1 this implies that $\frac{d\mu_{p_n,s}(f)}{ds} s \leqslant K$ for some constant K and some s in the threshold interval. Let us call this scaling factor s_c. Note that $s_c = \Theta(t)$, since s_c is in the threshold interval and t is an asymptotic threshold function. Due to Corollary 3.1 this means $\mathbf{I}[f] \leqslant 4 \cdot K$. For the corollary to hold, we have to assure $s_c < \left(\max_{i \in [N]}(q_{n,i})\right)^{-1}$. This follows due to our assumption

$$p_{n,\max} := \|\boldsymbol{p}_n\|_\infty = o\left(t^{-\frac{k}{2k-1}} \cdot \log^{-\frac{k-1}{2k-1}} t\right) = o\left(t^{-1/k}\right),$$

which implies

$$q_{n,\max}(s_c) := \max_{i \in [N]}\left(q_{n,i}(s_c)\right) = s_c \cdot \mathcal{O}\left(p_{n,\max}^k\right) = o(1). \tag{3.2}$$

Since f is $\{-1,1\}$-valued it holds that $\mathbb{E}[f] = 1 - 2 \cdot \mu_{p_n,s_c}(f)$ and $\mathbf{Var}[f] = 4 \cdot \mu_{p_n,s_c}(f)\left(1 - \mu_{p_n,s_c}(f)\right)$. Since we are in the threshold interval, it holds that $\mu_{p_n,s_c}(f)$ is constant and so are $\mathbb{E}[f]$ and $\mathbf{Var}[f]$.

Now we can use Theorem 3.1 to see that, at least with constant probability τ, our formulas have a subformula (or lack thereof) consisting of at most $\mathcal{O}(K) = \mathcal{O}(1)$ clauses, so that conditioning on the existence (or non-existence) of these clauses increases (or decreases) the probability that our random k-CNFs are unsatisfiable by at least $\tau/2$. The subformulas with these properties are the boosters. The theorem actually allows us to choose appropriate specific constants for τ and the upper bound on $|T|$.

Since the property of being unsatisfiable is monotone, it is not beneficial to forbid some clauses and demand others. We can therefore concentrate on the cases of either only forbidding or only enforcing clauses in our boosters. The following lemma shows that it suffices to concentrate on enforcing boosters. The idea is that every constant-sized subset of clauses a. a. s. does not exist in the formula, since clause probabilities are $o(1)$. Therefore, conditioning on the non-existence of such a subformula does not change the overall probability by too much.

Lemma 3.2. *Every booster which assumes the non-existence of clauses only boosts the probability to be satisfiable or unsatisfiable by $o(1)$.*

We can now concentrate on conditioning on the *existence* of clauses. Our goal is to show that no constant-sized boosters exist with constant probability.

Unsatisfiable Subformulas Are Too Improbable. A sure way to boost the probability of being unsatisfiable to one is to condition on the existence of an unsatisfiable subformula. To rule this case out, the next lemma shows that the probability that our formulas have an unsatisfiable subformula of constant size is smaller than any constant τ for sufficiently large n. The proof essentially shows that any minimally unsatisfiable subformula of constant size cannot exist with constant probability. This can be seen from the fact that such subformulas contain each variable in them at least twice and the probability for this can be bounded using $\|p_n\|_2^2$ and $\|p_n\|_\infty$.

Lemma 3.3. *Let $a, k \in \mathbb{N}$ be constants and let $(p_n)_{n \in \mathbb{N}}$ be an ensemble of variable probability distributions. If $\|p_n\|_\infty = o\left(s^{-1/k}\right)$ and $\|p_n\|_2^2 = \mathcal{O}\left(s^{-2/k}\right)$, then a random formula from $\mathcal{F}\left(n, k, (p_n)_{n \in \mathbb{N}}, s\right)$ has an unsatisfiable subformula of length at most a with probability $o(1)$.*

Maximally Quasi-Unsatisfiable Subformulas Provide the Second-Highest Boost. Since we ruled out unsatisfiable subformulas as the boosters we are looking for, we now turn our attention to satisfiable subformulas. Let Φ_T be the formula encoded by $x_T = (-1)^{|T|}$ and let $V(T) \subseteq \{X_1, \ldots, X_n\}$ be the variables in Φ_T. Note that $|V(T)|$ is constant since $|T|$ is constant and each clause contains k many variables. We call Φ_T *maximally quasi-unsatisfiable (mqu)* if it is satisfiable by only one of the $2^{|V(T)|}$ assignments over its variable set (quasi-unsatisfiable) and if adding any new clause with variables only from $V(T)$ makes it unsatisfiable (maximally satisfiable). The following lemma formalizes a statement by Friedgut [21] that the biggest possible boost any satisfiable subformula can give is achieved by mqu subformulas. The proof of the statement uses the fact that every satisfiable subformula can be extended to a mqu subformula over the same variables. It also uses positive correlation of increasing events [20] and the fact that we have a product probability space.

Lemma 3.4. *For every $T \subseteq [N]$ so that Φ_T is satisfiable, there is a $T' \supseteq T$ so that $\Phi_{T'}$ is maximally quasi-unsatisfiable and*

$$\Pr_{x \sim \pi} \left(f(x) = -1 \mid x_{T'} = (-1)^{|T'|} \right) \geq \Pr_{x \sim \pi} \left(f(x) = -1 \mid x_T = (-1)^{|T|} \right).$$

The Part of the Formula Containing only Variables from the Booster Is Still Satisfiable. We now turn to analyzing the boost maximally quasi-unsatisfiable subformulas can give. In the end will will show that they cannot boost the unsatisfiability probability by a constant. Lemma 3.4 implies that the same holds for all satisfiable subformulas, thus giving us the desired contradiction.

Let $T \subseteq [N]$ with Φ_T mqu. In order to see how big the boost by such a T can be, we split x into two parts, the part x_S, so that each clause in Φ_S only contains variables from $V(T)$, and the part $x_{\overline{S}}$, in which each encoded clause contains at least one variable from $\overline{V(T)} = \{X_1, \ldots, X_n\} \setminus V(T)$. Let $f(x_S)$ be -1 if Φ_S is unsatisfiable and 1 otherwise. The following lemma asserts that Φ_S can only be unsatisfiable with sub-constant probability. This is the case, because it is very unlikely to flip one of the constant number of clauses that can make the maximally satisfiable booster unsatisfiable.

Lemma 3.5. *It holds that $\Pr_{x \sim \pi} \left(f(x_S) = -1 \mid x_T = (-1)^{|T|} \right) = o(1)$.*

The Booster Adds Shorter Clauses to the Other Part of the Formula. We can now concentrate on the case that Φ_S is satisfiable. Since Φ_T is maximally unsatisfiable, it holds that $\Phi_S = \Phi_T$, and since Φ_T is quasi-unsatisfiable, Φ_S also only has one satisfying assignment.

We now want to create $x_{\overline{S}}$ under these conditions. To this end, we assume that the variables $V(t)$ take the one assignment that makes Φ_S satisfiable. For a clause containing both variables from $V(T)$ and variables from $\overline{V(T)}$ this means the clause is either satisfied or the variables from $V(T)$ can be eliminated as their literals are all set to false. Effectively, this means that we can have clauses over $\overline{V(T)}$ of length $0 < l < k$. The following lemma makes this statement more precise. Its proof is a simple application of the Markov bound.

Lemma 3.6. *If $p_{n,\max} = o\left(t^{-\frac{k}{2k-1}} \cdot \log^{-\frac{k-1}{2k-1}} t \right)$, then a mqu subformula of constant length spawns at most $D_l = o\left(\left(\frac{t}{\log t} \right)^{\frac{l}{k+l}} \right)$ clauses of length $l = 1, \ldots, k-1$ with probability $1 - o(1)$.*

We now want to create the resulting formula over variables from $\overline{V(T)}$ in two parts. First we create k-clauses over $\overline{V(T)}$ with the usual clause-flipping model, where the clause-probabilities are the same as in $\mathcal{F}(n, k, (\boldsymbol{p}_n)_{n \in \mathbb{N}}, s_c)$. Then, for each $l \in [k-1]$ we add D_l l-clauses over $\overline{V(T)}$ with the clause-drawing model. The probability q_c to add a clause $c = (\ell_1 \vee \ell_2 \vee \ldots \vee \ell_l)$ of size l is equal to the probability of flipping any clause which contains c and $k - l$ literals negated by the assignment of $\overline{V(T)}$:

$$q_c = C \frac{k! \cdot s_c}{2^k} \prod_{i=1}^{l} p_n(|\ell_i|) \cdot \sum_{J \in \mathcal{P}_{k-l}(V(T))} \prod_{X \in J} p_n(X). \tag{3.3}$$

We can now choose $q_c' = \frac{q_c/(1-q_c)}{\sum_{j \in [n]} q_c/(1-q_c)}$ as the probability to draw clause c. This helps us apply Lemma 2.1 to relate the resulting random formula $\hat{\Phi}$ to our original probability space. Furthermore, the following lemma also uses Lemma 3.6 and the fact that no clauses are drawn twice with probability $1 - o(1)$.

Lemma 3.7. *It holds that*

$$\Pr_{x \sim \pi} \left(f(x) = -1 \wedge f(x_S) = 1 \mid x_T = (-1)^{|T|} \right) \leqslant \Pr \left(\hat{\Phi} \ unsat \right) + o(1).$$

Shorter Clauses Can Be Substituted with k-Clauses. We now want to bound $\Pr(\hat{\Phi} \text{ unsat})$. To this end, let $\widetilde{\Phi}$ be the part of $\hat{\Phi}$ only consisting of k-clauses. Let us *assume* $\Pr(\hat{\Phi} \text{ unsat}) \geqslant \mu_{p_n,s_c}(f) + \delta$ for some constant $\delta > 0$. We know that $\widetilde{\Phi}$ is unsatisfiable with probability at most $\mu_{p_n,s_c}(f)$, since it is drawn from $\mathcal{F}(n, k, (p_n)_{n \in \mathbb{N}}, s_c)$ with the difference that only clauses over $\overline{V(T)}$ are flipped. This implies $\Pr(\hat{\Phi} \text{ unsat} \wedge \widetilde{\Phi} \text{ sat}) \geqslant \delta$. We now define a more general concept of coverability, analogously to Friedgut [21]. This will allow us to substitute l-clauses with k-clauses while maintaining the probability to make $\widetilde{\Phi}$ unsatisfiable.

Definition 3.3. *Let $D_1, \ldots, D_a \in \mathbb{N}$ and $l_1, \ldots, l_a \in \mathbb{N}$ and let $q_1, \ldots, q_a$ be probability distributions. For $A \subseteq \{0,1\}^n$, we define A to be $((d_1, l_1, q_1), (d_2, l_2, q_2), \ldots, (d_a, l_a, q_a), \varepsilon)$-coverable, if the union of d_i subcubes of co-dimension l_i chosen according to probability distribution q_i for $1 \leqslant i \leqslant a$ has a probability of at least ε to cover A.*

In contrast to Friedgut's definition, we allow subcubes of arbitrary co-dimension and with arbitrary probability distributions instead of only subcubes of co-dimension 1 with a uniform distribution. In the context of satisfiability we say that a specific formula (*not* a random formula) Φ is $((d_1, l_1, q_1), \ldots, (d_a, l_a, q_a), \varepsilon)$-coverable if the probability to make it unsatisfiable by adding d_i random clauses of size l_i chosen according to distribution q_i for $i = 1, 2, \ldots a$ is at least ε in total.

Now let $q_l = (q_c')_c$ for all clauses c of size l over $\overline{V(T)}$, where q_c' is the clause drawing probability we defined for $\hat{\Phi}$. It holds that with a sufficiently large constant probability $\hat{\Phi}$ is $((D_1, 1, q_1), \ldots, (D_{k-1}, k - 1, q_{k-1}), \delta)$-coverable. The following lemma shows that formulas with this property are also $((g(n), k, q_k), \delta')$-coverable for some function $g(n) = o(t)$ and any constant $\delta' < \delta$. Its proof is a more precise and general version of Friedgut's original proof.

Lemma 3.8. *Let q_k be our original clause probability distribution and let all other probability distributions be as described in Eq. (3.3) and let $D_1 \ldots D_{k-1}$ be as defined. If a concrete formula Φ is $((D_1, 1, q_1), \ldots, (D_{k-1}, k-1, q_{k-1}), \delta)$-coverable for some constant $\delta > 0$, it is also $((g(t), k, q_k), \delta')$-coverable for some function $g(t) = o(t)$ for any constant $0 < \delta' < \delta$.*

By substituting shorter clauses with k-clauses we lose at most an arbitrarily small additive constant from the probability $\mu_{p_n, s_c}(f) + \delta$ that $\hat{\Phi}$ is unsatisfiable. Thus, we still have a constant probability bigger than $\mu_{p_n, s_c}(f)$.

Bounding the Boost by Bounding the Slope of the Probability Function. We can now show that instead of adding $g(t)$ k-clauses, we can increase the scaling factor s_c of our original clause-flipping model to achieve the same probability. The proof of the following lemma uses Lemma 2.1 together with a Chernoff-Bound on the number of clauses added in the clause-flipping model.

Lemma 3.9. *For $g'(t) = g(t) + c \cdot \sqrt{t} \cdot \ln t = o(t)$ with $c > 0$ an appropriately chosen constant it holds that*

$$\Pr\left(\hat{\Phi} \text{ unsat}\right) \leqslant \mu_{p_n, s_c + g'(t)}(f) + o(1).$$

Under the assumption that $\Pr(\hat{\Phi} \text{ unsat}) \geqslant \mu_{p_n, s_c}(f) + \delta$, it follows that $\mu_{p_n, s_c + g'(t)}(f) \geqslant \mu_{p_n, s_c}(f) + \varepsilon$ for some constant $\varepsilon > 0$. We show that this cannot be the case under the assumption of a coarse threshold. The proof of this lemma is a simple application of Taylor's theorem and uses the fact that we evaluate the probability function at the point s_c, where $\frac{d\mu_{p,n_s}(f)}{ds} s \Big|_{s=s_c} \leqslant K$ due to Lemma 3.1.

Lemma 3.10. *It holds that $\mu_{p_n, s_c + g'(t)}(f) \leqslant \mu_{p_n, s_c}(f) + o(1)$.*

This contradicts our conclusion of $\mu_{p_n, s_c + g'(t)}(f) \geqslant \mu_{p_n, s_c}(f) + \varepsilon$ for some constant $\varepsilon > 0$. Therefore, our assumption $\Pr(\hat{\Phi} \text{ unsat}) \geqslant \mu_{p_n, s_c}(f) + \delta$ for some constant $\delta > 0$ has to be wrong, i.e. $\Pr(\hat{\Phi} \text{ unsat}) \leqslant \mu_{p_n, s_c}(f) + o(1)$. Now we can put all error probabilities together to see

$$\Pr_{x \sim \pi}\left(f(x) = -1 \mid x_T = (-1)^{|T|}\right) \leqslant \mu_{p_n, s_c}(f) + o(1).$$

This is smaller than $\mu_{p_n, s_c}(f) + \tau$ for sufficiently large values of n. This means, the maximally quasi-unsatisfiable subformula Φ_T cannot be a τ-booster for any constant $\tau > 0$. Due to Lemma 3.4 the boost by every satisfiable subformula is at most as big as the one by a mqu subformula. Thus, no T which encodes a satisfiable subformula can be a τ-booster. Since we already ruled out unsatisfiable subformulas, this means there are no τ-boosters which appear with probability at least $\tau/2$. This contradicts the implication of the Sharp Threshold Theorem and therefore the assumption of a coarse threshold, thus proving Theorem 3.3. $\square$

3.4   Relation to the Clause-Drawing Model

After proving the sharpness of the threshold for $\mathcal{F}(n, k, (\boldsymbol{p}_n)_{n\in\mathbb{N}}, s)$ in Theorem 3.2, it now remains to relate $\mathcal{F}(n, k, (\boldsymbol{p}_n)_{n\in\mathbb{N}}, s)$ to $\mathcal{D}(n, k, (\boldsymbol{p}_n)_{n\in\mathbb{N}}, m)$.

Usually, the satisfiability threshold is only determined for the clause-drawing model and not for the clause-flipping model. Nevertheless, the following lemma shows that for certain probability distribution ensembles $(\boldsymbol{p}_n)_{n\in\mathbb{N}}$ the asymptotic thresholds of both models are the same. This allows us to determine the asymptotic threshold function of the clause-flipping model and to apply Theorem 3.2. The proofs of Lemmas 3.11 and 3.12 use Lemma 2.1 and Chernoff Bounds.

Lemma 3.11. *Let $(\boldsymbol{p}_n)_{n\in\mathbb{N}}$ be an ensemble of variable probability distributions on n variables each and let $t = \omega(1)$ be an asymptotic threshold with respect to m for a monotone property P on $\mathcal{D}(n, k, (\boldsymbol{p}_n)_{n\in\mathbb{N}}, m)$. If $\|p_n\|_2^2 = o\left(t^{-1/k}\right)$, then $s_c = \Theta(t)$ is an asymptotic threshold with respect to s for P on $\mathcal{F}(n, k, (\boldsymbol{p}_n)_{n\in\mathbb{N}}, s)$.*

With the help of the former lemma, we can now prove Lemma 3.12.

Lemma 3.12. *Let $(\boldsymbol{p}_n)_{n\in\mathbb{N}}$ be an ensemble of variable probability distributions on n variables each and let $t = \omega(1)$ be an asymptotic threshold with respect to s for any monotone property P on $\mathcal{F}(n, k, (\boldsymbol{p}_n)_{n\in\mathbb{N}}, s)$. If $\|p_n\|_2^2 = o\left(t^{-1/k}\right)$ and if the threshold for P with respect to s on $\mathcal{F}(n, k, (\boldsymbol{p}_n)_{n\in\mathbb{N}}, s)$ is sharp, then P has a sharp threshold on $\mathcal{D}(n, k, \boldsymbol{p}, m)$ at $m_c = \Theta(t)$.*

Theorem 3.3, now follows from the two lemmas above and from Theorem 3.2.

Theorem 3.3. *Let $k \geqslant 2$, let $(\boldsymbol{p}_n)_{n\in\mathbb{N}}$ be an ensemble of variable probability distributions on n variables each and let $m_c = t(n) = \omega(1)$ be the asymptotic satisfiability threshold for $\mathcal{D}(n, k, (\boldsymbol{p}_n)_{n\in\mathbb{N}}, m)$ with respect to m. If $\|p_n\|_\infty = o\left(t^{-\frac{k}{2k-1}} \cdot \log^{-\frac{k-1}{2k-1}} t\right)$ and $\|p_n\|_2^2 = \mathcal{O}\left(t^{-2/k}\right)$, then satisfiability has a sharp threshold on $\mathcal{D}(n, k, (\boldsymbol{p}_n)_{n\in\mathbb{N}}, m)$ with respect to m.*

3.5   Example Application of the Theorem

We can now use Theorem 3.3 as a tool to show sharpness of the threshold for non-uniform random k-SAT with different probability distributions on the variables. As an example, we apply the theorem for an ensemble of power-law distributions.

Corollary 3.2. *Let $(\boldsymbol{p}_n)_{n\in\mathbb{N}}$ be an ensemble of general power-law distributions with the same power-law exponent $\beta \geqslant \frac{2k-1}{k-1} + 1 + \varepsilon$, where $\varepsilon > 0$ is a constant and $\boldsymbol{p}_n$ is defined over n variables. For $k \geqslant 2$ both $\mathcal{F}(n, k, (\boldsymbol{p}_n)_{n\in\mathbb{N}}, s)$ and $\mathcal{D}(n, k, (\boldsymbol{p}_n)_{n\in\mathbb{N}}, m)$ have a sharp threshold with respect to s and m, respectively.*

Proof. From [23] we know that the asymptotic threshold for $\mathcal{D}(n, k, (\boldsymbol{p}_n)_{n\in\mathbb{N}}, m)$ is at $m = \Theta(n)$ for $\beta \geqslant \frac{2k-1}{k-1} + \varepsilon$. It is now an easy exercise to see that

$$\|\boldsymbol{p}_n\|_2^2 = \sum_{i=1}^{n} p_{n,i}^2 = \begin{cases} \mathcal{O}\left(n^{-2(\beta-2)/(\beta-1)}\right) & ,\beta < 3 \\ \mathcal{O}\left(\frac{\ln n}{n}\right) & ,\beta = 3 \\ \mathcal{O}\left(n^{-1}\right) & ,\beta > 3 \end{cases}$$

and that $\|\boldsymbol{p}_n\|_\infty = \max_{i=1,2,\dots,n}(p_{n,i}) = \mathcal{O}(n^{-(\beta-2)/(\beta-1)})$. One can now verify $\|\boldsymbol{p}_n\|_2^2 = \mathcal{O}(n^{-2/k})$ and $\|\boldsymbol{p}_n\|_\infty = o(n^{-\frac{k}{2k-1}} \cdot \log^{-\frac{k-1}{2k-1}}(n))$ for $\beta > \frac{2k-1}{k-1} + 1 + \varepsilon$ and $k \geqslant 2$. Lemma 3.11 now states that the asymptotic satisfiability threshold for $\mathcal{F}(n, k, (\boldsymbol{p}_n)_{n\in\mathbb{N}}, s)$ is at $s = \Theta(n)$. Theorems 3.2 and 3.3 now imply a sharp threshold for $\mathcal{F}(n, k, (\boldsymbol{p}_n)_{n\in\mathbb{N}}, s)$ and $\mathcal{D}(n, k, (\boldsymbol{p}_n)_{n\in\mathbb{N}}, m)$.

4 Discussion of the Results

In this work we have shown sufficient conditions on the variable probability distribution of non-uniform random k-SAT for the satisfiability threshold to be sharp. The main theorems can readily be used to prove sharpness for a wide range of random k-SAT models with heterogeneous distributions on the variable occurrences: If the threshold function is known asymptotically, one only has to verify the two conditions on the variable distribution.

We suspect that it is possible to generalize the result to demanding only $\|\boldsymbol{p}\|_\infty = o\left(t^{-1/k}\right)$, since the additional factor is only needed in Lemma 3.8. In any case it would be interesting to complement the result with matching conditions on coarseness of the threshold.

We hope that our results make it possible to derive a proof in the style of Ding et al. [19] for certain variable probability ensembles with a sharp threshold, effectively proving the satisfiability threshold conjecture for these ensembles.

References

1. Achlioptas, D., Kirousis, L.M., Kranakis, E., Krizanc, D.: Rigorous results for random (2+p)-SAT. Theor. Comput. Sci. **265**(1–2), 109–129 (2001)
2. Alistarh, D., Sauerwald, T., Vojnović, M.: Lock-free algorithms under stochastic schedulers. In: 34th Symposium on Principles of Distributed Computing (PODC), pp. 251–260 (2015)
3. Ansótegui, C., Bonet, M.L., Giráldez-Cru, J., Levy, J.: The fractal dimension of SAT formulas. In: 7th International Joint Conference on Automated Reasoning (IJCAR), pp. 107–121 (2014)
4. Ansótegui, C., Bonet, M.L., Giráldez-Cru, J., Levy, J.: On the classification of industrial SAT families. In: 18th International Conference of the Catalan Association for Artificial Intelligence (CCIA), pp. 163–172 (2015)
5. Ansótegui, C., Bonet, M.L., Levy, J.: On the structure of industrial SAT instances. In: Gent, I.P. (ed.) CP 2009. LNCS, vol. 5732, pp. 127–141. Springer, Heidelberg (2009). https://doi.org/10.1007/978-3-642-04244-7_13

6. Ansótegui, C., Bonet, M.L., Levy, J.: Towards industrial-like random SAT instances. In: 21st International Joint Conference on Artificial Intelligence (IJCAI), pp. 387–392 (2009)
7. Ansótegui, C., Giráldez-Cru, J., Levy, J.: The community structure of SAT formulas. In: Cimatti, A., Sebastiani, R. (eds.) SAT 2012. LNCS, vol. 7317, pp. 410–423. Springer, Heidelberg (2012). https://doi.org/10.1007/978-3-642-31612-8_31
8. Bapst, V., Coja-Oghlan, A.: The condensation phase transition in the regular k-SAT model. In: Approximation, Randomization, and Combinatorial Optimization. Algorithms and Techniques, APPROX/RANDOM 2016, pp. 22:1–22:18 (2016)
9. Boufkhad, Y., Dubois, O., Interian, Y., Selman, B.: Regular random k-SAT: properties of balanced formulas. J. Autom. Reason. 35(1–3), 181–200 (2005)
10. Bradonjic, M., Perkins, W.: On sharp thresholds in random geometric graphs. In: Approximation, Randomization, and Combinatorial Optimization. Algorithms and Techniques, APPROX/RANDOM 2014, pp. 500–514 (2014)
11. Bringmann, K.: Why walking the dog takes time: Frechet distance has no strongly subquadratic algorithms unless SETH fails. In: 55th Symposium on Foundations of Computer Science (FOCS), pp. 661–670 (2014)
12. Chvatal, V., Reed, B.: Mick gets some (the odds are on his side). In: 33rd Symposium on Foundations of Computer Science (FOCS), pp. 620–627 (1992)
13. Coja-Oghlan, A.: The asymptotic k-SAT threshold. In: 46th Symposium on Theory of Computing (STOC), pp. 804–813 (2014)
14. Coja-Oghlan, A., Panagiotou, K.: The asymptotic k-SAT threshold. Adv. Math. 288, 985–1068 (2016)
15. Coja-Oghlan, A., Wormald, N.: The number of satisfying assignments of random regular k-SAT formulas. CoRR abs/1611.03236 (2016)
16. Cook, S.A.: The complexity of theorem-proving procedures. In: 3rd Symposium on Theory of Computing (STOC), pp. 151–158 (1971)
17. Cygan, M., Nederlof, J., Pilipczuk, M., Pilipczuk, M., van Rooij, J.M.M., Wojtaszczyk, J.O.: Solving connectivity problems parameterized by treewidth in single exponential time. In: 52nd Symposium on Foundations of Computer Science (FOCS), pp. 150–159 (2011)
18. Díaz, J., Kirousis, L.M., Mitsche, D., Pérez-Giménez, X.: On the satisfiability threshold of formulas with three literals per clause. Theoret. Comput. Sci. 410(30–32), 2920–2934 (2009)
19. Ding, J., Sly, A., Sun, N.: Proof of the satisfiability conjecture for large k. In: 47th Symposium on Theory of Computing (STOC), pp. 59–68 (2015)
20. Fortuin, C.M., Kasteleyn, P.W., Ginibre, J.: Correlation inequalities on some partially ordered sets. Commun. Math. Phys. 22(2), 89–103 (1971)
21. Friedgut, E.: Sharp thresholds of graph properties, and the k-SAT problem. J. Am. Math. Soc. 12(4), 1017–1054 (1999)
22. Friedgut, E.: Hunting for sharp thresholds. Random Struct. Algorithms 26(1–2), 37–51 (2005)
23. Friedrich, T., Krohmer, A., Rothenberger, R., Sauerwald, T., Sutton, A.M.: Bounds on the satisfiability threshold for power law distributed random SAT. In: 25th European Symposium on Algorithms (ESA), pp. 37:1–37:15 (2017)
24. Friedrich, T., Krohmer, A., Rothenberger, R., Sutton, A.M.: Phase transitions for scale-free SAT formulas. In: 31st Conference on Artificial Intelligence (AAAI), pp. 3893–3899 (2017)
25. Giráldez-Cru, J., Levy, J.: A modularity-based random SAT instances generator. In: 24th International Joint Conference on Artificial Intelligence (IJCAI), pp. 1952–1958 (2015)

26. Giráldez-Cru, J., Levy, J.: Locality in random SAT instances. In: 26th International Joint Conference on Artificial Intelligence (IJCAI), pp. 638–644 (2017)
27. Goerdt, A.: A threshold for unsatisfiability. J. Comput. Syst. Sci. **53**(3), 469–486 (1996)
28. Hajiaghayi, M.T., Sorkin, G.B.: The satisfiability threshold of random 3-SAT is at least 3.52. Technical report, RC22942, IBM, October 2003
29. Impagliazzo, R., Paturi, R.: On the complexity of k-SAT. J. Comput. Syst. Sci. **62**(2), 367–375 (2001)
30. Impagliazzo, R., Paturi, R., Zane, F.: Which problems have strongly exponential complexity? In: 39th Symposium on Foundations of Computer Science (FOCS), pp. 653–663 (1998)
31. Kaporis, A.C., Kirousis, L.M., Lalas, E.G.: The probabilistic analysis of a greedy satisfiability algorithm. Random Struct. Algorithms **28**(4), 444–480 (2006)
32. Karp, R.M.: Reducibility among combinatorial problems. In: Proceedings of a Symposium on the Complexity of Computer Computations, held 20–22 March 1972, at the IBM Thomas J. Watson Research Center, Yorktown Heights, New York, pp. 85–103 (1972)
33. Levin, L.A.: Universal sorting problems. Prob. Inf. Transm. **9**, 265–266 (1973)
34. Mézard, M., Parisi, G., Zecchina, R.: Analytic and algorithmic solution of random satisfiability problems. Science **297**(5582), 812–815 (2002)
35. Mitchell, D.G., Selman, B., Levesque, H.J.: Hard and easy distributions of SAT problems. In: 10th Conference on Artificial Intelligence (AAAI), pp. 459–465 (1992)
36. Monasson, R., Zecchina, R.: Statistical mechanics of the random k-satisfiability model. Phys. Rev. E **56**, 1357–1370 (1997)
37. Monasson, R., Zecchina, R., Kirkpatric, S., Selman, B., Troyansky, L.: Phase transition and search cost in the 2+ p-SAT problem. In: 4th Workshop on Physics and Computation, Boston, MA (1996)
38. Monasson, R., Zecchina, R., Kirkpatrick, S., Selman, B., Troyansky, L.: 2+p-SAT: relation of typical-case complexity to the nature of the phase transition. Random Struct. Algorithms **15**(3–4), 414–435 (1999)
39. Mull, N., Fremont, D.J., Seshia, S.A.: On the hardness of SAT with community structure. In: Creignou, N., Le Berre, D. (eds.) SAT 2016. LNCS, vol. 9710, pp. 141–159. Springer, Cham (2016). https://doi.org/10.1007/978-3-319-40970-2_10
40. Müller, T.: The critical probability for confetti percolation equals 1/2. Random Struct. Algorithms **50**(4), 679–697 (2017)
41. O'Donnell, R.: Analysis of Boolean Functions. Cambridge University Press, Cambridge (2014)
42. Rathi, V., Aurell, E., Rasmussen, L., Skoglund, M.: Bounds on threshold of regular random k-SAT. In: Strichman, O., Szeider, S. (eds.) SAT 2010. LNCS, vol. 6175, pp. 264–277. Springer, Heidelberg (2010). https://doi.org/10.1007/978-3-642-14186-7_22

In Between Resolution and Cutting Planes: A Study of Proof Systems for Pseudo-Boolean SAT Solving

Marc Vinyals[2], Jan Elffers[1], Jesús Giráldez-Cru[1], Stephan Gocht[1], and Jakob Nordström[1(✉)]

[1] KTH Royal Institute of Technology, Stockholm, Sweden
{elffers,giraldez,gocht,jakobn}@kth.se
[2] Tata Institute of Fundamental Research, Mumbai, India
marc.vinyals@tifr.res.in

Abstract. We initiate a proof complexity theoretic study of subsystems of cutting planes (CP) modelling proof search in conflict-driven pseudo-Boolean (PB) solvers. These algorithms combine restrictions such as that addition of constraints should always cancel a variable and/or that so-called saturation is used instead of division. It is known that on CNF inputs cutting planes with cancelling addition and saturation is essentially just resolution. We show that even if general addition is allowed, this proof system is still polynomially simulated by resolution with respect to proof size as long as coefficients are polynomially bounded.

As a further way of delineating the proof power of subsystems of CP, we propose to study a number of *easy* (but tricky) instances of problems in NP. Most of the formulas we consider have short and simple tree-like proofs in general CP, but the restricted subsystems seem to reveal a much more varied landscape. Although we are not able to formally establish separations between different subsystems of CP—which would require major technical breakthroughs in proof complexity—these formulas appear to be good candidates for obtaining such separations. We believe that a closer study of these benchmarks is a promising approach for shedding more light on the reasoning power of pseudo-Boolean solvers.

1 Introduction

The efficiency of modern Boolean satisfiability (SAT) solvers is one of the most fascinating success stories in computer science. The SAT problem lies at the foundation of the theory of NP-completeness [13], and as such is believed to be completely beyond reach from a computational complexity point of view. Yet solvers based on *conflict-driven clause learning (CDCL)* [4,38,40] are nowadays used routinely to solve instances with millions of variables.

From a theoretical point of view, it is an intriguing question *how to explain* the performance of state-of-the-art SAT solvers, and unfortunately our understanding of this remains quite limited. Perhaps the only tool currently available

© Springer International Publishing AG, part of Springer Nature 2018
O. Beyersdorff and C. M. Wintersteiger (Eds.): SAT 2018, LNCS 10929, pp. 292–310, 2018.
https://doi.org/10.1007/978-3-319-94144-8_18

for giving rigorous answers to such questions is provided by *proof complexity* [14], where one essentially ignores the question of algorithmic proof search and instead studies the power and limitations of the underlying method of reasoning.

Conflict-Driven Clause Learning and Resolution. It is well-known (see, e.g., [5]) that CDCL solvers search for proofs in the proof system *resolution* [7]. Ever since resolution-based SAT solvers were introduced in [16,17,46], subsystems of resolution corresponding to these algorithms, such as tree-like and regular resolution, have been studied. Exponential lower bounds for general resolution proofs were established in [11,29,51], and later it was proven that general resolution is exponentially stronger than regular resolution, which in turn is exponentially stronger than tree-like resolution (see [1,6,52] and references therein). More recently, CDCL viewed as a proof system was shown to simulate general resolution efficiently [3,43] (i.e., with at most a polynomial blow-up), though an algorithmic version of this result seems unlikely in view of [2].

A problem that is arguably even more intriguing than the analysis of CDCL solver performance is why attempts to build SAT solvers on stronger methods of reasoning than resolution have had such limited success so far. Resolution lies very close to the bottom in the hierarchy of proof systems studied in proof complexity, and even quite a limited extension of this proof system with algebraic or geometric reasoning holds out the prospect of exponential gains in performance.

Pseudo-Boolean Solving and Cutting Planes. In this paper we consider one such natural extension to *pseudo-Boolean (PB) solving* using linear inequalities over Boolean variables with integer coefficients, which is formalized in the proof system *cutting planes (CP)* [10,15,28]. By way of a brief overview, Hooker [31,32] considered generalizations of resolution to linear constraints and investigated the completeness of such methods. More general algorithms were implemented by Chai and Kuehlman [9], Sheini and Sakallah [49], and Dixon et al. [18–20]. The focus in all of these papers is mostly on algorithmic questions, however, and not on properties of the corresponding proof systems.

Papers on the proof complexity side have studied tree-like cutting planes [34] and CP with bounded constant terms in the inequalities [27], and resolution has been shown to simulate cutting planes when this upper bound is constant [30]. Exponential lower bounds for cutting planes with coefficients of polynomial magnitude were obtained in [8], and for general cutting planes with coefficients of arbitrary size strong lower bounds were proven in [45] and (very recently) in [26,33]. These papers consider more general derivation rules than are used algorithmically, however, and in contrast to the situation for resolution we are not aware of any work analysing the proof complexity of subsystems of CP corresponding to the reasoning actually being used in pseudo-Boolean solvers.

Our Contributions. We initiate a study of proof systems intended to capture the reasoning in pseudo-Boolean solvers searching for cutting planes proofs. In this work we focus on *cdcl-cuttingplanes* [21] and *Sat4j* [36,48], which are the two CP-based solvers that performed best in the relevant satisfiability problems category *DEC-SMALLINT-LIN* in the *Pseudo-Boolean Competition 2016* [44].[1] Our subsystems of CP combine algorithmically natural restrictions such as that addition should always cancel a variable and/or that *saturation* is used instead of the more expensive to implement division rule. We stress that these derivation rules are nothing new—indeed, the point is that they are already used in practice, and they are formally defined in, e.g., the excellent survey on pseudo-Boolean solving [47]. Our contribution is to initiate a systematic study of concrete combinations of these rules, using tools from proof complexity to establish concrete limitations on what solvers using these rules can achieve.

PB solvers typically perform poorly on inputs in conjunctive normal form (CNF), and it has been known at least since [31,32] that in this case CP with cancelling addition and saturation degenerates into resolution. We observe that strengthening just one of these rules is not enough to solve this problem: CP with cancelling addition and division is easily seen still to be resolution, and resolution can also polynomially simulate the saturation rule plus unrestricted additions as long as the coefficients are of polynomial magnitude. The issue here is that while all versions of CP we consider are *refutationally complete*, meaning that they can prove unsatisfiability of an inconsistent set of constraints, the subsystems of CP are not *implicationally complete*, i.e., even though some linear constraint is implied by a set of other constraints there might be no way of deriving it. This makes reasoning in these subsystems very sensitive to exactly how the input is encoded. Thus, a strong conceptual message of our paper is that in order to function robustly over a wide range of input formats (including, in particular, CNF), PB solvers will need to explore a stronger set of reasoning rules.

In a further attempt to understand the relative strength of these subsystems of cutting planes, we present some (to the best of our knowledge) new combinatorial formulas encoding NP-complete problems, but with the concrete instances chosen to be "obviously" unsatisfiable. We then investigate these formulas, as well as the even colouring formulas in [37], from the point of view of proof complexity. Most of these formulas have very short and simple proofs in general cutting planes, and these proofs are even tree-like. With some care the applications of addition in these proofs can also be made cancelling, but having access to the division rule rather than the saturation rule appears critical. Although we are not able to establish any formal separations between the subsystems of cutting planes that we study (other than for the special case of CNF inputs as noted above), we propose a couple of formulas which we believe are promising candidates for separations. Obtaining such results would require fundamentally new techniques, however, since the tools currently available for analysing CP cannot distinguish between subsystems defined in terms of different sets of syntactic rules.[2]

---

[1] There is now an updated version of *cdcl-cuttingplanes* called *RoundingSat* [24], but any theoretical claims we make in this paper hold for this new version also.

[2] Essentially all lower bound proofs for CP work for any semantically sound proof system operating on pseudo-Boolean constraints, completely ignoring the syntactic rules, and the one exception [25] that we are aware of uses a very specific trick to separate fully semantic (and non-algorithmic) CP from the syntactic version.

We also consider these formulas for other ranges of parameter values and show that for such values the formulas are very easy even for the weakest subsystems of CP that we consider. This would seem to imply that solving such instances should be well within reach for *cdcl-cuttingplanes* and *Sat4j*. However, as reported in [22] many of these instances are instead very challenging in practice. This suggests that in order to make significant advances in pseudo-Boolean solving one crucial aspect is to make full use of the division rule in cutting planes, and we believe that further study of these benchmarks is a promising approach for gaining a deeper understanding of the theoretical reasoning power of pseudo-Boolean solvers implementing conflict-driven proof search.

Organization of This Paper. We discuss conflict-driven proof search in resolution and cutting planes and give formal definitions of proof systems in Sect. 2. In Sect. 3 we prove simulation results for different subsystems of CP, and in Sect. 4 we present our new combinatorial formulas providing candidates for separations. We make some brief concluding remarks in Sect. 5. We refer the reader to the upcoming full-length version of the paper for all missing proofs.

2   Proof Systems for Pseudo-Boolean SAT Solving

Let us start by giving a more formal exposition of the proof systems studied in this paper. Our goal in this section is to explain to complexity theorists without much prior exposure to applied SAT solving how these proof systems arise naturally in the context of pseudo-Boolean (PB) solving, and to this end we start by reviewing resolution and conflict-driven clause learning (CDCL) solvers. By necessity, our treatment is very condensed, but an excellent reference for more in-depth reading on PB solving is [47], and more details on proof complexity material relevant to this paper can be found, e.g., in [41, 42].

We use the standard notation $\mathbb{N} = \{0, 1, 2, 3, \ldots\}$ and $\mathbb{N}^+ = \mathbb{N} \backslash \{0\}$ for natural numbers and positive natural numbers, respectively, and write $[n] = \{1, 2, \ldots, n\}$ and $[n, m] = \{n, n + 1, \ldots, m\}$ for $m, n \in \mathbb{N}^+$, $m > n$.

Resolution and Conflict-Driven Clause Learning. Throughout this paper we identify 1 with *true* and 0 with *false*. A *literal* over a Boolean variable x is either a *positive literal* x or a *negative* or *negated literal* $\bar{x}$. It will also be convenient to write x^σ, $\sigma \in \{0, 1\}$, to denote $x^1 = x$ and $x^0 = \bar{x}$. A *clause* $C = \ell_1 \vee \cdots \vee \ell_k$ is a disjunction of literals over pairwise disjoint variables. A *CNF formula* $F = C_1 \wedge \cdots \wedge C_m$ is a conjunction of clauses. We write *Vars*(F) to denote the set of variables appearing in a formula F. We think of clauses and formulas as sets, so that order is irrelevant and there are no repetitions.

We can represent a partial truth value assignment ρ as the set of literals set to true by ρ. We write $\rho(x^\sigma) = 1$ if $x^\sigma \in \rho$, $\rho(x^\sigma) = 0$ if $x^{1-\sigma} \in \rho$, and $\rho(x^\sigma) = *$ otherwise (i.e., when ρ does not assign any truth value to x). A clause C is satisfied by ρ if it contains some literal set to true by ρ; falsified if ρ sets all literals in C to false; and undetermined otherwise. The *restricted* clause $C{\restriction}_\rho$ is the trivial clause 1 if ρ satisfies C and otherwise C with all literals falsified by ρ removed, i.e., $C\!\restriction_\rho = C \setminus \{x^\sigma \mid x^{1-\sigma} \in \rho\}$. A *unit clause* is a clause with only one literal. We say that C is *unit under* ρ if $C\!\restriction_\rho = \{x^\sigma\}$ is a unit clause, and if so C is also said to *propagate* x^σ *under* ρ.

A *resolution refutation* π of F is a sequence of clauses $\pi = (D_1, D_2, \ldots, D_L)$ such that $D_L = \bot$ is the empty clause without literals and each D_i is either an *axiom clause* $D_i \in F$ or a *resolvent* on the form $D_i = B \vee C$ derived from $D_j = B \vee x$ and $D_k = C \vee \bar{x}$ for $j, k < i$ by the *resolution rule*

$$\frac{B \vee x \qquad C \vee \bar{x}}{B \vee C} \, . \tag{1}$$

It is sometimes convenient to add also a *weakening rule*

$$\frac{B}{B \vee C} \, , \tag{2}$$

which allows to derive any strictly weaker clause from an already derived clause, but it is not hard to show that any use of weakening in a resolution refutation can be eliminated without loss of generality. It is a standard fact that resolution is *implicationally complete*, meaning that a clause C can be derived from a formula F if and only if F semantically implies C.[3] In particular, F is unsatisfiable if and only if there exists a resolution refutation of F.

The *length* $L(\pi)$ of a refutation π is the number of clauses in it. Viewing the list of clauses π as annotated with explanations how they were obtained, we define an associated directed acyclic graph (DAG) G_π with vertices $\{v_1, v_2, \ldots, v_L\}$ labelled by the clauses $\{D_1, D_2, \ldots, D_L\}$ and with edges from resolved clauses to resolvents. We say that π is *tree-like* if G_π is a tree, or, equivalently, if every clause D_i is used at most once as a premise in the resolution rule (repetitions of clauses in π are allowed; i.e., different vertices can be labelled by the same clause). The *(clause) space* at step i in π is the number of clauses $D_j, j < i$ used to obtain resolvents $D_{j'}$, $j' \geq i$, plus 1 for the clause D_i itself, and the space $Sp(\pi)$ of the refutation is the maximal space at any step in π.

Turning next to CDCL solvers, we give a simplified description below that is sufficient for our needs—a more complete (theoretical) treatment can be found in [23]. In one sentence, a CDCL solver running on a CNF formula F repeatedly decides on variable assignments and propagates values that follow from such assignments until a clause is falsified, at which point a learned clause is added to the clause database $\mathcal{D}$ (where we always have $F \subseteq \mathcal{D}$) and the search backtracks. In a bit more detail, the solver maintains a current partial assignment ρ, where every assignment also has a *decision level* (the initial state is at decision level 0 with $\rho = \emptyset$ and $\mathcal{D} = F$). If there is a clause $C \in \mathcal{D}$ that is unit under ρ, the solver adds the propagated literal $x^\sigma = C\!\restriction_\rho$ to ρ with *reason clause* C and repeats the check for unit clauses until either (i) some clause $D \in \mathcal{D}$ is falsified

---

[3] In case the definition of resolution without weakening is used, the notion of implicational completeness is adapted in the natural way to mean that resolution can derive either C or some clause C' that *subsumes* C, i.e., such that $C' \subsetneq C$.

by the current assignment (referred to as a *conflict clause*), or else (ii) there are no propagating clauses. In the latter case the solver makes a *decision* $y = \nu$ and adds y^ν to ρ with decision level increased by 1 (unless there are no more variables left, in which case ρ is a satisfying assignment for F). In the former case, the solver instead performs a *conflict analysis* as described next.

Suppose for concreteness that the last propagated literal in ρ before reaching the conflict clause D was x^σ with reason clause $C = C^* \vee x^\sigma$. Since this propagation caused a conflict the variable x must appear with the opposite sign in D, which can hence be written on the form $D = D^* \vee x^{1-\sigma}$. The solver can therefore resolve $C^* \vee x^\sigma$ and $D^* \vee x^{1-\sigma}$ to get $D' = C^* \vee D^*$, after which x^σ is removed from ρ. We refer to D' as the new *conflict-side clause*. During conflict analysis the conflict-side clause D' is resolved in reverse chronological order with the reason clauses propagating literals in D' to false, and these literals are removed one by one from ρ. An important invariant during this process is that the current conflict-side clause is always falsified by the partial assignment ρ after removing the literal just resolved over. Therefore, every derived clause on the conflict side provides an "explanation" why the corresponding partial assignment fails.

The conflict analysis loop ends when the conflict-side clause contains only one literal from the current decision level at which point the solver *learns* this clause and adds it to the database $\mathcal{D}$. By the invariant, this learned clause is still falsified by ρ, and so the solver removes further literals from ρ in reverse chronological order until the decision level decreases to that of the second largest decision level of any literal in the learned clause. At this point the solver returns from conflict analysis and resumes the main loop described above. By design it now holds that the newly learned clause immediately causes unit propagation, flipping some previously assigned literal to the opposite value. Learned clauses having this property are called *asserting*, and a common feature of essentially all clause learning schemes used in practice is that they learn such asserting clauses.

The CDCL solver terminates either when it finds a satisfying assignment or when it detects unsatisfiability by learning the empty clause $\bot$ (or, more precisely, when it reaches a conflict at decision level 0, in which case the conflict analysis is guaranteed to derive the empty clause). There are, of course, lots of details that we are omitting above. The important conclusions, as we prepare to generalize the description of CDCL to a pseudo-Boolean context, is that the CDCL solver decides on variables and propagates values based on the clauses currently in the database, and that when a conflict is reached a new clause is added to the database obtained by a resolution derivation from the conflict and reason clauses. This means that from any run of CDCL on an unsatisfiable formula F we can extract a resolution refutation of F.

Cutting Planes and Pseudo-Boolean Solving. Recall that throughout this paper we are considering pseudo-Boolean constraints encoded as linear inequalities over Boolean variables with integral coefficients (and all linear inequalities discussed are assumed to be over $\{0, 1\}$-valued variables unless stated otherwise). In order to give a description of cutting planes that is suitable when we want to reason about pseudo-Boolean solvers, it is convenient to keep negated literals as objects in their own right, and to insist that all inequalities consist of positive linear combinations of literals. Therefore, we will write all linear constraints in *normalized form*

$$\sum_{i\in[n],\,\sigma\in\{0,1\}} a_i^\sigma x_i^\sigma \geq A \ , \tag{3}$$

where for all $a_i^\sigma \in \mathbb{N}$ with $i \in [n]$ and $\sigma \in \{0,1\}$ at least one of a_i^0 or a_i^1 equals 0. (variables occur only with one sign in any given inequality), and where the right-hand constant term $A \in \mathbb{N}$ is called the *degree of falsity* (or just *degree*). Note that the normalization is only a convenient form of representation and does not affect the strength of the proof system. If the input is a CNF formula F we just view every clause $C = x_1^{\sigma_1} \vee \cdots \vee x_w^{\sigma_w}$ as a linear constraint $x_1^{\sigma_1} + \cdots + x_w^{\sigma_w} \geq 1$, i.e., a constraint on the form (3) with $a_i^\sigma \in \{0,1\}$ and $A = 1$.

When generalizing CDCL to a pseudo-Boolean setting we want to build a solver that decides on variable values and propagates forced values until conflict, at which point a new linear constraint is learned and the solver backtracks. The main loop of a conflict-driven PB solver can be made identical to that of a CDCL solver, except that we change the word "clause" to "constraint." However, a naive generalization of the conflict analysis does not work. For an example of this, suppose we have $\rho = \{\overline{x}_1, x_2, \overline{x}_3\}$ under which $x_1 + 2\overline{x}_2 + x_3 + 2x_4 + 2x_6 \geq 3$ unit propagates x_6 to true, causing a conflict with $x_3 + 2x_5 + 2\overline{x}_6 \geq 3$. By analogy with the CDCL conflict analysis, we "resolve" (i.e., add and normalize) these two constraints to eliminate x_6, yielding $x_1 + 2\overline{x}_2 + 2x_3 + 2x_4 + 2x_5 \geq 3 + 3 - 2 = 4$ (since $2x_6 + 2\overline{x}_6 = 2$). But now the important invariant that the derived constraint is falsified by the current partial assignment fails, because the new constraint is not falsified by $\rho = \{\overline{x}_1, x_2, \overline{x}_3\}$! There are different ways of modifying the pseudo-Boolean conflict analysis to address this problem, and these different approaches are partly reflected in the different proof systems studied in this paper.

Starting with the most general version of the *cutting planes* proof system used in the proof complexity literature, using the normalized form (3) we can define the derivation rules[4] to be *literal axioms*

$$\frac{}{x_i^\sigma \geq 0} \ , \tag{4a}$$

linear combination

$$\frac{\sum_i a_i^\sigma x_i^\sigma \geq A \qquad \sum_i b_i^\sigma x_i^\sigma \geq B}{\sum_i (\alpha a_i^\sigma + \beta b_i^\sigma) x_i^\sigma \geq \alpha A + \beta B} \quad \alpha, \beta \in \mathbb{N}^+ \ , \tag{4b}$$

and *division*

$$\frac{\sum_i a_i^\sigma x_i^\sigma \geq A}{\sum_i \lceil a^\sigma / \alpha \rceil x_i^\sigma \geq \lceil A/\alpha \rceil} \quad \alpha \in \mathbb{N}^+ \ , \tag{4c}$$

---

[4] Attentive readers might note that division looks slightly stronger in our definition than the standard rule in the proof complexity literature, but the two versions are easily verified to be equivalent up to a linear factor in length. It is important to note that multiplication is only ever performed in combination with addition.

where in the linear combination rule we tacitly assume that the cancellation rule $x^\sigma + x^{1-\sigma} = 1$ is applied to bring the derived constraint into normalized form, as in the example we just saw. Just as in this example, for any linear combination that arises during conflict analysis it will be the case that there is a literal x_i^σ for which $\alpha a_i^\sigma = \beta b_i^{1-\sigma} > 0$. We say that this is an instance of *cancelling linear combination* since the variable x_i vanishes, and we also require for such linear combinations that α and β are chosen so that $\alpha a_i^\sigma = \beta b_i^{1-\sigma}$ is the least common multiple of a_i^σ and $b_i^{1-\sigma}$. We remark that this is also referred to as *generalized resolution* in the literature [31,32], since it is a natural generalization of (1) from disjunctive clauses to general linear constraints, and we will sometimes refer to the resulting constraint as a *(generalized) resolvent*.

We want to highlight that in the division rule (4c) we can divide *and round up* to the closest integer, since we are only interested in $\{0,1\}$-valued solutions. This division rule is where the power of cutting planes lies. And indeed, this is how it must be, since the other rules are sound also for real-valued variables, and so without the division rule we would not be able to distinguish sets of linear inequalities that have real-valued solutions but no $\{0,1\}$-valued solutions.

Pseudo-Boolean solvers such as *Sat4j* [36,48] and *cdcl-cuttingplanes* [21] do not implement the full set of cutting planes derivation rules as described above, however. In proofs generated by these solvers the linear combinations will always be *cancelling*. Division is used in *cdcl-cuttingplanes* only in a restricted setting to ensure that the learned constraint is always conflicting, and *Sat4j* omits this rule pretty much completely and instead applies the *saturation* rule

$$\frac{\sum_{(i,\sigma)} a_i^\sigma x_i^\sigma \geq A}{\sum_{(i,\sigma)} \min\{a_i^\sigma, A\} \cdot x_i^\sigma \geq A} , \tag{5a}$$

saying that no coefficient on the left need be larger than the degree on the right. (For instance, saturation applied to $3x_1 + x_2 + x_3 \geq 2$ yields that $2x_1 + x_2 + x_3 \geq 2$ holds.) As the division rule, the saturation rule is sound only for integral solutions. It is an interesting question how the division and saturation rules are related. Saturation can be simulated by division, but it is not clear whether this simulation can be made efficient in general. In the other direction, we give examples in this paper of when division is exponentially stronger than saturation.

We remark that another rule that is important in practice is *weakening*

$$\frac{\sum_{(i,\sigma)} a_i^\sigma x_i^\sigma \geq A}{\sum_{(i,\sigma) \neq (i^*,\sigma^*)} a_i^\sigma x_i^\sigma \geq A - a_{i^*}^{\sigma^*}} , \tag{5b}$$

which—perhaps somewhat counter-intuitively—is used during conflict analysis to maintain the invariant that the constraint being learned is conflicting with respect to the current partial assignment. In contrast to the weakening rule in resolution, the rule (5b) is crucial for pseudo-Boolean solvers, but since this rule can be implemented using (4a) and a cancelling linear combination we do not need to include it in our formal proof system definitions.

In order to try to understand the reasoning power of pseudo-Boolean solvers such as *cdcl-cuttingplanes* and *Sat4j*, in this paper we study the following four subsystems of cutting planes, where for brevity we will write just *cancellation* instead of *cancelling linear combination*:

General CP: Rules (4a), (4b), and (4c).
CP with saturation: Rules (4a), (4b), and (5a).
CP with saturation and cancellation: Rules (4a) and (5a) plus the cancelling version of (4b); essentially corresponding to *Sat4j*.
CP with division and cancellation: Rules (4a) and (4c) plus the cancelling version of (4b); strong enough to capture *cdcl-cuttingplanes*.

General cutting planes is refutationally complete in that it can disprove any inconsistent set of linear inequalities [28]: One can show that there is no $\{0, 1\}$-valued solution by using the cutting planes rules (4a)–(4c) to derive the contradiction $0 \geq A$ for some $A > 0$, which is the pseudo-Boolean equivalent of the empty clause, from the given linear inequalities. The *length* of such a cutting planes refutation is the total number of inequalities in it, and the *size* also sums the sizes of all coefficients (i.e., the bit size of representing them). We can also define a *line space* measure analogous to the clause space measure counting the number of inequalities in memory during a proof.

It is not hard to show—as we will argue shortly—that the three restricted versions of CP defined above are also refutationally complete. However, while general cutting planes is also implicationally complete [10], meaning that it can derive any inequality that is implied by a set of linear equations, the subsystems we consider are not even *weakly implicationally complete*.

Let us pause to explain what we mean by this terminology. For disjunctive clauses C and D it is not hard to see that the only way C can imply D is if $C \subseteq D$. In a pseudo-Boolean context, however, there are infinitely many ways to express a linear threshold function over the Boolean hypercube as a linear inequality (for instance, by multiplying the inequality by an arbitrary positive integer). We say, therefore, that a PB proof system is *weakly implicationally complete* if when some set of inequalities implies $\sum_{(i,\sigma)} a_i^\sigma x_i^\sigma \geq A$ it holds that the proof system can derive some potentially syntactically different inequality $\sum_{(i,\sigma)} b_i^\sigma x_i^\sigma \geq B$ implying $\sum_{(i,\sigma)} a_i^\sigma x_i^\sigma \geq A$, and that it is *(strongly) implicationally complete* if it can derive a constraint on the exact syntactic form $\sum_{(i,\sigma)} a_i^\sigma x_i^\sigma \geq A$.

Returning to our previous discussion, given the constraint $\sum_{i=1}^k x_i \geq d$ written as a set of disjunctive clauses $\{\sum_{i \in S} x_i \geq 1 \,|\, S \subseteq [k], |S| = k - d + 1\}$ (in pseudo-Boolean notation), it is not hard to see that there is no way CP with cancellation can derive any inequality implying the former encoding from the constraints in the latter encoding [31].[5] A slightly less obvious fact, which we

---

[5] This is so since the only possibility to apply cancelling linear combinations is to use literal axioms (4a) yielding (trivial) constraints on the form $\sum_{i \in S} x_i \geq 0$ for $|S| \geq 0$, and the set of such constraints is invariant under both division and saturation.

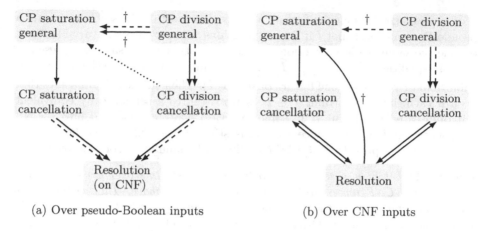

(a) Over pseudo-Boolean inputs   (b) Over CNF inputs

Fig. 1. Relations between proof systems. $A \longrightarrow B$: A polynomially simulates B; $A\text{-}\blacktriangleright B$: B cannot simulate A (there is an exponential separation); $A\cdots\blacktriangleright B$: candidate for a separation, †: known only for coefficients of polynomial magnitude.

shall prove in Sect. 3, is that even with general addition and saturation it is not possible to recover a cardinality constraint from its CNF encoding.

We want to emphasize again that we make no claims of originality when it comes to defining the derivation rules—they arise naturally in the context of pseudo-Boolean solving, and indeed all of them are described in [47]. However, we are not aware of any previous work defining and systematically studying the subsystems of CP described above from a proof complexity point of view, i.e., proving upper and lower bounds on proof resources. This is the purpose of the current paper, and we study the strength of these proof systems both for CNF inputs and general (linear) pseudo-Boolean inputs.

As a final remark for completeness, we want to point out that one further important rule, which is used, e.g., in [9], is *rounding to cardinality constraints* We leave as future work a study of formal proof systems using this rule.

3   Relations Between Subsystems of Cutting Planes

We now proceed to examine how having saturation instead of division and/or requiring linear combinations to be cancelling affects the reasoning power of cutting planes. The conclusions of this section are pictorially summarized in Fig. 1.

For starters, it is an easy observation that all the subsystems of cutting planes that we consider can simulate resolution when the input is in CNF, and we show that this is still the case when we start with a pseudo-Boolean input for cutting planes and the straightforward encoding into CNF of that input for resolution. This is immediate if we have the division rule, but in fact it is not hard to prove that the simulation also works with saturation.

Let us make these claims formal. We say that a set of clauses $\widehat{I}$ *represents* a pseudo-Boolean constraint I if both expressions are over the same variables[6] and encode the same Boolean function, and a CNF formula $\widehat{F}$ is said to represent a set of inequalities F if $\widehat{F} = \bigcup_{I \in F} \widehat{I}$ (where it is important to note that each CNF subformula $\widehat{I}$ represents one linear constraint I). Then the next lemma says that even if each linear constraint $I \in F$ is rewritten to a semantically equivalent but obfuscated constraint I' in some awkward way, but encoded into a CNF representation $\widehat{F}$ in some nice way, it is still the case that even the weakest version of CP applied to $F' = \bigcup I'$ can efficiently simulate resolution on $\widehat{F}$.

Lemma 1. *Let F be a set of pseudo-Boolean constraints over n variables and let $\widehat{F}$ be any CNF representation of F as described above. Then if there is a resolution refutation $\widehat{\pi}$ of $\widehat{F}$ in length L and clause space s, there is also a CP refutation π of F in length $O(nL)$ and line space $s + O(1)$ using only cancellation and saturation. If $\widehat{\pi}$ is tree-like, then π is also tree-like.*

It follows from Lemma 1 that CP with saturation is refutationally complete.

Corollary 2. *Any unsatisfiable set of pseudo-Boolean constraints over n variables has a tree-like CP refutation in length $O(n2^n)$ and line space $O(n)$ using cancellation and saturation.*

In the other direction, cutting planes with cancellation is equivalent to resolution when restricted to CNF inputs, and this is so regardless of whether division or saturation is used. The reason for this is that cancelling linear combinations of disjunctive clauses can only produce inequalities with degree of falsity 1, which are equivalent to clauses. This is essentially just an observation from [31] rewritten in the language of proof complexity, but let us state it here for the record.

Lemma 3. *If cutting planes with cancellation and either division or saturation can refute a CNF formula F in length L and line space s, then there is a resolution refutation of F in length L and clause space s.*

We can use this observation to show that systems allowing general linear combinations can be strictly stronger than systems with cancellation. To see this, consider *subset cardinality formulas* [39,50,53] defined in terms of $0/1$ $n \times n$ matrices $A = (a_{i,j})$, which have variables $x_{i,j}$ for all $a_{i,j} = 1$ and constraints claiming that in each row there is a majority of positive variables but in each column there is a majority of negative variables, i.e.,

$$\sum_{j \in R_i} x_{i,j} \geq \lceil |R_i|/2 \rceil \qquad\qquad i \in [n] \qquad\qquad (6a)$$

$$\sum_{i \in C_j} x_{i,j} \leq \lfloor |C_j|/2 \rfloor \qquad\qquad j \in [n] \qquad\qquad (6b)$$

where $R_i = \{j \mid a_{i,j} = 1\}$ and $C_j = \{i \mid a_{i,j} = 1\}$. In the case when all rows and columns have $2k$ variables, except for one row and column that have $2k + 1$

---

[6] We do not allow encodings with extension variables, since then formulas are no longer semantically equivalent and it becomes very hard to make meaningful comparisons.

variables, these formulas are unsatisfiable and are easily refutable in general CP, but if the matrix is *expanding* in a certain sense, then resolution proofs require exponential length [39]. This yields the following corollary of Lemma 3.

Corollary 4. *There are formulas on n variables that can be refuted in length $O(n)$ in general CP but require length $\exp(\Omega(n))$ in CP with cancellation.*

When it comes to comparing division versus saturation, it was observed in [9] that saturation can be simulated by repeated division. Working out the details, we obtain the following proposition.

Proposition 5. *If a set of pseudo-Boolean constraints has a CP refutation with saturation in length L and coefficients bounded by A, then there is a CP refutation with division in length AL.*

We remark that a direct simulation may lead to an exponential blow-up if the proof uses coefficients of exponential magnitude.

Our main contribution in this section is to show that when the input is in CNF, then cutting planes proofs with saturation and unrestricted addition can in fact be efficiently simulated by resolution assuming that all CP coefficients are of polynomial magnitude. Observe that this last condition also implies that the the the degree of falsity has polynomial magnitude, which is the slightly more precise assumption used in the next theorem.

Theorem 6. *If a CNF formula F has a CP refutation π with saturation in length L and every constraint in π has degree of falsity at most A, then there is a resolution refutation of F in length $O(AL)$.*

We can then use subset cardinality formulas again to separate CP with division from CP with saturation. The formal claim follows below, where the constant hidden in the asymptotic notation depends on the size of the coefficients.

Corollary 7. *There are formulas on n variables that can be refuted in length $O(n)$ in general CP but require length $\exp(\Omega(n))$ in CP with saturation if all coefficients in the proofs have polynomial magnitude.*

The idea behind the proof of Theorem 6 is to maintain for every inequality with degree of falsity A a set of A clauses that implies the inequality. We simulate linear combination steps by resolving the sets of clauses corresponding to the two inequalities over the variables that cancel, and we do not do anything for saturation steps.

Note that this approach does not work if the input is not in CNF. For instance, if we start with the pseudo-Boolean constraint $x + y + z \geq 2$ with degree of falsity 2, which is equivalent to the clauses $(x \vee y) \wedge (y \vee z) \wedge (x \vee z)$, then it is not possible to pick any 2 out of these 3 clauses that would imply the inequality.

We remark that we do not know of any separation between CP with saturation and division except those exhibited by CNF formulas. These separations are somewhat artificial in that they crucially use that the implicationally incomplete subsystems of CP cannot recover the the cardinality constraints "hidden" in the CNF encodings. In Sect. 4 we propose more natural candidates for separations between CP with division and cancellation and CP with saturation, where the difficulty would not be due to an "obfuscated" CNF encoding.

We conclude this section by the observation that any version of CP considered in this paper can easily refute any set of linear constraints that define an empty polytope over the reals, i.e., for which there is no real-valued solution. For general addition this is an immediate consequence of Farkas' lemma, and we can make the additions cancelling using the Fourier–Motzkin variable elimination procedure.

Lemma 8. *If a set of linear inequalities on n variables defines an empty polytope over the reals, then there is a tree-like CP refutation using only addition in length $O(n)$ and space $O(1)$, and a CP refutation using only cancelling addition in length $O(n^2)$ and space $O(n)$.*

As a consequence of Corollary 7 and Lemma 8 we obtain the following theorem.

Theorem 9. *CP with saturation is not (even weakly) implicationally complete.*

4   Tricky Formulas Based on Easy **NP** Instances

In this section we present candidates for achieving separations between the subsystems of cutting planes studied in this paper, and where these separations would not be a consequence of presenting pseudo-Boolean constraints as "obfuscated" CNF formulas but would highlight fundamental differences in pseudo-Boolean reasoning power between the proof systems.

All of our candidate formulas have short proofs for CP with division (and all refutations have constant-size coefficients unless stated otherwise), but for appropriately chosen parameter values it seems plausible that some of them are not possible to refute efficiently using the saturation rule. We also show that it is possible to chose other parameter values for these formulas to generate instances that are very easy in theory even for the weakest subsystem of CP that we consider. This is in striking contrast to what one can observe empirically when running pseudo-Boolean solvers on these instances, as reported in [22]—in practice, many of these theoretically easy instances appear to be very challenging.

Even Colouring. The even colouring formula $EC(G)$ [37] over a connected graph $G = (V, E)$ with all vertices of even degree consists of the constraints

$$\sum_{e \in E(v)} x_e = \deg(v)/2 \qquad\qquad v \in V \qquad\qquad (7)$$

(where $E(v)$ denotes the set of edges incident to v), claiming that each vertex has an equal number of incident 0- and 1-edges. The formula is unsatisfiable if and only if $|E|$ is odd, which we assume is always the case it what follows.

Even colouring formulas have short CP proofs: just add all positive and negative inequalities separately, divide by 2 and round up, and add the results. We can make the additions cancelling by processing inequalities in breadth-first order, alternating between positive and negative inequalities.

Proposition 10. *Tree-like CP with division and cancellation can refute $EC(G)$ in length $O(n)$ and space $O(1)$.*

If the graph is t-*almost bipartite*, by which we mean that removing t edges yields a bipartite graph, then we can make the proof work with saturation instead of division at the price of an exponential blow-up in t (which becomes a constant factor if t is constant).

Proposition 11. *If G is a t-almost bipartite graph then the formula $EC(G)$ can be refuted in length $O(2^t + n)$ and space $O(t)$ by CP with saturation and cancellation, and the refutation can be made tree-like in length $O(2^t n)$.*

An example of such graphs are rectangular $m \times n$ grids (where edges wrap around the borders to form a torus) and where we subdivide one edge into a degree-2 vertex to get an odd number of edges. If both m and n are even, then the graph is bipartite except for 1 edge, so we have cutting planes proofs with saturation and cancellation of length $O(mn)$, and if m is even and n is odd, then the graph is bipartite except for $m + 1$ edges, so we have proofs of length $O(2^m + mn)$. In all cases even colouring formulas on grids have resolution refutations of length $2^{O(m)} n$ and space $2^{O(m)}$ which we can simulate.

We conjecture that these formulas are exponentially hard for CP with saturation when the graph is a square grid of odd side length, i.e., $m = n = 2\ell + 1$ (so that the graph is far from bipartite), or is a $2d$-regular random graph.

Vertex Cover. Recall that a *vertex cover* of a graph $G = (V, E)$ is a subset of vertices $V' \subseteq V$ such that every edge $(u, v) \in E$ is incident to some vertex in V'. A graph G has a vertex cover of size at most $S \in \mathbb{N}^+$ if and only if the formula $VC(G, S)$ given by the constraints

$$x_u + x_v \geq 1 \qquad\qquad (u, v) \in E; \qquad\qquad (8a)$$

$$\sum_{v \in V} x_v \leq S \qquad\qquad\qquad\qquad\qquad\qquad (8b)$$

has a $\{0, 1\}$-valued solution.

We consider vertex cover instances over grid graphs $R_{m,n}$ Since a grid has degree 4 any cover must have size at least $mn/2$. This bound is not achievable when one dimension, say n, is odd, in which case the minimal cover size is $m\lceil n/2 \rceil$. We can choose the parameter S in (8b) in the interval $[mn/2, m\lceil n/2 \rceil - 1]$ to obtain unsatisfiable formulas with different levels of overconstrainedness.

Vertex cover formulas have short cutting planes proofs: add the horizontal edge inequalities (8a) for every row, divide by 2 (which rounds up the degree of falsity), and add all of these inequalities to find a contradiction with the upper bound (8b), and these additions can be reordered to be made cancelling.

Proposition 12. *CP with division and cancellation can refute $VC(R_{m,n}, S)$ with n odd and $S < m\lceil n/2 \rceil$ in length $O(mn)$ and space $O(1)$.*

A similar approach works with saturation instead of division, but since we cannot round up every row we need a stronger cover size constraint (8b).

Proposition 13. *Tree-like CP with saturation and cancellation is able to refute $VC(R_{m,n}, S)$ with n odd and $S \leq \lfloor mn/2 \rfloor$ in length $O(mn)$ and space $O(1)$.*

Alternatively, using what we find to be a rather nifty approach it turns out to be possible to derive all the 2^m clauses over the m variables corresponding to vertices in the first column, after which one can simulate a brute-force resolution refutation of this formula.

Proposition 14. *Tree-like CP with saturation and cancellation is able to refute $VC(R_{m,n}, S)$ with n odd and $S < m\lceil n/2 \rceil$ in length $O(2^m mn)$ and space $O(m)$.*

We conjecture that the exponential gap between Propositions 12 and 14 for $m = \Theta(n)$ and $S = m\lceil n/2 \rceil - 1$ is real and is due to the weakness of saturation.

Dominating Set. A *dominating set* of a graph $G = (V, E)$ is a subset of vertices $V' \subseteq V$ such that every vertex in $V \setminus V'$ has a neighbour in V'. G has a dominating set of size $S \in \mathbb{N}^+$ if and only if there is a $\{0, 1\}$-valued solution to the set of constraints $DS(G, S)$ defined as

$$x_v + \sum_{u \in N(v)} x_u \geq 1 \qquad\qquad v \in V; \qquad\qquad (9a)$$

$$\sum_{v \in V} x_v \leq S \qquad\qquad\qquad (9b)$$

We consider dominating set formulas over hexagonal grid graphs $H_{m,n}$, which can be visualized as brick walls. As it turns out these formulas have short proofs even in CP with saturation and cancellation, but the proofs are not obvious and the formulas have a surprisingly rich structure and present particularly challenging benchmarks in practice.

Since a hexagonal grid has degree 3, the minimum size of a dominating set is $\lceil |V|/4 \rceil = \lceil mn/4 \rceil$, so we set $S = \lfloor mn/4 \rfloor$. Whether these formulas are satisfiable depends on the largest power of 2 that divides m and n—also known as the 2-adic valuation or v_2. Formulas where $v_2(mn) = 1$ are unsatisfiable and can be refuted by adding all inequalities, and these additions can be made cancelling with some care.

Proposition 15. *Tree-like CP with cancellation can refute $DS(H_{m,n}, \lfloor mn/4 \rfloor)$ with $v_2(mn) = 1$ in length $O(mn)$ and space $O(1)$.*

Formulas where $v_2(mn) = 2$ are unsatisfiable and the proof follows by dividing the resulting inequalities in the previous proof by 2 and rounding up.

Proposition 16. *Tree-like CP with division and cancellation is able to refute $DS(H_{m,n}, mn/4)$ with $v_2(mn) = 2$ in length $O(mn)$ and space $O(1)$.*

When $v_2(n) \geq 2$ the dominating set must in fact define a *tiling* of the hexagonal grid. If furthermore $v_2(m) \geq 1$ then formulas are satisfiable. Among the remaining formulas some are satisfiable and some are not, and the next lemma sums up our knowledge in this matter.

Lemma 17. *Dominating set formulas over hexagonal grids are unsatisfiable if*

- $v_2(m) \geq 2$ *and* $v_2(n) = 1$, *or*
- $v_2(m) = 0$ *and* $v_2(n) \geq 3$ *and* $v_2(n) \leq v_2(4 \lfloor m/4 \rfloor)$, *or*
- $v_2(n) = 0$ *and* $v_2(m) \geq 3$ *and* $v_2(m) \leq v_2(4 \lfloor n/4 \rfloor)$.

We conjecture that Lemma 17 in fact provides an exact characterization.

To find CP refutations of the unsatisfiable dominating set instances, we can derive tiling constraints $x_v + \sum_{u \in N(v)} = 1$ for all vertices using only cancelling addition. CP with saturation and cancellation can then easily refute these formulas with tiling constraints in polynomial length.

Proposition 18. *If $DS(H_{m,n}, mn/4)$ is as in Lemma 17, then it can be refuted in length $O((nm)^2)$ in CP with saturation and cancellation.*

5 Concluding Remarks

In this paper, we investigate subsystems of cutting planes motivated by pseudo-Boolean proof search algorithms. Using tools from proof complexity, we differentiate between the reasoning power of different methods and show that current state-of-the-art pseudo-Boolean solvers are inherently unable to exploit the full strength of cutting planes even in theory, in stark contrast to what is the case for CDCL solvers with respect to resolution.

Some of these limitations are in some sense folklore, in that it is known that pseudo-Boolean solvers perform badly on input in CNF, but we show that this is true for all natural restrictions suggested by current solvers that fall short of full-blown cutting planes reasoning. Also, we propose a number of new crafted benchmarks as a way of going beyond CNF-based lower bounds to study the inherent limitations of solvers even when given natural pseudo-Boolean encodings. We show how the parameters for these benchmarks can be varied to yield versions that appear to be hard or easy for different subsystems of cutting planes.

Although we cannot establish any formal separations between the subsystems of cutting planes studied in this paper—this would seem to require the development of entirely new proof complexity techniques—it is our hope that further investigations of these benchmarks could yield more insights into the power and limitations of state-of-the-art pseudo-Boolean solvers.

Acknowledgements. We are most grateful to Daniel Le Berre for long and patient explanations of the inner workings of pseudo-Boolean solvers, and to João Marques-Silva for helping us get an overview of relevant references for pseudo-Boolean solving. We would like to thank Susanna F. de Rezende, Arnold Filtser, and Robert Robere for helpful discussions on polytopes. We also extend our gratitude to the *SAT 2018* anonymous reviewers for the many detailed comments that helped to improve the paper considerably.

Some empirical pseudo-Boolean solver experiments made within the context of this work were performed on resources provided by the Swedish National Infrastructure for Computing (SNIC) at the High Performance Computing Center North (HPC2N) at Umeå University. For these experiments we also used the tool CNFgen [12,35], for which we gratefully acknowledge Massimo Lauria.

The first author performed part of this work while at KTH Royal Institute of Technology. All authors were funded by the European Research Council under the European Union's Seventh Framework Programme (FP7/2007–2013) / ERC grant agreement no. 279611. The first author was also supported by the Prof. R Narasimhan post-doctoral award, and the fourth and fifth authors received support from Swedish Research Council grants 621-2012-5645 and 2016-00782.

References

1. Alekhnovich, M., Johannsen, J., Pitassi, T., Urquhart, A.: An exponential separation between regular and general resolution. Theory Comput. **3**(5), 81–102 (2007). preliminary version in STOC 2002
2. Alekhnovich, M., Razborov, A.A.: Resolution is not automatizable unless W[P] is tractable. SIAM J. Comput. **38**(4), 1347–1363 (2008). Preliminary version in FOCS 2001
3. Atserias, A., Fichte, J.K., Thurley, M.: Clause-learning algorithms with many restarts and bounded-width resolution. J. Artif. Intell. Res. **40**, 353–373 (2011). Preliminary version in SAT 2009
4. Bayardo Jr., R.J., Schrag, R.: Using CSP look-back techniques to solve real-world SAT instances. In: Proceedings of the 14th National Conference on Artificial Intelligence (AAAI 1997), pp. 203–208, July 1997
5. Beame, P., Kautz, H., Sabharwal, A.: Towards understanding and harnessing the potential of clause learning. J. Artif. Intell. Res. **22**, 319–351 (2004). Preliminary version in IJCAI 2003
6. Ben-Sasson, E., Impagliazzo, R., Wigderson, A.: Near optimal separation of tree-like and general resolution. Combinatorica **24**(4), 585–603 (2004)
7. Blake, A.: Canonical Expressions in Boolean Algebra. Ph.D. thesis, University of Chicago (1937)
8. Bonet, M., Pitassi, T., Raz, R.: Lower bounds for cutting planes proofs with small coefficients. J. Symbolic Logic **62**(3), 708–728 (1997). Preliminary version in STOC 1995
9. Chai, D., Kuehlmann, A.: A fast pseudo-Boolean constraint solver. IEEE Trans. Comput.-Aided Des. Integr. Circuits Syst. **24**(3), 305–317 (2005). Preliminary version in DAC 2003
10. Chvátal, V.: Edmonds polytopes and a hierarchy of combinatorial problems. Discrete Math. **4**(1), 305–337 (1973)
11. Chvátal, V., Szemerédi, E.: Many hard examples for resolution. J. ACM **35**(4), 759–768 (1988)
12. CNFgen: Combinatorial benchmarks for SAT solvers. https://github.com/MassimoLauria/cnfgen
13. Cook, S.A.: The complexity of theorem-proving procedures. In: Proceedings of the 3rd Annual ACM Symposium on Theory of Computing (STOC 1971), pp. 151–158 (1971)

14. Cook, S.A., Reckhow, R.: The relative efficiency of propositional proof systems. J. Symbolic Log. **44**(1), 36–50 (1979)
15. Cook, W., Coullard, C.R., Turán, G.: On the complexity of cutting-plane proofs. Discrete Appl. Math. **18**(1), 25–38 (1987)
16. Davis, M., Logemann, G., Loveland, D.: A machine program for theorem proving. Commun. ACM **5**(7), 394–397 (1962)
17. Davis, M., Putnam, H.: A computing procedure for quantification theory. J. ACM **7**(3), 201–215 (1960)
18. Dixon, H.E., Ginsberg, M.L., Hofer, D.K., Luks, E.M., Parkes, A.J.: Generalizing Boolean satisfiability III: implementation. J. Artif. Intell. Res. **23**, 441–531 (2005)
19. Dixon, H.E., Ginsberg, M.L., Luks, E.M., Parkes, A.J.: Generalizing Boolean satisfiability II: theory. J. Artif. Intell. Res. **22**, 481–534 (2004)
20. Dixon, H.E., Ginsberg, M.L., Parkes, A.J.: Generalizing Boolean satisfiability I: Background and survey of existing work. J. Artif. Intell. Res. **21**, 193–243 (2004)
21. Elffers, J.: CDCL-cuttingplanes: A conflict-driven pseudo-Boolean solver (2016). Submitted to the Pseudo-Boolean Competition 2016
22. Elffers, J., Giráldez-Cru, J., Nordström, J., Vinyals, M.: Using combinatorial benchmarks to probe the reasoning power of pseudo-Boolean solvers. In: Proceedings of the 21st International Conference on Theory and Applications of Satisfiability Testing (SAT 2018), July 2018. To appear
23. Elffers, J., Johannsen, J., Lauria, M., Magnard, T., Nordström, J., Vinyals, M.: Trade-offs between time and memory in a tighter model of CDCL SAT solvers. In: Creignou, N., Le Berre, D. (eds.) SAT 2016. LNCS, vol. 9710, pp. 160–176. Springer, Cham (2016). https://doi.org/10.1007/978-3-319-40970-2_11
24. Elffers, J., Nordström, J.: Divide and conquer: towards faster pseudo-Boolean solving. In: Proceedings of the 27th International Joint Conference on Artificial Intelligence (IJCAI-ECAI 2018), July 2018. To appear
25. Filmus, Y., Hrubeš, P., Lauria, M.: Semantic versus syntactic cutting planes. In: Proceedings of the 33rd International Symposium on Theoretical Aspects of Computer Science (STACS 2016). Leibniz International Proceedings in Informatics (LIPIcs), vol. 47, pp. 35:1–35:13, February 2016
26. Fleming, N., Pankratov, D., Pitassi, T., Robere, R.: Random $\theta(\log n)$-CNFs are hard for cutting planes. In: Proceedings of the 58th Annual IEEE Symposium on Foundations of Computer Science (FOCS 2017), pp. 109–120, October 2017
27. Goerdt, A.: The cutting plane proof system with bounded degree of falsity. In: Proceedings of the 5th International Workshop on Computer Science Logic (CSL 1991), pp. 119–133, October 1991
28. Gomory, R.E.: An algorithm for integer solutions of linear programs. In: Graves, R., Wolfe, P. (eds.) Recent Advances in Mathematical Programming, pp. 269–302. McGraw-Hill, New York (1963)
29. Haken, A.: The intractability of resolution. Theoret. Comput. Sci. **39**(2–3), 297–308 (1985)
30. Hirsch, E.A., Kojevnikov, A., Kulikov, A.S., Nikolenko, S.I.: Complexity of semi-algebraic proofs with restricted degree of falsity. J. Satisfiability Boolean Model. Comput. **6**, 53–69 (2008). Preliminary version in SAT 2005 and SAT 2006
31. Hooker, J.N.: Generalized resolution and cutting planes. Ann. Oper. Res. **12**(1), 217–239 (1988)
32. Hooker, J.N.: Generalized resolution for 0-1 linear inequalities. Ann. Math. Artif. Intell. **6**(1), 271–286 (1992)

33. Hrubeš, P., Pudlák, P.: Random formulas, monotone circuits, and interpolation. In: Proceedings of the 58th Annual IEEE Symposium on Foundations of Computer Science (FOCS 2017), pp. 121–131, October 2017

34. Impagliazzo, R., Pitassi, T., Urquhart, A.: Upper and lower bounds for tree-like cutting planes proofs. In: Proceedings of the 9th Annual IEEE Symposium on Logic in Computer Science (LICS 1994). pp. 220–228, July 1994

35. Lauria, M., Elffers, J., Nordström, J., Vinyals, M.: CNFgen: a generator of crafted benchmarks. In: Gaspers, S., Walsh, T. (eds.) SAT 2017. LNCS, vol. 10491, pp. 464–473. Springer, Cham (2017). https://doi.org/10.1007/978-3-319-66263-3_30

36. Le Berre, D., Parrain, A.: The SAT4J library, release 2.2. J. Satisfiability Boolean Model. Comput. **7**, 59–64 (2010)

37. Markström, K.: Locality and hard SAT-instances. J. Satisfiability Boolean Model. Comput. **2**(1–4), 221–227 (2006)

38. Marques-Silva, J.P., Sakallah, K.A.: GRASP: A search algorithm for propositional satisfiability. IEEE Trans. Comput. **48**(5), 506–521 (1999). Preliminary version in ICCAD 1996

39. Mikša, M., Nordström, J.: Long proofs of (seemingly) simple formulas. In: Sinz, C., Egly, U. (eds.) SAT 2014. LNCS, vol. 8561, pp. 121–137. Springer, Cham (2014). https://doi.org/10.1007/978-3-319-09284-3_10

40. Moskewicz, M.W., Madigan, C.F., Zhao, Y., Zhang, L., Malik, S.: Chaff: engineering an efficient SAT solver. In: Proceedings of the 38th Design Automation Conference (DAC 2001), pp. 530–535, June 2001

41. Nordström, J.: Pebble games, proof complexity and time-space trade-offs. Log. Methods Comput. Sci. **9**(3), 15:1–15:63 (2013)

42. Nordström, J.: On the interplay between proof complexity and SAT solving. ACM SIGLOG News **2**(3), 19–44 (2015)

43. Pipatsrisawat, K., Darwiche, A.: On the power of clause-learning SAT solvers as resolution engines. Artif. Intell. **175**(2), 512–525, February 2011. Preliminary version in CP 2009

44. Pseudo-Boolean competition 2016. http://www.cril.univ-artois.fr/PB16/, July 2016

45. Pudlák, P.: Lower bounds for resolution and cutting plane proofs and monotone computations. J. Symbol. Log. **62**(3), 981–998 (1997)

46. Robinson, J.A.: A machine-oriented logic based on the resolution principle. J. ACM **12**(1), 23–41 (1965)

47. Roussel, O., Manquinho, V.M.: Pseudo-Boolean and cardinality constraints. In: Biere, A., Heule, M.J.H., van Maaren, H., Walsh, T. (eds.) Handbook of Satisfiability, Frontiers in Artificial Intelligence and Applications, vol. 185, chap. 22, pp. 695–733. IOS Press, February 2009

48. SAT4J: The Boolean satisfaction and optimization library in Java. http://www.sat4j.org/

49. Sheini, H.M., Sakallah, K.A.: Pueblo: A hybrid pseudo-Boolean SAT solver. J. Satisfiability Boolean Model. Comput. **2**(1–4), 165–189, March 2006. Preliminary version in DATE 2005

50. Spence, I.: sgen1: A generator of small but difficult satisfiability benchmarks. J. Exp. Algorithmics **15**, 1.2:1–1.2:15, March 2010

51. Urquhart, A.: Hard examples for resolution. J. ACM **34**(1), 209–219 (1987)

52. Urquhart, A.: A near-optimal separation of regular and general resolution. SIAM J. Comput. **40**(1), 107–121 (2011). Preliminary version in SAT 2008

53. Van Gelder, A., Spence, I.: Zero-one designs produce small hard SAT instances. In: Strichman, O., Szeider, S. (eds.) SAT 2010. LNCS, vol. 6175, pp. 388–397. Springer, Heidelberg (2010). https://doi.org/10.1007/978-3-642-14186-7_37

Cops-Robber Games and the Resolution of Tseitin Formulas

Nicola Galesi[1], Navid Talebanfard[2], and Jacobo Torán[3(✉)]

[1] Universita di Roma La Sapienza, Rome, Italy
nicola.galesi@uniroma1.it
[2] Czech Academy of Sciences, Prague, Czech Republic
talebanfard@math.cas.cz
[3] Universität Ulm, Ulm, Germany
jacobo.toran@uni-ulm.de

Abstract. We characterize several complexity measures for the resolution of Tseitin formulas in terms of a two person cop-robber game. Our game is a slight variation of the one Seymour and Thomas used in order to characterize the tree-width parameter. For any undirected graph, by counting the number of cops needed in our game in order to catch a robber in it, we are able to exactly characterize the width, variable space and depth measures for the resolution of the Tseitin formula corresponding to that graph. We also give an exact game characterization of resolution variable space for any formula.

We show that our game can be played in a monotone way. This implies that the corresponding resolution measures on Tseitin formulas correspond exactly to those under the restriction of regular resolution.

Using our characterizations we improve the existing complexity bounds for Tseitin formulas showing that resolution width, depth and variable space coincide up to a logarithmic factor, and that variable space is bounded by the clause space times a logarithmic factor.

1 Introduction

Tseitin propositional formulas for a graph $G = (V, E)$ encode the combinatorial statement that the sum of the degrees of the vertices of G is even. Such formulas provide a great tool for transforming in a uniform way a graph into a propositional formula that inherits some of the properties of the graph. Tseitin formulas have been extensively used to provide hard examples for resolution or as benchmarks for testing SAT-solvers. To name just a few examples, they were used for proving exponential lower bounds on the minimal size required in tree-like and regular resolution [17], in general resolution [18] and for proving lower bounds on resolution proof measures as the width [7] and the space [9], or more recently for proving time-space trade-offs in resolution [5,6]. Due to the importance of these formulas, it is of great interest to find ways to understand how different parameters on the underlying graphs are translated as some complexity measures of

---

N. Talebanfard—Supported by ERC grant FEALORA 339691.

© Springer International Publishing AG, part of Springer Nature 2018
O. Beyersdorff and C. M. Wintersteiger (Eds.): SAT 2018, LNCS 10929, pp. 311–326, 2018.
https://doi.org/10.1007/978-3-319-94144-8_19

the corresponding Tseitin formula. This was the key of the mentioned resolution results. For example the expansion of the graph translated into resolution lower bounds for the corresponding formula in all mentioned lower bounds, while the carving-width or the cut-width of the graph were used to provide upper bounds for the resolution width and size in [2,5].

In this paper we obtain an exact characterization of the complexity measures of resolution width, variable space and depth for any Tseitin formula in terms of a cops-robber game played on its underlying graph. There exists a vast literature on such graph searching games (see eg. [10]). Probably the best known game of this kind is the one used by Seymour and Thomas [15] in order to characterize exactly the graph tree-width parameter. In the original game, a team of cops has to catch a robber that moves arbitrarily fast in a graph. Cops and robber are placed on vertices, and have perfect information of the positions of the other player. The robber can move any time from one vertex to any other reachable one but cannot go through vertices occupied by a cop. Cops are placed or removed from vertices and do not move. The robber is caught when a cop is placed on the vertex where she is standing. The value of the game for a graph G is the minimum number of cops needed to catch the robber on G. In [15] Seymour and Thomas also showed that this game is monotone in the sense that there is always an optimal strategy for the cops in which they never occupy the same vertex again after a cop has been removed from it. In a previous version of the game [13] the robber is invisible and the cops have to search the whole graph to be sure to catch her. The minimum number of cops needed to catch the robber in this game on G, characterizes exactly the path-width of G [8]. The invisible cop game is also monotone [13].

Our game is just a slight variation from the original game from [15]. The only differences are that the cops are placed on the graph edges instead of on vertices, and that the robber is caught when she is completely surrounded by cops. We show that the minimum number of cops needed to catch a robber on a graph G in this game, exactly characterizes the resolution width of the corresponding Tseitin formula. We also show that the number of times some cop is placed on an edge of G exactly coincides with the resolution depth of the Tseitin formula on G. Also, if we consider the version of the game with an invisible robber instead, we exactly obtain the resolution variable space of the Tseitin formula on G.

We also show that the ideas behind the characterization of variable space in terms of a game with an invisible robber, can in fact be extended to define a new combinatorial game to exactly characterize the resolution variable space of any formula (not necessarily a Tseitin formula). Our game is a non-interactive version of the Atserias and Dalmau game for characterizing resolution width [4].

An interesting consequence of the cops-robber game characterizations is that the property of the games being monotone can be used to show that for the corresponding complexity measures, the resolution proof can be regular without changing the bounds. As mentioned, the vertex-cops games are known to be monotone. This did not need to be true for our game. In fact, the robber-marshals game [11], another version of the game in which the cops are placed on the (hyper)edges, is know to be non-monotone [1]. We are able to show that the edge-cops game (for both cases of visible and invisible robber) is also monotone. This is done by reducing our edge game to the Seymour and Thomas vertex game. This fact immediately implies that in the context of Tseitin formulas, the width and variable space in regular resolution proofs is not worse that in general resolution[1]. A long standing open question from Urquhart [18] asks whether regular resolution can simulate general resolution on Tseitin formulas (in size). Our results show that this is true for the measures of width and variable space.

Finally we use the game characterization to improve the known relationships between different complexity measures on Tseitin formulas. In particular we show that for any graph G with n vertices, the resolution depth of the corresponding formula is at most its resolution width times $\log n$. From this follows that all the three measures width, depth and variable space are within a logarithmic factor in Tseitin formulas. Our results provide a family of a uniform class of propositional formulas where clause space is polynomially bounded in the variable space. No such result was known before as recently pointed to by Razborov in [14].

The paper is organized as follows. In Sect. 2, we have all the necessary preliminaries on resolution and its complexity measures. In Sect. 3 we present the characterization of variable space in resolution. In Sect. 4 we introduce our variants of the Cops-Robber games on graphs and we show the characterizations of width, variable space and depth of the Tseitin formula on G in terms of Cops-Robber games played on G. In Sect. 5 we focus on the monotone version of the games and we prove that all our characterizations can be made monotone. In the last Sect. 6 we use all our previous results to prove the new relationships between width, depth, variable space and clause space for Tseitin formulas. We finish with some conclusions and open questions.

2 Preliminaries

Let $[n] = \{1, 2, ..., n\}$. A *literal* is either a Boolean variable x or its negation $\bar{x}$. A *clause* is a disjunction (possibly empty) of literals. The empty clause will be denoted by $\square$. The set of variables occurring in a clause C, will be denoted by $\mathsf{Vars}(C)$. The *width* of a clause C is defined as $\mathsf{W}(C) := |\mathsf{Vars}(C)|$.

A CNF F_n over n variables $x_1, \ldots, x_n$ is a conjunction of clauses defined over $x_1, \ldots, x_n$. We often consider a CNF as a set of clauses and to simplify the notation in this section we omit the index n expressing the dependencies of F_n from the n variables. The width of a CNF F is $\mathsf{W}(F) := \max_{c \in F} \mathsf{W}(C)$. A CNF is a k-CNF if all clauses in it have width at most k.

The *resolution* proof system is a refutational propositional system for CNF formulas handling with clauses, and consisting of the only *resolution rule*:

$$\frac{C \vee x \qquad D \vee \bar{x}}{C \vee D}$$

---

[1] The resolution depth is well know to coincide with the regular resolution depth for any formula.

A *proof* π of a clause C from a CNF F (denoted by $F \vdash_\pi C$) is a sequence of clauses $\pi := C_1, \ldots, C_m$, $m \geq 1$ such that $C_m = C$ and each C_i in π is either a clause of F or obtained by the resolution rule applied to two previous clauses (called *premises*) in the sequence. When C is the empty clause $\square$, π is said to be a *refutation* of F. Resolution is a sound a complete system for unsatisfiable formulas in CNF.

Let $\pi := C_1, \ldots, C_m$ be a resolution proof from a CNF F. The *width* of π is defined as $W(\pi) := \max_{i \in [m]} W(C_i)$. The width needed to refute an unsatisfiable CNF F in resolution is $W(F \vdash) := \min_{F \vdash_\pi \square} W(\pi)$. The *size* of π is defined as $S(\pi) := m$. The size needed to refute an unsatisfiable CNF F in resolution is $S(F \vdash) := \min_{F \vdash_\pi \square} S(\pi)$.

Resolution proofs $F \vdash_\pi C$, can be represented also in two other notations: as directed acyclic graphs (DAG) or as sequences of set of clauses $\mathbb{M}$, called *(memory) configurations*. As a DAG, π is represented as follows: source nodes are labeled by clauses of F, the (unique) target node is labeled by C and each non-source node, labeled by a clause D, has two incoming edges from the (unique) nodes labeled by the premises of D in π. Using this notation the size of a proof π, is the number of nodes in the DAG representing π. The DAG notation allow to define other proof measures for resolution proofs. The *depth* of a proof π, $D(\pi)$ is the length of the longest path in the DAG representing π. The depth for refuting an unsatisfiable CNF F is $D(F \vdash) := \min_{F \vdash_\pi \square} D(\pi)$.

The representation of resolution proofs as configurations was introduced in [3,9] in order to define *space* complexity measures for resolution proofs. A proof π, $F \vdash_\pi C$, is a sequence $\mathbb{M}_1, \ldots, \mathbb{M}_s$ such that: $\mathbb{M}_1 = \emptyset$, $C \in \mathbb{M}_s$ and for each $t \in [s-1]$, $\mathbb{M}_{t+1}$ is obtained from $\mathbb{M}_t$, by one of the following rules:

[*Axiom Download*]: $\mathbb{M}_{t+1} = \mathbb{M}_t \cup \{D\}$, for D a clause in F;
[*Erasure*]: $\mathbb{M}_{t+1} \subset \mathbb{M}_t$;
[*Inference*]: $\mathbb{M}_{t+1} = \mathbb{M}_t \cup \{D\}$, if $A, B \in \mathbb{M}_t$ and $\frac{A \quad B}{D}$ is a valid resolution rule.

π is a refutation if C is $\square$.

The *clause space* of a configuration $\mathbb{M}$ is $Cs(\mathbb{M}) := |\mathbb{M}|$. The clause space of a refutation $\pi := \mathbb{M}_1, \ldots, \mathbb{M}_s$ is $Cs(\pi) := \max_{i \in [s]} Cs(\mathbb{M}_i)$. Finally the clause space to refute an unsatisfiable F is $Cs(F \vdash) := \min_{F \vdash_\pi \square} Cs(\pi)$. Analogously, we define the *variable space* and the *total space* of a configuration $\mathbb{M}$ as $Vs(\mathbb{M}) := |\bigcup_{C \in \mathbb{M}} Vars(C)|$ and $Ts(\mathbb{M}) := \sum_{C \in \mathbb{M}} W(C)$. Variable space and total space needed to refute an unsatisfiable F, are respectively $Vs(F \vdash) := \min_{F \vdash_\pi \square} Vs(\pi)$ and $Ts(F \vdash) := \min_{F \vdash_\pi \square} Ts(\pi)$.

An *assignment* for a set of variables X, specifies a truth-value ($\{0, 1\}$ value) for all variables in X. Variables, literals, clauses and CNFs are simplified under partial assignments (i.e. assignment to a subset of their defining variables) in the standard way.

2.1   Tseitin Formulas

Let $G = (V, E)$ be a connected undirected graph with n vertices, and let $\varphi :$ $V \to \{0, 1\}$ be an *odd* marking of the vertices of G, i.e. satisfying the property

$$\sum_{x \in V} \varphi(x) = 1 (\text{mod } 2).$$

For such a graph we can define an unsatisfiable formula in conjunctive normal form $\mathsf{T}(G, \varphi)$ in the following way: The formula has E as set of variables, and is a conjunction of the CNF translation of the formulas F_x for $x \in V$, where F_x expresses that $e_1(x) \oplus \cdots \oplus e_d(x) = \varphi(x)$ and $e_1(x) \ldots e_d(x)$ are the edges (variables) incident with vertex x.

$\mathsf{T}(G, \varphi)$ encodes the combinatorial principle that for all graphs the sum of the degrees of the vertices is even. $\mathsf{T}(G, \varphi)$ is unsatisfiable if and only if the marking φ is odd. For an undirected graph $G = (V, E)$, let $\Delta(G)$ denote its maximal degree. It is easy to see that $\mathsf{W}(\mathsf{T}(G, \varphi)) = \Delta(G)$.

The following fact was proved several times (see for instance [9,18]).

Fact 1. *For an odd marking φ, for every $x \in V$ there exists an assignment α_φ such that $\alpha_\varphi(F_x) = 0$, and $\alpha_\varphi(F_y) = 1$ for all $y \neq x$. Moreover if φ is an even marking, then $\mathsf{T}(G, \varphi)$ is satisfiable.*

Consider a partial truth assignment α of some of the variables of $\mathsf{T}(G, \varphi)$. We refer to the following process as applying α to (G, φ): Setting a variable $e = (x, y)$ in α to 0 corresponds to deleting the edge e in the graph G, and setting it to 1 corresponds to deleting the edge from the graph and toggling the values of $\varphi(x)$ and $\varphi(y)$ in G. Observe that the formula $\mathsf{T}(G', \varphi')$ resulting after applying α to (G, m) is still unsatisfiable.

3   A Game Characterization of Resolution Variable Space

We start by giving a new characterization of resolution variable space. This result holds for any CNF formula and is therefore quite independent of the rest of the paper. We include it at the beginning since it will be used to show that the invisible robber game characterizes variable space in Tseitin formulas.

The game is a *non-interactive* version of the Spoiler-Duplicator width game from Atserias and Dalmau [4]:

Given an unsatisfiable formula F in CNF with variable set V, Player 1 constructs step by step a finite list $L = L_0, L_1, \ldots, L_k$ of sets of variables, $L_i \subseteq V$. Starting by the empty set, $L_0 = \emptyset$, in each step he can either add variables to the previous set, or delete variables from it. The *cost* of the game is the size of the largest set in the list.

Once the Player 1 finishes his list, Player 2 has to construct dynamically a partial assignment for the set of variables in the list. In each step i, the domain of the assignment is the set of variables L_i in the list at this step. She starts giving some value to the first set of variables in the list, L_1, in a way that no clause of F is falsified. If variables are added to the set at any step, she has to extend the previous partial assignment to the new domain in any way, but again, no initial clause can be falsified. If a variable is kept from one set to the next one in the list, its value in the assignment remains. If variables are removed from the set at any step, the new partial assignment is the restriction of the previous one to the new domain.

If Player 2 manages to come to the end of the list without having falsified any clause of F at any point, she wins. Otherwise Player 1 wins.

Define $\mathsf{nisd}(F)$ to be the minimum cost of a winning game for Player 1 on F. We prove that for any unsatisfiable formula F the variable space of F coincides exactly with $\mathsf{nisd}(F)$.

Theorem 1. *Let F be an unsatisfiable formula, then* $\mathsf{nisd}(F) \leq \mathsf{Vs}(F \vdash)$.

Proof. (sketch) Consider a resolution proof Π of F as a list of configurations. The strategy of Player 1 consists in constructing a list L of sets of variables, that in each step i contains the variables present in the i-th configuration. The cost for this list is exactly $\mathsf{Vs}(\Pi)$.

We claim that any correct list of partial assignments of Player 2 that does not falsify any clause in F, has to satisfy simultaneously all the clauses at the configurations in each step. The argument is completed by observing that there must be a step in Π in which the clauses in the configuration are not simultaneously satisfiable.

Theorem 2. *Let F be an unsatisfiable formula, then* $\mathsf{Vs}(F \vdash) \leq \mathsf{nisd}(F)$.

Proof. (sketch) Let L be the list of sets of variables constructed by Player 1, containing at each step i a set L_i of at most $\mathsf{nisd}(F)$ variables. We consider for each step i a set of clauses $\mathcal{C}_i$ containing only the variables in L_i. Initially L_1 is some set of variables and $\mathcal{C}_1$ is the set of all clauses that can be derived by resolution (in any number of steps) from the clauses in F containing only variables in L_1. At any step i, if L_i is constructed by adding some new variables to L_{i-1}, $\mathcal{C}_i$ is defined to be the set of clauses that can be derived from the clauses in $\mathcal{C}_{i-1}$ and the clauses in F containing only variables in L_i. If L_i is constructed by subtracting some new variables from L_{i-1}, $\mathcal{C}_i$ is defined to be the set of clauses in $\mathcal{C}_{i-1}$ that only have variables in the set L_i. By definition $\mathcal{C}_i$ can be always be constructed from $\mathcal{C}_{i-1}$ by using only resolution steps, deletion or inclusion of clauses in F, and therefore this list of sets of clauses can be written as a resolution proof. At every step in this proof at most $\mathsf{nisd}(F)$ variables are present.

We claim that if L is a winning strategy for Player 1, then at some point i, $\mathcal{C}_i$ must contain the empty clause. This implies the result. To sketch a proof of this claim let us define at each step i the set A_i of partial assignments for the variables in L_i that satisfy all the clauses in $\mathcal{C}_i$, and the set B_i to be the set of partial assignments for the variables in L_i that do not falsify any initial clause and can be constructed by Player 2 following the rules of the game. It can be seen by induction on i that at each step, $A_i = B_i$. Since at some point i, Player 2 does not have any correct assignment that does not falsify a clause in F, it follows that $A_i = B_i = \emptyset$, which means that C_i in unsatisfiable and must contain the empty clause by the definition of C_i and the completeness of resolution.

4   Cops and Robber Games

We consider a slight variation of the Cops and Robber game from Seymour and Thomas [15] which they used to characterize exactly the tree-width of a graph. We call our version the *Edge* (Cops and Robber) *Game*.

Initially a robber is placed on a vertex of a connected graph G. She can move arbitrarily fast to any other vertex along the edges. The team of cops, directed by one person, want to capture her, and can always see where she is. They are placed on *edges* and do not move.

Definition 3. *(Edge Cops-Robber Game) Player 1 takes the role of the cops. At any stage he can place a cop on any unoccupied edge or remove a cop from and edge. The robber (Player 2) can then move to any vertex that is reachable from his actual position over a path without cops. Both teams have at any moment perfect information of the position of the other team. Initially no cop is on the graph. The game finishes when the robber is captured. This happens when the vertex she occupies is completely surrounded by cops.*

The value of the game is the maximum number of edge-cops present on the edges at any point in the game. We define $\mathsf{ec}(G)$ *as the minimum value in a finishing Edge Game on G.*

The only difference between our Edge Cops-Robber Game and the Cops-Robber game from Seymour and Thomas in that here the cops are placed on the edges, while in [15] they were placed on the vertices and that our game ends with the robber surrounded while in the Seymour-Thomas game a cop must occupy the same vertex as the robber.

4.1   The Cops-Robber Game Characterizes Width on Tseitin Formulas

The edge-cops game played on a connected graph G characterizes exactly the minimum width of a resolution refutation of $\mathsf{T}(G, \varphi)$ for any odd marking φ. In order to show this, we use the Atserias-Dalmau game [4] introduced to characterize resolution width. We prove that $\mathsf{ec}(G) = \mathsf{sd}(\mathsf{T}(G, \varphi))$ where $\mathsf{sd}(\mathsf{T}(G, \varphi))$ denotes the value of the Atserias-Dalmau game played on $\mathsf{T}(G, \varphi)$. We use the simplified explanation of the game from [16].

Spoiler and Duplicator play on a CNF formula F. Spoiler wants to falsify a clause of the formula, while Duplicator tries to prevent this from happening. During the game they construct a partial assignment α of the variables in F and the game ends when α falsifies a clause from F. Initially α is empty. At each step, Spoiler can select an unassigned variable x or forget (unassign) a variable from

α. In the first case the Duplicator assigns a value to x, in the second case she does not do anything. The value of a game is the maximum number of variables that are assigned in α at some point during the game. $\mathsf{sd}(F)$ is the minimum value of a finishing game on F.

Atserias and Dalmau [4] proved that this measure characterizes the width of a resolution refutation of any unsatisfiable F, $\mathsf{W}(F \vdash) = \max\{\mathsf{W}(F), \mathsf{sd}(F) - 1\}$[2]. Let us observe how the game goes when played on the formula $\mathsf{T}(G, \varphi)$. In a finishing game on $\mathsf{T}(G, \varphi)$ Spoiler and Duplicator construct a partial assignment α of the edges. Applying α to the variables of $\mathsf{T}(G, \varphi)$ a new graph G' and marking φ' are produced. Consider a partial truth assignment α of some of the variables. Assigning a variable $e = \{x, y\}$ in α to 0 corresponds to deleting the edge e in the graph, and setting it to 1 corresponds to deleting the edge from the graph and toggling the values of $\varphi(x)$ and $\varphi(y)$. The formula $\mathsf{T}(G', \varphi')$ resulting after applying α to (G, φ) is still unsatisfiable. We will call a connected component of G' for which the sum of the markings of its vertices is odd, an *odd component*. Initially G is an odd component under φ. By assigning an edge, an odd component can be divided in at most two smaller components, an odd one and an even one. The only way for Spoiler to end the game is to construct an assignment α that assigns values to all the edges of a vertex, contradicting its marking under α. This falsifies one of the clauses corresponding to the vertex.

Theorem 4. *For any connected graph G and any odd marking φ, $\mathsf{ec}(G) = \mathsf{sd}(\mathsf{T}(G, \varphi))$.*

Proof. In order to compare both games, the team of cops will be identified with the Spoiler and the robber will be identified with the Duplicator. Since the variables in $\mathsf{T}(G, \varphi)$ are the edges of G, the action of Spoiler selecting (forgetting) a variable in the Atserias-Dalmau game will be identified with placing (removing) a cop on that edge.

We show first that $\mathsf{ec}(G) \leq \mathsf{sd}(\mathsf{T}(G, \varphi))$. No matter what the strategy of Duplicator is, Spoiler has a way to play in which he spends at most $\mathsf{sd}(\mathsf{T}(G, \varphi))$ points at the Spoiler-Duplicator game on $\mathsf{T}(G, \varphi)$. In order to obtain a value smaller or equal than $\mathsf{sd}(\mathsf{T}(G, \varphi))$ in the Edge Game, the cops just have to imitate Spoiler's strategy on $\mathsf{T}(G, \varphi)$. At the same time, they compute a strategy for Duplicator that simulates the position of the robber. This is done by considering a Duplicator assigning values in such a way that there is always a unique odd component which corresponds to the subgraph of G isolated by cops where the robber is. At any step in the Edge Game, we the following invariant is kept:

The partial assignment produced in the Spoiler-Duplicator game on $\mathsf{T}(G, \varphi)$ defines a unique odd component corresponding to the component of the robber.

If in a step of the Spoiler-Duplicator game the edge selected by Spoiler does not cut the component where the robber is, Player 1 can simulate Duplicator's assignment for this variable in a way in which a unique odd component is kept

---

[2] In the original paper [4] it is stated that $\mathsf{W}(F \vdash) = \mathsf{sd}(F) - 1$, by inspecting the proof it can be seen that the formulation involving the width of F is the correct one.

and continue with the next decision of Spoiler. At a step right after the component of the robber is cut by the cops, Player 1 can compute an assignment of Duplicator for the last occupied edge, which would create a labeling that identifies the component where the robber as the unique odd component of the graph. This is always possible. Then Player 1 just needs to continue the imitation of Spoiler's strategy for the assignment produced by Duplicator.

At the end of the game Spoiler falsifies an initial clause, and the vertex corresponding to this clause is the unique odd component under the partial assignment. Therefore the cops will be on the edges of a falsified clause, thus catching the robber on the corresponding vertex.

The proof of $ec(G) \geq sd(T(G, \varphi))$ is very similar. Now we consider that there is a strategy for Player 1 in the Edge Game using at most $ec(G)$ cops, and we want to extract from it a strategy for the Spoiler. He just needs to select (remove) variables is the same way as the cops are being placed (removed). This time, all through the game we have the following invariant:

The component isolated by cops in which the robber is, is an odd component in the Spoiler-Duplicator game.

When the variable (edge) selected does not cut the component where the robber is, he does not need to do anything. When the last selected variable cuts the component of the robber, by choosing a value for this variable Duplicator decides which one of the two new components is the odd one. Spoiler figures that the robber has gone to the new odd component and asks the cops what to do next in this situation. When the robber is caught, this will be in an odd component of size 1 which all its edges assigned. This partial assignment falsifies the corresponding clause in $T(G, \varphi)$.

Corollary 5. *For any connected graph G and any odd marking φ,*

$$W(T(G, \varphi) \vdash) = \max\{\Delta(G), ec(G) - 1\}.$$

4.2 An Invisible Robber Characterizes Variable Space on Tseitin Formulas

Consider now the cops game in which the robber is invisible. That means that the cops strategy cannot depend on the robber and the cops have to explore the whole graph to catch her. As in the visible version of the game, the robber is caught if all the edges around the vertex in which she is, are occupied by cops. For a graph G let $iec(G)$ be the minimum number of edge-cops needed to catch an invisible robber in G. Let $T(G, \varphi)$ be the Tseitin formula corresponding to G. We show that $iec(G)$ corresponds exactly with $Vs(T(G, \varphi))$.

Theorem 6. $Vs(T(G, \varphi)) = iec(G)$.

Proof. (sketch)

(i) $Vs(T(G, \varphi) \vdash) \leq iec(G)$. We use the game characterization of variable space. Consider the strategy of the cops. At each step the set of variables constructed by Spoiler corresponds to the set of edges (variables) where the cops are. Now consider any list of partial assignments that Player 2 might construct. Any such assignment can be interpreted as deleting some edges and moving the robber to an odd component in the graph. But the invisible robber is caught at some point, no matter what she does, and this corresponds to a falsified initial clause.

(ii) $\mathsf{iec}(G) \leq \mathsf{Vs}(\mathsf{T}(G, \varphi) \vdash)$. Now we have a strategy for Spoiler, and the cops just need to be placed on the edges corresponding to the variables selected by Player 1. If the robber could escape, by constructing a list of partial assignments mimicking the robber moves (that is, each time the cops produce a new cut in the component where the robber is, she sets the value of the last assigned variable to make odd the new component where the robber has moved to), Player 2 never falsifies a clause in $\mathsf{T}(G, \varphi)$.

4.3   A Game Characterization of Depth on Tseitin Formulas

We consider now a version of the game in which the cops have to remain on their edges until the end of the game and cannot be reused.

Definition 7. *For a graph G let $\mathsf{iec}(G)$ be the minimum number of edge-cops needed in order to catch a visible robber on G, in the cops-robber game, with the additional condition that the cops once placed, cannot be removed from the edges until the end of the game.*

Theorem 8. *For any connected undirected graph G and any odd marking φ of G, $\mathsf{D}(\mathsf{T}(G, \varphi) \vdash) = \mathsf{iec}(G)$.*

Proof. (sketch)

(i) $\mathsf{D}(\mathsf{T}(G, \varphi) \vdash) \leq \mathsf{iec}(G)$. Based on the strategy of the cops, we construct a regular resolution proof tree of $\mathsf{T}(G, \varphi)$ in which the variables are resolved in the order (from the empty clause) as the cops are being placed on the edges. Starting at the node in the tree corresponding to the empty clause, in each step when a cop is placed on edge e we construct two parent edges, one labeled by e and the other one by $\bar{e}$. A node in the tree is identified by the partial assignment defined by the path going from the empty clause to this node. Each time the cops produce a cut in G, such an assignment defines two different connected components in G, one with odd marking and one with even marking. We consider at this point the resolution of the component with the odd marking, following the cop strategy for the case in which the robber did go to this component.

(ii) $\mathsf{iec}(G) \leq \mathsf{D}(\mathsf{T}(G, \varphi) \vdash)$. Consider a resolution proof Π of $\mathsf{T}(G, \varphi)$. Starting by the empty clause, the cops are placed on the edges corresponding to the variables being resolved. At the same time a partial assignment is being constructed (by the robber) that defines a path in the resolution that goes through the clauses that are negated by the partial assignment. If removing these edges where the cops are produces a cut in G, the cops continue from a node in the resolution proof corresponding to an assignment for the last chosen variable that gives odd value to the component where the robber has moved. At the end a clause in $\mathsf{T}(G, \varphi)$ is falsified, which corresponds to the cops being placed in the edges around the robber. The number of cops needed is at most the resolution depth.

5   Regular Resolution and Monotone Games

We show in this section that the fact that the games can be played in a monotone way, implies that width and variable space in regular resolution are as good as in general resolution in the context for Tseitin formulas.

We need some further notation. For a set S and $k > 0$, we denote the set of subsets of S of size at most k by S^k.

5.1   The Visible Robber

We recall the game of [15]. Let $G = (V, E)$ be a simple graph and let $Y \subseteq V$. A Y-flap is the vertex set of a connected component in $G \setminus Y$. A position in this game is a pair (Y, Q) where $Y \subseteq V$ and Q is an Y-flap. The game starts in position $(\emptyset, V)$. Assume that position (Y_i, Q_i) is reached. The cops-player chooses Y_{i+1} such that either $Y_i \subseteq Y_{i+1}$ or $Y_{i+1} \subseteq Y_i$. Then the robber-player chooses a Y_{i+1}-flap Q_{i+1} such that $Q_i \subseteq Q_{i+1}$ or $Q_{i+1} \subseteq Q_i$. The cops-player wins when $Q_i \subseteq Y_{i+1}$. We say that a sequence of positions $(Y_0, Q_0), \ldots, (Y_t, Q_t)$ is *monotone* if for all $0 \leq i \leq j \leq k \leq t$, $Y_i \cap Y_k \subseteq Y_j$. The main result of Seymour and Thomas is that if k cops can win the game, they can also win monotonically. We will use this result to prove an analogous statement about our games where we put the cops on edges.

We extend the framework of Seymour and Thomas to talk about edges. Now we have $X \subseteq E$. An X-flap is the edge set of a connected component in $G \setminus X$. A position is a pair (X, R) with $X \subseteq E$ and R an X-flap. Assume that a position (X_i, R_i) is reached. The cops-player chooses X_{i+1} such that either $X_i \subseteq X_{i+1}$ or $X_{i+1} \subseteq X_i$. Then the robber-player chooses an X_{i+1}-flap R_{i+1} such that either $R_i \subseteq R_{i+1}$ or $R_{i+1} \subseteq R_i$. The cops win when $R_i \subseteq X_{i+1}$. Note that under this definition if some X isolates more than one vertex, then we will have multiple empty sets as X-flaps. However if the robber moves to such an X-flap she will immediately lose as in the next round the cops remain where they are and $\emptyset \subseteq X$.

Similarly a sequence of positions $(X_0, R_0), \ldots, (X_t, R_t)$ is monotone if for all $0 \leq i \leq j \leq k \leq t$, $X_i \cap X_k \subseteq X_j$.

Given a graph $G = (V, E)$ the *line graph* of G is $L(G) = (V', E')$ defined as follows: for every edge $e \in E$ we put a vertex $w_e \in V'$. We then set

$$E' = \{\{w_{e_1}, w_{e_2}\} : e_1, e_2 \in E, e_1 \cap e_2 \neq \emptyset\}.$$

For $X \subseteq E$ define $L(X) := \{w_e : e \in X\}$ and for $Y \subseteq V'$ define $L^{-1}(Y) = \{e : w_e \in Y\}$.

Proposition 9. *Let $G = (V, E)$ be a graph and let $X \subseteq E$. It follows that $R \subseteq E$ is an X-flap if and only $L(R)$ is an $L(X)$-flap.*

Proof. It is enough to show that any $e_1, e_2 \in E \setminus X$ are reachable from each other in $G \setminus X$ if and only if w_{e_1} and w_{e_2} are reachable from each other in $L(G) \setminus L(X)$. Let $P = e_1, f_1, \ldots, f_t, e_2$ be a path in $G \setminus X$ connecting e_1 and e_2. By construction we have a path $w_{e_1}, w_{f_1}, \ldots, w_{f_t}, w_{e_2}$ in $L(G) \setminus L(X)$.

Conversely let $w_{e_1}, w_{f_1}, \ldots, w_{f_t}, w_{e_2}$ be a path of minimum length between w_{e_1} and w_{e_2} in $L(G) \setminus L(X)$. It is easy to see that $e_1, f_1, \ldots, f_t, e_2$ is a path between e_1 and e_2 in $G \setminus X$.

Theorem 10. *Assume that there is a strategy for the edge-cops game on G with k cops. Then there exists a strategy for the vertex-cops game in $L(G)$ with k cops.*

Proof. Fix a strategy σ for the edge-cops on G, i.e., for every $X \in E^k$ and every X-flap R, $\sigma(X, R) \in E^k$ which guarantees that the robber will eventually be captured. We will inductively construct a sequence $\{(Y_i, Q_i)\}$ of positions in the vertex game on $L(G)$, where Q_is are the responses of the robber, while keeping a corresponding sequence $\{(X_i, R_i)\}$ for the edge game on G. The vertex game starts in position $(Y_0, Q_0) = (\emptyset, V')$ and the edge game starts in $(X_0, R_0) = (\emptyset, E)$. We have $X_1 = \sigma(X_0, R_0)$. In general we set $Y_i = L(X_i)$ and after the robber has responded with Q_i we define $R_i = L^{-1}(Q_i)$, from which we construct $X_{i+1} = \sigma(X_i, R_i)$ and so on. That R_i is an X_i-flap follows immediately from Proposition 9. To see that this is indeed a winning strategy, note that at some point we reach a position with $R_i \subseteq X_{i+1}$. This happens only when $Q_i \subseteq Y_{i+1}$.

Theorem 11. *Assume that there is a monotone strategy for the vertex-cops game in $L(G)$ with k cops. Then there exists a monotone strategy with k cops for the edge-cops game in G.*

Proof. We will construct a sequence $\{(X_i, R_i)\}$ of positions in the edge game on G while keeping a corresponding sequence $\{(Y_i, Q_i)\}$ of positions in the vertex game on $L(G)$. Note that R_i will be the response of the robber on G. Let σ be a monotone strategy with k vertex-cops on $L(G)$. We will inductively construct $X_i = \{e : w_e \in Y_i\}$ and after the robber has responded with R_i we define $Q_i = L(R_i)$. Proposition 9 implies that Q_i is a Y_i-flap. Since σ is a winning strategy at some point we reach a position with $Q_i \subseteq Y_{i+1}$. This happens only when $R_i \subseteq X_{i+1}$. The monotonicity of the strategy follows immediately.

5.2   The Invisible Robber

In a similar way as we did with the visible robber game, we can reduce the edge-game with an invisible robber to the invisible robber vertex-game of Kirousis and Papadimitriou [12] (we will call this game KP). In their game cops are placed on vertices. An edge is *cleared* if both its endpoints have cops. An edge can be *recontaminated* if it is connected to an uncleared edge passing through no cops. It is shown in [12] that the cops can optimally clear all the edges without occupying any vertex twice.

Theorem 12. *Assume that k cops can win the edge-game capturing an invisible robber on G. Then k cops can capture the robber in KP game on $L(G)$.*

Theorem 13. *Assume that k cops can monotonically capture the robber in KP game on $L(G)$. Then k cops can monotonically capture the invisible robber in the edge-game on G.*

Corollary 14. *Let $G = (V, E)$ be a simple connected graph and let φ be any odd marking of G. Assume that there exist a resolution refutation of $\mathsf{T}(G, \varphi)$ of variable space at most k. Then there exists a regular resolution refutation of $\mathsf{T}(G, \varphi)$ of variable space at most k.*

6 New Relations Between Complexity Measures for Tseitin Formulas

For any unsatisfiable formula F the following inequalities hold:

$$\mathsf{W}(F \vdash) \leq \mathsf{Vs}(F \vdash) \tag{1}$$
$$\mathsf{Vs}(F \vdash) \leq \mathsf{D}(F \vdash) \tag{2}$$
$$\mathsf{Cs}(F \vdash) \leq \mathsf{D}(F \vdash) + 1 \tag{3}$$
$$\mathsf{Cs}(F \vdash) \geq \mathsf{W}(F \vdash) - \mathsf{W}(F) + 1 \tag{4}$$

Here Eq. 1 follows by definition, Eq. 2 is proved in [19], Eq. 4 is the Atserias-Dalmau [4] width-space inequality and Eq. 3 follows from the following two observations:

1. Any resolution refutation π can be transformed, doubling subproofs, in a tree-like refutation with the same depth of the original proof π.
2. The clause space of a treelike refutation is at most as large as its depth+1 [9].

In general the relationship between variable space and clause space is not clear. It is also an open problem to know whether variable space and depth are polynomially related (see [14,19]) and if clause space is polynomially bounded in variable space (see Razborov in [14], Open problems). In this section we answer this questions in the context of Tseitin formulas. We show in Corollary 17 below that for any Tseitin formula $\mathsf{T}(G, \varphi)$ corresponding to a graph G with n vertices,

$$\mathsf{D}(\mathsf{T}(G, \varphi) \vdash) \leq \mathsf{W}(\mathsf{T}(G, \varphi) \vdash) \log n \tag{5}$$

From this and the inequalities above it follow the following new relations:

$$\mathsf{D}(\mathsf{T}(G, \varphi) \vdash) \leq \mathsf{Vs}(\mathsf{T}(G, \varphi) \vdash) \log n \tag{6}$$
$$\mathsf{Cs}(\mathsf{T}(G, \varphi) \vdash) \leq \mathsf{Vs}(\mathsf{T}(G, \varphi) \vdash) \log n + 1 \tag{7}$$
$$\mathsf{Vs}(\mathsf{T}(G, \varphi) \vdash) \leq (\mathsf{Cs}(\mathsf{T}(G, \varphi) \vdash) + \Delta(G) - 1) \log n. \tag{8}$$

Where the last equation follow since $\mathsf{W}(\mathsf{T}(G, \varphi)) = \Delta(G)$.

That is, in the context of Tseitin formulas $\mathsf{T}(G, \varphi)$:

1. If G is a graph of bounded degree, the width, depth, variable space and clause space for refuting $\mathsf{T}(G, \varphi)$ differ by at most a $\log n$ factor.

2. For any graph G the clause space of refuting $\mathsf{T}(G,\varphi)$ is bounded above by the a $\log n$ factor of the variable space of refuting $\mathsf{T}(G,\varphi)$.

To prove our results, we need two preliminary lemmas.

Lemma 15. *Let $\mathsf{T}(G,\varphi)$ be a Tseitin formula and Π be a width k resolution refutation of $\mathsf{T}(G,\varphi)$. From Π it is possible to find in linear time in $|\Pi|$ a set W of at most $k+1$ variables such that any assignment of these variables when applied to G in the usual way, defines a graph G' and a labeling φ' in which there is some odd connected component with at most $\lceil \frac{|V|}{2} \rceil$ vertices.*

Proof. We use again the Spoiler and Duplicator game from [4]. A way for Spoiler to pay at most $k+1$ points on the game on $\mathsf{T}(G,\varphi)$ is to use the structure of Π starting at the empty clause and query each time the variable that is being resolved at the parent clauses. When Duplicator assigns a value to this variable, Spoiler moves to the parent clause falsified by the partial assignment and deletes from this assignment any variables that do not appear in the parent clause. In this way he always reaches at some point an initial clause, falsifying it and thus winning the game. At any point at most $k+1$ variables have to be assigned. To this strategy of Spoiler, Duplicator can oppose the following strategy: She applies the partial assignment being constructed to the initial graph G producing a subgraph G' and a new labeling φ'. Every time a variable e has to be assigned, if e does not produce a new cut in G' she gives to e an arbitrary value. If e cuts an odd component in G' she assigns e with the value that makes the largest of the two new components an odd component. In case e cuts an even component in two, Duplicator gives to e the value which keeps both components even. Observe that with this strategy there is always a unique odd component. Even when Spoiler releases the value of some assigned variable he cannot create more components, he either keeps the same number of components or connects two of them.

While playing the game on $\mathsf{T}(G,\varphi)$ with these two strategies, both players define a path from the empty clause to an initial one. There must be a first clause K along this path in which the constructed partial assignment constructed in the game at the point t in which K is reached, when applied to G, defines a unique odd component of size at most $\lceil \frac{|V|}{2} \rceil$. This is so because the unique odd component initially has size $|V|$ while at the end has size 1. This partial assignment has size at most $k+1$. Not only the odd component, but any component produced by the partial assignment has size at most $\lceil \frac{|V|}{2} \rceil$. This is because at the point before t the odd component was larger than $\lceil \frac{|V|}{2} \rceil$ and therefore any other component had to be smaller than this. At time t Spoiler chooses a variable that when assigned cuts the odd component in two pieces. Duplicator assigns it in such a way that the largest of these two components is odd and has size at most $\lceil \frac{|V|}{2} \rceil$. Therefore the other new component must have at most this size.

Any other assignment of these variables also produces an odd component of size at most $\lceil \frac{|V|}{2} \rceil$. They correspond to other strategies and they all produce the same cuts and components in the graph, just different labellings of the components. Since the initial formula was unsatisfiable there must always be at least one odd component. In order to find the set W of variables, one just has to move on refutation Π simulating Spoiler and Duplicator strategies. This can be done in linear time in the size of Π.

Theorem 16. *There is an algorithm that on input a connected graph $G = (V, E)$ with an odd labeling φ and a resolution refutation Π of $\mathsf{T}(G, \varphi)$ with width k, produces a tree-like resolution refutation Π' of $\mathsf{T}(G, \varphi)$ of depth $k \log(|V|)$.*

Proof. Let $W = \{e_1, \ldots e_{|W|}\}$ be a set of variables producing an odd connected component of size at most $\lceil \frac{|V|}{2} \rceil$, as guaranteed by Lemma 15. We can construct a tree-like resolution of depth $|W|$ of the complete formula F_W with $2^{|W|}$ clauses, each containing all variables in W but with a different sign combination.

By the Lemma, each assignment of the variables, when applied to G produces a subgraph G_i and a labeling φ_i with an odd component with at most $\lceil \frac{|V|}{2} \rceil$ vertices. The problem of finding a tree-like refutation for $\mathsf{T}(G, \varphi)$ has been reduced to finding a tree-like resolution refutation for each of the formulas $\mathsf{T}(G_i, \varphi_i)$. But each of the graphs G_i have an odd component with at most $\lceil \frac{|V|}{2} \rceil$ vertices and the problem is to refute the Tseitin formulas corresponding to these components. After at most $\log(|V|)$ iterations we reach Tseitin formulas with just two vertices that can be refuted by trees of depth one. Since W has width at most $k + 1$ literals, in each iteration the refutation trees have depth at most k. Putting everything together we get a tree-like refutation of depth at most $k \log(|V|)$. $\qquad \square$

Corollary 17. *For any graph $G = (V, E)$ and any odd labeling φ,*
$$\mathsf{D}(\mathsf{T}(G, \varphi) \vdash) \leq \mathsf{W}(\mathsf{T}(G, \varphi) \vdash) \log(|V|).$$

Corollary 18. *For any graph $G = (V, E)$ and any odd labeling φ*
$$\mathsf{Cs}(\mathsf{T}(G, \varphi) \vdash) \leq \mathsf{Vs}(\mathsf{T}(G, \varphi) \vdash) \log(|V|).$$

7   Conclusions and Open Problems

We have shown that the measures of width, depth and variable space in the resolution of Tseitin formulas can be exactly characterized in terms of a graph searching game played on the underlying graph. Our game is a slight modification of the well known cops-robber game from Seymour and Thomas. The main motivation for this characterization is the fact that some results in graph searching can be used to solve questions in proof complexity. Using the monotonicity properties of the Seymour and Thomas game, we have proven that the measures of width and variable space in regular resolution coincide exactly with those of general resolution in the context of Tseitin formulas. Previously it was only known that for Tseitin formulas, regular width was within a constant factor of the width in general resolution [2]. The game characterization also inspired new relations between the three resolution measures on Tseitin formulas and we proved that they are all within a logarithmic factor.

We have also obtained a game characterization of variable space for the resolution of general CNF formulas, as a non-interactive version of the Atserias and Dalmau game [4] for resolution width.

Still open is whether for Tseitin formulas, regular resolution can also simulate general resolution in terms of size, as asked by Urquhart [18]. Also game characterizations for other resolution measures like size or space, either for Tseitin or general formulas, would be a very useful tool in proof complexity.

Acknowledgments. The authors would like to thank Osamu Watanabe and the ELC project were this research was started. We are also grateful to Dimitrios Thilikos and to the anonymous referees for helpful comments.

References

1. Adler, I.: Marshals, monotone marshals, and hypertree width. J. Gr. Theory **47**, 275–296 (2004)
2. Alekhnovich, M., Razborov, A.A.: Satisfiability, branch-width and Tseitin tautologies. Comput. Complex. **20**(4), 649–678 (2011)
3. Alekhnovich, M., Ben-Sasson, E., Razborov, A.A., Wigderson, A.: Space complexity in propositional calculus SIAM. J. Comput. **31**(4), 1184–1211 (2002)
4. Atserias, A., Dalmau, V.: A combinatorial characterization of resolution width. In: 18th IEEE Conference on Computational Complexity, pp. 239–247 (2003)
5. Beame, P., Beck, C., Impagliazzo, R.: Time-space trade-offs in resolution: super-polynomial lower bounds for superlinear space. SIAM J. Comput. **49**(4), 1612–1645 (2016)
6. Beck, C., Nordström, J., Tang, B.: Some trade-off results for polynomial calculus: extended abstract. In: Proceedings of the 45th ACM Symposium on the Theory of Computing, pp. 813–822 (2013)
7. Ben-Sasson, E., Wigderson, A.: Short proofs are narrow - resolution made simple. J. ACM **48**(2), 149–169 (2001)
8. Ellis, J.A., Sudborough, I.H., Turner, J.S.: The vertex separation and search number of a graph. Inf. Computat. **113**(1), 50–79 (1994)
9. Esteban, J.L., Torán, J.: Space bounds for resolution. Inf. Comput. **171**(1), 84–97 (2001)
10. Fomin, F.V., Thilikos, D.: An annotated bibliography on guaranteed graph searching. Theor. Comput. Sci. **399**, 236–245 (2008)
11. Gottlob, G., Leone, N., Scarello, F.: Robbers, marshals and guards: game theoretic and logical characterizations of hypertree width. J. Comput. Syst. Sci. **66**, 775–808 (2003)
12. Kirousis, L.M., Papadimitriou, C.H.: Searching and pebbling. Theor. Comput. Sci. **47**(3), 205–218 (1986)
13. LaPaugh, A.S.: Recontamination does not help to search a graph. Technical report Electrical Engineering and Computer Science Department. Princeton University (1883)
14. Razborov, A.: On space and depth in resolution. Comput. Complex., 1–49 (2017). https://doi.org/10.1007/s00037-017-0163-1
15. Seymour, P.D., Thomas, R.: Graph searching and a min-max theorem of tree-width. J. Comb. Theory Ser. B **58**, 22–35 (1993)
16. Torán, J.: Space and width in propositional resolution. Comput. Complex. Column Bull. EATCS **83**, 86–104 (2004)
17. Tseitin, G.S.: On the complexity of derivation in propositional calculus. In: Studies in Constructive Mathematics and Mathematical Logic, Part 2, pp. 115–125. Consultants Bureau (1968)
18. Urquhart, A.: Hard examples for resolution. J. ACM **34**, 209–219 (1987)
19. Urquhart, A.: The depth of resolution proofs. Stud. Logica **99**, 349–364 (2011)

Minimally Unsatisfiable Sets

Minimal Unsatisfiability and Minimal Strongly Connected Digraphs

Hoda Abbasizanjani[✉] and Oliver Kullmann[✉]

Swansea University, Swansea, UK
{Hoda.Abbasizanjani,O.Kullmann}@Swansea.ac.uk

Abstract. Minimally unsatisfiable clause-sets (MUs) are the hardest unsatisfiable clause-sets. There are two important but isolated characterisations for nonsingular MUs (every literal occurs at least twice), both with ingenious but complicated proofs: Characterising 2-CNF MUs, and characterising MUs with deficiency 2 (two more clauses than variables). Via a novel connection to Minimal Strong Digraphs (MSDs), we give short and intuitive new proofs of these characterisations, revealing an underlying common structure.

1 Introduction

This paper is about understanding basic classes of minimally unsatisfiable CNFs, short MUs. The most basic MUs are those with only *one more clause* than variables, i.e., with deficiency $\delta = 1$. This whole class is explained by the expansion rule, which replaces a single clause C by two clauses $C' \cup \{v\}, C'' \cup \{\overline{v}\}$ for $C' \cup C'' = C$ and a new variable v, starting with the empty clause. So in a sense only trivial reasoning takes place here. Somewhat surprisingly, this covers all Horn cases in MU ([5]). At the next level, there are two classes, namely *two more clauses* than variables ($\delta = 2$), and *2-CNF*. Characterisations have been provided in the seminal paper [12] for the former class, and in the technical report [15] for the latter. Both proofs are a tour de force. We introduce in this paper a new unifying reasoning scheme, based on graph theory.

This reasoning scheme considers MUs with two parts. The clauses of the "core" represent AllEqual, that is, all variables are equal. The two "full monotone clauses", a disjunction over all positive literals and a disjunction over all negative literals, represent the negation of AllEqual. This is the new class FM ("full monotone") of MUs, which still, though diluted, is as complex as all of MU. So we demand that the reasoning for AllEqual is graph theoretical, arriving at the new class DFM ("D" for digraph).

Establishing AllEqual on the variables happens via SDs, "strong digraphs", where between any two vertices there is a path. For minimal reasoning we use MSDs, minimal SDs, where every arc is necessary. Indeed, just demanding to have an MU with two full monotone clauses, while the rest are binary clauses, is enough to establish precisely MSDs. The two most fundamental classes of MSDs are the *(directed) cycles* C_n and the *dipaths*, the directed versions $D(P_n)$ of the

© Springer International Publishing AG, part of Springer Nature 2018
O. Beyersdorff and C. M. Wintersteiger (Eds.): SAT 2018, LNCS 10929, pp. 329–345, 2018.
https://doi.org/10.1007/978-3-319-94144-8_20

undirected paths P_n, where every undirected edge is replaced by two directed arcs, for both directions. The cycles are at the heart of MUs with $\delta = 2$, while the dipaths are at the heart of MUs in 2-CNF.

To connect MSDs (that is, DFMs) with more general MUs, two transformations are used. First, *singular variables*, occurring in one sign only once, are eliminated by singular DP-reduction, yielding *nonsingular* MUs. This main (poly-time) reduction removes "trivialities", and indeed deficiency 1 consists purely of these trivialities (as the above generation process shows). Second we need to add "missing" literal occurrences, non-deterministically, to clauses, as long as one stays still in MU. This process is called *saturation*, yielding saturated MUs. As it turns out, the nonsingular MUs of deficiency 2 are already saturated and are already of the form DFM, while the nonsingular 2-CNFs have to be (partially) saturated to reach the form DFM.

Before continuing with the overview, we introduce a few basic notations. The class of MUs as clause-sets is formally denoted by $\mathcal{MU}$, while the nonsingular elements are denoted by $\mathcal{MU}' \subset \mathcal{MU}$ (every variable occurs at least twice positively and twice negatively). The number of clauses of a clause-set F is $c(F)$, the number of (occurring) variables is $n(F)$, and the deficiency is $\delta(F) := c(F) - n(F)$. The basic fact is $\delta(F) \geq 1$ for $F \in \mathcal{MU}$ ([1]), and that deficiency is a good complexity parameter ([6]). We use indices for subclassing in the obvious way, e.g., $\mathcal{MU}_{\delta=1} = \{F \in \mathcal{MU} : \delta(F) = 1\}$. Furthermore, like in the DIMACS file format for clause-sets, we use natural numbers in $\mathbb{N} = \{1, 2, \ldots\}$ for variables, and the non-zero integers for literals. So the clause $\{-1, 2\}$ stands for the usual clause $\{\overline{v_1}, v_2\}$, where we just got rid off the superfluous variable-symbol "v". In propositional calculus, this would mean $\neg v_1 \vee v_2$, or, equivalently, $v_1 \rightarrow v_2$.

The Two Fundamental Examples. After this general overview, we now state the central two families of MUs for this paper, for deficiency 2 and 2-CNF. The MUs $\mathcal{F}_n := \{\{1, \ldots, n\}, \{-1, \ldots, -n\}, \{-1, 2\}, \ldots, \{-(n-1), n\}, \{-n, 1\}\}$ of deficiency 2 have been introduced in [12]. It is known, and we give a proof in Lemma 7, that the $\mathcal{F}_n$ are saturated. As shown in [12], the elements of $\mathcal{MU}'_{\delta=2}$ are exactly (up to isomorphism, of course) the formulas $\mathcal{F}_n$. The elimination of singular variables by singular DP-reduction is not confluent in general for MUs. However in [18] it is shown, that we have confluence up to isomorphism for deficiency 2. These two facts reveal that the elements of $\mathcal{MU}_{\delta=2}$ contain a unique "unadorned reason" for unsatisfiability, namely the presence of a complete cycle over some variables (of unique length) together with the requirement that these variables do not have the same value. In the report [15], as in [20] (called "$F^{(2)}$"), the 2-CNF MUs $\mathcal{B}_n := \{\{-1, 2\}, \{1, -2\}, \ldots, \{-(n-1), n\}, \{n-1, -n\}, \{-1, -n\}, \{1, n\}\}$ have been introduced, which are 2-uniform (all clauses have length 2). In [15] it is shown that the nonsingular MUs in 2-CNF are exactly the $\mathcal{B}_n$. By [18] it follows again that we have confluence modulo isomorphism of singular DP-reduction on 2-CNF-MUs. Thus a 2-CNF-MU contains, up to renaming, a complete path of equivalences of variables, where the length of the path is unique; this path establishes the equivalence of all these variables, and then there is the equivalence of the starting point and the negated end point, which yields the contradiction.

Background We have referred above to the fundamental result about singular DP-reduction (sDP) in [18], that for $F \in \mathcal{MU}$ and any $F', F'' \in \mathcal{MU}'$ obtained from F by sDP we have $n(F') = n(F'')$. So we can define $\mathbf{nst}(F) := n(F') \in \mathbb{N}_0$ (generalising [18, Definition 75]). We have $0 \leq \mathrm{nst}(F) \leq n(F)$, with $\mathrm{nst}(F) = 0$ iff $\delta(F) = 1$, and $\mathrm{nst}(F) = n(F)$ iff F is nonsingular. The "nonsingularity type" $\mathrm{nst}(F)$ provides basic information about the isomorphism type of MUs, after (completed) sDP-reduction, and suffices for deficiency 2 and 2-CNF.

We understand a class $\mathcal{C} \subseteq \mathcal{MU}$ "fully" if we have a full grasp on its elements, which should include a complete understanding of the isomorphism types involved, that is, an easily accessible catalogue of the essentially different elements of $\mathcal{C}$. The main conjecture is that the nonsingular cases of fixed deficiency have polytime isomorphism decision, and this should be extended to "all basic classes". Singular DP-reduction is essential here, since already Horn-MU, which has deficiency one, is GI-complete (graph-isomorphism complete; [14]).

Before giving an overview on the main proof ideas, we survey the relevant literature on 2-CNF. Irredundant 2-CNF is studied in [21], mostly concentrating on satisfiable cases, while we are considering only unsatisfiable cases. As mentioned, the technical report [15] contains the proof of the characterisation of 2-CNF-MU, while [20] has some bounds, and some technical details are in [2]. MUSs (MU-sub-clause-sets) of 2-CNF are considered in [3], showing how to compute shortest MUSs in polytime, while in [4] MUSs with shortest resolution proofs are determined in polytime. It seems that enumeration of all MUSs of a 2-CNF has not been studied in the literature. However, in the light of the strong connection to MSDs established in this paper, for the future [11] should become important, which enumerates all MSDs of an SD in incremental polynomial time.

Two Full Clauses. The basic new class is $\mathcal{FM} \subset \mathcal{MU}$, which consists of all $F \in \mathcal{MU}$ containing the full positive clause (all variables) and the full negative clause (all complemented variables). Using "monotone clauses" for positive and negative clauses, "$\mathcal{FM}$" reminds of "full monotone". Let A_n be the basic MUs with n variables and 2^n full clauses; so we have $A_n \in \mathcal{FM}$ for all $n \geq 0$. The trivial cases of $\mathcal{FM}$ are A_0 and A_1, while a basic insight is that $\mathcal{FM}' := \mathcal{FM} \cap \mathcal{MU}'$ besides $\{\bot\}$ contains precisely all the nontrivial elements of $\mathcal{FM}$. In this sense it can be said that $\mathcal{FM}$ studies only nonsingular MUs. We expect the class $\mathcal{FM}_{\delta=k}$ at least for $\delta = 3$ to be a stepping stone towards understanding $\mathcal{MU}_{\delta=3}$ (the current main frontier). The most important new class for this paper is $\mathcal{DFM} \subset \mathcal{FM}$, which consists of all $F \in \mathcal{FM}$ such that besides the monotone clauses all other clauses are binary. Indeed graph isomorphisms for MSDs is still GI-complete ([23]), and thus so is isomorphism for $\mathcal{DFM}$.

After having now DFM at our disposal, we gain a deeper understanding how the seminal characterisations of the basic nonsingular F, that is, $F \in \mathcal{MU}'$, $n(F) > 0$, work: [I] From $\delta(F) = 2$ follows $F \cong \mathcal{F}_{n(F)}$ ([12]; see Corollary 1). [II] From $F \in 2\text{-}\mathcal{CLS}$ follows $F \cong \mathcal{B}_{n(F)}$ ([15]; see Corollary 2). The main step is to make the connection to the class $\mathcal{DFM}$: [I] In case of $\delta(F) = 2$, up to renaming it actually already holds that $F \in \mathcal{DFM}$. The main step here to show is the existence of the two full monotone clauses — that the rest then is in 2-CNF follows by the minimality of the deficiency. [II] In case of $F \in 2\text{-}\mathcal{CLS}$ there must exist exactly one positive and one negative clause and these can be saturated to full positive resp. full negative clauses, and so we obtain $F' \in \mathcal{DFM}$. Once the connection to $\mathcal{DFM}$ is established, graph-theoretical reasoning does the remaining job: [I] The MSDs of minimal deficiency 0 are the cycles. [II] The only MSDs G such that the corresponding DFMs can be obtained as partial saturations of *nonsingular* 2-CNF are the dipaths, since we can only have two *linear vertices* in G, vertices of in- and out-degree one.

An overview on the main results of this paper is given in Fig. 1.

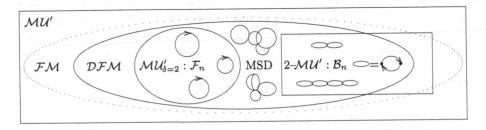

Fig. 1. Directed cycles at the heart of $\mathcal{MU}_{\delta=2}$, and dipaths at the heart of $2\text{-}\mathcal{MU}$.

2   Preliminaries

We use clause-sets F, finite sets of clauses, where a clause is a finite set of literals, and a literal is either a variable or a negated/complemented variable. The set of all variables is denoted by $\mathcal{VA}$ (we use variables also as vertices in graphs), and we assume $\mathbb{N} = \{1, 2, \ldots\} \subseteq \mathcal{VA}$. This makes creating certain examples easier, since we can use integers different from zero as literals (as in the DIMACS format). The set of clause-sets is denoted by $\mathcal{CLS}$, the empty clause-set by $\top := \emptyset \in \mathcal{CLS}$ and the empty clause by $\bot := \emptyset$. Clause-sets are interpreted as CNFs, conjunctions of disjunction of literals. A clause-set F is uniform resp. k-uniform, if all clauses of F have the same length resp. length k. This paper is self-contained, if however more background is required, then the Handbook chapter [13] can be consulted.

Clauses C do not contain clashes (conflicts), i.e., they are "non-tautological", which formally is denoted by $C \cap \overline{C} = \emptyset$, where for a set L of literals by $\overline{L}$ we denoted elementwise complementation. With $\mathrm{var}(F)$ we denote the set of variables occurring in F, while by $\mathrm{lit}(F) := \mathrm{var}(F) \cup \overline{\mathrm{var}(F)}$ we denote the possible literals of F (one of the two polarities of a literal in $\mathrm{lit}(F)$ must occur in F). Since the union $\bigcup F$ is the set of occurring literals, we have $\mathrm{lit}(F) = (\bigcup F) \cup \overline{\bigcup F}$, while $\mathrm{var}(F) = \mathrm{lit}(F) \cap \mathcal{VA}$. A clause C is *positive* if $C \subset \mathcal{VA}$, while C is *negative* if $C \subset \overline{\mathcal{VA}}$, and C is *mixed* otherwise; a non-mixed clause is called *monotone*. A **full clause** of a clause-set F is some $C \in F$ with $\mathrm{var}(C) = \mathrm{var}(F)$. A *full clause-set* is an $F \in \mathcal{CLS}$ where all $C \in F$ are full. By $\boldsymbol{A_n}$ we denote the full clause-set consisting of the 2^n full clauses over variables $1, \ldots, n$ for $n \in \mathbb{N}_0$.

So $A_0 = \{\bot\}$, $A_1 = \{\{-1\}, \{1\}\}$, and $A_2 = \{\{-1, -2\}, \{1, 2\}, \{-1, 2\}, \{1, -2\}\}$. For $F \in \mathcal{CLS}$ we use $n(F) := |\mathrm{var}(F)| \in \mathbb{N}_0$ for the number of (occurring) variables, $c(F) := |F| \in \mathbb{N}_0$ for the number of clauses, and $\delta(F) := c(F) - n(F) \in \mathbb{Z}$ for the **deficiency**. $p\text{-}\mathcal{CLS}$ is the set of $F \in \mathcal{CLS}$ such that for all $C \in F$ holds $|C| \leq p$. The application of partial assignments φ to $F \in \mathcal{CLS}$, denoted by $\varphi * F$, yields the clause-set obtained from F by removing clauses satisfied by φ, and removing falsified literals from the remaining clauses. Contractions can occur, since we are dealing with clause-*sets*, i.e., previously unequal clauses may become equal, and so more clauses might disappear than expected. Also more variables than just those in φ might disappear, since we consider only occurring variables. $\mathcal{SAT}$ is the set of satisfiable clause-sets, those $F \in \mathcal{CLS}$ where there is a partial assignment φ with $\varphi * F = \top$. $\mathcal{CLS}$ is partitioned into $\mathcal{SAT}$ and the set of unsatisfiable clause-sets. A clause-set F is irredundant iff for every $C \in F$ there exists a total assignment φ which satisfies $F \setminus \{C\}$ (i.e., $\varphi * (F \setminus \{C\}) = \top$) while falsifying C (i.e., $\varphi * \{C\} = \{\bot\}$). Every full clause-set is irredundant.

Isomorphism of clause-sets $F, G \in \mathcal{CLS}$ is denoted by $F \cong G$, that is, there exists a complement-preserving bijection from $\mathrm{lit}(F)$ to $\mathrm{lit}(G)$ which induces a bijection from the clauses of F to the clauses of G. For example for an unsatisfiable full clause-set F we have $F \cong A_{n(F)}$. $\mathcal{RHO}$ is the set of renamable Horn clause-sets, i.e., $F \in \mathcal{CLS}$ with $F \cong G$ for some Horn clause-set G (where every clause contains at most one positive literal, i.e., $\forall C \in G : |C \cap \mathcal{VA}| \leq 1$).

The **DP-operation** (sometimes also called "variable elimination") for $F \in \mathcal{CLS}$ and a variable v results in $\mathrm{DP}_v(F) \in \mathcal{CLS}$, which replaces all clauses in F containing variable v (positively or negatively) by their resolvents on v. Here for clauses C, D with $C \cap \overline{D} = \{x\}$ the resolvent of C, D on $\mathrm{var}(x)$ is $(C \setminus \{x\}) \cup (D \setminus \{\overline{x}\})$ (note that clauses can only be resolved if they contain *exactly* one clashing literal, since clauses are non-tautological).

We conclude by recalling some notions from graph theory: A graph/digraph G is a pair (V, E), with $V(G) := V$ a finite set of "vertices", while $E(G) := E$ is the set of "edges" resp. "arcs", which are two-element subsets $\{a, b\} \subseteq V$ resp. pairs $(a, b) \in V^2$ with $a \neq b$. An isomorphism between two (di)graphs is a bijection between the vertex sets, which induces a bijection on the edges/arcs. Isomorphism of clause-sets can be naturally reduced in polytime to graph isomorphism, and *GI-completeness* of such isomorphism problems means additionally that also the graph isomorphism problem can be reduced to it.

3   Review on Minimal Unsatisfiability (MU)

$\mathcal{MU}$ is the set of unsatisfiable clause-sets such that every strict sub-clause-set is satisfiable. For $F \in \mathcal{CLS}$ holds $F \in \mathcal{MU}$ iff F is unsatisfiable and irredundant. We note here that "$\mathcal{MU}$" is the *class* of MUs, while "MU" is used in text in a substantival role. $\mathcal{MU}' \subset \mathcal{MU}$ is the set of **nonsingular** MUs, that is, $F \in \mathcal{MU}$ such that every literal occurs at least twice. We use $2\text{-}\mathcal{MU} := \mathcal{MU} \cap 2\text{-}\mathcal{CLS}$ and $2\text{-}\mathcal{MU}' := \mathcal{MU}' \cap 2\text{-}\mathcal{CLS}$. **Saturated** MUs are those unsatisfiable $F \in \mathcal{CLS}$, such that for every $C \in F$ and every clause $D \supset C$ we have $(F \setminus \{C\}) \cup \{D\} \in$

$\mathcal{SAT}$. For $F \in \mathcal{MU}$ a **saturation** is some saturated $F' \in \mathcal{MU}$ where there exists a bijection $\alpha : F \to F'$ with $\forall C \in F : C \subseteq \alpha(C)$; by definition, every MU can be saturated. If we just add a few literal occurrences (possibly zero), staying with each step within $\mathcal{MU}$, then we speak of a **partial saturation** (this includes saturations); we note that the additions of a partial saturation can be arbitrarily permuted. Dually there is the notion of **marginal** MUs, those $F \in \mathcal{MU}$ where removing any literal from any clause creates a redundancy, that is, some clause following from the others. For $F \in \mathcal{MU}$ a **marginalisation** is some marginal $F' \in \mathcal{MU}$ such that there is a bijection $\alpha : F' \to F$ with $\forall C \in F' : C \subseteq \alpha(C)$; again, every MU can be marginalised, and more generally we speak of a **partial marginalisation**. As an example all $A_n \in \mathcal{MU}$, $n \in \mathbb{N}_0$, are saturated and marginal, while A_n is nonsingular iff $n \neq 1$.

For $F \in \mathcal{MU}$ and a variable $v \in \text{var}(F)$, we define **local saturation** as the process of adding literals $v, \overline{v}$ to some clauses in F (not already containing $v, \overline{v}$), until adding any additional v or $\overline{v}$ yields a satisfiable clause-set. Then the result is **locally saturated on** v. For a saturated $F \in \mathcal{MU}$, as shown in [17, Lemma C.1], assigning any (single) variable in F (called "splitting") yields MUs (for more information see [19, Subsect. 3.4]). The same proof yields in fact, that for a locally saturated $F \in \mathcal{MU}$ on a variable v, splitting on v maintains minimal unsatisfiability:

Lemma 1. *Consider $F \in \mathcal{MU}$ and a variable $v \in \text{var}(F)$. If F is locally saturated for variable v, then we have $\langle v \to \varepsilon \rangle * F \in \mathcal{MU}$ for both $\varepsilon \in \{0, 1\}$.*

In general, application of the DP-operation to some MU may or may not yield another MU. A positive example for $n \in \mathbb{N}$ and $v \in \{1, \ldots, n\}$ is $\text{DP}_v(A_n) \cong A_{n-1}$. A special case of DP-reduction, guaranteed to stay inside MU, is **singular DP-reduction**, where v is a singular variable in F. In this case, as shown in [18, Lemma 9], no tautological resolvents can occur and no contractions can take place (recall that we are using clause-*sets*, where as a result of some operations previously different clause can become equal – a "contraction"). So even each $\mathcal{MU}_{\delta=k}$ is stable under singular DP-reduction. We use $\textbf{sDP}(F) \subset \mathcal{MU}'_{\delta=\delta(F)}$, $F \in \mathcal{MU}$, for the set of all clause-sets obtained from F by singular DP-reduction. By [18, Corollary 64] for any $F', F'' \in \text{sDP}(F)$ holds $n(F') = n(F'')$. So we can define for $F \in \mathcal{MU}$ the **nonsingularity type** $\text{nst}(F) := n(F') \in \mathbb{N}_0$ via any $F' \in \text{sDP}(F)$. Thus $\text{nst}(F) = n(F)$ iff F is nonsingular.

MU(1) A basic fact is that $F \in \mathcal{MU} \setminus \{\{\bot\}\}$ contains a variable occurring positively and negatively each at most $\delta(F)$ times ([17, Lemma C.2]). So the minimum variable degree (the number of occurrences) is $2\delta(F)$ (sharper bounds are given in [19]). This implies that $F \in \mathcal{MU}_{\delta=1}$ has a **1-singular** variable (i.e., degree 2). It is well-known that for $F \in \mathcal{RHO}$ there exists an input-resolution tree T yielding $\{\bot\}$ ([10]); in the general framework of [9], these are those T with the Horton-Strahler number $\text{hs}(T)$ at most 1. W.l.o.g. we can assume all these trees to be regular, that is, along any path no resolution variable is repeated. This implies that for $F \in \mathcal{MU} \cap \mathcal{RHO}$ holds $\delta(F) = 1$, and all variables in F are singular. By [17] all of $\mathcal{MU}_{\delta=1}$ is described by a binary tree T, which just describes the expansion process as mentioned in the Introduction, and which is basically the same as a resolution tree refuting F (T is not unique). Since the variables in the tree are all unique (the creation process does not reuse variables), any two clauses clash in at most one variable. For $F \in \mathcal{MU}_{\delta=1}$ with exactly one 1-singular variable holds $\text{hs}(T) = 1$ (and so $F \in \mathcal{RHO}$), since $\text{hs}(T) \geq 2$ implies that there would be two nodes whose both children are leaves, and so F would have two 1-singular variables. Furthermore if $F \in \mathcal{MU} \cap \mathcal{RHO}$ has a full clause C, then C is on top of T and so the complement of its literals occur only once:

Lemma 2. *Consider $F \in \mathcal{MU}$. If $\delta(F) = 1$, and F has only one 1-singular variable, then $F \in \mathcal{RHO}$. If $F \in \mathcal{RHO}$, then all variables are singular, and if F has a full clause C, then for every $x \in C$ the literal $\overline{x}$ occurs only once in F.*

The Splitting Ansatz. The main method for analysing $F \in \mathcal{MU}$ is "splitting": choose an appropriate variable v in $F \in \mathcal{MU}$, apply the partial assignments $\langle v \rightarrow 0 \rangle$ and $\langle v \rightarrow 1 \rangle$ to F, obtain F_0, F_1, analyse them, and lift the information obtained back to F. An essential point here is to have $F_0, F_1 \in \mathcal{MU}$. In general this does not hold. The approach of Kleine Büning and Zhao, as outlined in [16, Sect. 3], is to remove clauses appropriately in F_0, F_1, and study various conditions. Our method is based on the observation, that if a clause say in F_0 became redundant, then $\overline{v}$ can be added to this clause in F, while still remaining MU, and so the assignment $v \rightarrow 0$ then takes care of the removal. This is the essence of *saturation*, with the advantage that we are dealing again with MUs. A saturated MU is characterised by the property, that for any variable, splitting yields two MUs. For classes like $2\text{-}\mathcal{CLS}$, which are not stable under saturation, we introduced *local saturation*, which only saturates the variable we want to split on. In our application, the local saturation uses all clauses, and this is equivalent to a "disjunctive splitting" as surveyed [2, Definition 8]. On the other hand, for deficiency 2 the method of saturation is more powerful, since we have stability under saturation, and the existence of a variable occurring twice positively and twice negatively holds *after* saturation. Splitting needs to be done on nonsingular variables, so that the deficiency becomes strictly smaller in F_0, F_1 — we want these instances "to be easy", to know them well. In both our cases we obtain indeed renamable Horn clause-sets. For deficiency 2 we exploit, that the splitting involves the minimal number of clauses, while for 2-CNF we exploit that the splitting involves the maximal number of clauses after local saturation. In order to get say F_0 "easy", while F is "not easy", the part which gets removed, which is related to F_1, must have special properties.

4   MU with Full Monotone Clauses (FM)

We now introduce formally the main classes of this paper, $\mathcal{FM} \subset \mathcal{MU}$ (Definition 1) and $\mathcal{DFM} \subset \mathcal{FM}$ (Definition 4). Examples for these classes showed up in the literature, but these natural classes haven't been studied yet.

Definition 1. *Let $\mathcal{FM}$ be the set of $F \in \mathcal{MU}$ such that there is a full positive clause $P \in F$ and a full negative clause $N \in F$ (that is, $\mathrm{var}(P) = \mathrm{var}(N) = \mathrm{var}(F)$, $P \subset \mathcal{VA}$, $N \subset \overline{\mathcal{VA}}$). More generally, let $\mathcal{FC}$ be the set of $F \in \mathcal{MU}$ such that there are full clauses $C, D \in F$ with $D = \overline{C}$.*

The closure of $\mathcal{FM}$ under isomorphism is $\mathcal{FC}$. In the other direction, for any $F \in \mathcal{FC}$ and any pair $C, D \in F$ of full clauses with $D = \overline{C}$ (note that in general such a pair is not unique), flip the signs so that C becomes a positive clause (and thus D becomes a negative clause), and we obtain an element of $\mathcal{FM}$. As usual we call the subsets of nonsingular elements $\mathcal{FM}'$ resp. $\mathcal{FC}'$. The *trivial elements* of $\mathcal{FM}$ and $\mathcal{FC}$ are the MUs with at most one variable: $\mathcal{FM}_{n\leq 1} = \mathcal{FM}_{\delta=1} = \mathcal{FC}_{n\leq 1} = \mathcal{FC}_{\delta=1} = \{\{\bot\}\} \cup \{\{v\}, \{\overline{v}\} : v \in \mathcal{VA}\}$. The singular cases in $\mathcal{FM}$ and $\mathcal{FC}$ are just these cases with only one variable:

Lemma 3. $\mathcal{FM}' = \mathcal{FM}_{\delta \geq 2} \cup \{\{\bot\}\}$, $\mathcal{FC}' = \mathcal{FC}_{\delta \geq 2} \cup \{\{\bot\}\}$.

Proof. Assume that there is a singular $F \in \mathcal{FC}$ with $n(F) \geq 2$. Let C, D be full complementary clauses in F. W.l.o.g. we can assume that there is $x \in C$ (so $\overline{x} \in D$) such that literal x only occurs in C. Consider now some $y \in D\backslash\{\overline{x}\}$ (exists due to $n(F) \geq 2$). There exists a satisfying assignment φ for $F' := F\backslash\{D\}$, and it must hold $\varphi(x) = 1$ and $\varphi(y) = 0$ (otherwise F would be satisfiable). Obtain φ' by flipping the value of x. Now φ' still satisfies F', since the only occurrence of literal x is C, and this clause contains $\overline{y}$ — but now φ' satisfies F. $\square$

So the study of $\mathcal{FM}$ is about special nonsingular MUs. In general we prefer to study $\mathcal{FM}$ over $\mathcal{FC}$, since here we can define the "core" as a sub-clause-set:

Definition 2. *For $F \in \mathcal{FM}$ there is exactly one positive clause $P \in F$, and exactly one negative clause $N \in F$ (otherwise there would be subsumptions in F), and we call $F \setminus \{P, N\}$ the* **core** *of F.*

We note that cores consist only of mixed clauses, and in general any mixed clause-set (consisting only of mixed clauses) has always at least two satisfying assignments, the all-0 and the all-1 assignments. The decision complexity of $\mathcal{FM}$ is the same as that of $\mathcal{MU}$ (which is the same as $\mathcal{MU}'$), which has been determined in [22, Theorem 1] as complete for the class D^P, whose elements are differences of NP-classes (for example "MU = Irredundant minus SAT"):

Theorem 1. *For $F \in \mathcal{CLS}$, the decision whether "$F \in \mathcal{FM}$?" is D^P-complete.*

Proof. The decision problem is in D^P, since $F \in \mathcal{FM}$ iff F is irredundant with full monotone clauses and $F \notin \mathcal{SAT}$. For the reduction of $\mathcal{MU}$ to $\mathcal{FM}$, we consider $F \in \mathcal{CLS}$ with $n := n(F) \geq 2$, and first extend F to F', forcing a full positive clause, by taking a new variable v, adding literal v to all clauses of $\mathcal{F}_n$ and adding literal $\overline{v}$ to all clauses of F. Then we force additionally a full negative clause, extending F' to F'' in the same way, now using new variable w, and adding w to all clauses of F' and adding $\overline{w}$ to all clauses of $\mathcal{F}_{n+1}$. We have $F \in \mathcal{MU}$ iff $F'' \in \mathcal{MU}$. $\square$

We now turn to the semantics of the core:

Definition 3. *For $V \subset \mathcal{VA}$ the **AllEqual** function on V is the boolean function which is true for a total assignment of V if all variables are assigned the same value, and false otherwise. A **CNF-realisation** of AllEqual on V is a clause-set F with $\text{var}(F) \subseteq V$, which is as a boolean function the AllEqual function on V.*

The core of every FM F realises AllEqual on $\text{var}(F)$ irredundantly, and this characterises $\mathcal{FM}$, yielding the ALLEQUAL THEOREM:

Theorem 2. *Consider $F \in \mathcal{CLS}$ with a full positive clause $P \in F$ and a full negative clause $N \in F$, and let $F' := F \setminus \{P, N\}$. Then $F \in \mathcal{FM}$ if and only if F' realises AllEqual on $\text{var}(F)$, and F' is irredundant.*

5   FM with Binary Clauses (DFM)

Definition 4. $\mathcal{DFM}$ *is the subset of $\mathcal{FM}$ where the core is in $2\text{-}\mathcal{CLS}$, while $\mathcal{DFC}$ is the set of $F \in \mathcal{FC}$, such that there are full complementary clauses $C, D \in F$ with $F \setminus \{C, D\} \in 2\text{-}\mathcal{CLS}$.*

The core of DFMs consists of clauses of length exactly 2. $\mathcal{DFC}$ is the closure of $\mathcal{DFM}$ under isomorphism.

Definition 5. *For $F \in \mathcal{DFM}$ the **positive implication digraph** $\text{pdg}(F)$ has vertex set $\text{var}(F)$, i.e., $V(\text{pdg}(F)) := \text{var}(F)$, while the arcs are the implications on the variables as given by the core F' of F, i.e., $E(\text{pdg}(F)) := \{(a, b) : \{\overline{a}, b\} \in F', a, b \in \text{var}(F)\}$. This can also be applied to any mixed binary clause-set F (note that the core F' is such a mixed binary clause-set).*

The essential feature of mixed clause-sets $F \in 2\text{-}\mathcal{CLS}$ is that for a clause $\{\overline{v}, w\} \in F$ we only need to consider the "positive interpretation" $v \to w$, not the "negative interpretation" $\overline{w} \to \overline{v}$, since the positive literals and the negative literals do not interact. So we do not need the (full) implication digraph. Via the positive implication digraphs we can understand when a mixed clause-set realises AllEqual. We recall that digraph G is a **strong digraph** (SD), if G is strongly connected, i.e., for every two vertices a, b there is a path from a to b. A **minimal strong digraph** (MSD) is an SD G, such that for every arc $e \in E(G)$ holds that $(V(G), E(G) \setminus \{e\})$ is not strongly connected. Every digraph G with $|V(G)| \leq 1$ is an MSD. We are ready to formulate the CORRESPONDENCE LEMMA:

Lemma 4. *A mixed binary clause-set F is a CNF-realisation of AllEqual iff $\text{pdg}(F)$ is an SD, where F is irredundant iff $\text{pdg}(F)$ is an MSD.*

Proof. The main point here is that the resolution operation for mixed binary clauses $\{\overline{a}, b\}, \{\overline{b}, c\}$, resulting in $\{\overline{a}, c\}$, corresponds exactly to the formation of transitive arcs, i.e., from $(a, b), (b, c)$ we obtain (a, c). So the two statements of the lemma are just easier variations on the standard treatment of logical reasoning for 2-CNFs via "path reasoning". $\qquad\square$

As explained before, $F \mapsto \mathrm{pdg}(F)$ converts mixed binary clause-sets with full monotone clauses to a digraph. Also the reverse direction is easy:

Definition 6. *For a finite digraph G with $V(G) \subset \mathcal{VA}$, the clause-set $\mathbf{mcs}(G) \in \mathcal{CLS}$ ("m" like "monotone") is obtained by interpreting the arcs $(a, b) \in E(G)$ as binary clauses $\{\overline{a}, b\} \in \mathrm{mcs}(G)$, and adding the two full monotone clauses $\{V(G), \overline{V(G)}\} \subseteq \mathrm{mcs}(G)$.*

For the map $G \mapsto \mathrm{mcs}(G)$, we use the vertices of G as the variables of $\mathrm{mcs}(G)$. An arc (a, b) naturally becomes a mixed binary clause $\{\overline{a}, b\}$, and we obtain the set F' of mixed binary clauses, where by definition we have $\mathrm{pdg}(F') = G$. This yields a bijection between the set of finite digraphs G with $V(G) \subset \mathcal{VA}$ and the set of mixed binary clause-sets. By the Correspondence Lemma 4, minimal strong connectivity of G is equivalent to F' being an irredundant AllEqual-representation. So there is a bijection between MSDs and the set of mixed binary clause-sets which are irredundant AllEqual-representation. We "complete" the AllEqual-representations to MUs, by adding the full monotone clauses, and we get the DFM $\mathrm{mcs}(G)$. We see, that DFMs and MSDs are basically the "same thing", only using different languages, which is now formulated as the CORRESPONDENCE THEOREM (with obvious proofs left out):

Theorem 3. *The two formations $F \mapsto \mathrm{pdg}(F)$ and $G \mapsto \mathrm{mcs}(G)$ are inverse to each other, i.e., $\mathrm{mcs}(\mathrm{pdg}(F)) = F$ and $\mathrm{pdg}(\mathrm{mcs}(G)) = G$, and they yield inverse bijections between DFMs and MSDs: For every $F \in \mathcal{DFM}$ the digraph $\mathrm{pdg}(F)$ is an MSD, and for every MSD G with $V(G) \subset \mathcal{VA}$ we have $\mathrm{mcs}(G) \in \mathcal{DFM}$.*

The Correspondence Theorem 3 can be considerably strengthened, by including other close relations, but here we formulated only what we need. For a DFM $F \neq \{\bot\}$ and an MSD $G \neq (\emptyset, \emptyset)$ we obtain $\delta(\mathrm{pdg}(F)) = \delta(F) - 2$ and $\delta(\mathrm{mcs}(G)) = \delta(G) + 2$, where we define the **deficiency** of a digraph G as $\boldsymbol{\delta(G)} := |E(G)| - |V(G)|$. Concerning isomorphisms there is a small difference between the two domains, since the notion of clause-set isomorphism includes flipping of variables, which for DFMs can be done all at once (flipping "positive" and "negative") — this corresponds in $\mathrm{pdg}(F)$ to the reversal of the direction of all arcs. For our two main examples, cycles and dipaths, this yields an isomorphic digraph, but this is not the case in general.

Marginalisation of DFMs concerns only the full monotone clauses and not the binary clauses, formulated as the MARGINALISATION LEMMA:

Lemma 5. *Consider a clause-set F obtained by partial marginalisation of a non-trivial DFM F'. Then F has no unit-clause and its formation did not touch binary clauses but only shortened its monotone clauses.*

Proof. By definition, partial marginalisation can be arbitrarily reordered. If some binary clause would be shortened, then, put first, this would yield unit-clauses, subsuming some full monotone clauses. □

Deciding $F \in \mathcal{DFM}$ can be done in polynomial time: Check whether we have the two full monotone clauses, while the rest are binary clauses, if yes, translate the binary clauses to a digraph and decide whether this digraph is an MSD (which can be done in quadratic time; recall that deciding the SD property can be done in linear time) — if yes, then $F \in \mathcal{DFM}$, otherwise $F \notin \mathcal{DFM}$. We now come to the two simplest example classes, cycles and "di-paths". Let $C_n := (\{1,\ldots,n\}, \{(1,2),\ldots,(n-1,n),(n,1)\})$ for $n \geq 2$ be the directed cycle of length n. The directed cycles C_n have the minimum deficiency zero among MSDs. We obtain the basic class $\mathcal{F}_n$, as already explained in the Introduction:

Definition 7. Let $\mathcal{F}_n := \mathrm{mcs}(C_n) \in \mathcal{DFM}$ for $n \geq 2$ (Definition 6).

Lemma 6. For $F \in \mathcal{DFM}_{\delta=2}$ holds $F \cong \mathcal{F}_{n(F)}$.

Proof. By the Correspondence Theorem 3, $\mathrm{pdg}(F)$ is an MSD with the deficiency $\delta(F) - 2 = 0$, and thus is a directed cycle of length $n(F)$. □

Lemma 7. For every $n \geq 2$, $\mathcal{F}_n$ is saturated.

Proof. We show that adding a literal x to any $C \in \mathcal{F}_n$ introduces a satisfying assignment, i.e., $\mathcal{F}_n$ is saturated. The monotone clauses are full, and saturation can only touch the mixed clauses. Recall $\mathrm{var}(\mathcal{F}_n) = \{1,\ldots,n\}$. Due to symmetry assume $C = \{-n, 1\}$, and we add $x \in \{2,\ldots,n-1\}$ to C. Let φ be the total assignment setting all variables $2,\ldots,n$ to true and 1 to false. Then φ satisfies the monotone clauses and the new clause $\{-n, 1, x\}$. Recall that every literal occurs only once in the core of $\mathcal{F}_n$. So literal 1 occurs only in C. Thus φ satisfies also every mixed clause in $F \setminus \{C\}$ (which has a positive literal other than 1). □

For a tree G (a finite connected acyclic graph with at least one vertex) we denote by $D(G) := (V(G), \{(a,b), (b,a) : \{a,b\} \in E(G)\})$ the directed version of G, converting every edge $\{a,b\}$ into two arcs $(a,b), (b,a)$; in [7] these are called "directed trees", and we use **ditree** here. For every tree G the ditree $D(G)$ is an MSD. Let $P_n := (\{1,\ldots,n\}, \{\{1,2\},\ldots,\{n-1,n\}\})$, $n \in \mathbb{N}_0$, be the pathgraph.

Definition 8. Let $\mathbf{DB}_n := \mathrm{mcs}(D(P_n)) \in \mathcal{DFM}$ $(n \in \mathbb{N}_0)$ (Definition 6).

So $\mathrm{DB}_n = A_n$ for $n \leq 2$, while in general $n(\mathrm{DB}_n) = n$, and for $n \geq 1$ holds $c(\mathrm{DB}_n) = 2 + 2(n-1) = 2n$, and $\delta(\mathrm{DB}_n) = n$. DB_n for $n \neq 1$ is nonsingular, and every variable in $\mathrm{var}(\mathrm{DB}_n) \setminus \{1, n\}$ is of degree 6 for $n \geq 2$, while the variables $1, n$ (the endpoints of the dipath) have degree 4. Among ditrees, only dipaths can be marginalised to nonsingular 2-uniform MUs, since dipaths are the only ditrees with exactly two *linear vertices* (i.e., vertices with indegree and outdegree equal to 1). The unique marginal MUs obtained from dipaths are as follows:

Definition 9. For $n \geq 1$ obtain the uniform $\mathcal{B}_n \in 2\text{-}\mathcal{MU}$ from DB_n by replacing the full positive/negative clause with $\{1, n\}$ resp. $\{-1, -n\}$, i.e., $\mathcal{B}_n = \{\{-1, -n\}, \{1, n\}, \{-1, 2\}, \{1, -2\}, \ldots, \{-(n-1), n\}, \{n-1, -n\}\}$; $\mathcal{B}_0 := \mathrm{DB}_0$.

6   Deficiency 2 Revisited

We now come to the first main application of the new class $\mathcal{DFM}$, and we give a new and relatively short proof, that the $\mathcal{F}_n$ are precisely the nonsingular MUs of deficiency 2. The core combinatorial-logical argument is to show $\mathcal{MU}'_{\delta=2} \subseteq \mathcal{FC}_{\delta=2}$, i.e., every $F \in \mathcal{MU}'_{\delta=2}$ must have two full complementary clauses $C, D \in F$. The connection to the "geometry" then is established by showing $\mathcal{FM}_{\delta=2} \subseteq \mathcal{DFM}_{\delta=2}$, i.e., if an FM F has deficiency 2, then it must be a DFM, i.e., all clauses besides the full monotone clauses are binary. The pure geometrical argument is the characterisation of $\mathcal{DFM}_{\delta=2}$, which has already been done in Lemma 6.

The proof of the existence of full clauses $D = \overline{C}$ in F is based on the Splitting Ansatz, as explained in Sect. 3. Since $\mathcal{MU}_{\delta=2}$ is stable under saturation, we can start with a saturated F, and can split on any variable (though later an argument is needed to undo saturation). There must be a variable v occurring at most twice positively as well as negatively (otherwise the basic lemma $\delta(F) \geq 1$ for any MU F would be violated), and due to nonsingularity v occurs exactly twice positively and negatively. The splitting instances F_0, F_1 have deficiency 1. So they have at least one 1-singular variable. There is very little "space" to reduce a nonsingular variable in F to a 1-singular variable in F_0 resp. F_1, and indeed those two clauses whose vanishing in F_0 do this, are included in F_1, and vice versa. Since clauses in $\mathcal{MU}_{\delta=1}$ have at most one clash, F_0, F_1 have exactly one 1-singular variable. And so by the geometry of the structure trees (resp. their Horton-Strahler numbers), both F_0, F_1 are in fact renamable Horn! Thus every variable in F_0, F_1 is singular, and F_0, F_1 must contain a unit-clause. Again considering both sides, it follows that the (two) positive occurrences of v must be a binary clause (yielding the unit-clause) and a full clause C (whose vanishing yields the capping of all variables to singular variables), and the same for the (two) negative occurrences, yielding D. So $F_0, F_1 \in \mathcal{RHO}$ both contain a full clause and we know that the complements of the literals in the full clause occur exactly once in F_0 resp. F_1. Thus in fact C resp. D have the "duty" of removing each others complement, and we get $D = \overline{C}$.

Now consider $F \in \mathcal{FM}_{\delta=2}$ with monotone full clauses $C, D \in F$. Transform the core F' within F into an equivalent F'', by replacing each clause in F' by a contained prime implicate of F', which, since the core means that all variables are equal (semantically), is binary. So we arrive in principle in $\mathcal{DFM}$, but we could have created redundancy — and this can not happen, since an MSD has minimum deficiency 0. The details are as follows:

Theorem 4. $\mathcal{DFC}_{\delta=2} = \mathcal{FC}_{\delta=2} = \mathcal{MU}'_{\delta=2}$.

Proof. By definition and Lemma 3 we have $\mathcal{DFC}_{\delta=2} \subseteq \mathcal{FC}_{\delta=2} \subseteq \mathcal{MU}'_{\delta=2}$. First we show $\mathcal{MU}'_{\delta=2} \subseteq \mathcal{FC}_{\delta=2}$, i.e., every $F \in \mathcal{MU}'_{\delta=2}$ has two full complementary clauses. Recall that F has a variable $v \in \text{var}(F)$ of degree 4, which by nonsingularity is the minimum variable degree. So v has two positive occurrences in clauses $C_1, C_2 \in F$ and two negative occurrences in clauses $D_1, D_2 \in F$. We assume that F is saturated (note that saturation maintains minimal unsatisfiability and deficiency). By the Splitting Ansatz, $F_0 := \langle v \to 0 \rangle * F \in \mathcal{MU}_{\delta=1}$ and $F_1 := \langle v \to 1 \rangle * F \in \mathcal{MU}_{\delta=1}$. So F_0 removes D_1, D_2 and shortens C_1, C_2, while F_1 removes C_1, C_2 and shortens D_1, D_2. Both F_0, F_1 contain a 1-singular variable (i.e., of degree 2), called a resp. b. We obtain $\{a, \bar{a}\} \subseteq D_1 \cup D_2$, since F has no singular variable and only by removing D_1, D_2 the degree of a decreased to 2. Similarly $\{b, \bar{b}\} \subseteq C_1 \cup C_2$. In $\mathcal{MU}_{\delta=1}$ any two clauses have at most one clash, and thus indeed F_0, F_1 have each exactly one 1-singular variable. Now $F_0, F_1 \in \mathcal{MU}_{\delta=1}$ with exactly one 1-singular variable are renamable Horn clause-sets (Lemma 2). Since $F_0, F_1 \in \mathcal{RHO} \cap \mathcal{MU}$ contain unit-clauses, created by clause-shortening, one of C_1, C_2 and one of D_1, D_2 are binary. W.l.o.g. assume C_1, D_1 are binary. Furthermore by Lemma 2 all variables in F_0, F_1 are singular, while F has no singular variable. So in F_0 all singularity is created by the removal of D_1, D_2, and in F_1 all singularity is created by the removal of C_1, C_2. Thus C_2, D_2 are full clauses. For a full clause in F_0, F_1, the complement of its literals occur only once (recall Lemma 2). Thus C_2 and D_2 have the duty of eliminating each others complements, and so we obtain $C_2 = \overline{D_2}$. To finish the first part, we note that the literals in clauses C, D each occurs exactly twice by the previous argumentation, and thus, since F was nonsingular to start with, indeed the initial saturation did nothing.

We turn to the second part of the proof, showing $\mathcal{FM}_{\delta=2} \subseteq \mathcal{DFM}_{\delta=2}$, i.e., the core F' of every $F \in \mathcal{FM}_{\delta=2}$ contains only binary clauses. By the characterisation of FMs, the AllEqual Theorem 2, F' realises AllEqual over the variables of F. The deficiency of F' is $\delta(F') = \delta(F) - 2 = 0$. Obtain F'' by replacing each $C \in F'$ by a prime implicate $C'' \subseteq C$ of F', where every prime implicate is binary. Now F'' is logically equivalent to F', and we can apply the Correspondence Lemma 4 to F'', obtaining an MSD $G := \mathrm{pdg}(F'')$ with $\delta(G) = \delta(F'')$. Due to the functional characterisation of F' we have $\mathrm{var}(F'') = \mathrm{var}(F') = \mathrm{var}(F)$. Using that MSDs have minimal deficiency 0, thus $\delta(G) = 0$, and so G is the cycle of length $n(F)$, and thus F'' is isomorphic to $\mathcal{F}_{n(F)}$. Now $\mathcal{F}_{n(F)}$ is saturated (Lemma 7), and thus indeed $F'' = F'$.   □

Corollary 1 ([12]). *For $F \in \mathcal{MU}'_{\delta=2}$ holds $F \cong \mathcal{F}_{n(F)}$.*

7 MU for 2-CNF

Lemma 8. *$F \in 2\text{-}\mathcal{MU}$ with a unit-clause is in $\mathcal{RHO}$, and has at most two unit-clauses ([2]). In every $F \in 2\text{-}\mathcal{MU}$ each literal occurs at most twice ([15]).*

Lemma 9. *In $F \in 2\text{-}\mathcal{MU}$ with exactly two unit-clauses, every literal occurs exactly once. Both unit-clauses can be partially saturated to a full clause (yielding two saturations), and these two full clauses are complementary.*

Proof. By Lemma 8, $F \in \mathcal{RHO}$ with $\delta(F) = 1$. Since F is uniform except of two unit-clauses, the number of literal occurrences is $2c(F) - 2 = 2n(F)$, and so every literal in F occurs only once (F is marginal). Consider an underlying tree T according to Sect. 3, in the form of an input-resolution tree T. The key is that any of the two unit-clauses can be placed at the top, and thus can be saturated (alone) to a full clause. Since in an input-resolution tree, at least one unit-clause is needed at the bottom, to derive the empty clause, we see that the two possible saturations yield complementary clauses.   □

We now come to the main results of this section, characterising the nonsingular MUs in 2-CNF. First the combinatorial part of the characterisation: the goal is to show that $F \in 2\text{-}\mathcal{MU}'$ can be saturated to a DFM, up to renaming, i.e., there exist a positive clause and a negative clause which can be partially saturated to full positive and negative clauses. The proof is based on the Splitting Ansatz. Unlike $\mathcal{MU}'_{\delta=2}$, 2-CNF MUs are not stable under saturation. So we use *local* saturation on a variable $v \in \text{var}(F)$, where we get splitting instances $F_0, F_1 \in 2\text{-}\mathcal{MU}$. Then we show F_0, F_1 are indeed in $\mathcal{RHO}$ with exactly 2 unit-clauses, and we apply that any of these unit-clauses can be saturated to a full clause. W.l.o.g. we saturate any of the two unit-clauses in F_0 to a full positive clause. Now one of the two unit-clauses in F_1 can be saturated to a full negative clause, and the two full monotone clauses can be lifted to F. This yields a DFM which is a partial saturation of F. The details are as follows:

Theorem 5. *Every* $2\text{-}\mathcal{MU}'$ *can be partially saturated to some* $\mathcal{DFC}$.

Proof. We show $F \in 2\text{-}\mathcal{MU}'$ contains, up to flipping of signs, exactly one positive and one negative clause, and these can be saturated to full monotone clauses. F has no unit-clause and is 2-uniform. By Lemma 8 every literal in F has degree 2. Let $F' \in \mathcal{MU}$ be a clause-set obtained from F by locally saturating $v \in \text{var}(F)$. So $F_0 := \langle v \to 0 \rangle * F'$ and $F_1 := \langle v \to 1 \rangle * F'$ are in $2\text{-}\mathcal{MU}$ (Lemma 1) and each has exactly two unit-clauses (obtained precisely from the clauses in F containing $v, \overline{v}$). By Lemma 8 holds $F_0, F_1 \in \mathcal{RHO} \cap \mathcal{MU}_{\delta=1}$. And by Lemma 9 all variables are 1-singular and in each of F_0, F_1, both unit-clauses can be partially saturated to a full clause. These full clauses can be lifted to the original F (by adding v resp. $\overline{v}$) while maintaining minimal unsatisfiability (if both splitting results are MU, so is the original clause-set; see [19, Lemma 3.15, Part 1]). Now we show that for a full clause in F_0, F_1 adding v or $\overline{v}$ yields a full clause in F, i.e., only v vanished by splitting. All variables in F_0, F_1 are 1-singular, while F has no singular variable. If there would be a variable w in F_0 but not in F_1, then the variable degree of w would be 2 in F, a contradiction. Thus $\text{var}(F_0) \subseteq \text{var}(F_1)$. Similarly we obtain $\text{var}(F_1) \subseteq \text{var}(F_0)$. So $\text{var}(F_0) = \text{var}(F_1) = \text{var}(F) \setminus \{v\}$.

It remains to show that we can lift w.l.o.g. a full positive clause from F_0 and a full negative clause from F_1. Let $C_1, C_2 \in F$ be the clauses containing v and $D_1, D_2 \in F$ be the clauses containing $\overline{v}$. Assume the unit-clause $C_1 \setminus \{v\} \in F_0$ can be saturated to a full positive clause. This implies that every $C \in F \setminus \{C_1\}$ has a negative literal (since $F \setminus \{C_1\}$ is satisfied by setting all variables to false). Then by Lemma 9 the unit-clause $C_2 \setminus \{v\}$ can be saturated to a full negative clause in F_0. Similarly we obtain that every clause in $F \setminus \{C_2, D_1, D_2\}$ has a positive literal. So F has exactly one positive clause C_1 and all binary clauses in F_0, F_1 are mixed. Since $c(F_1) = n(F_1) + 1 = (n(F) - 1) + 1 = n(F)$ and there are

$n(F) - 1$ occurrences of each literal in F_1, w.l.o.g. D_1 is a negative clause and D_2 is mixed. Recall that in $\mathcal{MU}_{\delta=1}$ every two clauses have at most one clash, and so $D_1 \setminus \{\overline{v}\} \in F_1$ can be saturated to a full negative clause (otherwise there would be a clause with more than one clash with the full clause). So we obtain a DFM which is a partial saturation of F. □

By [7, Theorem 4], every MSD with at least two vertices has at least two linear vertices. We need to characterise a special case of MSDs with exactly two linear vertices. This could be derived from the general characterisation by [8, Theorem 7], but proving it directly is useful and not harder than to derive it:

Lemma 10. *An MSD G with exactly two linear vertices, where every other vertex has indegree and outdegree both at least 2, is a dipath.*

Proof. We show that G is a dipath by induction on $n := |V(G)|$. For $n = 2$ clearly G is MSD iff G is a dipath. So assume $n \geq 3$. Consider a linear vertex $v \in V(G)$ with arcs (w, v) and (v, w'), where $w, w' \in V(G)$. If $w \neq w'$ would be the case, then the MSD obtained by removing v and adding the arc (w, w') had only one linear vertex (since the indegree/outdegree of other vertices are unchanged). So we have $w = w'$. Let G' be the MSD obtained by removing v. Now w is a linear vertex in G' (since every MSD has at least two linear vertices). By induction hypothesis G' is a dipath, and the assertion follows. □

By definition, for a mixed binary clause-set F a 1-singular variable is a linear vertex in $\text{pdg}(F)$. So by the Correspondence Theorem, a variable v in a DFM F has degree 4 (i.e., degree 2 in the core) iff v is a linear vertex in $\text{pdg}(F)$.

Theorem 6. *$F \in \mathcal{DFC}$ can be partially marginalised to some nonsingular element of 2-$\mathcal{CLS}$ if and only if $F \cong \text{DB}_{n(F)}$.*

Proof. Since $\mathcal{B}_n$ is a marginalisation of DB_n (obviously then the unique nonsingular one), it remains to show that a DFM F, which can be partially marginalised as in the assertion, is isomorphic to $\text{DB}_{n(F)}$. We show that $\text{pdg}(F)$ has exactly two linear vertices, while all other vertices have indegree and outdegree at least two, which proves the statement by Lemma 10. Consider a nonsingular $G \in 2\text{-}\mathcal{MU}$ obtained by marginalisation of F. Recall that by the Marginalisation Lemma 5 the mixed clauses are untouched. $\text{pdg}(F)$ has at least two linear vertices, so the mixed clauses in G have at least two 1-singular variables. Indeed the core of F has exactly two 1-singular variables, since these variables must occur in the positive and negative clauses of G, which are of length two. The other vertices have indegree/outdegree at least two due to nonsingularity. □

By Theorems 5, 6 we obtain a new proof for the characterisation of nonsingular MUs with clauses of length at most two:

Corollary 2 ([15]). *For $F \in 2\text{-}\mathcal{MU}'$ holds $F \cong \mathcal{B}_{n(F)}$.*

8   Conclusion

We introduced the novel classes $\mathcal{FM}$ and $\mathcal{DFM}$, which offer new conceptual insights into MUs. Fundamental for $\mathcal{FM}$ is the observation, that the easy syntactical criterion of having both full monotone clauses immediately yields the complete understanding of the semantics of the core. Namely that the satisfying assignments of the core are precisely the negations of the full monotone clauses, and so all variables are either all true or all false, i.e., all variables are equivalent. $\mathcal{DFM}$ is the class of FMs where the core is a 2-CNF. This is equivalent to the clauses of the core, which must be mixed binary clauses $\{\overline{v}, w\}$, constituting an MSD via the arcs $v \rightarrow w$. Due to the strong correspondence between DFMs and MSDs, once we connect a class of MUs to $\mathcal{DFM}$, we can use the strength of graph-theoretical reasoning. As a first application of this approach, we provided the known characterisations of $\mathcal{MU}'_{\delta=2}$ and $2\text{-}\mathcal{MU}'$ in an accessible manner, unified by revealing the underlying graph-theoretical reasoning.

References

1. Aharoni, R., Linial, N.: Minimal non-two-colorable hypergraphs and minimal unsatisfiable formulas. J. Comb. Theor. Ser. A **43**(2), 196–204 (1986). https://doi.org/10.1016/0097-3165(86)90060-9
2. Kleine Büning, H., Wojciechowski, P., Subramani, K.: On the computational complexity of read once resolution decidability in 2CNF formulas. In: Gopal, T.V., Jäger, G., Steila, S. (eds.) TAMC 2017. LNCS, vol. 10185, pp. 362–372. Springer, Cham (2017). https://doi.org/10.1007/978-3-319-55911-7_26
3. Buresh-Oppenheim, J., Mitchell, D.: Minimum witnesses for unsatisfiable 2CNFs. In: Biere, A., Gomes, C.P. (eds.) SAT 2006. LNCS, vol. 4121, pp. 42–47. Springer, Heidelberg (2006). https://doi.org/10.1007/11814948_6
4. Buresh-Oppenheim, J., Mitchell, D.: Minimum 2CNF resolution refutations in polynomial time. In: Marques-Silva, J., Sakallah, K.A. (eds.) SAT 2007. LNCS, vol. 4501, pp. 300–313. Springer, Heidelberg (2007). https://doi.org/10.1007/978-3-540-72788-0_29
5. Davydov, G., Davydova, I., Büning, H.K.: An efficient algorithm for the minimal unsatisfiability problem for a subclass of CNF. Ann. Math. Artif. Intell. **23**(3–4), 229–245 (1998). https://doi.org/10.1023/A:1018924526592
6. Fleischner, H., Kullmann, O., Szeider, S.: Polynomial-time recognition of minimal unsatisfiable formulas with fixed clause-variable difference. Theor. Comput. Sci. **289**(1), 503–516 (2002). https://doi.org/10.1016/S0304-3975(01)00337-1
7. García-López, J., Marijuán, C.: Minimal strong digraphs. Discrete Math. **312**(4), 737–744 (2012). https://doi.org/10.1016/j.disc.2011.11.010
8. García-López, J., Marijuán, C., Pozo-Coronado, L.M.: Structural properties of minimal strong digraphs versus trees. Linear Algebra Appl. **540**, 203–220 (2018). https://doi.org/10.1016/j.laa.2017.11.027
9. Gwynne, M., Kullmann, O.: Generalising unit-refutation completeness and SLUR via nested input resolution. J. Autom. Reasoning **52**(1), 31–65 (2014). https://doi.org/10.1007/s10817-013-9275-8
10. Henschen, L.J., Wos, L.: Unit refutations and Horn sets. J. Assoc. Comput. Mach. **21**(4), 590–605 (1974). https://doi.org/10.1145/321850.321857

11. Khachiyan, L., Boros, E., Elbassioni, K., Gurvich, V.: On enumerating minimal dicuts and strongly connected subgraphs. Algorithmica **50**, 159–172 (2008). https://doi.org/10.1007/s00453-007-9074-x
12. Büning, H.K.: On subclasses of minimal unsatisfiable formulas. Discrete Appl. Math. **107**(1–3), 83–98 (2000). https://doi.org/10.1016/S0166-218X(00)00245-6
13. Büning, H.K., Kullmann, O.: Minimal unsatisfiability and autarkies. In: Biere, A., Heule, M.J.H., van Maaren, H., Walsh, T. (eds.) Handbook of Satisfiability, vol. 185 of Frontiers in Artificial Intelligence and Applications, Chap. 11, pp. 339–401. IOS Press, February 2009. https://doi.org/10.3233/978-1-58603-929-5-339
14. Büning, H.K., Xu, D.: The complexity of homomorphisms and renamings for minimal unsatisfiable formulas. Ann. Math. Artif. Intell. **43**(1–4), 113–127 (2005). https://doi.org/10.1007/s10472-005-0422-8
15. Büning, H.K., Zhao, X.: Minimal unsatisfiability: results and open questions. Technical report tr-ri-02-230, Series Computer Science, University of Paderborn, University of Paderborn, Department of Mathematics and Computer Science (2002). http://wwwcs.uni-paderborn.de/cs/ag-klbue/de/research/MinUnsat/index.html
16. Büning, H.K., Zhao, X.: On the structure of some classes of minimal unsatisfiable formulas. Discrete Appl. Math. **130**(2), 185–207 (2003). https://doi.org/10.1016/S0166-218X(02)00405-5
17. Kullmann, O.: An application of matroid theory to the SAT problem. In: Proceedings of the 15th Annual IEEE Conference on Computational Complexity, pp. 116–124, July 2000. See also TR00-018, Electronic Colloquium on Computational Complexity (ECCC), March 2000. https://doi.org/10.1109/CCC.2000.856741
18. Kullmann, O., Zhao, X.: On Davis-Putnam reductions for minimally unsatisfiable clause-sets. Theor. Comput. Sci. **492**, 70–87 (2013). https://doi.org/10.1016/j.tcs.2013.04.020
19. Kullmann, O., Zhao, X.: Bounds for variables with few occurrences in conjunctive normal forms. Technical report arXiv:1408.0629v5 [math.CO], arXiv, January 2017. http://arxiv.org/abs/1408.0629
20. Lee, C.: On the size of minimal unsatisfiable formulas. Electron. J. Combinatorics **16**(1) (2009). Note #N3. http://www.combinatorics.org/Volume_16/Abstracts/v16i1n3.html
21. Liberatore, P.: Redundancy in logic II: 2CNF and Horn propositional formulae. Artif. Intell. **172**(2–3), 265–299 (2008). https://doi.org/10.1016/j.artint.2007.06.003
22. Papadimitriou, C.H., Wolfe, D.: The complexity of facets resolved. J. Comput. Syst. Sci. **37**(1), 2–13 (1988). https://doi.org/10.1016/0022-0000(88)90042-6
23. Brendan McKay (https://mathoverflow.net/users/9025/brendan-mckay). Answer: Graph isomorphism problem for minimally strongly connected digraphs. MathOverflow, August 2017. https://mathoverflow.net/q/279299 (version: 2017-08-23); question asked by Hoda Abbasizanjani (Swansea University).https://mathoverflow.net/a/279386

Finding All Minimal Safe Inductive Sets

Ryan Berryhill[1](✉), Alexander Ivrii[3], and Andreas Veneris[1,2]

[1] Department of Electrical and Computer Engineering,
University of Toronto, Toronto, Canada
{ryan,veneris}@eecg.utoronto.ca
[2] Department of Computer Science, University of Toronto, Toronto, Canada
[3] IBM Research Haifa, Haifa, Israel
ALEXI@il.ibm.com

Abstract. Computing minimal (or even just small) certificates is a central problem in automated reasoning and, in particular, in automated formal verification. For unsatisfiable formulas in CNF such certificates take the form of Minimal Unsatisfiable Subsets (MUSes) and have a wide range of applications. As a formula can have multiple MUSes that each provide different insights on unsatisfiability, commonly studied problems include computing a smallest MUS (SMUS) or computing all MUSes (AllMUS) of a given unsatisfiable formula. In this paper, we consider certificates to safety properties in the form of Minimal Safe Inductive Sets (MSISes), and we develop algorithms for exploring such certificates by computing a smallest MSIS (SMSIS) or computing all MSISes (AllMSIS) of a given safe inductive invariant. More precisely, we show how the well-known MUS enumeration algorithms CAMUS and MARCO can be adapted to MSIS enumeration.

1 Introduction

Computing minimal (or even just small) certificates is a central problem in automated reasoning, and, in particular, in Model Checking. Given an unsatisfiable Boolean formula in conjunctive normal form (CNF), a *minimal unsatisfiable subset* (MUS) is a subset of the formula's clauses that is itself unsatisfiable. MUSes have a wide range of applicability, including Proof-Based Abstraction [18], improved comprehension of verification results through vacuity [22], and much more. It is not surprising that a large body of research is dedicated to efficiently computing MUSes. As a formula can have multiple MUSes, each of which may provide different insights on unsatisfiability, several algorithms have been developed to extract all MUSes from an unsatisfiable formula (AllMUS) [2,14,15,19,21], and in particular a smallest MUS of an unsatisfiable formula (SMUS) [11]. For a recent application of AllMUS and SMUS to Model Checking, see [8].

R. Berryhill—This work was completed while Ryan Berryhill was an intern at IBM Research Haifa.

© Springer International Publishing AG, part of Springer Nature 2018
O. Beyersdorff and C. M. Wintersteiger (Eds.): SAT 2018, LNCS 10929, pp. 346–362, 2018.
https://doi.org/10.1007/978-3-319-94144-8_21

For safety properties, certificates come in the form of safe inductive invariants. A recent trend, borrowing from the breakthroughs in Incremental Inductive Verification (such as IMC [17], IC3 [6], and PDR [7]), is to represent such invariants as a conjunction of simple lemmas. Lemmas come in the form of clauses encoding facts about reachable states, and hence the invariant is represented in CNF. The problem of efficiently minimizing the set of such lemmas, and especially constructing a *minimal safe inductive subset* (MSIS) of a given safe inductive invariant has applications to SAT-based model checking [5,10], and has been further studied in [12].

By analogy to MUS extraction, in this paper we consider the problem of computing all MSISes of a given safe inductive invariant (AllMSIS), and in particular finding the smallest MSIS (SMSIS). The problem of minimizing safe inductive invariants appears on its surface to share many commonalities with minimizing unsatisfiable subsets. However, a key aspect of MUS extraction is *monotonicity*: adding clauses to an unsatisfiable formula always yields an unsatisfiable formula. On the other hand, MSIS extraction seems to lack this monotonicity: adding clauses to a safe inductive formula always yields a safe formula, but it may not yield an inductive one. In spite of non-monotonicity, this paper lifts existing MUS enumeration algorithms to the problem of MSIS enumeration.

The CAMUS [15] algorithm solves the AllMUS problem using a well-known hitting set duality between MUSes and *minimal correction subsets* (MCSes). This work defines analogous concepts for safe inductive sets called *support sets* and *collapse sets*, and, using a hitting set duality between them, lifts CAMUS to MSIS extraction. When considering MSIS (resp., MUS) extraction, the algorithm works by enumerating the collapse sets (resp. MCSes) and then exploiting the duality to enumerate MSISes (resp. MUSes) in ascending order of size (*i.e.,* from smallest to largest). When considering the AllMSIS problem, the collapse set enumeration step fundamentally limits the algorithm's anytime performance, as MSIS discovery can only begin after that step. However, when considering SMSIS, the ability to discover the smallest MSIS first is a significant advantage.

MARCO [14], another significant AllMUS algorithm, addresses this limitation by directly exploring the power set of a given unsatisfiable CNF formula. In this work, we translate MSIS extraction to a monotone problem and demonstrate how to solve it with MARCO. This improves anytime performance when considering the AllMSIS problem. However, when considering SMSIS, MARCO may have to compute every MSIS before concluding that it has found the smallest one.

Towards the goal of better understanding the complexity of MSIS problems, we also lift some well-studied MUS-based decision problems to their MSIS analogs and demonstrate complexity results for those problems. Specifically, we prove that the MSIS identification problem "is this subset of a CNF formula an MSIS?" is D^P-complete (*i.e.,* it can be expressed as the intersection of an NP-complete language and a co-NP-complete language). Further, the MSIS existence problem "does this inductive invariant contain an MSIS with k or fewer clauses?" is found to be Σ_2^P-complete. Both of these results match the corresponding MUS problems' complexities.

Experiments are presented on hardware model checking competition benchmarks. On 200 benchmarks, it is found that the CAMUS-based algorithm can find all MSISes within 15 min for 114 benchmarks. The most successful MARCO-based algorithm is almost as successful, finding all MSISes on 110 benchmarks within the time limit. Further, the CAMUS-based algorithm solves the SMSIS problem within 15 min on 156 benchmarks.

The rest of this paper is organized as follows. Section 2 introduces necessary background material. Section 3 formulates the AllMSIS and SMSIS problems. Section 4 presents the CAMUS-inspired MSIS algorithm while Sect. 5 presents the MARCO-based one. Section 6 introduces complexity results for MSIS problems. Section 7 presents experimental results. Finally, Sect. 8 concludes the paper.

2 Preliminaries

2.1 Basic Definitions

The following terminology and notation is used throughout this paper. A literal is either a variable or its negation. A clause is a disjunction of literals. A Boolean formula in Conjunctive Normal Form (CNF) is a conjunction of clauses. It is often convenient to treat a CNF formula as a set of clauses. For a CNF formula φ, $c \in \varphi$ means that clause c appears in φ. A Boolean formula φ is satisfiable (SAT) if there exists an assignment to the variables of φ such that φ evaluates to 1. Otherwise it is unsatisfiable (UNSAT).

2.2 MUSes, MCSes and Hitting Set Duality

If φ is UNSAT, an UNSAT subformula $\varphi_1 \subseteq \varphi$ is called an UNSAT core of φ. If the UNSAT core is minimal or irreducible (*i.e.*, every proper subset of the core is SAT) it is called a Minimal Unsatisfiable Subset (MUS). A subset $C \subseteq \varphi$ is a Minimal Correction Subset (MCS) if $\varphi \setminus C$ is SAT, but for every proper subset $D \subsetneq C$, $\varphi \setminus D$ is UNSAT. In other words, an MCS is a minimal subset such that its removal would render the formula satisfiable.

The *hitting set duality* between MUSes and MCSes states that a subset C of φ is a MUS if and only if C is a minimal hitting set of $MCSes(\varphi)$, and vice versa. For example, if C is a hitting set of $MCSes(\varphi)$, then C contains at least one element from every MCS and therefore corresponds to an UNSAT subset of φ. Moreover, if C is minimal, then removing any element of C would result in at least one MCS not being represented. Therefore, the resulting formula would be SAT implying that C is in fact a MUS. For more details, see Theorem 1 in [15].

2.3 Safe Inductive Invariants and MSIS

Consider a finite transition system with a set of state variables $\mathcal{V}$. The primed versions $\mathcal{V}' = \{v' | v \in \mathcal{V}\}$ represent the next-state functions. For each $v \in \mathcal{V}$, v' is a Boolean function of the current state and input defining the next state for v.

For any formula F over $\mathcal{V}$, the primed version F' represents the same formula with each $v \in \mathcal{V}$ replaced by v'.

A model checking problem is a tuple $P = (Init, Tr, Bad)$ where $Init(\mathcal{V})$ and $Bad(\mathcal{V})$ are CNF formulas over $\mathcal{V}$ that represent the initial states and the unsafe states, respectively. States that are not unsafe are called safe states. The transition relation $Tr(\mathcal{V}, \mathcal{V}')$ is a formula over $\mathcal{V} \cup \mathcal{V}'$. It is encoded in CNF such that $Tr(v, v')$ is satisfiable iff state v can transition to state v'.

A model checking instance is UNSAFE iff there exists a natural number N such that the following formula is satisfiable:

$$Init(v_0) \wedge \left(\bigwedge_{i=0}^{N-1} Tr(v_i, v_{i+1}) \right) \wedge Bad(v_N) \tag{1}$$

The instance is SAFE iff there exists a formula $Inv(\mathcal{V})$ that meets the following conditions:

$$Init(v) \Rightarrow Inv(v) \tag{2}$$

$$Inv(v) \wedge Tr(v, v') \Rightarrow Inv(v') \tag{3}$$

$$Inv(v) \Rightarrow \neg Bad(v) \tag{4}$$

A formula satisfying (2) satisfies *initiation*, meaning that it contains all initial states. A formula satisfying (3) is called *inductive*. An inductive formula that satisfies initiation contains all reachable states and is called an *inductive invariant*. A formula satisfying (4) is *safe*, meaning that it contains only safe states. A safe inductive invariant contains all reachable states and contains no unsafe states, so it is a certificate showing that P is SAFE.

For a model checking problem P with a safe inductive invariant Inv_0 in CNF, a subset $Inv_1 \subseteq Inv_0$ is called a *Safe Inductive Subset* (SIS) of Inv_0 relative to P if Inv_1 is also a safe inductive invariant. Furthermore, if no proper subset of Inv_1 is a SIS, then Inv_1 is called a *Minimal Safe Inductive Subset* (MSIS).

2.4 Monotonicity and MSMP

Let $\mathcal{R}$ denote a reference set, and let $p : 2^{\mathcal{R}} \mapsto \{0, 1\}$ be a predicate defined over elements of the power set of $\mathcal{R}$. The predicate p is *monotone* if $p(\mathcal{R})$ holds and for every $\mathcal{R}_0 \subseteq \mathcal{R}_1 \subseteq \mathcal{R}$, $p(\mathcal{R}_0) \Rightarrow p(\mathcal{R}_1)$. In other words, adding elements to a set that satisfies the predicate yields another set that satisfies the predicate.

Many computational problems involve finding a minimal subset that satisfies a monotone predicate. Examples include computing prime implicants, minimal models, minimal unsatisfiable subsets, minimum equivalent subsets, and minimal corrections sets [16]. For example, for MUS extraction the reference set is the original formula φ, and for a subset $C \subseteq \varphi$, the monotone predicate is $p(C) = 1$ iff C is UNSAT. The *Minimal Set over a Monotone Predicate* problem (MSMP) [16] generalizes all of these notions to the problem of finding a subset $M \subseteq \mathcal{R}$ such that $p(M)$ holds, and for any $M_1 \subsetneq M$, $p(M_1)$ does not hold. State-of-the-art MSMP algorithms heavily rely on monotonicity.

On the other hand, MSIS extraction does not appear to be an instance of MSMP. The natural choice of predicate is "$p(Inv) = 1$ iff Inv is a SIS." This predicate is not monotone, as adding clauses to a safe inductive invariant can yield a non-inductive formula.

3   Problem Formulation: AllMSIS and SMSIS

Modern safety checking algorithms (such as IC3 [6] and PDR [7]) return safe inductive invariants represented as a conjunction of clauses, and hence in CNF. In general there is no guarantee that these invariants are simple or minimal. On the other hand, some recent SAT-based model-checking algorithms [5,10] benefit from simplifying and minimizing these invariants. Given a safe inductive invariant Inv_0 in CNF, some common techniques include removing literals from clauses of Inv_0 [5] and removing clauses of Inv_0 [5,12].

In this paper, we address the problem of minimizing the set of clauses in a given safe inductive invariant. We are interested in computing a smallest safe inductive subset or computing all minimal safe inductive subsets, as stated below.

Enumeration of All Minimal Safe Inductive Subsets (AllMSIS): Given a model checking problem $P = (Init, Tr, Bad)$ and safe inductive invariant Inv_0, enumerate all MSISes of Inv_0.

Finding a Smallest-Sized Safe Inductive Subset (SMSIS): Given a model checking problem $P = (Init, Tr, Bad)$ and safe inductive invariant Inv_0, find a minimum-sized MSIS of Inv_0.

On the surface, computing minimal safe inductive subsets of an inductive invariant appears closely related to computing minimal unsatisfiable subsets of an unsatisfiable formula. However, we are not aware of a direct simple translation from SMSIS and AllMSIS to the analogous MUS problems. This may be due to the lack of monotonicity noted in the previous subsection.

4   MSIS Enumeration Using Hitting Set Duality

In this section we examine precise relationships between different clauses in a safe inductive invariant. We define the notions of a *support set* and a *collapse set* of an individual clause in the invariant, which are somewhat analogous to MUSes and MCSes, respectively. A hitting set duality is identified between support and collapse sets and used to develop an MSIS enumeration algorithm. The algorithm is based on CAMUS [15], a well-known algorithm for MUS enumeration. We present a detailed example to illustrate the concepts and algorithm.

4.1   Inductive Support and Collapse Sets

For a clause c in an inductive invariant Inv, c is inductive relative to Inv by definition. However, it may be the case that c is inductive relative to a small subset of Inv. The notions of *support sets*, borrowed from [4], and *minimal support sets* formalize this concept:

Definition 1. *Given a model checking problem $P = (Init, Tr, Bad)$, a safe inductive invariant Inv, and a clause $c \in Inv$, a support set Γ of c is a subset of clauses of Inv relative to which c is inductive (i.e., the formula $\Gamma \wedge c \wedge Tr \wedge \neg c'$ is UNSAT). A minimal support set Γ of c is a support set of c such that no proper subset of Γ is a support set of c.*

Intuitively, minimal support sets of $c \in Inv$ correspond to MUSes of $Inv \wedge c \wedge Tr \wedge \neg c'$ (where the minimization is done over Inv). Thus support sets provide a more refined knowledge of why a given clause is inductive. Note that as c appears unprimed in the formula, it never appears in any of its minimal support sets. Support sets have various applications, including MSIS computations [12] and a recent optimization to IC3 [3]. The set of all minimal support sets of c is denoted $\mathtt{MinSups}(c)$, and $\mathtt{MinSup}(c)$ denotes a specific minimal support set of c.

Inspired by the duality between MUSes and MCSes, we also consider sets of clauses that cannot be simultaneously removed from a support set. *Collapse sets* and *minimal collapse sets*, defined below, formalize this concept.

Definition 2. *Given a model checking problem $P = (Init, Tr, Bad)$, a safe inductive invariant Inv, and a clause $c \in Inv$, a collapse set Ψ of c is a subset of clauses of Inv such that $Inv \setminus \Psi$ is not a support set of c. A minimal collapse set Ψ of c is a collapse set such that no proper subset of Ψ is a collapse set of c.*

We denote by $\mathtt{MinCols}(c)$ the set of all collapse sets of c. Somewhat abusing the notation, we define the *support sets* and *collapse sets* of $\neg Bad$ as related to safety of P. Formally, $\Gamma \subseteq Inv_0$ is a support set of $\neg Bad$ iff $\Gamma \wedge Bad$ is UNSAT. The set $\Psi \subseteq Inv_0$ is a collapse set of $\neg Bad$ if $Inv_0 \setminus \Psi$ is not a support set of $\neg Bad$. Minimal support sets and minimal collapse sets of $\neg Bad$ are defined accordingly. The following lemma summarizes the relations between various definitions.

Lemma 1. *Let Inv_1 be a SIS of Inv_0.*

1. *There exists $\Gamma \in \mathtt{MinSups}(\neg Bad)$ such that $\Gamma \subseteq Inv_1$;*
2. *For each $c \in Inv_1$, there exists $\Gamma \in \mathtt{MinSups}(c)$ such that $\Gamma \subseteq Inv_1$;*
3. *For each $\Psi \in \mathtt{MinCols}(\neg Bad)$ we have that $\Psi \cap Inv_1 \neq \emptyset$;*
4. *For each $c \in Inv_1$ and for each $\Psi \in \mathtt{MinCols}(c)$ we have that $\Psi \cap Inv_1 \neq \emptyset$.*

The following example illustrates the concept of support sets, which are somewhat analogous to MUSes. The example is extended throughout the paper to illustrate additional concepts and algorithms.

Running Example: Let us suppose that $Inv = \{c_1, c_2, c_3, c_4, c_5, c_6\}$ is a safe inductive invariant for P. Omitting the details on the actual model checking problem, let us suppose that the minimal support sets are given as follows: $\mathtt{MinSups}(\neg Bad) = \{\{c_1, c_2\}, \{c_1, c_3\}\}$, $\mathtt{MinSups}(c_1) = \{\emptyset\}$, $\mathtt{MinSups}(c_2) = \{\{c_4\}, \{c_6\}\}$, $\mathtt{MinSups}(c_3) = \{\{c_5\}\}$, $\mathtt{MinSups}(c_4) = \{\{c_2, c_5\}\}$, $\mathtt{MinSups}(c_5) = \{\{c_3\}\}$, $\mathtt{MinSups}(c_6) = \{\emptyset\}$. In particular, all the following formulas are unsatisfiable: $c_1 \wedge c_2 \wedge \neg Bad$, $c_1 \wedge c_3 \wedge \neg Bad$, $c_1 \wedge Tr \wedge \neg c_1'$, $c_4 \wedge c_2 \wedge Tr \wedge \neg c_2'$, $c_6 \wedge c_2 \wedge Tr \wedge \neg c_2'$,

$c_5 \wedge c_3 \wedge Tr \wedge \neg c_3'$, $c_2 \wedge c_5 \wedge c_4 \wedge Tr \wedge \neg c_4'$, $c_3 \wedge c_5 \wedge Tr \wedge \neg c_5'$, $c_6 \wedge Tr \wedge \neg c_6'$. Conversely, the following formulas are satisfiable: $c_1 \wedge \neg Bad$, $c_2 \wedge \neg Bad$, $c_3 \wedge \neg Bad$, $c_2 \wedge Tr \wedge \neg c_2'$, $c_3 \wedge Tr \wedge \neg c_3'$, $c_2 \wedge c_4 \wedge Tr \wedge \neg c_4'$, $c_5 \wedge c_4 \wedge Tr \wedge \neg c_4'$, $c_5 \wedge Tr \wedge \neg c_5'$.

Further, the following example illustrates the "dual" concept of collapse sets, which are somewhat analogous to MCSes.

Running Example (cont.): The minimal collapse sets are given as follows: $\texttt{MinCols}(\neg Bad) = \{\{c_1\}, \{c_2, c_3\}\}$, $\texttt{MinCols}(c_1) = \{\emptyset\}$, $\texttt{MinCols}(c_2) = \{\{c_4, c_6\}\}$, $\texttt{MinCols}(c_3) = \{\{c_5\}\}$, $\texttt{MinCols}(c_4) = \{\{c_2\}, \{c_5\}\}$, $\texttt{MinCols}(c_5) = \{\{c_3\}\}$, $\texttt{MinCols}(c_6) = \{\emptyset\}$.

One way to construct a (not necessarily minimal) SIS of Inv is to choose a minimal support for each clause in the invariant, and then, starting from the support of $\neg Bad$, recursively add all the clauses participating in the supports. In the running example, we only need to make the choices for $\texttt{MinSup}(\neg Bad)$ and for $\texttt{MinSup}(c_2)$, as all other minimal supports are unique. The following example illustrates three different possible executions of such an algorithm, demonstrating that such an approach does not necessarily lead to an MSIS.

Running Example (cont.): Fixing $\texttt{MinSup}(\neg Bad) = \{c_1, c_2\}$ and $\texttt{MinSup}(c_2) = \{c_4\}$ leads to $Inv_1 = \{c_1, c_2, c_3, c_4, c_5\}$. The clauses c_1 and c_2 are chosen to support $\neg Bad$, c_4 is chosen to support c_2, c_5 is needed to support c_4, and c_3 is needed to support c_5. A second possibility fixes $\texttt{MinSup}(\neg Bad) = \{c_1, c_2\}$ and $\texttt{MinSup}(c_2) = \{c_6\}$, which leads to $Inv_2 = \{c_1, c_2, c_6\}$. A third possibility chooses $\texttt{MinSup}(\neg Bad) = \{c_1, c_3\}$ and leads to the $Inv_3 = \{c_1, c_3, c_5\}$, regardless of the choice for $\texttt{MinSup}(c_2)$.

We can readily see that certain choices for minimal supports to do not produce a minimal safe inductive invariant. Indeed, Inv_3 is minimal but Inv_1 is not. The problem has exactly two MSISes represented by Inv_2 and Inv_3, and both also happen to be smallest minimal inductive invariants.

4.2 CAMUS for MSIS Extraction

Our MSIS enumeration algorithm is strongly motivated by CAMUS [15], which enumerates all MUSes of an unsatisfiable formula in CNF. Given an unsatisfiable formula φ, CAMUS operates in two phases. The first enumerates all MCSes of φ using a MaxSAT-based algorithm. The second phase enumerates all MUSes of φ based on the hitting set duality between MCSes and MUSes. Our algorithm performs similar operations involving the analogous concepts of support and collapse sets.

4.3 The CAMSIS Algorithm

Given a model checking problem $P = (Init, Tr, Bad)$ and safe inductive invariant Inv_0 for P, the algorithm also operates in two phases. The first phase iterates over all $c \in Inv_0 \cup \{\neg Bad\}$ and computes the set $\texttt{MinCols}(c)$ of all minimal collapse sets of c. This is analogous to the first phase of CAMUS and is done very similarly. Indeed, collapse sets of c are enumerated by computing an UNSAT core of $Inv_0 \wedge c \wedge Tr \wedge \neg c'$, while minimizing with respect to the clauses of Inv_0.

Now, one possibility is to enumerate all minimal support sets of each $c \in Inv_0 \cup \{\neg Bad\}$, based on the duality between MinCols(c) and MinSups(c), and then to enumerate all MSISes of Inv_0 using a dedicated algorithm that chooses support sets in a way to produce minimal invariants. Instead, we suggest Algorithm 1 to enumerate MSISes of Inv_0 directly, only based on MinCols (and without computing MinSups first). One can think of this as a SAT-based algorithm for hitting-set duality "with a twist." It uses the last two statements in Lemma 1 to construct a formula in which satisfying assignments correspond to SISes, and then finding all satisfying assignments that correspond to MSISes.

Algorithm 1. CAMSIS

Input: $Inv_0 = \{c_1, \ldots, c_n\}$, MinCols($\neg Bad$), MinCols($c$) for every $c \in Inv_0$
Output: MSISes(Inv_0) relative to P

1: introduce new variable s_c for each $c \in Inv_0$
2: $\vartheta_1 = \bigwedge_{\{d_1,\ldots,d_k\} \in \text{MinCols}(\neg Bad)} (s_{d_1} \vee \cdots \vee s_{d_k})$
3: $\vartheta_2 = \bigwedge_{c \in Inv_0} \bigwedge_{\{d_1,\ldots,d_k\} \in \text{MinCols}(c)} (\neg s_c \vee s_{d_1} \vee \cdots \vee s_{d_k})$
4: $\vartheta = \vartheta_1 \wedge \vartheta_2$
5: $j \leftarrow 1$
6: **loop**
7:     **while** $(\vartheta \wedge \text{AtMost}(\{s_{c_1}, \ldots, s_{c_n}\}, j))$ is SAT (with model M) **do**
8:         Let $Inv = \{c_i \mid M \models (s_{c_i} = 1)\}$
9:         $\vartheta \leftarrow \vartheta \wedge (\bigvee_{c_i \in Inv} \neg s_{c_i})$
10:         MSISes $\leftarrow$ MSISes $\cup \{Inv\}$
11:     **end while**
12:     **break if** ϑ is UNSAT
13:     $j \leftarrow j + 1$
14: **end loop**
15: **return** MSISes

The algorithm accepts the initial safe inductive invariant $Inv_0 = \{c_1, \ldots, c_n\}$, the set of minimal collapse sets of $\neg Bad$, and the set of minimal collapse sets for each clause in the invariant. All SAT queries use an incremental SAT solver. On line 1, an auxiliary variable s_c is introduced for each clause c. The intended meaning is that $s_c = 1$ iff c is selected as part of the MSIS. On lines 2–4, the algorithm constructs a formula ϑ that summarizes Lemma 1. First, for each minimal collapse set Ψ of $\neg Bad$, at least one clause of Ψ must be in the invariant. This ensures that the invariant is safe. Further, for each selected clause c (i.e., where $s_c = 1$) and for each minimal collapse set Ψ of c, at least one clause of Ψ must be in the invariant. This ensures that each selected clause is inductive relative to the invariant, thereby ensuring that the resulting formula is inductive.

The algorithm uses the `AtMost` cardinality constraint to enumerate solutions from smallest to largest. The loop on line 6 searches for MSISes using ϑ. It starts by seeking MSISes of cardinality 1 and increases the cardinality on each iteration. Each time an MSIS is found, all of its supersets are blocked by adding a clause on line 9. Line 12 checks if all MSISes have been found using ϑ without any `AtMost` constraint. This check determines if any MSISes of *any* size remain, and if not the algorithm exits the loop. The following example illustrates an execution of the algorithm.

Running Example (cont.): Initially, $\vartheta = (s_1) \wedge (s_2 \vee s_3) \wedge (\neg s_2 \vee s_4 \vee s_6) \wedge (\neg s_3 \vee s_5) \wedge (\neg s_4 \vee s_2) \wedge (\neg s_4 \vee s_5) \wedge (\neg s_5 \vee s_3)$. Let $S = \{s_1, \ldots, s_6\}$. It is easy to see that both $\vartheta \wedge \texttt{AtMost}(S, 1)$ and $\vartheta \wedge \texttt{AtMost}(S, 2)$ are UNSAT. Suppose that the first solution returned for $\vartheta \wedge \texttt{AtMost}(S, 3)$ is $s_1 = 1, s_2 = 0, s_3 = 1, s_4 = 0, s_5 = 1, s_6 = 0$. It corresponds to a (minimum-sized) safe inductive invariant $\{c_1, c_3, c_5\}$. It is recorded and ϑ is modified by adding the clause $(\neg s_1 \vee \neg s_3 \vee \neg s_5)$. Rerunning on $\vartheta \wedge \texttt{AtMost}(S, 3)$ produces another solution $s_1 = 1, s_2 = 1, s_3 = 0, s_4 = 0, s_5 = 0, s_6 = 1$, corresponding to the MSIS $\{c_1, c_2, c_6\}$. It is recorded and ϑ is modified by adding $(\neg s_1 \vee \neg s_2 \vee \neg s_6)$. Now $\vartheta \wedge \texttt{AtMost}(S, 3)$ is UNSAT. In addition, ϑ is UNSAT and the algorithm terminates.

We now prove the algorithm's completeness and soundness. The proof relies on the fact that satisfying assignments of the formula ϑ constructed on line 4 are safe inductive subsets of the given inductive invariant. The theorem below demonstrates this fact.

Theorem 1. *Each satisfying assignment $M \models \vartheta$ corresponds to a SIS of Inv_0.*

Proof. ϑ is the conjunction of ϑ_1 and ϑ_2. Each clause of ϑ_2 relates to a clause $c \in Inv_0$ and collapse set Ψ of c. It requires that either c is not selected or an element of Ψ is selected. ϑ_2 contains all such constraints, so it requires that for each clause $c \in Inv_0$, either c is not selected or a member of every minimal collapse set of c is selected. By duality, this is equivalent to requiring a support set of c is selected. Therefore ϑ_2 requires that an inductive formula is selected.

ϑ_1 encodes the additional constraint that a support set of $\neg Bad$ is selected. In other words, it requires that a safe formula is selected. Since each clause of Inv_0 must satisfy initiation by definition, a satisfying assignment of ϑ corresponds to a safe inductive invariant contained within Inv_0. □

Corollary 1. *Algorithm 1 returns the set of all MSISes of Inv_0.*

Proof. The algorithm finds only minimal models of ϑ and finds all such models.

Several simple optimizations are possible. The technique in [12] describes an algorithm to identify certain clauses that appear in every inductive invariant. If such a set $\mathcal{N}$ is known in advance, it is sound to add constraints (s_c) to ϑ for each $c \in \mathcal{N}$ and start the search from cardinality $|\mathcal{N}|$. Further, it is possible to start by finding collapse sets only for the clauses in $\mathcal{N}$, and then find collapse sets for the clauses in those collapse sets, and so on until a fixpoint is reached.

5 MARCO and MSIS Extraction

In this section we show how the MARCO algorithm for MUS enumeration [14] can be adapted for MSIS enumeration. In Sect. 5.1 we present the MARCO algorithm from [14] trivially extended to a more general class of *monotone predicate* problems [16]. In Sect. 5.2 we describe a monotone reformulation of the MSIS extraction problem and fill in the missing details on the special functions used by the MARCO algorithm.

5.1 MARCO Algorithm for MSMP

Algorithm 2 displays the basic MARCO algorithm from [14] trivially extended to the more general class of monotone predicates [16]. The algorithm accepts a monotone predicate p and a set F satisfying $p(F) = 1$. It returns the set of all minimal subsets of F satisfying p. Recall that the *monotonicity* of p means that $p(F_0) \Rightarrow p(F_1)$ whenever $F_0 \subseteq F_1$.

Algorithm 2. MARCO for MSMP

Input: monotone predicate p, formula F in CNF s.t. $p(F) = 1$
Output: set M of all minimal subsets of F that satisfy p

1: $map \leftarrow \top$
2: **while** map is SAT **do**
3:     $seed \leftarrow$ getUnexplored(map)
4:     **if** $p(seed) = 0$ **then**
5:         $mss \leftarrow$ grow$(seed)$
6:         $map \leftarrow map \wedge$ blockDown(mss)
7:     **else**
8:         $mus \leftarrow$ shrink$(seed)$
9:         $M \leftarrow M \cup \{mus\}$
10:         $map \leftarrow map \wedge$ blockUp(mus)
11:     **end if**
12: **end while**
13: **return** M

MARCO directly explores the power set lattice of the input set F. In greater detail, it operates as follows. *Seeds* are selected using a Boolean formula called the map, where each satisfying assignment corresponds to an unexplored element of the power set. This is handled by the getUnexplored procedure on line 3. The map has a variable for each element of F, such that the element is selected as part of the seed iff the variable is assigned to 1. Initially, the map is empty and the first seed is chosen arbitrarily.

If $p(seed) = 0$, the grow procedure attempts to expand it to a larger set mss that also does not satisfy p (line 5). This can be accomplished by adding elements of $F \setminus seed$ and checking if the result satisfies p. If so, the addition of the element is backed out, otherwise it is kept. Once every such element has been tried, the result it a maximal set that does not satisfy p. Since any subset of such a set does not satisfy p, the algorithm blocks mss and all of its subsets from consideration as future seeds by adding a new clause to the map (line 6).

Conversely, if $p(seed) = 1$, it is shrunk to a minimal such set (an MSMP) by removing clauses in a similar fashion using the `shrink` procedure (line 8). Subsequently, the minimal set and all of its supersets are blocked by adding a clause to the map (line 10). This is because a strict superset of such a set is not minimal, and therefore not an MSMP.

MUS enumeration is a concrete instantiation of this algorithm. The natural choice of predicate is $p(F) = 1$ iff F is unsatisfiable. The `shrink` subroutine returns a MUS of $seed$. The `grow` subroutine returns a maximal satisfiable subset of F containing $seed$.

5.2 A Monotone Version of MSIS Enumeration

Suppose that we are given a safe inductive invariant Inv_0. In order to extract MSISes of Inv_0 with MARCO, it is necessary to construct a monotone predicate such that the minimal subsets satisfying this monotone predicate are MSISes. As we saw before, the predicate $p(F) =$ "is F a SIS of Inv_0?" is not monotone. However, let us define the predicate $p_0(F) =$ "does F contain a SIS of Inv_0?"

Lemma 2. *The predicate p_0 defined above is monotone. Furthermore, minimal subsets of Inv_0 satisfying p_0 are MSISes of Inv_0.*

Proof. To show monotonicity of p_0, suppose that $F_0 \subseteq F_1 \subseteq Inv_0$ and suppose that $G \subseteq F_0$ is a SIS of F_0. Then G is also a SIS of F_1. For the second property, note that a *minimal* set that contains a SIS must be a SIS itself. □

In order to apply p_0 for computing MSISes, we need to specify the missing subroutines of the MARCO algorithm, or equivalently we need to show how to compute $p_0(seed)$, and how to implement `shrink` and `grow`.

In order to compute $p_0(seed)$, we need check whether $seed$ contains a SIS. We accomplish this using the algorithm `MaxIndSubset` that computes a *maximal inductive subset* of a potentially non-inductive set of clauses. Following [12], we compute `MaxIndSubset`(R) of a set of clauses R by repeatedly removing those clauses of R that are not inductive with respect to R, and we check whether the fixpoint R_0 of this procedure is safe using a SAT solver. In particular we can replace the condition $p_0(F) =$ "does F contain a SIS" by an equivalent condition "is `MaxIndSubset`(F) safe?"

The `shrink` procedure involves finding an MSIS of `MaxIndSubset`$(seed)$. A basic algorithm that finds a single MSIS is presented in [5]. Given a safe inductive invariant R, this algorithm repeatedly selects a clause c in R and checks whether `MaxIndSubset`$(R \setminus \{c\})$ is safe. If so, then R is replaced by `MaxIndSubset`$(R \setminus \{c\})$. For more details and optimizations, refer to [12].

The `grow` procedure expands a seed that does not contain a SIS to a maximal subset of Inv_0 that does not contain a SIS. A basic algorithm repeatedly selects a clause $c \in Inv_0 \setminus seed$ and checks whether $p_0(seed \cup \{c\}) = 0$. If so, then $seed$ is replaced by $seed \cup \{c\}$. The following continuation of the running example demonstrates several iterations of MARCO.

Running Example (cont.): Assume that grow and shrink are implemented as described above, and the clauses are always processed in the order of their index. On the first iteration, suppose getUnexplored(map) returns $seed = \emptyset$. The grow procedure initially sets $mss = seed = \emptyset$ and makes the following queries and updates:

- MaxIndSubset($\{c_1\}$) = $\{c_1\}$ is not safe; $mss \leftarrow mss \cup \{c_1\}$;
- MaxIndSubset($\{c_1, c_2\}$) = $\{c_1\}$ is not safe; $mss \leftarrow mss \cup \{c_2\}$;
- MaxIndSubset($\{c_1, c_2, c_3\}$) = $\{c_1\}$ is not safe; $mss \leftarrow mss \cup \{c_3\}$;
- MaxIndSubset($\{c_1, c_2, c_3, c_4\}$) = $\{c_1\}$ is not safe; $mss \leftarrow mss \cup \{c_4\}$;
- MaxIndSubset($\{c_1, c_2, c_3, c_4, c_5\}$) = $\{c_1, c_2, c_3, c_4, c_5\}$ is safe;
- MaxIndSubset($\{c_1, c_2, c_3, c_4, c_6\}$) = $\{c_1, c_2, c_6\}$ is safe.

Thus at the end we obtain $mss = \{c_1, c_2, c_3, c_4\}$. Next, map is updated to $map \wedge \texttt{blockDown}(\{c_1, c_2, c_3, c_4\})$, forcing $seed$ to include either c_5 or c_6 from thereon. On the second iteration, let's suppose that getUnexplored(map) returns $seed = \{c_1, c_2, c_3, c_4, c_5, c_6\}$. The shrink procedure initially sets $mus = $ MaxIndSubset($seed$) = $\{c_1, c_2, c_3, c_4, c_5, c_6\}$ and makes the following queries and updates:

- MaxIndSubset($mus \setminus \{c_1\}$) = $\{c_2, c_3, c_4, c_5, c_6\}$ is not safe;
- MaxIndSubset($mus \setminus \{c_2\}$) = $\{c_1, c_3, c_5, c_6\}$ is safe; $mus \leftarrow \{c_1, c_3, c_5, c_6\}$;
- MaxIndSubset($mus \setminus \{c_3\}$) = $\{c_1, c_6\}$ is not safe;
- MaxIndSubset($mus \setminus \{c_5\}$) = $\{c_1, c_6\}$ is not safe;
- MaxIndSubset($mus \setminus \{c_6\}$) = $\{c_1, c_3, c_5\}$ is safe; $mus \leftarrow \{c_1, c_3, c_5\}$.

Hence at the end we obtain $mus = \{c_1, c_3, c_5\}$. This allows to update map to $map \wedge \texttt{blockUp}\{c_1, c_3, c_5\}$, forcing $seed$ to exclude either c_1, c_3 or c_5 thereafter.

It is important to note that in practice grow and shrink are implemented using additional optimizations. The example uses the simple versions for ease-of-understanding.

6   Complexity of MSIS and MUS

This section briefly summarizes complexity results for the MUS and MSIS identification and existence problems. The results for MUS are well-known while, as far as we know, the results for MSIS are novel. Note that the algorithms presented in Sects. 4 and 5 solve the *function problems* of AllMSIS and SMSIS. The complexity of related MUS problems has been studied in works such as [13]. In this section, we study closely-related *decision problems*, which are also solved implicitly by the presented algorithms. We present the problems and their known complexity classes below.

Lemma 3. *The MUS existence problem "does F_0 have a MUS of size k or less?" is Σ_2^P-complete.*

Lemma 4. *The MUS identification problem "is F_1 a MUS of F_0?" is D^P-complete.*

Proofs of the above lemmas are presented in [9,20], respectively. The novel result for the MSIS existence problem relies on a reduction from the MUS existence problem. The problem is similarly stated as "does Inv_0 have an MSIS of size k or less?" To see that it is in Σ_2^P, notice that it can be solved by a non-deterministic Turing machine that guesses a subset of Inv_0 and checks if it is a SIS, which requires only a constant number of satisfiability queries. We demonstrate that it is Σ_2^P-Hard by reduction from the MUS existence problem.

Theorem 2. *MUS existence problem $\leq_m^P$ MSIS existence problem.*

Proof. Let (C, k) be an instance of the MUS existence problem. Construct an MSIS existence instance as follows:

$$Inv_C = \{(c_i \vee \neg Bad) : c_i \in C\}$$
$$Init = \neg Bad$$
$$Tr = \{(v_i' = v_i) : v_i \in Vars(C)\}$$

where Bad is a new variable that does not appear in C. Inv_C is a safe inductive invariant because:

1. $Init \Rightarrow Inv_C$ since $\neg Bad$ satisfies every clause
2. Inv_C is inductive because every formula is inductive for Tr
3. $Inv_C \Rightarrow \neg Bad$ since $Inv_C \wedge Bad$ is equi-satisfiable with C, which is UNSAT.

Next, we show that every MSIS of Inv_C corresponds to a MUS of C. For any $D \subseteq C$, let $Inv_D = \{(c_i \vee \neg Bad) : c_i \in D\}$. Note that:

1. Every Inv_D satisfies initiation
2. Every Inv_D is inductive
3. $Inv_D \wedge Bad$ is equi-satisfiable with D. It is UNSAT iff D is UNSAT.

The three points above imply Inv_D is a SIS iff D is an UNSAT core, so Inv_D is an MSIS iff D is a MUS. This implies that Inv_C contains an MSIS of size k or less iff C contains a MUS of size k or less. $\square$

Corollary 2. *MUS identification problem $\leq_m^P$ MSIS identification problem.*

Proof. Follows from the same reduction used to prove Theorem 2.

We now present the proof that the MSIS identification problem is D^P-complete. It is D^P-Hard due to Corollary 2. The proof that it is in D^P is presented in Theorem 3 below.

Theorem 3. *The MSIS identification problem "is Inv_1 an MSIS of Inv_0?" is in D^P.*

Proof. Let Inv_0 be an safe inductive invariant for $P = (Init, Tr, Bad)$. Let $\mathcal{L}_1$ be the language "subsets of Inv_0 that do not (strictly) contain a SIS." $\mathcal{L}_1$ is in NP. Given a $C_0 \subseteq Inv_0$, MaxIndSubset(C_0) [12] would execute at most $|C_0|$ satisfiability queries to determine that C_0 does not contain a SIS. All of the queries return SAT in this case. Given a $C \in \mathcal{L}_1$, executing MaxIndSubset on the $|C|$ strict subsets of cardinality $|C| - 1$ yields $O(|C|^2)$ satisfying assignments. They form a certificate for a positive instance with size polynomial in $|C|$.

Let $\mathcal{L}_2$ be the language "subsets of Inv_0 that are SISes." $\mathcal{L}_2$ is in co-NP. This follows from the fact that $C \subseteq Inv$ is a SIS if $C \wedge Tr \wedge \neg C'$ and $C \wedge Bad$ are both UNSAT. The satisfying assignment for the disjunction of those two formulas forms a certificate for a negative instance (where C is not a SIS).

The language "MSISes of Inv_0" is $\mathcal{L}_1 \cap \mathcal{L}_2$, so it is in D^P.   □

7   Experimental Results

This section presents empirical results for the presented algorithms and the MSIS algorithm of [12] on safe single-property benchmarks from the 2011 Hardware Model Checking Competition [1]. This particular benchmark set was chosen because it has a large number of problems solved during the competition. Experiments are executed on a 2.00 GHz Linux-based machine with an Intel Xeon E7540 processor and 96 GB of RAM. In order to generate inductive invariants for minimization, our implementation of IC3 is run with a 15 min time limit, which produces invariants in 280 cases. In 77 cases, IC3 generates a minimal invariant (including cases where the given property is itself inductive). These benchmarks are removed from further consideration, as are 3 additional benchmarks for which none of the minimization algorithms terminated within the time limit. For each of the 200 remaining testcases, CAMSIS (Sect. 4.3), MSIS [12], and MARCO (Sect. 5) are used to minimize the inductive invariant. Motivated by [14], we consider 3 slightly different versions of MARCO, by forcing getUnexplored to return either any seed satisfying the map (MARCO-ARB), the seed of smallest possible cardinality (MARCO-UP), or the seed of largest possible cardinality (MARCO-DOWN). In this way, MARCO-UP favors earlier detection of *mss*es, while MARCO-DOWN favors earlier detection of *mus*es (using the terminology of Algorithm 2). Each of the above techniques is run with a time limit of 15 min, not including the time required to run IC3.

Table 1 summarizes the results. For each technique and for several different values of k, the first line reports the number of testcases on which the technique is able to find k MSISes (or all MSISes if this number does not exceed k). The second line reports the total time, in seconds. In addition, for the CAMSIS algorithm, the column "preparation" reports the number of testcases in which it was able to enumerate all collapse sets, and the total time for doing so.

First we note that while CAMSIS spends a significant amount of time to compute the collapse sets, it is the winning algorithm when computing all or a large

Table 1. Summary of results

	Preparation	k = 1	k = 5	k = 10	k = 100	k = ALL
MSIS		200				
		5,944				
CAMSIS	165	156	155	155	**155**	114
	34,212	42,907	43,842	43,853	**44,047**	**79,908**
MARCO-UP		146	145	145	142	110
		58,904	60,461	60,725	62,021	83,934
MARCO-ARB		143	139	138	127	101
		56,175	60,297	61,687	71,747	93,251
MARCO-DOWN		**199**	**184**	**176**	145	100
		7,917	**21,007**	**29,305**	57,217	92,970

number of MSISes. It is also the winning algorithm for computing the guaranteed smallest MSIS, succeeding in 156 cases. In contrast, the best MARCO-based approach for computing all or the smallest MSIS only succeeds in 110 cases. It is interesting to note that this is the approach favoring earlier detection of *msses* rather than *muses*. On the other hand, the MARCO-DOWN approach, which is tailored towards finding *muses*, shows much better anytime behavior, prevailing over the other algorithms when computing a small number of MSISes (such as 1, 5 or 10). We note that the result for MARCO-DOWN for $k = 1$ is not surprising, as in this case the first assignment returned by getUnexplored returns the original invariant as the seed, so MARCO-DOWN simply reduces to finding any arbitrary MSIS of the original invariant. Finally, we note that the MARCO-ARB algorithm is in general worse than either MARCO-DOWN or MARCO-UP.

To give some intuition on the nature of the problems considered, Table 2 shows the number of MSISes for the 115 testcases solved by at least one AllMSIS algorithm. The largest number of MSISes was 149280. Incidentally, this the only of the 115 benchmarks for which CAMSIS could not find every MSIS.

Table 2. Total number of MSISes (115 benchmarks)

MSISes	1	2–10	11–100	101–1000	> 1000
Frequency	55	20	17	11	12

A more in-depth analysis shows that while on average the MSIS technique from [12] is significantly better for finding a single MSIS than CAMSIS for finding a smallest-size MSIS, there are several cases where CAMSIS significantly outperforms MSIS. In other words, in some cases first finding all the collapse sets and then finding a minimum inductive invariant using hitting set duality is faster than looking for a minimal inductive invariant directly.

8   Conclusion and Future Work

This work lifts the MUS extraction algorithms of CAMUS and MARCO to the non-monotone problem of MSIS extraction. The former is accomplished by identifying a hitting set duality between support sets and collapse sets, which is analogous to the MCS/MUS duality exploited by CAMUS. The latter is accomplished by converting MSIS extraction to a monotone problem and applying MARCO directly. Further, complexity results are proven demonstrating that MSIS identification is D^P-complete and the MSIS existence problem is Σ_2^P-complete, both of which match the corresponding MUS problems.

The work of [14] suggests many optimizations to MARCO algorithm, it would be interesting to explore these in our context. It would also be of interest to determine if the predicates used to convert the non-monotone MSIS problems into a monotone one suitable for use with MARCO can be applied in CAMSIS. Further, we intend to lift other MUS extraction algorithms such as dualize-and-advance (DAA) [2] to MSIS problems. Finally, further study of the application of MARCO to non-monotone problems and the complexity of doing so is a natural extension of this work.

References

1. Hardware Model Checking Competition (2011). http://fmv.jku.at/hwmcc11
2. Bailey, J., Stuckey, P.J.: Discovery of minimal unsatisfiable subsets of constraints using hitting set dualization. In: Hermenegildo, M.V., Cabeza, D. (eds.) PADL 2005. LNCS, vol. 3350, pp. 174–186. Springer, Heidelberg (2005). https://doi.org/10.1007/978-3-540-30557-6_14
3. Berryhill, R., Ivrii, A., Veira, N., Veneris, A.: Learning support sets in IC3 and Quip: The good, the bad, and the ugly. In: 2017 Formal Methods in Computer Aided Design (FMCAD) (2017)
4. Berryhill, R., Veira, N., Veneris, A., Poulos, Z.: Learning lemma support graphs in Quip and IC3. In: Proceedings of the 2nd International Verification and Security Workshop, IVSW 2017 (2017)
5. Bradley, A.R., Somenzi, F., Hassan, Z., Zhang, Y.: An incremental approach to model checking progress properties. In: 2011 Formal Methods in Computer-Aided Design (FMCAD), pp. 144–153, October 2011
6. Bradley, A.: SAT-based model checking without unrolling. In: International Conference on Verification, Model Checking, and Abstract Interpretation, VMCAI 2011 (2011)
7. Eén, N., Mishchenko, A., Brayton, R.: Efficient implementation of property directed reachability. In: Proceedings of the International Conference on Formal Methods in Computer-Aided Design, FMCAD 2011 (2011)
8. Ghassabani, E., Whalen, M., Gacek, A.: Efficient generation of all minimal inductive validity cores. In: 2017 Formal Methods in Computer Aided Design (FMCAD) (2017)
9. Gupta, A.: Learning Abstractions for Model Checking. Ph.D. thesis, Pittsburgh, PA, USA (2006), aAI3227784

10. Hassan, Z., Bradley, A.R., Somenzi, F.: Incremental, inductive CTL model checking. In: Madhusudan, P., Seshia, S.A. (eds.) CAV 2012. LNCS, vol. 7358, pp. 532–547. Springer, Heidelberg (2012). https://doi.org/10.1007/978-3-642-31424-7_38

11. Ignatiev, A., Previti, A., Liffiton, M., Marques-Silva, J.: Smallest MUS extraction with minimal hitting set dualization. In: Pesant, G. (ed.) CP 2015. LNCS, vol. 9255, pp. 173–182. Springer, Cham (2015). https://doi.org/10.1007/978-3-319-23219-5_13

12. Ivrii, A., Gurfinkel, A., Belov, A.: Small inductive safe invariants. In: Proceedings of the 14th Conference on Formal Methods in Computer-Aided Design, FMCAD 2014 (2014)

13. Janota, M., Marques-Silva, J.: On the query complexity of selecting minimal sets for monotone predicates. Artif. Intell. **233**, 73–83 (2016)

14. Liffiton, M.H., Previti, A., Malik, A., Marques-Silva, J.: Fast, flexible MUS enumeration. Constraints **21**(2), 223–250 (2016). https://doi.org/10.1007/s10601-015-9183-0

15. Liffiton, M.H., Sakallah, K.A.: Algorithms for computing minimal unsatisfiable subsets of constraints. J. Autom. Reasoning **40**(1), 1–33 (2008). https://doi.org/10.1007/s10817-007-9084-z

16. Marques-Silva, J., Janota, M., Belov, A.: Minimal Sets over Monotone Predicates in Boolean Formulae (2013). https://doi.org/10.1007/978-3-642-39799-8_39

17. McMillan, K.L.: Interpolation and SAT-based model checking. In: Hunt, W.A., Somenzi, F. (eds.) CAV 2003. LNCS, vol. 2725, pp. 1–13. Springer, Heidelberg (2003). https://doi.org/10.1007/978-3-540-45069-6_1

18. McMillan, K.L., Amla, N.: Automatic abstraction without counterexamples. In: Garavel, H., Hatcliff, J. (eds.) TACAS 2003. LNCS, vol. 2619, pp. 2–17. Springer, Heidelberg (2003). https://doi.org/10.1007/3-540-36577-X_2

19. Nadel, A.: Boosting minimal unsatisfiable core extraction. In: Proceedings of the 2010 Conference on Formal Methods in Computer-Aided Design, FMCAD 2010, pp. 221–229 (2010)

20. Papadimitriou, C.H., Wolfe, D.: The complexity of facets resolved. J. Comput. Syst. Sci. **37**(1), 2–13 (1988)

21. Previti, A., Marques-Silva, J.: Partial MUS enumeration. In: Proceedings of the Twenty-Seventh AAAI Conference on Artificial Intelligence, AAAI 2013 (2013)

22. Simmonds, J., Davies, J., Gurfinkel, A., Chechik, M.: Exploiting resolution proofs to speed up LTL vacuity detection for BMC. In: FMCAD, pp. 3–12. IEEE Computer Society (2007)

Satisfiability Modulo Theories

Effective Use of SMT Solvers for Program Equivalence Checking Through Invariant-Sketching and Query-Decomposition

Shubhani Gupta[✉], Aseem Saxena, Anmol Mahajan, and Sorav Bansal

Indian Institute of Technology Delhi, New Delhi, India
shubhani@cse.iitd.ac.in

Abstract. Program equivalence checking is a fundamental problem in computer science with applications to translation validation and automatic synthesis of compiler optimizations. Contemporary equivalence checkers employ SMT solvers to discharge proof obligations generated by their equivalence checking algorithm. Equivalence checkers also involve algorithms to infer invariants that relate the intermediate states of the two programs being compared for equivalence. We present a new algorithm, called *invariant-sketching*, that allows the inference of the required invariants through the generation of counter-examples using SMT solvers. We also present an algorithm, called *query-decomposition*, that allows a more capable use of SMT solvers for application to equivalence checking. Both invariant-sketching and query-decomposition help us prove equivalence across program transformations that could not be handled by previous equivalence checking algorithms.

1 Introduction

The general problem of program equivalence checking is undecidable. Several previous works have tackled the problem for applications in (a) translation validation, where the equivalence checker attempts to automatically generate a proof of equivalence across the transformations (translations) performed by a compiler [2,3]; and (b) program synthesis, where the equivalence checker is tasked with determining if the optimized program proposed by the synthesis algorithm is equivalent to the original program specification [4,5]. For both these applications, soundness is critical, i.e., if the equivalence checker determines the programs to be equivalent, then the programs are guaranteed to have equivalent runtime behaviour. On the other hand, completeness may not always be achievable (as the general problem is undecidable), i.e., it is possible that the equivalence checker is unable to prove the programs equivalent, even if they are actually equivalent. For example, recent work on black-box equivalence checking [6] involves comparing the unoptimized (O0) and optimized (O2/O3) implementations of the same C programs in x86 assembly. While their algorithm guarantees soundness, it does not guarantee completeness; their work reported that they could

© Springer International Publishing AG, part of Springer Nature 2018
O. Beyersdorff and C. M. Wintersteiger (Eds.): SAT 2018, LNCS 10929, pp. 365–382, 2018.
https://doi.org/10.1007/978-3-319-94144-8_22

```
C0  int g[144];                          A0  example0_compiled:
C1  int example0() {                      A1    r1 := 0;      //sum'
C2    int sum = 0;                        A2    r2 := 144;    //loop index(i')
C3    for (int i = 0; i < 144; i++)       A3    r3 := 0;      //array index(a')
C4    {                                   A4  loop:
C5      sum = sum + g[i];                 A5      r1 := r1 + [ base_g + 4*r3];
C6    }                                   A6      r2 := r2 - 1;
C7    retval = sum/144; //return          A7      r3 := r3 + 1;
C8  }                                     A8    if (r2 != 0) goto loop
                                          A9    rax := mul-shift-add(r1,144)
```

Fig. 1. C-function `example0()` and abstracted version of its compiled assembly (as produced by `gcc -O2`). We use a special keyword `retval` to indicate the location that holds the return value of the function. In assembly, sum and i variables are register allocated to `r1` and `r2` respectively and r3 is an iterator for indexing the array g. Division operation in C program is optimized to mul-shift-add instructions in assembly. base_g represents the base address of array g in memory. [x] is short-hand for 4 bytes in memory at address x.

prove equivalence only across 72–76% of the functions in real-world programs, across transformations produced by modern compilers like GCC, LLVM, ICC, and CompCert. Our work aims to reason about equivalence in scenarios where these previous algorithms would fail.

To understand the problem of equivalence checking and the general solution, we discuss the proof of equivalence across the example pair of programs in Fig. 1. The most common approach to proving that this pair of programs is equivalent involves the construction of a *simulation relation* between them. If $Prog_A$ represents the C language specification and $Prog_B$ represents the optimized x86 implementation, a simulation relation is represented as a table, where each row is a tuple $((L_A, L_B), P)$ such that L_A and L_B are program locations in $Prog_A$ and $Prog_B$ respectively, and P is a set of invariants on the live program variables[1] at locations L_A and L_B. Program locations represent the next instruction (PC values) to be executed in a program and the live program variables are identified by performing liveness analysis at every program location. A tuple $((L_A, L_B), P)$ represents that the invariants P hold whenever the two programs are at L_A and L_B respectively. A simulation relation is valid if the invariants at each location pair are inductively provable from the invariants at all its predecessor location pairs. Invariants at the entry location (the pair of entry locations of the two programs) represent the equivalence of program inputs ($Init$) and form the base case of this inductive proof. If we can thus inductively prove equivalence of return values at the exit location (the pair of exits of the two programs), we establish the equivalence of the two programs. For C functions, the return values include return registers (e.g., `rax` and `rdx`) and the state of the heap and global variables. Formally, a simulation relation is valid if:

$$Init \Leftrightarrow invariants_{(Entry_A, Entry_B)} \tag{1}$$

$$\forall_{(L'_A, L'_B) \to (L_A, L_B)} invariants_{(L'_A, L'_B)} \Rightarrow_{(L'_A, L'_B) \to (L_A, L_B)} invariants_{(L_A, L_B)} \tag{2}$$

---

[1] For assembly code, variables represent registers, stack and memory regions.

Table 1. Simplified simulation relation for the programs in Fig. 1. M_A and M_B are memory states in $Prog_A$ and $Prog_B$ respectively. $=_\Delta$ represents equivalent arrays except for Δ, where Δ represents the stack region.

Location	Invariants (P)
(C0, A0)	`g` $\mapsto$ `base_g`
(C3, A4)	`144-i = i'`, `sum = sum'`, `i = a'`, `g` $\mapsto$ `base_g`
(C7, A9)	`(retval`$_C$` = rax`$_A$`)`, `g` $\mapsto$ `base_g`
Init: $g \mapsto base\_g$, $M_A =_\Delta M_B$	

Here, $invariants_{(L_A, L_B)}$ is the same as P and represents the conjunction of invariants in the simulation relation for the location pair (L_A, L_B), L'_A and L'_B are predecessors of L_A and L_B in programs $Prog_A$ and $Prog_B$ respectively, and $\Rightarrow_{(L'_A, L'_B) \to (L_A, L_B)}$ represents implication over the paths $L'_A \to L_A$ and $L'_B \to L_B$ in programs $Prog_A$ and $Prog_B$ respectively.

Almost all compiler optimizations are *similarity* preserving, i.e. the optimized program simulates the original program, and hence approaches that rely on the construction of a simulation relation usually suffice for computing equivalence across compiler optimizations. There have been proposals in previous work [7] to handle transformations that do not preserve similarity (but preserve equivalence), but we do not consider them in this paper. In our experience, modern compilers rarely (if ever) produce transformations that do not preserve similarity. Table 1 shows a simulation relation that proves the equivalence between the two programs in Fig. 1. Given a valid simulation relation, proving equivalence is straight-forward; however the construction of a simulation relation is undecidable in general.

Static approaches to equivalence checking attempt to construct a simulation relation purely through static analysis. On the other hand, *data-driven* approaches [8,9] extract information from the program execution traces to infer a simulation relation. In either case, the construction of a simulation relation involves the inference of the *correlation* between program locations (i.e., the first column of the simulation relation table) and the invariants at each correlated pair of locations (i.e., the second column of the simulation relation table).

A data-driven approach to inferring a correlation involves identifying program locations in the two programs where the control-flow is correlated across multiple execution runs, *and* where the number and values of the live variables in the two programs are most similar. Also, the inference of invariants in these data-driven approaches is aided by the availability of actual runtime values of the live program variables.

On the other hand, static approaches usually employ an algorithm based on the guess-and-check strategy. We discuss a static algorithm for automatic construction of a simulation relation in Sect. 2. Essentially it involves an incremental construction of a simulation relation, where at each incremental step, the invariants at the currently correlated program locations are inferred (using a guess-and-check strategy) and future correlations are guided through the invariants inferred thus far. These guess-and-check based approaches are able to infer only simple forms of invariants and run into scalability bottlenecks while trying to infer more complex invariants. This is why data-driven approaches are more powerful because they can sidestep these scalability limitations by being able to infer more expressive invariants using real execution data.

However, data-driven approaches cannot work in the absence of a sufficient number of execution traces. Further, these approaches fail if certain portions (or states) of the program remain unexercised (uncovered) through the execution runs. We provide a counter-example guided strategy for invariant inference that allows us to scale our guess-and-check procedure much beyond existing approaches to guess-and-check. Our strategy resembles previous data-driven approaches; however, we are able to do this without access to execution traces, and only through counter-examples provided by modern SMT solvers. In particular, our algorithm involves *sketch generation* through syntax-guided synthesis, where a *sketch* is a template for an invariant. We use counter-examples to try and *fill* the sketch to arrive at a final invariant. To our knowledge, this is the first sketching-based approach to invariant inference and is the first contribution of our work. We call our algorithm *invariant-sketching*.

Several steps during the construction of a simulation relation involve proof-obligations (or checks) that can be represented as SMT satisfiability queries and discharged to an SMT solver. We find that modern SMT solvers face tractability limitations while computing equivalence across several compiler transformations. This is primarily due to two reasons:

1. Often, equivalence across these types of transformations are not captured in higher-order decision procedures in SMT solvers, and it appears that modern SMT solvers resort to expensive exponential-time algorithms to decide equivalence in these cases;
2. Even if these transformations are captured in SMT solvers, the composition of multiple such transformations across relatively large program fragments makes it intractable for the SMT solver to reason about them.

For the transformations that are not readily supported by modern SMT solvers, we employ *simplification passes* that can be applied over SMT expression DAGs[2], before submitting them to the SMT solver. Simplification passes involve rewriting expression DAGs using pattern-matching rules. We find several cases where the discharge of certain proof obligations during equivalence computation is tractable only after our simplification passes have been applied. We believe that our observations could inform SMT solvers and guide their optimization strategy.

The latter scalability limitation is due to the composition of multiple compiler transformations in a single program fragment. To tackle this, we propose a novel algorithm called *Query-decomposition*. Query decomposition involves breaking down a larger query into multiple sub-queries: we find that while an SMT solver may find it hard to reason about one large query, it may be able to discharge

---

[2] A DAG is a more compact representation of an expression tree where identical subtrees in the same tree are merged into one canonical node.

tens of smaller queries in much less total time. Further, we find that counter-examples obtained from previous SMT solver queries can be used to significantly prune the number of required smaller queries. Using query-decomposition with counter-example based pruning allows us to decide more proof-obligations than were previously possible, in turn allowing us to compute equivalence across a larger class of transformations/programs. The simplification passes and our query-decomposition algorithm with counter-example based pruning are the second contribution of this paper.

In both these contributions, we make use of the `get-model` feature [10] available in modern SMT solvers to obtain counter-examples. Previous approaches to equivalence checking have restricted their interaction with SMT solvers to a one-bit SAT/UNSAT answer; we demonstrate algorithms that can scale equivalence checking procedures beyond what was previously possible through the use of solver-generated counter-examples.

Paper Organization: Section 2 provides background on automatic construction of a simulation relation - our work builds upon this previous work to improve its scalability and robustness. Section 3 presents a motivating example for our work. Section 4 describes our novel sketching-based invariant inference procedure. Section 5 focusses on some important limitations of SMT solvers while reasoning about compiler optimizations, and discusses our simplification passes and the query-decomposition algorithm in this context. Section 6 discusses the experiments and results. Section 7 summarizes previous work and concludes.

2   Background: Automatic Generation of Simulation Proof

Automatic construction of a provable simulation relation between a program and its compiled output has been the subject of much research with several motivating applications. Our algorithm resembles previous work on black-box equivalence checking [6], in that it attempts to construct a simulation relation incrementally as a *joint transfer function graph* (JTFG). The JTFG is a graph with nodes and edges, and represents the partial simulation relation computed so far. A JTFG node (L_A, L_B) represents a pair of program nodes L_A and L_B and indicates that $Prog_A$ is at L_A and $Prog_B$ is at L_B. Similarly, a JTFG edge $(L_A', L_B') \to (L_A, L_B)$ represents a pair of transitions $L_A' \to L_A$ and $L_B' \to L_B$ in $Prog_A$ and $Prog_B$ respectively. Each JTFG node (L_A, L_B) contains invariants relating the live variables at locations L_A and L_B in the two programs respectively. Further, for each JTFG edge, the *edge conditions* (*edgecond*) of its two individual constituent edges ($L_A' \to L_A$ and $L_B' \to L_B$) should be equivalent. An edge condition represents the condition under which that edge is taken, as a function of the live variables at the source location of that edge.

The algorithm for constructing a JTFG is succinctly presented in Algorithm 1. Section 3 describes the running of this algorithm on an example pair of programs. The JTFG is initialized with a single node, representing the pair of

---

Algorithm 1. Algorithm to construct the JTFG (simulation relation). $edges_B$ is a list of edges in $Prog_B$ in depth-first search order. The `AddEdge()` function returns a new JTFG `jtfg'`, formed by adding the edge to the old JTFG `jtfg`.

Function *CorrelateEdges(jtfg, edges$_B$)*
    if *edges$_B$ is empty* **then**
       | **return** LiveValuesAtExitAreEq(jtfg)
    end
    edge$_B$ ← RemoveFirst(edges$_B$)
    fromPC$_B$ ← GetFromPC(edge$_B$)
    fromPC$_A$ ← FindCorrelatedPC(jtfg, fromPC$_B$)
    edges$_A$ ← GetEdgesTillUnroll(Prog$_A$, fromPC$_A$, μ)
    foreach *edge$_A$ in edges$_A$* **do**
       jtfg' = AddEdge(jtfg, edge$_A$, edge$_B$)
       PredicatesGuessAndCheck(jtfg')
       if *IsEqualEdgeConditions(jtfg')* ∧ *CorrelateEdges(jtfg', edges$_B$)* **then**
       | **return** true
       end
    end
    return false

---

entry locations of the two programs. It also has the associated invariants encoding equivalence of program values at entry (base case). The loop heads, function calls and exit locations in $Prog_B$ are then chosen as the program points that need to be correlated with a location in $Prog_A$. All other program points in $Prog_B$ are *collapsed* by composing their incoming and outgoing edges into *composite edges*. The `CorrelateEdges()` function picks one (composite) $Prog_B$ edge, say $edge_B$, at a time and tries to identify paths in the source program ($Prog_A$) that have an equivalent *path condition* to $edge_B$'s edge condition. Several candidate paths are attempted up to an unroll factor μ (`GetEdgesTillUnroll()`). All candidate paths must originate from a $Prog_A$ location (`fromPC_A`) that has already been correlated with the source location of $edge_B$ (`fromPC_B`). The unroll factor μ allows equivalence computation across transformations involving loop unrolling. The path condition of a path is formed by appropriately composing the edge conditions of the edges belonging to that path. The edge $edge_B$ is chosen in depth-first search order from $Prog_B$, and also dictates the order of incremental construction of the JTFG. The equivalence of the edge condition of $Prog_B$ with the path condition of $Prog_A$ is computed based on the invariants inferred so far at the already correlated JTFG nodes (`IsEqualEdgeConditions()`). These invariants, inferred at each step of the algorithm, are computed through a Houdini-style [11] guess-and-check procedure. The guesses are synthesized from a grammar, through the syntax-guided synthesis of invariants (`PredicatesGuessAndCheck`). These correlations for each edge ($edge_B$) are determined recursively to allow backtracking (the recursive call to `CorrelateEdges()`). If at any stage, an edge ($edge_B$) cannot be correlated with any path in $Prog_A$, the function returns with a failure, prompting the caller frame in this recursion stack, to try another correlation for a previously correlated edge.

PredicatesGuessAndCheck() synthesizes invariants through the following grammar of guessing: $\mathbb{G} = \{ \star_A \otimes \star_B \}$, where operator $\otimes \in \{<, >, =, \leq, \geq\}$ and $\star_A$ and $\star_B$ represent the live program values (represented as symbolic expressions) appearing in $Prog_A$ and $Prog_B$ respectively. The guesses are formed through a cartesian product of values in $Prog_A$ and $Prog_B$ using the patterns in $\mathbb{G}$. Our checking procedure is a sound fixed-point computation which keeps eliminating the unprovable predicates until only provable predicates remain (similar to Houdini). At each step of the fixed point computation, for each guessed predicate at each node, we try to prove it from current invariants at every predecessor node (as also done in the final simulation relation validity check in Eq. 2).

It is worth noting that we need to keep our guessing procedure simple to keep this procedure tractable; it currently involves only conjunctions of equality and inequality relations between the variables of the two programs. We find that this often suffices for the types of transformations produced by production compilers. In general, determining the right guesses for completing the proof is undecidable: a simple guessing grammar keeps the algorithm tractable, increasing grammar complexity significantly increases the runtime of the equivalence procedure. In our work, we augment the guessing procedure to generate *invariant-sketches* and use counter-examples to fill the sketches to arrive at the final invariants.

3   Running Example

Figure 2 shows the abstracted versions (*aka* transfer function graphs or TFGs) of the original C specification and its optimized assembly implementation for the program in Fig. 1. TFG nodes represent program locations and TFG edges indicate control flow. Each TFG edge is associated with its *edge-condition* and its *transfer function* (labelled as Cond and TF respectively in the figure). Notably, TFG and JTFG representations are almost identical. Across the two TFGs,

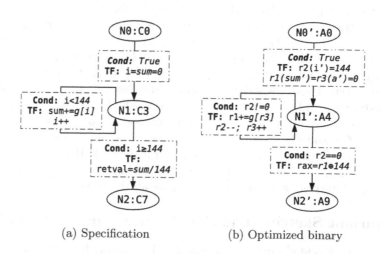

(a) Specification   (b) Optimized binary

Fig. 2. TFGs of C program and its optimized implementation for the program in Fig. 1.

the program has undergone multiple compiler transformations, namely (a) loop reversal (i counts from [0..144) in the original program but counts backwards from [144..0) in the optimized program), (b) strength-reduction of the expensive division operation to a cheaper combination of multiply-shift-add (represented by $\oplus$), and (c) register allocation of variables sum and i. All these are common optimizations produced by modern compilers, and failing to prove equivalence across any of these (or their composition) directly impacts the robustness of an equivalence checker. (As an aside, while the general loop-reversal transformation does not preserve similarity, it preserves similarity in this example. In general, we find that modern compilers perform loop-reversal only if similarity is preserved).

Applying Algorithm 1 to this pair of TFGs, we begin with a JTFG with one node representing the start node (N0,N0'). Our first correlation involves correlating the loop heads of the two programs by adding the node (N1,N1') to the JTFG. At this point, we need to infer the invariants at (N1,N1'). While all other invariants can be inferred using the grammar presented in the procedure PredicatesGuessAndCheck in Sect. 2, the invariant i' = 144-i is not generated by our grammar. This is so because the grammar only relates the variable values computed in the two programs, but the value 144 - i is never computed in the body of the source program.

One approach to solving this problem is to generalize our guessing grammar such that it also generates invariants of the shape $\{C_A \star_A + C_B \star_B + C_1 = 0\}$, where $\star_A$ and $\star_B$ represent program values (represented as symbolic expressions) appearing in $Prog_A$ and $Prog_B$ respectively, and C_A, C_B, and C_1 are the coefficients of $\star_A$, $\star_B$, and 1 respectively in this linear equality relation. C_A, C_B, and C_1 could be arbitrary constants. We call this extended grammar $\mathbb{G}'$, the grammar of *linear-equality relations*. The problem, however, is that this grammar explodes the potential number of guessed invariants, as the number of potential constant coefficients is huge. In contrast, a data-driven approach may identify the exact linear-relation through the availability of run-time values. We present an algorithm to tractably tackle this in a static setting through the generation of *invariant-sketches* in our grammar. An invariant-sketch is similar to an invariant, except that certain parts of the sketch are left unspecified. e.g., in our case, the constant coefficients C_A, C_B, and C_1 are left unspecified in the generated sketch. These unspecified constants are also called *holes* in the sketch.

We restrict C_A to be 1 and $C_B \in \{-1, 0, 1\}$ and find that this suffices for the types of transformations we have encountered in modern compilers. Generalizing to arbitrary C_A and C_B requires careful reasoning about bitvector arithmetic and associated overflows, and we leave that for future work. However, notice that we place no restrictions on C_1—e.g., for 64-bit arithmetic, $C_1 \in [-2^{63}..2^{63} - 1]$, making it prohibitively expensive to enumerate all the possibilities.

4   Invariant Sketches and the Use of Counter-Examples

We now discuss a syntax-guided invariant-sketching algorithm that uses counter-examples generated by SMT solvers to fill the sketches. The guessing grammar

G' generates invariant-sketches, in addition to the invariants generated by G. For example, one of the guesses generated using G' for our running example will be $(i + C_B i' + C_1 = 0)$. Recall that i represents a variable in $Prog_A$ and i' represents a variable in $Prog_B$ and C_B, C_1 represent holes in the generated sketch. Notice that we omit C_A as it is restricted to be $= 1$ in G'.

Algorithm 2. InvSketch algorithm to infer invariant between var_y of $Prog_A$ and var_x of $Prog_B$ at Node N.

Function *InvSketch(N, e, var_x, var_y)*
    $N1_A$ = QuerySatAssignment(e, true)
    if $N1_A$ *is empty* **then**
     | return (Inv $\mapsto$ {False})
    end
    $N2_A$ = QuerySatAssignment(e, (var_x, var_y) != $N1_A$)
    if $N2_A$ *is empty* **then**
     | return (Inv $\mapsto$ {$(var_x, var_y) = N1_A$})
    end
    Coeff$_{C_B,C_1}$ = InferLinearRelation($N1_A$, $N2_A$)
    Inv = FillSketch(Coeff$_{C_B,C_1}$)
    return Inv

For each invariant-sketch, we try to infer the potential values of the holes by querying the SMT solver for a *satisfying assignment* for the variables at the current node. A satisfying assignment N_A at a node N represents a mapping from program variables to some constants; this mapping should be satisfiable, assuming that the invariants at a predecessor node P and the edge condition for the edge $P \rightarrow N$ are true. For example, if a predecessor node P has an inferred invariant x=y and the edge condition and transfer function across the edge $P \rightarrow N$ are {true} and {x=x+1, y=y+2} respectively, then the assignment x=3,y=3 is *not* satisfiable at N. On the other hand, the assignment x=3,y=4 is satisfiable at N. To obtain satisfying assignments for variables at a node N, we first obtain a satisfying assignment P_A for the invariants at P and the edge condition for the edge $P \rightarrow N$ through an SMT query. We then apply the transfer function of the edge $P \rightarrow N$ on P_A to obtain N_A.

We define a procedure called QuerySatAssignment($e = P \rightarrow N$, cond$_{extra}$). This procedure generates a satisfying assignment (if it exists) for N, given the current invariants at P and the edge-condition of edge $e = P \rightarrow N$. Further, the satisfying assignment must satisfy the extra conditions encoded by the second argument cond$_{extra}$. Algorithm 2 presents our algorithm to infer the invariant, given an invariant-sketch, using satisfying assignments generated through calls to QuerySatAssignment(). The algorithm infers the values of the holes C_B and C_1 (if they exist) in a given invariant-sketch. At a high level, the algorithm first obtains two satisfying assignments, ensuring that the second satisfying assignment is distinct from the first one. Given two assignments, we can substitute these assignments in the invariant-sketch to obtain two linear equations in two unknowns, namely C_B and C_1. Based on these two linear equations, we can infer the potential values of C_B and C_1 using standard linear-algebra methods. If no satisfying assignment exists through any of the predecessors of N, we simply emit the invariant `false` indicating that this node is unreachable given the current invariants at the predecessors (this should happen only if the programs contain dead-code). Similarly, if we are able to generate only one satisfying assignment (i.e., the second SMT query fails to generate another distinct satisfying assignment), we simply generate the invariant encoding that the variables have constant values (equal to the ones generated by the satisfying assignment). If a node N has multiple predecessors, we can try generating satisfying assignments through *either* of the predecessors.

Thus, for each invariant-sketch generated at each step of our algorithm, we check to see if the satisfying assignments for program variables at that node result in a valid invariant. If so, we add the invariant to the pool of invariants generated by our guessing procedure. Notice that we need at most two SMT queries per invariant-sketch; in practice, the same satisfying assignment may be re-usable over multiple invariant-sketches drastically reducing the number of SMT queries required. For our running example, we will first obtain a satisfying assignment at node (N1,N1') using invariants at node (N0,N0'): i=0,i'=144. However we will be unable to obtain a second satisfying assignment at this stage, and so we will generate invariants i=0,i'=144 at (N1,N1'). In the next step of the algorithm, the edge N1'→N1' will be correlated with the corresponding *Prog*$_A$ edge N1→N1. At this stage, we will again try the same invariant-sketch, and this time we can obtain two distinct satisfying assignments at (N1,N1'): {i=0,i'=144} (due to the edge (N0,N0')→(N1,N1')) and {i=1,i'=143} (due to the edge (N1,N1')→(N1,N1')). Using these two satisfying assignments, and using standard linear-algebra techniques (to solve for two unknowns through two linear equations), we can infer that $C_B = 1$ and $C_1 = -144$, which is our required invariant guess. Notice that the output of our invariant-sketching algorithm is an invariant *guess*, which may subsequently be eliminated by our sound fixed-point procedure for checking the inductive validity of the simulation relation. The latter check ensures that our equivalence checking algorithm remains sound.

5   Efficient Discharge of Proof Obligations

In our running example, after the edges (N1→N1) and (N1'→N1') have been correlated, the algorithm will infer the required invariants correlating the program values at Node (N1,N1'). After that, the edge (N1'→N2') will get correlated with the edge (N1→N2) and we would be interested in proving that the final return values are identical. This will involve discharging the proof obligation of the form: (sum=sum')⇒((sum/144)=(sum'⊕144)). It turns out that SMT solvers find it hard to reason about equivalence under such transformations; as we discuss in Sect. 6, modern SMT solvers do not answer this query even after several hours. We find that this holds for some common types of compiler transformations such as: (a) transformations involving mixing of multi-byte arithmetic operations with select/store operations on arrays, (b) transformation of the division operator to a multiply-shift-add sequence of operations, (c) complex bitvector shift/extract operations mixed with arithmetic operations, etc.

We implement *simplification passes* to enable easier reasoning for such patterns by the SMT solvers. A simplification pass converts a pattern in the expression to its simpler canonical form. We discuss the "simplification" of the division operator into a sequence of multiply-shift-add operators to illustrate this. Given a dividend n and a constant divisor d, we convert it to:

$$n \div d \equiv (n + (n \times C_{mul}) \gg 32) \gg C_{shift},$$

where $0 < d < 2^{32}$ and $0 \le n < 2^{32}$ and C_{mul}, C_{shift} represent two constants dependent on d and are calculated using a method given in Hacker's Delight [12]. This simplification ensures that if the compiler performs a transformation that resembles this pattern, then both the original program and the transformed program will be simplified to the same canonical expression structure, which will enable the SMT solvers to reason about them more easily. We find that there exist more patterns that exhibit SMT solver time-outs by default, but their simplified versions (through our custom simplification passes) become tractable for solving through modern SMT solvers.

Even after applying the simplification passes, we find that several SMT queries still time-out because SMT solvers find it difficult to reason about equivalence in the face of several composed transformations performed by the compiler. We observe that while SMT solvers can easily compute equivalence across a smaller set of transformations, they often time out if the number of composed transformations is too many or too intertwined. Taking a cue from this observation, we propose the *query-decomposition* algorithm.

The general form of proof queries in an equivalence checker is: $Precond \Rightarrow (LHS = RHS)$, where $Precond$ represents a set of pre-conditions (e.g., x=y) and LHS and RHS expressions (e.g., x+1 and y+2) are obtained through symbolic execution of the C specification ($Prog_A$) and the optimized program ($Prog_B$) respectively. The RHS expression may contain several composed transformations for the computation performed in the LHS expression. Query-decomposition involves breaking this one large proof query into multiple smaller queries by using the following steps:

1. We walk the expression DAGs of LHS and RHS and collect all unique sub-expressions in LHS and RHS into two different sets, say $\{lhsSubExprs\}$ and $\{rhsSubExprs\}$.
2. We check the equivalence of each $lhsSubExpr \in \{lhsSubExprs\}$ with each $rhsSubExpr \in \{rhsSubExprs\}$ (assuming $Precond$), in increasing order of size of LHS sub expressions. i.e., $Precond \Rightarrow (lhsSubExpr = rhsSubExpr)$. The size of an expression is obtained by counting the number of operators in its expression DAG. If there are m unique sub-expressions in $\{lhsSubExprs\}$ and n unique sub-expressions in $\{rhsSubExprs\}$, we may need to perform $m * n$ equivalence checks.
3. For any check that is successful in step 2, we learn a substitution map from the bigger expression to the smaller expression. For example, if $lhsSubExpr$

was smaller in size than $rhsSubExpr$, we learn a substitution mapping $rhsSubExpr \mapsto lhsSubExpr$. We maintain a set of substitution mappings thus learned.

4. For any check that is unsuccessful in step 2, we learn a counter-example that satisfies $Precond$ but represents a variable-assignment that shows that $lhsSubExpr = rhsSubExpr$ is not provable. We add all such counter-examples to a set $\{counterExamples\}$.

5. In all future equivalence checks (of the total $m * n$ checks) of type $Precond \Rightarrow (lhsSubExpr = rhsSubExpr)$, we first check the set $\{counterExamples\}$ to see if any $counterExample \in \{counterExamples\}$ disproves the query. If so, we have already decided the equivalence check as false. If not, we rewrite both expressions $lhsSubExpr$ and $rhsSubExpr$ using the substitution-map learned in step 3. For a substitution mapping $e1 \rightarrow e2$, we replace every occurrence of $e1$ in an expression with $e2$ during this rewriting. After the rewriting procedure reaches a fixed-point, we use an off-the-shelf SMT solver to decide the rewritten query.

This decomposition of a larger expression into sub-expressions, and the substitution of equivalent sub-expressions while deciding equivalence of larger expressions ensures that the queries submitted to SMT solvers are simpler than the original query. In other words, through this strategy, the LHS' and RHS' expressions that are submitted to an SMT solver are more similar to each other, as multiple composed transformations have likely been decomposed into fewer transformations in each individual query. Because we only replace provably-equivalent sub-expressions during decomposition, the overall equivalence checking algorithm remains sound.

This bottom-up strategy of decomposing a larger query into several smaller queries would be effective only if (a) we expect equivalent sub-expressions to appear across LHS and RHS, and (b) the total time to decide equivalence for multiple sub-expression pairs is smaller than the time to decide equivalence for a single larger expression-pair. We find that both these criteria often hold while comparing distinct expressions that differ in the transformations performed by a compiler. Figure 3 illustrates this with an example. In this example, the proof query involves deciding equivalence between the top-level expressions $E1$ and $E5$. Also, it turns out that the sub-expressions $E3$ and $E4$ (on the left) are equivalent to the sub-expressions $E7$ and $E8$ (on the right) respectively. This query gets generated when comparing the unoptimized and optimized implementations generated by GCC for a fragment of a real-world program. Notably, modern SMT solvers like Z3/Yices/Boolector time-out even after several hours for such a query. On the other hand, they are able to decide the equivalence of individual sub-expressions ($E3 = E7$ and $E4 = E8$) within a few seconds. Experimentally, we have observed that if we substitute $E7$ and $E8$ with $E3$ and $E4$ respectively (in the expression DAG of $E5$), the resulting equivalence check between $E1$ and the rewritten $E5$ (as shown in Fig. 3(b)) also completes within a fraction of a second. In Sect. 6, we discuss more real-world functions and our results with transformations produced by multiple compilers, to demonstrate the

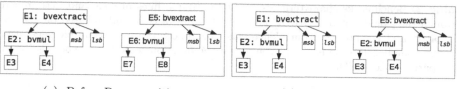

(a) Before Decomposition   (b) After Decomposition

Fig. 3. Original and result of query-decomposition for a fragment of expression DAGs for two SMT-expressions.

effectiveness of our query-decomposition procedure. We also evaluate the fraction of intermediate sub-expression queries that can be pruned through the use of the counter-example set.

6   Experiments

We evaluate invariant-sketching and query-decomposition algorithms by studying their effectiveness in a black-box equivalence checker across LLVM IR and x86 assembly code. We compile a C program using LLVM's clang-3.6 to generate unoptimized (O0) LLVM IR bitcode and using GCC, LLVM, ICC (Intel C Compiler), and CComp (CompCert [13]) with O2 optimization to generate the x86 binary executable code. We have written symbolic executors for LLVM bitcode and x86 instructions to convert the programs to their logical QF_AUFBV SMT-like representation. In this representation, program states including the state of LLVM variables, x86 registers and memory are modelled using bit-vectors and byte-addressable arrays respectively. Function calls are modelled through uninterpreted functions. The black-box equivalence checking tool employs the algorithm discussed in Sect. 2 with $\mu = 1$. The tool also models undefined-behaviour semantics of the C language [14] for improved precision in equivalence checking results. Proof obligations are discharged using Yices [15] (v2.5.4) and Z3 [16] (commit 0b6a836eb2) SMT solvers running in parallel: each proof obligation is submitted to both solvers, and the result is taken from the solver that finishes first. We use a time-out value of five hours for each proof obligation.

Benchmarks and Results: For evaluation, we use C functions from the SPEC CPU Integer benchmarking suite [17] that contain loops and cannot be handled by previous equivalence checking algorithms. Previous work on black box equivalence checking [6] fails to compute equivalence on all these functions. We also include the benchmarks used by previous work on data-driven equivalence checking [8] in our evaluation; we are able to statically compute equivalence for these benchmarks, where previous work relied on execution data for the same. The selected functions along with their characteristics and results obtained for each function-compiler pair are listed in Table 2.

The results in bold-red typeface depict the function-compiler pairs for which previous work fails to prove equivalence statically. Computing equivalence for

Table 2. Benchmarks characteristics. SLOC is source lines of code and determined through the sloc count tool. ALOC is assembly lines of code and is based on gcc-O0 compilation. T_X represents equivalence checking time taken for executable generated by "X" compiler in seconds. ✗ represents that the function could not be compiled with that particular compiler.

S.No.	Benchmark	Function	SLOC	ALOC	Checking time (sec)			
					T_{gcc}	T_{clang}	T_{icc}	T_{ccomp}
B1	knucleotide	ht_hashcode	5	28	17	18	215	10
B2	nsieve	main	11	39	1343	1687	2265	868
B3	sha1	do_bench	11	49	338	320	385	383
B4	DDEC	lerner1a	12	22	37	13	36	12
B5	twolf	controlf	8	16	73	79	75	✗
B6	gzip	display_ratio	21	78	738	121	677	570
B7	vpr	is_cbox	8	48	24	25	24	24
B8	vpr	get_closest_seg_start	14	57	27	27	27	27
B9	vpr	get_seg_end	16	63	28	29	29	30
B10	vpr	is_sbox	16	79	34	33	32	Fail
B11	vpr	toggle_rr	7	25	131	121	99	✗
B12	bzip2	makeMaps	11	39	217	240	214	221

these programs requires either sophisticated guessing procedures (which we address through our invariant-sketching algorithm) or/and involves complex proof queries that would time-out on modern SMT solvers (addressed by our simplification and query-decomposition procedures). The results in non-bold face depict the function-compiler pairs for which equivalence can be established even without our algorithms—in these cases, the transformations performed by the compiler require neither sophisticated guessing nor do their proof obligations time-out. For most cases, by employing our algorithms, the execution time for establishing equivalence is reasonably small. In general, we observe that the equivalence checking time depends on the size of the C program and the number and complexity of transformations performed by the compiler. For one of the benchmarks (is_sbox compiled through CComp), equivalence could not be established even after employing our invariant-sketching, simplification and query-decomposition algorithms. We next evaluate the improvements obtained by using counter-examples to prune the number of queries discharged to SMT solver. Recall that additional queries are generated by both invariant-sketching and query-decomposition algorithms. Also, the query-decomposition algorithm maintains a set of counter-examples and a substitution-map learned so far, to reduce the time required to discharge a query. For each query, we first check to see if the query can be answered through the set of currently-available counter-examples. If not, the second step involves rewriting the query through simplification passes and the currently-available substitution-map to see if the resulting query can be answered syntactically. Finally, if equivalence is not decidable

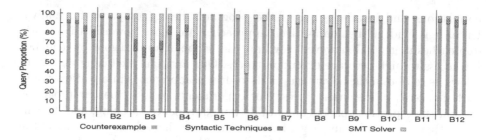

Fig. 4. Proof queries statistics. The bar represents the percentage of queries solved by each strategy for each benchmark-compiler pair in the same order as in Table 2.

even after simplification and substitution, we submit the simplified query to the SMT solver. Figure 4 provides a break-down of the fraction of queries answered by counter-examples, by syntactic simplification and substitution, and by the SMT solver. The counter-example set is able to answer more than 85% of the total proof queries, including the ones generated by our invariant-sketching and query-decomposition algorithms. Similarly, syntactic simplification and substitution are able to answer 3% of the queries, while the remaining 12% of the queries are answered by the SMT solver. Recall that the simplification and substitution passes help ensure that the 12% queries can be answered by the SMT solver efficiently; the SMT solver would often time-out without these simplifications and substitutions.

7   Related Work and Conclusions

Combinational equivalence checking through SAT-based algorithms has been studied extensively both in hardware verification [18] and in software verification [19]. Equivalence checking for sequential programs (e.g., programs with loops) has also been studied extensively in the context of translation validation [3,20], design and verification of compiler optimizations [1,21], and program synthesis and superoptimization [4,5,22–25]. Modern SMT solvers have further facilitated these applications (over traditional SAT solvers) by raising the level of abstraction at the interface of the SMT solver and the equivalence checker. Improving the capabilities of SMT solvers for various practical applications remains an important field of research [26,27]. For example, Limaye and Seshia [27] describe word-level simplification of queries before submitting them to SMT solvers; our simplification passes are similar to such previous work.

Our work studies the effective utilization of SMT solvers for the problem of equivalence checking for sequential programs containing loops. We demonstrate techniques that allow an equivalence checker to decide equivalence across a wider variety of programs and transformations; our invariant-sketching and query-decomposition algorithms are novel contributions in this context.

Data-driven equivalence checking and data-driven invariant inference are recent approaches [8,9,28,29] that utilize the information obtained through running the programs on real inputs (execution traces), for inferring the required correlations and invariants across the programs being compared for equivalence. It is evident that data-driven approaches are more powerful than static approaches in general; however they limit the scope of applications by demanding access to high-coverage execution traces. Our invariant-sketching algorithm allows us to obtain the advantages of data-driven approaches in a static setting, with no access to execution traces. Our experiments include the test programs used in these previous papers on data-driven equivalence checking, and demonstrate that a counter-example guided invariant-sketching scheme can achieve the same effect without access to execution traces. Further, some of the data-driven techniques, such as CEGIR [29] and Daikon [30], are unsound, i.e., they may return invariants that are not inductively provable but are only good enough for a given set of execution traces or the capabilities of a given verification tool. Unsound strategies are not useful for several applications of equivalence checking, such as translation validation and program synthesis. Both invariant-sketching and query-decomposition algorithms preserve soundness.

Recent work on synthesizing models for quantified SMT formulas [31] involves a similar computational structure to our invariant-sketching technique; the primary differences are in our use of a linear interpolation procedure (`InferLinearRelation`), and consequently the small number of invariant-synthesis attempts (at most two) for each invariant-sketch. These techniques make our procedure tractable, in contrast to the approach of synthesizing models for general quantified SMT formulas outlined in [31]. Invariant-sketching also has a parallel with previous approaches on counter-example guided abstraction refinement, such as recent work on worst-case execution time analysis [32]. From this perspective of abstraction refinement, our invariant-sketching algorithm refines an invariant from $(\text{Inv} \mapsto \{\text{False}\})$ to $(\text{Inv} \mapsto \{(\text{var}_x, \text{var}_y) = \text{N1}_A\})$ to the final linearly-interpolated invariant based on the invariant-sketch. This counter-example guided refinement is aided by invariant-sketches involving linear relations, that are designed to capture the underlying structure of the equivalence checking problem.

The query-decomposition algorithm for effective utilization of SMT solvers is based on our experiences with multiple SMT solvers. It is indeed interesting to note that SMT solvers can decide many smaller queries in much less time than one equivalent bigger query. This observation has motivated our decomposition algorithm, and our experiments show its efficacy in deciding equivalence across programs, where previous approaches would fail.

References

1. Necula, G.C.: Translation validation for an optimizing compiler. In: Proceedings of the ACM SIGPLAN 2000 Conference on Programming Language Design and Implementation, PLDI 2000, pp. 83–94. ACM, New York (2000)
2. Tristan, J.B., Govereau, P., Morrisett, G.: Evaluating value-graph translation validation for LLVM. In: Proceedings of the 32nd ACM SIGPLAN Conference on Programming Language Design and Implementation, PLDI 2011, pp. 295–305. ACM, New York (2011)

3. Tate, R., Stepp, M., Tatlock, Z., Lerner, S.: Equality saturation: a new approach to optimization. In: POPL 2009: Proceedings of the 36th Annual ACM SIGPLAN-SIGACT Symposium on Principles of Programming Languages, pp. 264–276. ACM, New York (2009)
4. Bansal, S., Aiken, A.: Automatic generation of peephole superoptimizers. In: Proceedings of the 12th International Conference on Architectural Support for Programming Languages and Operating Systems, ASPLOS XII, pp. 394–403. ACM, New York (2006)
5. Schkufza, E., Sharma, R., Aiken, A.: Stochastic superoptimization. In: Proceedings of the Eighteenth International Conference on Architectural Support for Programming Languages and Operating Systems, ASPLOS 2013, pp. 305–316. ACM, New York (2013)
6. Dahiya, M., Bansal, S.: Black-box equivalence checking across compiler optimizations. In: Chang, B.-Y.E. (ed.) APLAS 2017. LNCS, vol. 10695, pp. 127–147. Springer, Cham (2017). https://doi.org/10.1007/978-3-319-71237-6_7
7. Zuck, L., Pnueli, A., Goldberg, B., Barrett, C., Fang, Y., Hu, Y.: Translation and run-time validation of loop transformations. Form. Methods Syst. Des. **27**(3), 335–360 (2005)
8. Sharma, R., Schkufza, E., Churchill, B., Aiken, A.: Data-driven equivalence checking. In: Proceedings of the 2013 ACM SIGPLAN International Conference on Object Oriented Programming Systems Languages & Applications, OOPSLA 2013, pp. 391–406. ACM, New York (2013)
9. Padhi, S., Sharma, R., Millstein, T.: Data-driven precondition inference with learned features. In: Proceedings of the 37th ACM SIGPLAN Conference on Programming Language Design and Implementation, PLDI 2016, pp. 42–56. ACM, New York (2016)
10. Barrett, C., Stump, A., Tinelli, C.: The SMT-LIB standard - version 2.0. In: Proceedings of the 8th International Workshop on Satisfiability Modulo Theories, Edinburgh, Scotland (2010)
11. Flanagan, C., Leino, K.R.M.: Houdini, an annotation assistant for ESC/Java. In: Oliveira, J.N., Zave, P. (eds.) FME 2001. LNCS, vol. 2021, pp. 500–517. Springer, Heidelberg (2001). https://doi.org/10.1007/3-540-45251-6_29
12. Warren, H.S.: Hacker's Delight. Addison-Wesley Longman Publishing Co., Inc., Boston (2002)
13. Leroy, X.: Formal certification of a compiler back-end, or: programming a compiler with a proof assistant. In: 33rd ACM Symposium on Principles of Programming Languages, pp. 42–54. ACM Press (2006)
14. Dahiya, M., Bansal, S.: Modeling undefined behaviour semantics for checking equivalence across compiler optimizations. In: Strichman, O., Tzoref-Brill, R. (eds.) HVC 2017. LNCS, vol. 10629, pp. 19–34. Springer, Cham (2017). https://doi.org/10.1007/978-3-319-70389-3_2
15. Dutertre, B.: Yices 2.2. In: Biere, A., Bloem, R. (eds.) CAV 2014. LNCS, vol. 8559, pp. 737–744. Springer, Cham (2014). https://doi.org/10.1007/978-3-319-08867-9_49
16. de Moura, L., Bjørner, N.: Z3: an efficient SMT solver. In: Ramakrishnan, C.R., Rehof, J. (eds.) TACAS 2008. LNCS, vol. 4963, pp. 337–340. Springer, Heidelberg (2008). https://doi.org/10.1007/978-3-540-78800-3_24
17. Henning, J.L.: SPEC CPU2000: measuring CPU performance in the new millennium. Computer **33**(7), 28–35 (2000)
18. Popek, G.J., Goldberg, R.P.: Formal requirements for virtualizable third generation architectures. Commun. ACM **17**(7), 412–421 (1974)

19. Cadar, C., Dunbar, D., Engler, D.: KLEE: unassisted and automatic generation of high-coverage tests for complex systems programs. In: Proceedings of the 8th USENIX Conference on Operating Systems Design and Implementation, OSDI 2008, pp. 209–224. USENIX Association, Berkeley (2008)

20. Kundu, S., Tatlock, Z., Lerner, S.: Proving optimizations correct using parameterized program equivalence. In: Proceedings of the 2009 ACM SIGPLAN Conference on Programming Language Design and Implementation, PLDI 2009, pp. 327–337. ACM, New York (2009)

21. Tate, R., Stepp, M., Lerner, S.: Generating compiler optimizations from proofs. In: Proceedings of the 37th Annual ACM SIGPLAN-SIGACT Symposium on Principles of Programming Languages, POPL 2010, pp. 389–402. ACM, New York (2010)

22. Bansal, S., Aiken, A.: Binary translation using peephole superoptimizers. In: Proceedings of the 8th USENIX Conference on Operating Systems Design and Implementation, OSDI 2008, pp. 177–192. USENIX Association, Berkeley (2008)

23. Massalin, H.: Superoptimizer: a look at the smallest program. In: ASPLOS 1987: Proceedings of the Second International Conference on Architectural Support for Programming Languages and Operating Systems, pp. 122–126 (1987)

24. Phothilimthana, P.M., Thakur, A., Bodik, R., Dhurjati, D.: Scaling up superoptimization. In: Proceedings of the Twenty-First International Conference on Architectural Support for Programming Languages and Operating Systems, ASPLOS 2016, pp. 297–310. ACM, New York (2016)

25. Sharma, R., Schkufza, E., Churchill, B., Aiken, A.: Conditionally correct superoptimization. In: Proceedings of the 2015 ACM SIGPLAN International Conference on Object-Oriented Programming, Systems, Languages, and Applications, OOPSLA 2015, pp. 147–162. ACM, New York (2015)

26. Sulflow, A., Kuhne, U., Fey, G., Grosse, D., Drechsler, R.: WoLFram- a word level framework for formal verification. In: 2009 IEEE/IFIP International Symposium on Rapid System Prototyping, pp. 11–17, June 2009

27. Jha, S., Limaye, R., Seshia, S.A.: Beaver: engineering an efficient SMT solver for bit-vector arithmetic. In: Bouajjani, A., Maler, O. (eds.) CAV 2009. LNCS, vol. 5643, pp. 668–674. Springer, Heidelberg (2009). https://doi.org/10.1007/978-3-642-02658-4_53

28. Sharma, R., Gupta, S., Hariharan, B., Aiken, A., Liang, P., Nori, A.V.: A data driven approach for algebraic loop invariants. In: Felleisen, M., Gardner, P. (eds.) ESOP 2013. LNCS, vol. 7792, pp. 574–592. Springer, Heidelberg (2013). https://doi.org/10.1007/978-3-642-37036-6_31

29. Nguyen, T., Antonopoulos, T., Ruef, A., Hicks, M.: Counterexample-guided approach to finding numerical invariants. In: Proceedings of the 2017 11th Joint Meeting on Foundations of Software Engineering, ESEC/FSE 2017, pp. 605–615. ACM, New York (2017)

30. Ernst, M.D., Perkins, J.H., Guo, P.J., McCamant, S., Pacheco, C., Tschantz, M.S., Xiao, C.: The Daikon system for dynamic detection of likely invariants. Sci. Comput. Program. 69(1–3), 35–45 (2007)

31. Preiner, M., Niemetz, A., Biere, A.: Counterexample-guided model synthesis. In: Legay, A., Margaria, T. (eds.) TACAS 2017. LNCS, vol. 10205, pp. 264–280. Springer, Heidelberg (2017). https://doi.org/10.1007/978-3-662-54577-5_15

32. Černý, P., Henzinger, T.A., Kovács, L., Radhakrishna, A., Zwirchmayr, J.: Segment abstraction for worst-case execution time analysis. In: Vitek, J. (ed.) ESOP 2015. LNCS, vol. 9032, pp. 105–131. Springer, Heidelberg (2015). https://doi.org/10.1007/978-3-662-46669-8_5

Experimenting on Solving Nonlinear Integer Arithmetic with Incremental Linearization

Alessandro Cimatti[1], Alberto Griggio[1], Ahmed Irfan[1,2(✉)], Marco Roveri[1], and Roberto Sebastiani[2]

[1] Fondazione Bruno Kessler, Trento, Italy
{cimatti,griggio,irfan,roveri}@fbk.eu
[2] DISI, University of Trento, Trento, Italy
{ahmed.irfan,roberto.sebastiani}@unitn.it

Abstract. Incremental linearization is a conceptually simple, yet effective, technique that we have recently proposed for solving SMT problems over nonlinear real arithmetic constraints. In this paper, we show how the same approach can be applied successfully also to the harder case of nonlinear integer arithmetic problems. We describe in detail our implementation of the basic ideas inside the MathSAT SMT solver, and evaluate its effectiveness with an extensive experimental analysis over all nonlinear integer benchmarks in SMT-LIB. Our results show that Math-SAT is very competitive with (and often outperforms) state-of-the-art SMT solvers based on alternative techniques.

1 Introduction

The field of Satisfiability Modulo Theories (SMT) has seen tremendous progress in the last decade. Nowadays, powerful and effective SMT solvers are available for a number of quantifier-free theories[1] and their combinations, such as equality and uninterpreted functions (UF), bit-vectors (BV), arrays (AX), and linear arithmetic over the reals (LRA) and the integers (LIA). A fundamental challenge is to go beyond the linear case, by introducing nonlinear polynomials – theories of nonlinear arithmetic over the reals (NRA) and the integers (NIA). Although the expressive power of nonlinear arithmetic is required by many application domains, dealing with nonlinearity is a very hard challenge. Going from SMT(LRA) to SMT(NRA) yields a complexity gap that results in a computational barrier in practice – most available complete solvers rely on Cylindrical Algebraic Decomposition (CAD) techniques [8], which require double exponential time in worst case. Adding integrality constraints exacerbates the problem even further, because reasoning on NIA has been shown to be undecidable [16].

This work was funded in part by the H2020-FETOPEN-2016-2017-CSA project SC[2] (712689).

[1] In the following, we only consider quantifier-free theories, and we abuse the accepted notation by omitting the "QF_" prefix in the names of the theories.

© Springer International Publishing AG, part of Springer Nature 2018
O. Beyersdorff and C. M. Wintersteiger (Eds.): SAT 2018, LNCS 10929, pp. 383–398, 2018.
https://doi.org/10.1007/978-3-319-94144-8_23

Recently, we have proposed a conceptually simple, yet effective approach for dealing with the quantifier-free theory of nonlinear arithmetic over the reals, called *Incremental Linearization* [4–6]. Its underlying idea is that of trading the use of expensive, exact solvers for nonlinear arithmetic for an abstraction-refinement loop on top of much less expensive solvers for linear arithmetic and uninterpreted functions. The approach is based on an abstraction-refinement loop that uses SMT(UFLRA) as abstract domain. The uninterpreted functions are used to model nonlinear multiplications, which are incrementally axiomatized, by means of linear constraints, with a lemma-on-demand approach.

In this paper, we show how incremental linearization can be applied successfully also to the harder case of nonlinear integer arithmetic problems. We describe in detail our implementation of the basic ideas, performed within the MATHSAT [7] SMT solver, and evaluate its effectiveness with an extensive experimental analysis over all NIA benchmarks in SMT-LIB. Our results show that MATHSAT is very competitive with (and often outperforms) state-of-the-art SMT solvers based on alternative techniques.

Related Work. Several SMT solvers supporting nonlinear integer arithmetic (e.g., Z3 [10], SMT-RAT [9]) rely on the bit-blasting approach [12], in which a nonlinear integer satisfiability problem is iteratively reduced to a SAT problem by first bounding the integer variables, and then encoding the resulting problem into SAT. If the SAT problem is unsatisfiable then the bounds on the integer variables are increased, and the process is repeated. This approach is geared towards finding models, and it cannot prove unsatisfiability unless the problem is bounded.

In [3], the SMT(NIA) problem is approached by reducing it to SMT(LIA) via linearization. The linearization is performed by doing case analysis on the variables appearing in nonlinear monomials. Like the bit-blasting approach, the method aims at detecting satisfiable instances. If the domain of the problem is bounded, the method generates an equisatisfiable linear SMT formula. Otherwise, it solves a bounded problem and incrementally increases the bounds of some (heuristically chosen) variables until it finds a solution to the linear problem. In some cases, it may also detect (based on some heuristic) the unsatisfiability of the original problem.

The CVC4 [1] SMT solver uses a hybrid approach, in which a variant of incremental linearization (as presented in [5,17]) is combined with bit-blasting.

Recent works presented in [13] and [15] have considered a method that combines solving techniques for SMT(NRA) with branch and bound. The main idea is to relax the NIA problem by interpreting the variables over the reals, and apply NRA techniques for solving it. Since the relaxed problem is an over-approximation of the original problem, the unsatisfiability of the NIA problem is implied by the unsatisfiability of the NRA problem. If the NRA-solver finds a non-integral solution a to a variable x, then a lemma $(x \leq \lfloor a \rfloor \vee x \geq \lceil a \rceil)$ is added to the NRA problem. Otherwise, an integral solution is found for the NIA problem. In [13], the Cylindrical Algebraic Decomposition (CAD) procedure (as presented in [14]) is combined with branch and bound in the MCSAT framework. This is the method used by the YICES [11] SMT solver. In [15], the authors show how to combine CAD and virtual substitution with the branch-and-bound method in the DPLL(T) framework.

Contributions. Compared to our previous works on incremental linearization [4–6], we make the following contributions. First, we give a significantly more detailed description of our implementation (in the SMT solver MATHSAT), showing pseudo-code for all its major components. Second, we evaluate the approach over NIA problems, both by comparing it with the state of the art, and by evaluating the contributions of various components/heuristics of our procedure to its overall performance.

Structure of the Paper. This paper is organized as follows. In §2 we provide some background on the ideas of incremental linearization. In §3 we describe our implementation in detail. In §4 we present our experimental evaluation. Finally, in §5 we draw some conclusions and outline directions for future work.

2   Background

We assume the standard first-order quantifier-free logical setting and standard notions of theory, satisfiability, and logical consequence.

We denote with $\mathbb{Z}$ the set of integer numbers. A *monomial* in variables $v_1, v_2, \ldots, v_n$ is a product $v_1^{\alpha_1} * v_2^{\alpha_2} * \ldots * v_n^{\alpha_n}$, where each α_i is a non-negative integer called exponent of the variable v_i. When clear from context, we may omit the multiplication symbol $*$ and simply write $v_1^{\alpha_1} v_2^{\alpha_2} \ldots v_n^{\alpha_n}$. A *polynomial* p is a finite linear combination of monomials with coefficients in $\mathbb{Z}$, i.e., $p \stackrel{\text{def}}{=} \Sigma_{i=0}^n c_i m_i$ where each $c_i \in \mathbb{Z}$ and each m_i is a monomial. A *polynomial constraint* or simply *constraint* P is of the form $p \bowtie 0$ where p is a polynomial and $\bowtie \in \{<, \leq, >, \geq\}$.[2]

Satisfiability Modulo Theories (SMT) is the problem of deciding the satisfiability of a first-order formula with respect to some theory or combination of theories. Most SMT solvers are based on the lazy/DPLL(T) approach [2], where a SAT solver is tightly integrated with a T-solver, that is demanded to decide the satisfiability of a list of constraints (treated as a conjunction of constraints) in the theory T. There exist several theories that the modern SMT solvers support. In this work we are interested in the following theories: *Equality and Uninterpreted Functions* (UF), *Linear Arithmetic* and *Nonlinear Arithmetic* over the integers (LIA and NIA, resp.), and in their combinations thereof.

We denote formulas with φ, lists of constraints with ϕ, terms with t, variables with v, constants with a, b, c, monomials with w, x, y, z, polynomials with p, functions with f, each possibly with subscripts. If μ is a model and v is a variable,

---

[2] In the rest of the paper, for simplifying the presentation we assume that an equality constraint is written as a conjunction of weak inequality constraints, and an inequality constraint is written as a disjunction of strict inequality constraints.

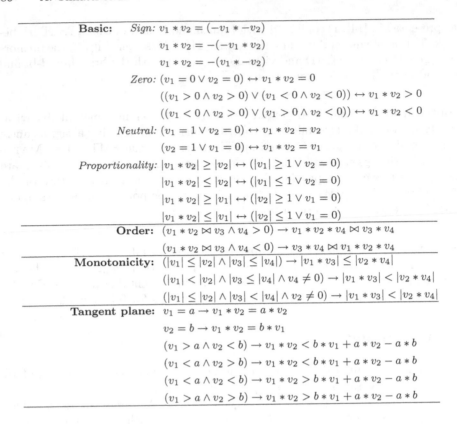

Fig. 1. Axioms of the multiplication function.

we write $\mu[v]$ to denote the value of v in μ, and we extend this notation to terms and formulas in the usual way. If ϕ is a list of constraints, we write $\bigwedge \phi$ to denote the formula obtained by taking the conjunction of all its elements.

We call a monomial m a *toplevel monomial* in a polynomial $p \overset{\text{def}}{=} \Sigma_{i=0}^{n} c_i m_i$ if $m = m_j$ for $0 \leq j \leq n$. Similarly, a monomial m is a *toplevel monomial* in φ if there exists a polynomial p in φ such that m is a toplevel monomial in p. Given φ, we denote with $\widehat{\varphi}$ the formula obtained by replacing every nonlinear multiplication between two monomials $x * y$ occurring in φ by a binary uninterpreted function $f_*(x, y)$.

We assume that the polynomials in φ are normalized by applying the distributivity property of multiplication over addition, and by sorting both the monomials and the variables in each monomial using a total order (e.g. lexicographic). Moreover, we always rewrite negated polynomial constraints into negation-free polynomial constraints by pushing the negation to the arithmetic relation (e.g., we write $\neg(p \leq 0)$ as $(p > 0)$).

```
result CHECK-NIA (φ : constraint list):
 1.    res = CHECK-UFLIA(φ̂):
 2.    if RES-IS-FALSE(res):
 3.        return res
 4.    μ = RES-GET-MODEL (res)
 5.    to_refine = ∅
 6.    φ' = {c | c ∈ φ and EVAL-MODEL(μ, c) = ⊥}
 7.    for each x * y in φ':
 8.        if EVAL-MODEL(μ, x * y) ≠ μ[x̂ * y]:
 9.            to_refine = to_refine ∪ {x * y}
10.    if to_refine = ∅:
11.        return ⟨TRUE, μ⟩
12.    res = CHECK-SAT(φ, μ)
13.    if RES-IS-TRUE(res):
14.        return res
15.    lemmas = ∅
16.    for round in ⟨1, 2, 3⟩:
17.        for each x * y in to_refine:
18.            L = GENERATE-LEMMAS(x * y, μ, round, to_refine, φ)
19.            lemmas = lemmas ∪ L
20.        if lemmas ≠ ∅:
21.            return ⟨UNDEF, lemmas⟩
22.    return ⟨UNKNOWN⟩
```

Fig. 2. The top-level NIA theory solver procedure.

Overview of Incremental Linearization. The main idea of incremental linearization is to trade the use of expensive, exact solvers for nonlinear arithmetic for an abstraction-refinement loop on top of much less expensive solvers for linear arithmetic and uninterpreted functions. First, the input SMT(NIA) formula φ is abstracted to the SMT(UFLIA) formula $\widehat{\varphi}$ (called its UFLIA-abstraction). Then the loop begins by checking the satisfiability of $\widehat{\varphi}$. If the SMT(UFLIA) check returns false then the input formula is unsatisfiable. Otherwise, the model μ for $\widehat{\varphi}$ is used to build an UFLIA underapproximation $\widehat{\varphi}^*$ of φ, with the aim of finding a model for the original NIA formula φ. If the SMT check for $\widehat{\varphi}^*$ is satisfiable, then φ is also satisfiable. Otherwise, a conjunction of linear *lemmas* that is sufficient to rule out the spurious model μ is added to $\widehat{\varphi}$, thus improving the precision of the abstraction, and another iteration of the loop is performed. The lemmas added are instances of the axioms of Fig. 1 obtained by replacing the free variables with terms occurring in φ, selected among those that evaluate to false under the current spurious model μ.

3   Implementing Incremental Linearization in a Lazy SMT Solver

We now describe in detail our implementation of the basic incremental linearization ideas as a theory solver inside an SMT prover based on the lazy/DPLL(T)

---

value EVAL-MODEL (μ : model, t : term):
1.   **match** t **with**
2. $x \bowtie y$:   **return** (EVAL-MODEL(μ, x) $\bowtie$ EVAL-MODEL(μ, y) ? $\top$: $\bot$)
3. $x * y$:   **return** EVAL-MODEL(μ, x) $*$ EVAL-MODEL(μ, y)
4. $x + y$:   **return** EVAL-MODEL(μ, x) $+$ EVAL-MODEL(μ, y)
5. $c * x$:   **return** $c *$ EVAL-MODEL(μ, x)
6. v :   **return** $\mu[v]$
7. c :   **return** c

Fig. 3. Recursive model evaluation.

---

result CHECK-SAT (ϕ : constraint list, μ : UFLIA-model):
1. $\varphi = \bigwedge \phi$
2.   **for each** $x * y$ **in** φ:
3. c_x = EVAL-MODEL(μ, x)
4. c_y = EVAL-MODEL(μ, y)
5. $\varphi = \varphi \wedge ((x * y = c_x * y \wedge x = c_x) \vee (x * y = c_y * x \wedge y = c_y))$
6.   **return** SMT-UFLIA-SOLVE ($\widehat{\varphi}$)

Fig. 4. Searching for a model via linearization.

approach. The pseudo-code for the toplevel algorithm is shown in Fig. 2. The algorithm takes as input a list of constraints ϕ, corresponding to the NIA constraints in the partial assignment that is being explored by the SAT search, and it returns a result consisting of a status flag plus some additional information that needs to be sent back to the SAT solver. If the status is TRUE, then ϕ is satisfiable, and a model μ for it is also returned. If the status is FALSE, then ϕ is unsatisfiable, and a conflict set $\phi' \subseteq \phi$ (serving as an explanation for the inconsistency of ϕ) is also returned. If the status is UNDEF, the satisfiability of ϕ cannot be determined yet. In this case, the returned result contains also a set of lemmas to be used by the SAT solver to refine its search (i.e. those lemmas are learnt by the SAT solver, and the search is resumed). Finally, a status of UNKNOWN means that the theory solver can neither determine the satisfiability of ϕ nor generate additional lemmas[3]; in this case, the search is aborted.

CHECK-NIA starts by invoking a theory solver for UFLIA on the abstract version $\widehat{\phi}$ of the input problem (lines 1–4), in which all nonlinear multiplications are treated as uninterpreted functions. The unsatisfiability of $\widehat{\phi}$ immediately implies that ϕ is inconsistent. Otherwise, the UFLIA solver generates a model μ for $\widehat{\phi}$. μ is then used (lines 5–9) to determine the set of nonlinear multiplications that need to be refined. This is done by collecting all nonlinear multiplication terms $x * y$ which have a wrong value in μ; that is, for which the value of the abstraction $\widehat{x * y}$ is different from the value obtained by fully evaluating the multiplication under μ (using the EVAL-MODEL function shown in Fig. 3). It is

---

[3] This can happen when the tangent lemmas (see Fig. 8) are not used.

```
lemma set GENERATE-LEMMAS (x * y : term, μ : model, r : int, to_refine : term set,
                           φ : constraint list):
 1.  if r = 1:
 2.      return GENERATE-BASIC-LEMMAS (x * y, μ)
 3.  else:
 4.      if r = 2:
 5.          L = GENERATE-ORDER-LEMMAS(x * y, μ, φ)
 6.          if L ≠ ∅:
 7.              return L
 8.          toplevel = true
 9.      else:
10.          toplevel = false
11.      L = GENERATE-MONOTONICITY-LEMMAS(x * y, μ, to_refine, toplevel)
12.      if L ≠ ∅:
13.          return L
14.      return GENERATE-TANGENT-LEMMAS(x * y, μ, toplevel)
```

Fig. 5. Main lemma generation procedure.

important to observe that here we can limit the search for multiplications to refine only to those that appear in constraints that evaluate to false under μ (line 6). In fact, if all the constraints evaluate to true, then by definition μ is a model for them, and we can immediately conclude that ϕ is satisfiable (line 10).

Even when μ is spurious, it can still be the case that there exists a model for ϕ that is "close" to μ. This is the idea behind the CHECK-SAT procedure of Fig. 4, which uses μ as a guide in the search for a model of ϕ. CHECK-SAT works by building an UFLIA-underapproximation of ϕ, in which all multiplications are forced to be linear. The resulting formula $\widehat{\varphi}$ can then be solved with an SMT solver for UFLIA. Although clearly incomplete, this procedure is cheap (since the Boolean structure of $\widehat{\varphi}$ is very simple) and, as our experiments will show, surprisingly effective.

When CHECK-SAT fails, we proceed to the generation of lemmas for refining the spurious model μ (lines 15–21). Our lemma generation strategy works in three rounds: we invoke the GENERATE-LEMMAS function (Fig. 5) on the multiplication terms $x * y$ that need to be refined using increasing levels of effort, stopping at the earliest successful round – i.e., a round in which lemmas are generated. In the first round, only basic lemmas encoding simple properties of multiplications (sign, zero, neutral, proportionality in Fig. 1) are considered (GENERATE-BASIC-LEMMAS). In the second round, we consider also "order" lemmas (GENERATE-ORDER-LEMMAS, Fig. 6), i.e. lemmas obtained via (a restricted, model-driven) application of the order axioms of $\mathbb{Z}$. If GENERATE-ORDER-LEMMAS fails, we proceed to generating monotonicity (GENERATE-MONOTONICITY-LEMMAS, Fig. 7) and tangent plane (GENERATE-TANGENT-LEMMAS, Fig. 8) lemmas, restricting the instantiation however to only toplevel monomials. Finally, in the last round, we repeat the generation of monotonicity and tangent lemmas, considering this time also non-toplevel monomials.

lemma set GENERATE-ORDER-LEMMAS ($x * y$: term, μ : model, ϕ : constraint list):
1.   **for each** variable v in $x * y$:
2.   monomials = GET-MONOMIALS (v, ϕ)
3.   bounds = GET-BOUNDS (v, ϕ)
4.   **for each** ($w * v \bowtie p$) in bounds:
5.   **for each** $t * v$ in monomials:
6.   **if** $t * w * v$ in ϕ **and** $t * p$ in ϕ:
7. n = EVAL-MODEL(t)
8.   **if** $n = 0$:
9.   **continue**
10.   **else if** $n > 0$:
11. $\psi = ((w * v \bowtie p) \wedge t > 0) \rightarrow (t * w * v \bowtie t * p)$
12.   **else**:
13. $\psi = ((w * v \bowtie p) \wedge t < 0) \rightarrow (t * p \bowtie t * w * v)$
14.   **if** EVAL-MODEL(μ, ψ) = $\bot$:
15.   **return** $\{\psi\}$
16.   **return** $\emptyset$

Fig. 6. Generation of order lemmas.

Lemma Generation Procedures. We now describe our lemma generation procedures in detail. The pseudo-code is reported in Figs. 5, 6, 7 and 8. All procedures share the following two properties: (i) the lemmas generated do not contain any nonlinear multiplication term that was not in the input constraints ϕ; and (ii) all the generated lemmas evaluate to false ($\bot$) in the current model μ.

The function GENERATE-BASIC-LEMMAS, whose pseudo-code is not reported, simply instantiates all the basic axioms for the input term $x * y$ that satisfy points (i) and (ii) above.

The function GENERATE-ORDER-LEMMAS (Fig. 6) uses the current model and asserted constraints to produce lemmas that are instances of the order axiom for multiplication. It is based on ideas that were first implemented in the CVC4 [1] SMT solver.[4] It works by combining, for each variable v in the input term $x * y$, the monomials $t * v$ in which v occurs (retrieved by GET-MONOMIALS) with the predicates of the form ($w * v \bowtie p$) (where $\bowtie \in \{<, >, \leq, \geq\}$ and p is a polynomial) that are induced by constraints in ϕ (which are collected by the GET-BOUNDS function). The (non-constant) coefficient t of v in the monomial $t * v$ is used to generate the terms $t * w * v$ and $t * p$: if both occur[5] in the input constraints ϕ, then an instance of the order axiom is produced, using the current model μ as a guide (lines 7–15).

The function GENERATE-MONOTONICITY-LEMMAS (Fig. 7) returns instances of monotonicity axioms relating the current input term $x * y$ with other monomials that occur in the set of terms to refine. In the second round of lemma generation, only toplevel monomials are considered.

---

[4] We are grateful to Andrew Reynolds for fruitful discussions about this.

[5] It is important to stress here that we keep the monomials in a normal form by reordering their variables, although this is not shown explicitly in the pseudo-code.

lemma set GENERATE-MONOTONICITY-LEMMAS $(x * y : \text{term}, \mu : \text{model}, \text{to\_refine} : \text{term set},$
$\qquad\qquad\qquad\qquad\qquad\qquad\qquad\qquad \text{toplevel} : \textbf{bool})$:

1.   **if** toplevel $\neq$ IS-TOPLEVEL-MONOMIAL$(x * y)$:
2.   **return** $\emptyset$
3. $L = \emptyset$
4.   **for each** $w * z$ **in** to_refine:
5.   **if not** toplevel **or** IS-TOPLEVEL-MONOMIAL $(z * w)$:
6. $\psi_1 = (|x| \leq |w| \wedge |y| \leq |z|) \rightarrow |x * y| \leq |w * z|$
7. $\psi_2 = (|x| \leq |z| \wedge |y| \leq |w|) \rightarrow |x * y| \leq |w * z|$
8. $\psi_3 = (|x| < |w| \wedge |y| \leq |z| \wedge z \neq 0) \rightarrow |x * y| < |w * z|$
9. $\psi_4 = (|x| < |z| \wedge |y| \leq |w| \wedge w \neq 0) \rightarrow |x * y| < |w * z|$
10. $\psi_5 = (|x| \leq |w| \wedge |y| < |z| \wedge w \neq 0) \rightarrow |x * y| < |w * z|$
11. $\psi_6 = (|x| \leq |z| \wedge |y| < |w| \wedge z \neq 0) \rightarrow |x * y| < |w * z|$
12. $L = L \cup \{\psi_i \mid \text{EVAL-MODEL}(\mu, \psi_i) = \bot\}$
13.   **return** L

Fig. 7. Generation of monotonicity lemmas.

Finally, the function GENERATE-TANGENT-LEMMAS (Fig. 8) produces instances of the tangent plane axioms. In essence, the function instantiates all the clauses of the tangent plane lemma using the two factors x and y of the input multiplication term $x * y$ and their respective values a and b in μ, returning all the instances that are falsified by μ. This is done in lines 15–21 of Fig. 8. In our actual implementation, however, we do not use the model values a and b directly to generate tangent lemmas, but we instead use a heuristic that tries to reduce the number of tangent lemmas generated for each $x * y$ term to refine. More specifically, we keep a 4-value tuple $\langle l_x, l_y, u_x, u_y \rangle$ associated with each $x * y$ term in the input problem (which we call *frontier*) consisting of the smallest and largest of the previous model values for x and y for which a tangent lemma has been generated, and for each frontier we maintain an invariant that whenever x is in the interval $[l_x, u_x]$ or y is in the interval $[l_y, u_y]$, then $x * y$ has both an upper and a lower bound. This condition is achieved by adding tangent lemmas for the following four points of each frontier: $(l_x, l_y), (l_x, u_y), (u_x, l_y), (u_x, u_y)$ (the function UPDATE-TANGENT-FRONTIER in Fig. 8 generates those lemmas). If the current model values a and b for x and y are outside the intervals $[l_x, u_x]$ and $[l_y, u_y]$ respectively, we try to adjust them with the goal of enlarging the frontier as much as possible whenever we generate a tangent plane. Intuitively, this can be seen as a form of lemma generalisation. The procedure is shown in lines 6–14 of Fig. 8: the various PUSH-TANGENT-POINT* functions try to move the input points along the specified directions (either 'U'p, by increasing a value, or 'D'own, by decreasing it) as long as the tangent plane passing through (a, b) still separates the multiplication curve from the spurious value c.[6]

---

[6] In our implementation we use a bounded (dichotomic) search for this. For example, for the 'UU' direction we try increasing both a and b until either the tangent plane passing through (a, b) cannot separate the multiplication curve from the bad point c anymore, or we reach a maximum bound on the number of iterations.

---

lemma set GENERATE-TANGENT-LEMMAS ($x * y$: term, μ : model, toplevel : **bool**):

1. **if** toplevel $\neq$ IS-TOPLEVEL-MONOMIAL($x * y$):
2.   **return** $\emptyset$
3. $a = $ EVAL-MODEL(μ, x)
4. $b = $ EVAL-MODEL(μ, y)
5. $c = \mu[\widehat{x * y}]$
6. $l_x, l_y, u_x, u_y = $ GET-TANGENT-FRONTIER($x * y$)
7.   **if** $a < l_x$ **and** $b < l_y$: $a, b = $ PUSH-TANGENT-POINTS-DD($x * y, a, b, c$)
8.   **else if** $a < l_x$ **and** $b > u_y$: $a, b = $ PUSH-TANGENT-POINTS-DU($x * y, a, b, c$)
9.   **else if** $a > u_x$ **and** $b > u_y$: $a, b = $ PUSH-TANGENT-POINTS-UU($x * y, a, b, c$)
10. **else if** $a > u_x$ **and** $b < l_y$: $a, b = $ PUSH-TANGENT-POINTS-UD($x * y, a, b, c$)
11. **else if** $a < l_x$: $a = $ PUSH-TANGENT-POINT1-D($x * y, a, b, c$)
12. **else if** $a > u_x$: $a = $ PUSH-TANGENT-POINT1-U($x * y, a, b, c$)
13. **else if** $b < l_y$: $b = $ PUSH-TANGENT-POINT2-D($x * y, a, b, c$)
14. **else if** $b > u_y$: $b = $ PUSH-TANGENT-POINT2-U($x * y, a, b, c$)
15. $\psi_1 = (x = a \rightarrow x * y = a * y)$
16. $\psi_2 = (y = b \rightarrow x * y = b * x)$
17. $\psi_3 = (x > a \wedge y < b) \rightarrow (x * y < b * x + a * y - a * b)$
18. $\psi_4 = (x < a \wedge y > b) \rightarrow (x * y < b * x + a * y - a * b)$
19. $\psi_5 = (x < a \wedge y < b) \rightarrow (x * y > b * x + a * y - a * b)$
20. $\psi_6 = (x > a \wedge y > b) \rightarrow (x * y > b * x + a * y - a * b)$
21. $L = \{\psi_i \mid$ EVAL-MODEL(μ, ψ_i) $= \perp\}$
22. **if** $L \neq \emptyset$:
23. $L = L \cup$ UPDATE-TANGENT-FRONTIER($x * y, a, b$)
24. **return** L

Fig. 8. Generation of tangent lemmas.

Example 1 (Tangent frontier enlargement – Fig. 9). Let $\langle -3, -1, 5, 2 \rangle$ be the current frontier of $x * y$ during the search. Suppose the abstract model gives: $\mu[x] = a = 15$, $\mu[y] = b = 5$, and $\mu[\widehat{x * y}] = c = 48$. This model is spurious because $15 * 5 \neq 48$. Notice that the point $(15, 5)$ is outside of the frontier, because 15 is not in $[-3, 5]$ and 5 is not in $[-1, 2]$. So, during the tangent lemmas generation, the function PUSH-TANGENT-POINTS-UU can return $a = 20$ and $b = 10$, as one of the constraints of the tangent lemma instantiated at that point is violated by the current model, i.e., we can obtain the following clauses from the tangent lemma:

$$x > 20 \wedge y < 10 \rightarrow x * y < 10 * x + 20 * y - 200$$
$$x < 20 \wedge y > 10 \rightarrow x * y < 10 * x + 20 * y - 200$$
$$x < 20 \wedge y < 10 \rightarrow x * y > 10 * x + 20 * y - 200$$
$$x > 20 \wedge y > 10 \rightarrow x * y > 10 * x + 20 * y - 200$$

by plugging in the values $x = 15$, $y = 5$, and $x * y = 48$, then we obtain a conflict in the third clause because $15 < 20$ and $5 < 10$, but $48 \not> 10 * 15 + 20 * 5 - 200$. This means that the tangent lemma instantiated at point $(20, 10)$ can be used for refinement (Fig. 9(c)).

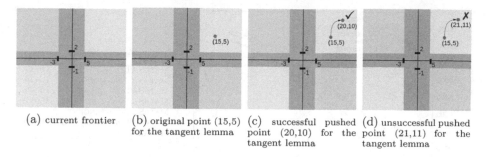

(a) current frontier   (b) original point (15,5)   (c) successful pushed   (d) unsuccessful pushed
                         for the tangent lemma   point (20,10) for the   point (21,11) for the
                                                     tangent lemma   tangent lemma

Fig. 9. Illustration of the strategy for adjusting the refinement point for the tangent lemma.

However, if we use $(21, 11)$ for the tangent lemma instantiation, we get the following clauses:

$$x > 21 \land y < 11 \rightarrow x * y < 11 * x + 21 * y - 231$$
$$x < 21 \land y > 11 \rightarrow x * y < 11 * x + 21 * y - 231$$
$$x < 21 \land y < 11 \rightarrow x * y > 11 * x + 21 * y - 231$$
$$x > 21 \land y > 11 \rightarrow x * y > 11 * x + 21 * y - 231$$

Notice that, all these clauses are satisfied if we plug in the values $x = 15$, $y = 5$, and $x * y = 48$. Therefore, we cannot use them for refinement (Fig. 9(d)).

4   Experimental Analysis

We have implemented our incremental linearization procedure in our SMT solver MATHSAT [7]. In this section, we experimentally evaluate its performance. Our implementation and experimental data are available at https://es.fbk.eu/people/irfan/papers/sat18-data.tar.gz.

Setup and Benchmarks. We have run our experiments on a cluster equipped with 2.6 GHz Intel Xeon X5650 machines, using a time limit of 1000 s and a memory limit of 6 Gb.

For our evaluation, we have used all the benchmarks in the QF_NIA category of SMT-LIB [18], which at the present time consists of 23876 instances. All the problems are available from the SMT-LIB website.

Our evaluation is composed of two parts. In the first, we evaluate the contribution of different parts of our procedure to the overall performance of MATHSAT, by comparing different configurations of the solver. In the second part, we compare our best configuration against the state of the art in SMT solving for nonlinear integer arithmetic.

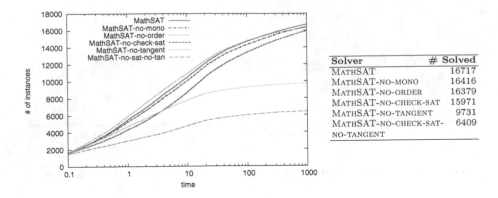

Solver	# Solved
MATHSAT	16717
MATHSAT-NO-MONO	16416
MATHSAT-NO-ORDER	16379
MATHSAT-NO-CHECK-SAT	15971
MATHSAT-NO-TANGENT	9731
MATHSAT-NO-CHECK-SAT-NO-TANGENT	6409

Fig. 10. Comparison among different configurations of MATHSAT.

Comparison of Different Configurations. We evaluate the impact of the main components of our procedure, by comparing five different configurations of MATHSAT:

- The standard configuration, using all the components described in the previous section (simply denoted MATHSAT);
- a configuration with CHECK-SAT disabled (denoted MATHSAT-NO-CHECK-SAT);
- a configuration with GENERATE-ORDER-LEMMAS disabled (denoted MATH-SAT-NO-ORDER);
- a configuration with GENERATE-MONOTONICITY-LEMMAS disabled (denoted MATHSAT-NO-MONO);
- a configuration with GENERATE-TANGENT-LEMMAS disabled (denoted MATH-SAT-NO-TANGENT); and finally
- a configuration with both CHECK-SAT and GENERATE-TANGENT-LEMMAS disabled (denoted MATHSAT-NO-CHECK-SAT-NO-TANGENT).

The results are presented in Fig. 10. The plot on the left shows, for each configuration, the number of instances that could be solved (on the y axis) within the given time limit (on the x axis). The table on the right shows the ranking of the configurations according to the number of instances solved. From Fig. 10, we can see that all components of our procedure contribute to the performance of MATHSAT. As expected, tangent lemmas are crucial, but it is also interesting to observe that the cheap satisfiability check by linearization is very effective, leading to an overall performance boost and to the successful solution of 746 additional benchmarks that could not be solved by MATHSAT-NO-CHECK-SAT. Finally, although the addition of order axioms (by GENERATE-ORDER-LEMMAS) does not pay off for simpler problems, its impact is clearly visible for harder instances, allowing MATHSAT to solve 338 more benchmarks than MATHSAT-NO-ORDER.

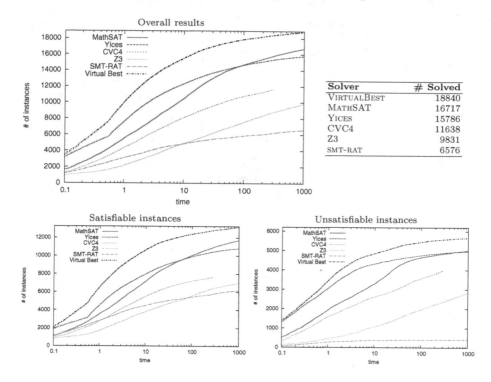

Fig. 11. Comparison with state-of-the-art SMT solvers for NIA.

Solver	# Solved
VIRTUALBEST	18840
MATHSAT	16717
YICES	15786
CVC4	11638
Z3	9831
SMT-RAT	6576

Comparison with the State of the Art. In the second part of our experiments, we compare MATHSAT with the state-of-the-art SMT solvers for NIA. We consider CVC4 [1], SMT-RAT [9], YICES [11] and Z3 [10]. Figures 11 and 12 show a summary of the results (with separate plots for satisfiable and unsatisfiable instances in addition to the overall plot), whereas Fig. 13 shows a more detailed comparison between MATHSAT and YICES. Additional information about the solved instances for each benchmark family is given in Table 1. From the results, we can see that the performance of MATHSAT is very competitive: not only it solves more instances than all the other tools, but it is also faster than CVC4, SMT-RAT and Z3. On the other hand, YICES is much faster than MATHSAT in the majority of cases, especially on easy unsatisfiable instances (solved in less than 10 s). However, the two tools are very complementary, as shown by Fig. 13: MATHSAT can solve 2436 instances for which YICES times out, whereas YICES can successfully handle 1505 instances that MATHSAT is unable to solve. Moreover, MATHSAT overall solves 931 more problems (915 satisfiable and 16 unsatisfiable) than YICES in the given resource limits.

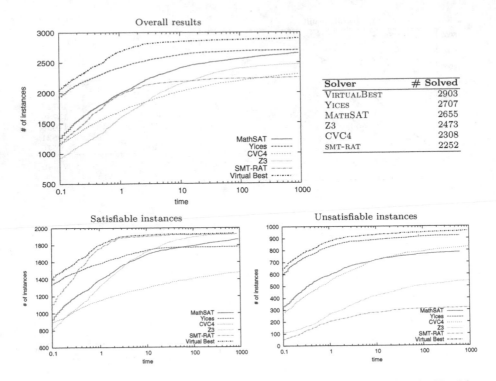

Fig. 12. Comparison with state-of-the-art SMT solvers for NIA – without the VeryMax benchmarks.

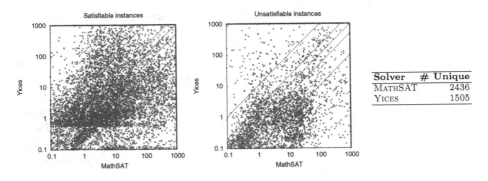

Fig. 13. Detailed comparison between MATHSAT and YICES.

Table 1. Summary of the comparison with the state of the art.

	Total (23876)	AProVE (2409)	Calypto (177)	Lasso Ranker (106)	LCTES (2)	Leipzig (167)	MCM (186)	Ultimate Automizer (7)	Ultimate Lasso-Ranker (32)	VeryMax (20790)
MATHSAT	**11723/4993**	1642/561	**79/89**	**4/100**	0/1	126/2	12/0	0/7	**6/26**	**9854/4207**
YICES	10808/4977	1595/**708**	**79/97**	4/84	0/0	92/1	8/0	0/7	**6/26**	9024/4054
CVC4	7653/3984	1306/608	77/89	4/94	0/1	84/2	6/0	0/6	**6/26**	6170/3158
Z3	6993/2837	1656/325	78/96	4/92	0/0	**162**/0	20/1	0/7	**6/26**	5067/2290
SMT-RAT	6161/414	**1663**/184	**79**/89	3/20	0/0	160/0	**21**/0	0/1	**6/26**	4229/94
VIRTUALBEST	13169/5669	1663/724	79/97	4/101	0/1	162/2	23/1	0/7	6/26	11232/4710

Each column shows a family of benchmarks in the QF_NIA division of SMT-LIB. For each solver, the table shows the number of sat/unsat results in each family. The best performing tools (in terms of # of results) are reported in boldface.

5 Conclusions

We have presented a solver for satisfiability modulo nonlinear integer arithmetic based on the incremental linearization approach. Our empirical analysis of its performance over all the nonlinear integer benchmarks in the SMT-LIB library shows that the approach is very competitive with the state of the art: our solver MATHSAT can solve many problems that are out of reach for other tools, and overall it solves the highest number of instances. Our evaluation has however also shown that current approaches for SMT(NIA) are very complementary, with no tool that always outperforms all the others. This suggests the investigation of hybrid approaches that combine multiple methods as a very promising direction for future work.

References

1. Barrett, C., Conway, C.L., Deters, M., Hadarean, L., Jovanovic, D., King, T., Reynolds, A., Tinelli, C.: CVC4. In: CAV. pp. 171–177 (2011)
2. Barrett, C.W., Sebastiani, R., Seshia, S.A., Tinelli, C.: Satisfiability modulo theories. In: Handbook of Satisfiability, Frontiers in Artificial Intelligence and Applications, vol. 185, pp. 825–885. IOS Press (2009)
3. Borralleras, C., Lucas, S., Oliveras, A., Rodríguez-Carbonell, E., Rubio, A.: Sat modulo linear arithmetic for solving polynomial constraints. J. Autom. Reason. **48**(1), 107–131 (2012)
4. Cimatti, A., Griggio, A., Irfan, A., Roveri, M., Sebastiani, R.: Incremental Linearization for Satisfiability and Verification Modulo Nonlinear Arithmetic and Transcendental Functions. Under Submission (2017), available at https://es.fbk.eu/people/irfan/papers/inclin-smt-vmt-nl-tf.pdf
5. Cimatti, Alessandro, Griggio, Alberto, Irfan, Ahmed, Roveri, Marco, Sebastiani, Roberto: Invariant Checking of NRA Transition Systems via Incremental Reduction to LRA with EUF. In: Legay, Axel, Margaria, Tiziana (eds.) TACAS 2017. LNCS, vol. 10205, pp. 58–75. Springer, Heidelberg (2017). https://doi.org/10.1007/978-3-662-54577-5_4

6. Cimatti, Alessandro, Griggio, Alberto, Irfan, Ahmed, Roveri, Marco, Sebastiani, Roberto: Satisfiability Modulo Transcendental Functions via Incremental Linearization. In: de Moura, Leonardo (ed.) CADE 2017. LNCS (LNAI), vol. 10395, pp. 95–113. Springer, Cham (2017). https://doi.org/10.1007/978-3-319-63046-5_7

7. Cimatti, Alessandro, Griggio, Alberto, Schaafsma, Bastiaan Joost, Sebastiani, Roberto: The MathSAT5 SMT Solver. In: Piterman, Nir, Smolka, Scott A. (eds.) TACAS 2013. LNCS, vol. 7795, pp. 93–107. Springer, Heidelberg (2013). https://doi.org/10.1007/978-3-642-36742-7_7

8. Collins, G.E.: Quantifier Elimination for Real Closed Fields by Cylindrical Algebraic Decomposition-preliminary Report. SIGSAM Bull. 8(3), 80–90 (1974)

9. Corzilius, F., Loup, U., Junges, S., Ábrahám, E.: SMT-RAT: An SMT-compliant nonlinear real arithmetic toolbox. In: SAT. pp. 442–448. Springer (2012)

10. De Moura, L., Bjørner, N.: Z3: An efficient SMT solver. In: TACAS. pp. 337–340. Springer (2008)

11. Dutertre, B.: Yices 2.2. In: Biere, A., Bloem, R. (eds.) Computer-Aided Verification (CAV'2014). LNCS, vol. 8559, pp. 737–744. Springer (July 2014)

12. Fuhs, Carsten, Giesl, Jürgen, Middeldorp, Aart, Schneider-Kamp, Peter, Thiemann, René, Zankl, Harald: SAT Solving for Termination Analysis with Polynomial Interpretations. In: Marques-Silva, João, Sakallah, Karem A. (eds.) SAT 2007. LNCS, vol. 4501, pp. 340–354. Springer, Heidelberg (2007). https://doi.org/10.1007/978-3-540-72788-0_33

13. Jovanović, Dejan: Solving Nonlinear Integer Arithmetic with MCSAT. In: Bouajjani, Ahmed, Monniaux, David (eds.) VMCAI 2017. LNCS, vol. 10145, pp. 330–346. Springer, Cham (2017). https://doi.org/10.1007/978-3-319-52234-0_18

14. Jovanović, D., De Moura, L.: Solving non-linear arithmetic. In: IJCAR. pp. 339–354. Springer (2012)

15. Kremer, Gereon, Corzilius, Florian, Ábrahám, Erika: A Generalised Branch-and-Bound Approach and Its Application in SAT Modulo Nonlinear Integer Arithmetic. In: Gerdt, Vladimir P., Koepf, Wolfram, Seiler, Werner M., Vorozhtsov, Evgenii V. (eds.) CASC 2016. LNCS, vol. 9890, pp. 315–335. Springer, Cham (2016). https://doi.org/10.1007/978-3-319-45641-6_21

16. Matiyasevich, Y.V.: Hilbert's Tenth Problem. MIT Press, Foundations of computing (1993)

17. Reynolds, Andrew, Tinelli, Cesare, Jovanović, Dejan, Barrett, Clark: Designing Theory Solvers with Extensions. In: Dixon, Clare, Finger, Marcelo (eds.) FroCoS 2017. LNCS (LNAI), vol. 10483, pp. 22–40. Springer, Cham (2017). https://doi.org/10.1007/978-3-319-66167-4_2

18. SMT-LIB, The Satisfiability Modulo Theories Library. http://smtlib.org

Tools and Applications

XOR-Satisfiability Set Membership Filters

Sean A. Weaver$^{(\boxtimes)}$ ⓘ, Hannah J. Roberts, and Michael J. Smith

Information Assurance Research Group, U.S. National Security Agency,
9800 Savage Rd., Suite 6845, Ft. George G. Meade, MD 20755, USA
saweave@tycho.ncsc.mil

Abstract. Set membership filters are used as a primary test for whether large sets contain given elements. The most common such filter is the Bloom filter [6]. Most pertinent to this article is the recently introduced Satisfiability (SAT) filter [31]. This article proposes the *XOR-Satisfiability filter*, a variant of the SAT filter based on random k-XORSAT. Experimental results show that this new filter can be more than 99% efficient (i.e., achieve the information-theoretic limit) while also having a query speed comparable to the standard Bloom filter, making it practical for use with very large data sets.

1 Introduction

To support timely computation on large sets, and in cases where being certain is not necessary, a quick, probabilistic test of a *set membership filter* is often used. A set membership filter is constructed from a set and queried with elements from the corresponding domain. Being probabilistic, the filter will return either *Maybe* or *No*. That is, the filter can return false positives, but never false negatives. The most well-known set membership filter is the Bloom filter [6]. Though many other set membership filters have been proposed, the most important to this work is the SAT filter [31].

A *SAT filter* is a set membership filter constructed using techniques based on SAT [5]. In [31], the authors describe the process of building a SAT filter as follows. First, each element in a set of interest is translated into a CNF clause (disjunction of literals). Next, every clause is logically conjoined into a CNF formula. Finally, solutions to the resulting formula are found using a SAT solver. These solutions constitute a SAT filter. To query a SAT filter, an element is translated into a clause (using the same method as during filter building) and if the clause is satisfied by all of the stored solutions, the element may be in the original set, otherwise it is definitely not in the original set. Parameters for tailoring certain aspects of the SAT filter such as false positive rate, query speed, and amount of long term storage are described in [31].

O. Beyersdorff and C. M. Wintersteiger (Eds.): SAT 2018, LNCS 10929, pp. 401–418, 2018.
https://doi.org/10.1007/978-3-319-94144-8_24

This article describes a new, practical variant of the generic SAT filter where clauses are considered to be the XORs of Boolean variables, rather than the traditional inclusive OR (disjunction) of literals. This approach (mentioned as possible future research in [31]) offers many advantages over a disjunction-based SAT filter such as practically near perfect filter efficiency [30], faster build and query times, and support for metadata storage and retrieval.

In terms of related work, there are other filter constructions that attempt to achieve high efficiency (e.g. via compression) (e.x., see [7–9,16,24]). Most similar to the XORSAT filter construction introduced here are Matrix filters [13,26]. Insofar as XORSAT equations are equivalent to linear equations over $GF(2)$, there are two obvious (and independent) ways to generalize such a linear system: either by considering equations over larger fields like $GF(2^s)$ (Matrix filters), or remaining over $GF(2)$ and working with s right-hand sides (XORSAT filters). In both constructions, the solutions can be used to store probabilistic membership in sets, as well as values corresponding to keys, but the XORSAT filter construction is motivated by some clear computational advantages.

First, Matrix filters require a hash function that yields elements over $GF(2^s)^n$, whereas hash functions for XORSAT filters yield elements over $GF(2)^n$ — an s-fold improvement in the data required. Also, Matrix filters require arithmetic over $GF(2^s)$, whereas XORSAT filters work entirely over $GF(2)$ and as such are more naturally suited to highly-optimized implementations; all computations devolve to simple and fast word operations (like AND and XOR) and bit-parity computations which are typically supported on modern computers. This article also proposes some simple and more practical methods for bucketing and handling sparse variants, which likewise correspond to efficiency and performance improvements.

2 XORSAT Filters

This section briefly describes XORSAT and the XORSAT filter.

2.1 XORSAT

Construction and query of an XORSAT filter rely heavily on properties of random k-XORSAT, a variant of SAT where formulas are expressed as conjunctions of random XOR clauses, i.e. the exclusive OR of Boolean variables.

Definition 1. *An XOR clause is an expression of the form*

$$v_{i_1} \oplus \ldots \oplus v_{i_{k_i}} \equiv b_i,$$

where the symbol $\oplus$ represents XOR, the symbol $\equiv$ represents logical equivalence, each v_i is a Boolean variable and each b_i (right-hand side) is a constant, either 0 (for False) or 1 (for True).

Definition 2. *A width k XOR clause has exactly k distinct variables.*

Definition 3. *A random k-XORSAT instance is a set of XOR clauses drawn uniformly, independently, and with replacement from the set of all width k XOR clauses* [20].

As with random k-SAT [1], a random k-XORSAT instance is likely to be satisfiable if its clauses-to-variables ratio is less than a certain threshold α_k, and likely to be unsatisfiable if greater than α_k [25]. Experimental results have established approximate values of α_k for small values of k, though it asymptotically approaches 1. Experimental values are given next and are reproduced from [11,12].

Table 1. Various α_k values for random k-XORSAT

k	2	3	4	5	6	7
α_k	0.5	0.917935	0.976770	0.992438	0.997379	0.999063

Polynomial time algorithms exist for reducing random k-XORSAT instances into reduced row echelon form. For example, Gaussian elimination can solve such instances in $\mathcal{O}(n^3)$ steps and the 'Method of Four Russians' [4,29] in $\mathcal{O}(\frac{n^3}{\log_2 n})$. Once in this reduced form, collecting random solutions (kernel vectors) is trivial—assign random values to all of the free variables (those in the identity submatrix), and backsolve for the dependent ones.

2.2 XORSAT Filter Construction

This section presents the basic XORSAT filter construction. Later sections provide enhancements which enable such filters to be used in practice. The XORSAT filter is built and queried in a manner very similar to the SAT filter. Provided below are updated algorithms for construction and query, analogous to those in [31], where deeper discussion on how to construct and query SAT filters can be found.

Building an XORSAT Filter. Being a variant of the the SAT filter, the XORSAT filter has similar properties. Building an XORSAT filter for a data set $Y \subseteq D$ (where D is a domain) is one-time work. The XORSAT filter is an offline filter, so, once built, it is not able to be updated. To build an XORSAT filter, all elements $y \in Y$ are transformed into width k XOR clauses that, when conjoined, constitute a random k-XORSAT instance. If the instance is unsatisfiable, a filter cannot be constructed for the given data set and parameters. Otherwise, a solution for that instance and acts as a filter for Y.

Algorithm 1 shows how to transform an element $e \in D$ into a width k XOR clause using a set of hash functions. Algorithm 2 shows how to build an XORSAT filter from a given set $Y \subseteq D$.

Algorithm 1. ELEMENTTOXORCLAUSE($e \in D, n, k, h_0, \ldots, h_{k-1}, h_b$)

e is the element used to generate an XOR clause
n is the number of variables per XORSAT instance
k is the number of variables per XOR clause
$h_0, \ldots, h_{k-1}$ are functions that map elements of D to $[0, n)$
h_b is a function that maps elements of D to $[0, 1]$

1: $nonce := 0$
2: **repeat**
3:     $V := \{\}$
4:     **for** $i := 0$ **to** $k - 1$ **do**
5:         $v := h_i(e, nonce)$, hash e to generate variable v
6:         $V := V \cup \{v\}$, add v to the XOR clause
7:     **end for**
8:     $nonce := nonce + 1$
9: **until** all variables of V are distinct
10: $b := h_b(e)$, hash e to generate the right-hand side
11: **return** (V, b)

Algorithm 2. BUILDXORSATFILTER($Y \subseteq D, n, k, h_0, \ldots, h_{k-1}, h_b$)

Y is the set used to build an XORSAT filter
n is the number of variables per XORSAT instance
k is the number of variables per XOR clause
$h_0, \ldots, h_{k-1}$ are functions that map elements of D to $[0, n)$
h_b is a function that maps elements of D to $[0, 1]$

1: $\mathcal{X}_Y := \{\}$, the empty formula
2: **for** each element $y \in Y$ **do**
3:     $(V_y, b_y) :=$ ELEMENTTOXORCLAUSE($y, n, k, h_0, \ldots, h_{k-1}, h_b$)
4:     $\mathcal{X}_Y := \mathcal{X}_Y \cup \{(V_y, b_y)\}$
5: **end for**
6: **if** the random k-XORSAT instance $\mathcal{X}_Y$ is unsatisfiable **then**
7:     **return** failure
8: **else**
9:     Let F_Y be a single solution to $\mathcal{X}_Y$
10:     **return** F_Y
11: **end if**

Querying a SAT Filter. Querying an XORSAT filter with an element $x \in D$ is very similar to querying a SAT filter. First, x is transformed into a k width XOR clause. Then, if the clause is satisfied by the solution generated by Algorithm 2 for a set Y, x is *maybe* in Y, otherwise x is definitely not in Y. Algorithm 3 shows how to query an XORSAT filter.

Algorithm 3. QUERYXORSATFILTER($F_Y, x \in D, n, k, h_0, \ldots, h_{k-1}, h_b$)
x is the element used to query the XORSAT filter F_Y
n is the number of variables per XORSAT instance
k is the number of variables per XOR clause
$h_0, \ldots, h_{k-1}$ are functions that map elements of D to $[0, n)$
h_b is a function that maps elements of D to $[0, 1]$

1: $(V_x, b_x) := $ ELEMENTTOXORCLAUSE($x, n, k, h_0, \ldots, h_{k-1}, h_b$)
2: **for** each variable $v \in V_x$ **do**
3: $\quad$ $b_x := b_x \oplus F_Y(v)$
4: **end for**
5: **if** $b_x = 0$ **then**
6: $\quad$ **return** *Maybe*
7: **end if**
8: **return** *No*

2.3 False Positive Rate, Query Time, and Storing Multiple Solutions

The false positive rate of an XORSAT filter is the probability that the XOR clause generated by the query is satisfied by the stored solution. This is equal to the probability that a random width k XOR clause is satisfied by a random solution, i.e., $\frac{1}{2}$. As with the SAT filter, the false positive rate can be improved by either storing multiple solutions to multiple XORSAT instances or storing multiple *uncorrelated* solutions to a single XORSAT instance. For SAT filters, this second method is preferred because querying is much faster (only one clause needs to be built, so the hash functions are called fewer times), but the challenge of finding uncorrelated solutions to a single instance has yet to be overcome, though recent work seems promising [3,14,17,21].

Fortunately, moving from SAT to XORSAT also moves past this difficulty. Since the XORSAT solving method used here, reduction to echelon form, is agnostic to the type of the elements in the matrix being reduced, s XORSAT instances can be encoded by treating the variables and right-hand side of each XOR clause as vectors of Booleans[1]. Then, the transformation to reduced row echelon form uses bitwise XOR on vectors (during row reduction) rather than Boolean XOR on single bits. Hence, s XORSAT instances can be solved in parallel, and just as fast as solving a single instance[2]. Also, since the XORSAT instances have random right-hand sides, the s solutions, one for each instance, will be uncorrelated.

The solutions are stored in the same manner as the SAT filter, that is, all s solution bits corresponding to a variable are stored together (the transpose of the

---

[1] The intuition for this idea came from Bryan Jacobs' work on isomorphic k-SAT filters and work by Heule and van Maaren on parallelizing SAT solvers using bitwise operators [19].

[2] As long as s is not greater than the native register size of the machine on which the solver is running.

array of solutions). If the solutions are stored this way, querying the s-wide filter is just as efficient as querying a filter created from a single instance. Moreover, the false positive rate is improved to $\frac{1}{2^s}$ because, during XORSAT filter query, s different right-hand side bits are generated from each element and each have to be satisfied by the corresponding solution.

2.4   Dictionaries

A small modification to the XORSAT filter construction can enable it to produce filters that also store and retrieve metadata d associated with each element y. To insert the tuple (y, d), a key-value pair, into the filter, append a bitwise representation of d, say r bits wide, to the right-hand side of the clause for y. Now every variable is treated as an $s+r$ wide vector of Booleans and the resulting instance is solved using word-level operations[3]. When querying, the first s bits of the right-hand side act as a check (to determine if the element passes the filter) and, if so (and not a false positive) the last r bits will take on the values of the bitwise representation of d.

On a purely practical note, the instances generated during build need not be entirely random k-XORSAT instances. By removing the check for duplicate variables, XOR clauses with less than k variables can be generated because duplicate variables in an XOR clause simply cancel out. In practice, this only slightly decreases efficiency (increases the size of the filter), but moderately decreases query time.

Algorithm 4 shows how to create a bit-packed sequence of XOR clauses, including support for dictionaries. Algorithm 5 shows how to query using the new Algorithm 4. To use these new algorithms to do purely filtering, set r to 0. For a pure dictionary, set s to 0. If both $r > 0$ and $s > 0$, the stored metadata will only be returned when an element passes the filter. If the element is a false positive, the returned metadata will be random.

2.5   Blocked XORSAT Filters

XORSAT filters suffer from the same size problem as SAT filters, namely, it is not practical to build filters for large sets. The reason being that the time it takes a modern solver to find a solution to an instance (with say millions of variables) is often too long for common applications. The natural way to overcome this with SAT filters is to increase the number of variables in the random k-SAT problem, decreasing efficiency, but also making the SAT problem easier by backing off of the k-SAT threshold [23]. This technique is not applicable to random k-XORSAT instances, that is, increasing the number of variables does not make significantly easier instances.

XORSAT (and SAT) filter build time can be decreased by first hashing elements into blocks (or buckets) and then building one filter for each block of elements, a process that is trivially parallelizable. This is a tailoring of a Blocked

---
[3] Adding an extra r bits of metadata means that the filter now has r more solutions.

Algorithm 4. ELEMENTTOXORCLAUSES($e \in D, d, k, h_0, \ldots, h_{k-1}, h_b$)

e is the element used to generate s XOR clauses
n is the number of variables per XORSAT instance
d is data to be stored, a bit-vector in $[0, 2^r)$
k is the number of variables per XOR clause
$h_0, \ldots, h_{k-1}$ are functions that map elements of D to $[0, n)$
h_b is a function that maps elements of D to bit-vectors in $[0, 2^s)$

1: $V := \{\}$
2: **for** $i := 0$ **to** $k - 1$ **do**
3:     $v := h_i(e)$, hash e to generate variable v
4:     $V := V \cup \{v\}$, add v to the XOR clause
5: **end for**
6: $b := h_b(e)$, hash e to generate the right-hand side
7: **return** $(V, b||d)$, append d to the right-hand side

Algorithm 5. QUERYXORSATDICTIONARY($F_Y, x \in D, n, k, s, r, h_0, \ldots, h_{k-1}, h_b$)

x is the element used to query the XORSAT filter F_Y
n is the number of variables per XORSAT instance
k is the number of variables per XOR clause
s is the number of solutions to be found
r is the number of bits of metadata stored with each element
$h_0, \ldots, h_{k-1}$ are functions that map elements of D to $[0, n)$
h_b is a function that maps elements of D to bit-vectors in $[0, 2^s)$

1: $(V_x, b_x = [b_0, \ldots, b_{s+r-1}]) :=$ ELEMENTTOXORCLAUSES($x, 0, k, h_0, \ldots, h_{k-1}, h_b$)
2: **for** each variable $v \in V_x$ **do**
3:     $b_x := b_x \oplus F_Y(v)$
4: **end for**
5: **if** $[b_0, \ldots, b_{s-1}] = 0$ **then**
6:     **return** ($Maybe$, $[b_s, \ldots, b_{s+r-1}]$)
7: **end if**
8: **return** No

Bloom filter [22, 28] to SAT filters of any constraint variation. The number of blocks can be determined by the desired runtime of the build process; the more blocks the faster the build process. The issue here is that, given a decent random hash function, elements are distributed into blocks according to a Poisson distribution [15], that is, some blocks will likely have a few more elements than others. Hence, to store the solutions for each block, one also needs to store some information about the number of variables in each block so that they can be accessed during query. Depending on the technique used, this is roughly a small number of extra bits per block. Otherwise, the blocks can be forced to a uniform size by setting the number variables for each block to be the maximum number of variables needed to make the largest block satisfiable. Either way, the long-term storage of the filter has slightly increased, slightly decreasing efficiency at the benefit of a (potentially much) shorter build-time. So, here is one trade-off between build time and efficiency that can make SAT filters practical for large datasets. Also, blocking can increase query speed since, depending on block size, the k lookups will be relatively near each other in a computer's memory, giving the processor an opportunity to optimize the lookups. In fact, this is the original motivation of Blocked Bloom Filters; it's simply advantageous that the idea can also be used to drastically decrease the build time of SAT filters.

The next section provides the mathematics needed to choose appropriate parameters for the XORSAT filter construction.

3   Filter Efficiency

As introduced in [30], given a filter with false positive rate p, n bits of memory, and $m = |Y|$, the efficiency of the filter is

$$\mathcal{E} = \frac{-\log_2 p}{n/m}.$$

Efficiency is a measure of how well a filter uses the memory available to it. The higher the efficiency, the more information packed into a filter. A filter with a fixed size can only store so much information. Hence, efficiency has an upper-bound, i.e., the *information-theoretic limit*, namely

$$\mathcal{E} = 1.$$

Since "$m/n = 1$ remains a *sharp* threshold for satisfiability of constrained[4] k-XORSAT for every $k \geq 3$" [25], the XORSAT filter construction, like the SAT

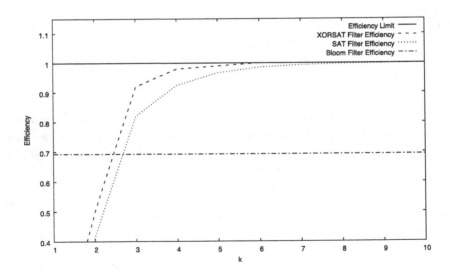

Fig. 1. Theoretically achievable XORSAT filter efficiency for various k.

---

[4] A *constrained* model is one where every variable appears in at least two equations.

filter construction, theoretically achieves $\mathcal{E} = 1$. In other words, it is possible to build an XORSAT filter for a given data set and false positive rate that uses as little long-term storage as possible. XORSAT filter efficiency tends to 1 faster than that of SAT filters, and the corresponding satisfiability threshold is much sharper. This means that, since there are diminishing returns as k grows, a small k (five or six) can give near optimal efficiency (see Fig. 1), and, unlike the SAT filter, these high efficiencies are able to be achieved in practice.

4   XORSAT Filter Parameters

This section discusses the selection of parameters for XORSAT filters (see Table 2).

Table 2. XORSAT filter parameters

p	The false positive rate of an XORSAT filter
s	The number of XORSAT instances
r	The number of bits of metadata stored with each element
n	The number of variables per XORSAT instance
m	The number of XOR clauses per XORSAT instance
k	The number of variables per XOR clause

As with the SAT filter, a value for k should be selected first. A small k (five or six) is sufficient to achieve near perfect efficiency (see Fig. 1). Larger k are undesirable as efficiency will not notably increase and query speed will significantly decrease.

The value of m is the number of elements being stored in the filter. Since the satisfiability threshold $\alpha_k = \frac{m}{n}$ is *sharp* for random k-XORSAT and tends quickly to 1, n should be set equal to, or slightly larger than m. For example, if $k = 3$, m should be roughly 91% of n (see Table 1 for precise calculations). Setting n much larger than m will cause a drop in efficiency without any advantage. This is not true for the SAT filter because random k-SAT problems become harder the closer they are to the satisfiability threshold [23], so, increasing n decreases build time. This is not the case with XORSAT filters and is the main reason they can practically achieve near perfect efficiency. Finally, a value for either s or p should be selected. These parameters determine the false positive rate $p = \frac{1}{2^s}$ and the amount of long-term storage (sn) of the filter.

To give an example set of parameters, an XORSAT filter for $m = 2^{16}$ elements with a false positive rate of $p \approx \frac{1}{2^7}$ needs $s = 7$ solutions to be stored and $n = 2^{16} + \epsilon$. Such an XORSAT filter, with $k = 6$, can be built and will use $sn \approx 460000$ bits of long-term storage, a 30% reduction over an optimal Bloom filter's long-term storage $\approx$660000 bits. See Sect. 6 for metrics on different size data sets, efficiencies, and query times.

5   Detailed Example

This section presents a detailed example of how to build and query an XORSAT filter, including details on how to use the filter to store and retrieve metadata. For the sole purpose of this example, let the set of interest be $Y = [(\text{"cat"}, 0),$ $(\text{"fish"}, 1), (\text{"dog"}, 2)]$. Here, Y is a list of three tuples where each tuple contains a word and an integer representing the tuple's index in the list.

5.1   Building the Filter

The first step in building an XORSAT filter for Y is to decide on parameters (see Sect. 4). For this example, let k, the number of variables per XOR clause, be three. Since there are three elements, m will be three. Let the number of variables per XORSAT instance be a number slightly larger than m to ensure the instances are satisfiable, say $n = 4$. Let p, the desired false positive rate, be $\frac{1}{2^3}$. This fixes s, the number of XORSAT instances, to three. Since there are three indices, only two bits of metadata, r, are needed to represent an index.

The next step is to create a list of hashes corresponding to each of the three words. This example will make use of the 32-bit xxHash algorithm [10]. Let the list of hashes be

$$H = [\text{xxHash}(\text{"cat"}), \text{xxHash}(\text{"fish"}), \text{xxHash}(\text{"dog"})]$$
$$= [\texttt{0xb85c341a}, \texttt{0x87024bb7}, \texttt{0x3fa6d2df}].$$

Next, the hashes are used to generate XOR clauses, one per hash. For the purpose of this example a scheme needs to be devised that will transform a hash into an XOR clause. One simple method is to first treat the hash as a bit-vector, then split the vector into parts and let each part represent a new variable in the XOR clause. Here, let the hashes be split into 4-bit parts, as $2^4 > n$ and it will be easy to see the split (represented in hexadecimal). The list of split hashes is

$$SH = [[\texttt{0xb}, \texttt{0x8}, \texttt{0x5}, \texttt{0xc}, \texttt{0x3}, \texttt{0x4}, \texttt{0x1}, \texttt{0xa}],$$
$$[\texttt{0x8}, \texttt{0x7}, \texttt{0x0}, \texttt{0x2}, \texttt{0x4}, \texttt{0xb}, \texttt{0xb}, \texttt{0x7}],$$
$$[\texttt{0x3}, \texttt{0xf}, \texttt{0xa}, \texttt{0x6}, \texttt{0xd}, \texttt{0x2}, \texttt{0xd}, \texttt{0xf}]].$$

The next step is to use the split hashes to create XOR clauses. This is done here by treating the groupings of 4 bits (under proper modulus) as variable indices and right-hand side of each clause. The variable indices and right-hand side for each clause would be

$$\mathcal{I}_{Y.0} = [[SH_{00}(\text{mod } n), SH_{01}(\text{mod } n), SH_{02}(\text{mod } n), SH_{03}(\text{mod } 2^s)],$$
$$[SH_{10}(\text{mod } n), SH_{11}(\text{mod } n), SH_{12}(\text{mod } n), SH_{13}(\text{mod } 2^s)],$$
$$[SH_{20}(\text{mod } n), SH_{21}(\text{mod } n), SH_{22}(\text{mod } n), SH_{23}(\text{mod } 2^s)]]$$
$$= [[\text{0xb}(\text{mod } 4), \text{0x8}(\text{mod } 4), \text{0x5}(\text{mod } 4), \text{0xc}(\text{mod } 8)],$$
$$[\text{0x8}(\text{mod } 4), \text{0x7}(\text{mod } 4), \text{0x0}(\text{mod } 4), \text{0x2}(\text{mod } 8)],$$
$$[\text{0x3}(\text{mod } 4), \text{0xf}(\text{mod } 4), \text{0xa}(\text{mod } 4), \text{0x6}(\text{mod } 8)]]$$
$$= [[3, 0, 1, 4],$$
$$[0, 3, 0, 2],$$
$$[3, 3, 2, 6]].$$

In practice, these first few steps are the bottleneck in terms of query speed and need to be heavily optimized. The simple scheme presented here is purely for demonstration purposes. A more practical but complex scheme is given in Sect. 6. As well, this scheme does not guarantee width k XOR clauses are generated because duplicates may arise. However, duplicate variables in XOR clauses simply cancel each other out, so, for the purpose of this example, this simplified scheme is enough to demonstrate the main concepts. Also, for specific applications, duplicate detection and removal may be too computationally expensive to outweigh any benefit gained in efficiency.

The three XORSAT instances are encoded as follows:

$$\mathcal{X}_{Y.0} = [x_3 \oplus x_0 \oplus x_1 \equiv [1, 0, 0],$$
$$x_0 \oplus x_3 \oplus x_0 \equiv [0, 1, 0],$$
$$x_3 \oplus x_3 \oplus x_2 \equiv [1, 1, 0]]$$
$$= [x_0 \oplus x_1 \oplus x_3 \equiv [1, 0, 0],$$
$$x_3 \equiv [0, 1, 0],$$
$$x_2 \equiv [1, 1, 0]].$$

Next, append each element's two bits of metadata to the right-hand side of each corresponding XOR clause, creating $s + r = 5$ instances.

$$\mathcal{X}_Y = [x_0 \oplus x_1 \oplus x_3 \equiv [1, 0, 0] \;||\; [0, 0],$$
$$x_3 \equiv [0, 1, 0] \;||\; [0, 1],$$
$$x_2 \equiv [1, 1, 0] \;||\; [1, 0]]$$
$$= [x_0 \oplus x_1 \oplus x_3 \equiv [1, 0, 0, 0, 0],$$
$$x_3 \equiv [0, 1, 0, 0, 1],$$
$$x_2 \equiv [1, 1, 0, 1, 0]].$$

The final steps are to solve and store $s + r = 5$ solutions, one for each of the XORSAT instances encoded by $\mathcal{X}_Y$. Though there are many different solutions to these instances, five such solutions are

$$S_Y = [[x_0 = 1, x_1 = 0, x_2 = 1, x_3 = 0],$$
$$[x_0 = 0, x_1 = 1, x_2 = 1, x_3 = 1],$$
$$[x_0 = 0, x_1 = 0, x_2 = 0, x_3 = 0],$$
$$[x_0 = 1, x_1 = 1, x_2 = 1, x_3 = 0],$$
$$[x_0 = 1, x_1 = 0, x_2 = 0, x_3 = 1]].$$

The filter F_Y is the transpose of the solutions S_Y, along with those parameters necessary for proper querying, namely

$$F_Y = ([[1, 0, 0, 1, 1],$$
$$[0, 1, 0, 1, 0],$$
$$[1, 1, 0, 1, 0],$$
$$[0, 1, 0, 0, 1]],$$
$$n = 3, k = 3, s = 3, r = 2).$$

The filter F_Y is now complete and the next part of the example will demonstrate how to query it.

5.2   Querying the Filter

The process to query F_Y with an example element $x =$ "horse" follows many of the same steps as building the filter. First, the same hash scheme from above is used to generate an XOR clause for "horse".

$$H = \mathsf{xxHash}(\text{``horse''})$$
$$= \mathsf{0x3f37a1a7}.$$

Next, the hash is split into groups of 4 bits.

$$SH = [\mathsf{0x3, 0xf, 0x3, 0x7, 0xa, 0x1, 0xa, 0x7}].$$

Then, three clause indices and a right-hand side are generated from the hash.

$$\mathcal{I} = [SH_0(\bmod\ n), SH_1(\bmod\ n), SH_2(\bmod\ n), SH_3(\bmod\ 2^s)]$$
$$= [\mathsf{0x3}(\bmod\ 4), \mathsf{0xf}(\bmod\ 4), \mathsf{0x3}(\bmod\ 4), \mathsf{0x7}(\bmod\ 8)]$$
$$= [3, 3, 3, 7].$$

Finally the clause is created from the indices and right-hand side and two bits (all True) are appended to support metadata retrieval.

$$C = x_3 \oplus x_3 \oplus x_3 \equiv [1, 1, 1]\ ||\ [1, 1]$$
$$= \qquad\qquad x_3 \equiv [1, 1, 1, 1, 1].$$

In Algorithm 5, the right-hand side metadata bits are all set to False and the terminal equivalence ($\equiv$) is treated as an XOR ($\oplus$). That choice was made purely for presentation of the algorithm. This example demonstrates that either way is acceptable.

Now that the clause C has been built, it can be tested against the filter F_Y. To do so, assign the variables in C their values in the stored solutions of F_Y and evaluate the resulting equation.

$$
\begin{aligned}
C_{F_Y} &= F_Y(3) \equiv [1,1,1,1,1] \\
&= [0,1,0,0,1] \equiv [1,1,1,1,1] \\
&= [0,1,0,0,1].
\end{aligned}
$$

Since the first three bits of C_{F_Y} are not all True, the element does not pass the filter. Hence, the string "horse" is definitely not in Y.

The final part of this example demonstrates a query that passes and returns stored metadata. Specifically, F_Y will be queried with $x =$ "cat". Again, the same hash scheme from above is used to generate an XOR clause for "cat". Since this was already demonstrated in the previous section, the details will not be repeated. Instead, the clause C is simply stated next, including the two True bits appended to support metadata retrieval.

$$
C = x_0 \oplus x_1 \oplus x_3 \equiv [1,0,0,1,1].
$$

Evaluating C against F_Y produces

$$
\begin{aligned}
C_{F_Y} &= F_Y(0) \oplus F_Y(1) \oplus F_Y(3) \equiv [1,0,0,1,1] \\
&= [1,0,0,1,1] \oplus [0,1,0,1,0] \oplus [0,1,0,0,1] \equiv [1,0,0,1,1] \\
&= [1,0,0,0,0] \equiv [1,0,0,1,1] \\
&= [1,1,1,0,0].
\end{aligned}
$$

Since the first three bits of C_{F_Y} are all True, the element passes the filter. Hence, "cat" is in Y with a $\frac{1}{2^3}$ chance of being a false positive. The last two bits of C_{F_Y}, $[0,0]$, represent the stored metadata, namely, the index 0.

6   Experimental Results

This section serves to demonstrate that it is practical to build efficient XORSAT filters for very large data sets. To do so, a research-grade XORSAT solver and XORSAT filter construction were implemented in the C language. The solver performs the 'Method of the Four Russians' [29].

As a proof of concept, seventeen dictionaries were built consisting of $2^{10}, \ldots,$ and 2^{26} random 16-byte strings. To ensure that random k-XORSAT instances were generated, the strings were transformed into XOR clauses using the 64-bit xxHash hash algorithm [10]. Each string was fed into a single call of xxHash and the output was used to seed a linear feedback shift register (LFSR) with 16-bit elements and primitive polynomial $1 + x^2 + x^3 + x^4 + x^8$. XOR clauses were produced by stepping the LFSR k times. Duplicate removal was not considered as searching for duplicates drastically increases query performance yet only marginally increases efficiency. All of the following results were collected using a late 2013 MacBook Pro with a 2.6-GHz Intel Core i7 and 16 GB of RAM. All times are reported in seconds.

Table 3. Achieved efficiency and seconds taken to build *non-blocked* XORSAT filters with $p = \frac{1}{2^{10}}$. m is the number of elements in the data set being stored and k is the number of variables per XOR clause.

m	$k = 3$	$k = 4$	$k = 5$	$k = 6$
2^{10}	(88%, <1)	(93%, <1)	(93%, <1)	(93%, <1)
2^{11}	(89%, <1)	(97%, <1)	(97%, <1)	(97%, <1)
2^{12}	(90%, <1)	(97%, <1)	(98%, <1)	(98%, <1)
2^{13}	(91%, 1)	(97%, 1)	(98%, 1)	(99%, 1)
2^{14}	(91%, 2)	(97%, 3)	(99%, 4)	(99%, 5)
2^{15}	(89%, 17)	(97%, 21)	(98%, 28)	(98%, 36)

Table 3 presents the achieved efficiency and time taken to build *non-blocked* XORSAT filters, that is, for each filter, $s = 10$ XORSAT instances were generated and one solution was found for each. The instances were solved in parallel using a single call to the XORSAT solver.

Unlike SAT filters, the number of solutions found does not affect either efficiency or runtime so long as s is less than the word-size of the computer (typically 64 bits, a very reasonable assumption in practice given that $s > 64$ would mean building a filter with a false positive rate less than $\frac{1}{2^{64}}$). Efficiency is not affected because the s right-hand sides are all uncorrelated. Runtime is not affected because all s instances are solved in parallel using bit-packing and word-level operations.

The XORSAT filters in Tables 3 and 4 achieve the desired false positive rate. This was verified experimentally by querying each XORSAT filter with 2^{23} 4-byte elements and using the results to calculate the achieved false positive rate and, for Table 4, query speed as well.

In terms of efficiency, the experimental results match the theoretical results from Table 1. And, if the number of XOR clauses per instance is above 2^{14}, filters can be practically built that are very close to the optimal efficiency possible for each given k.

The results also hint correctly that it is not practical to build *non-blocked* XORSAT filters for very large data sets as runtime will grow and become prohibitive in practice. It is likely that filter build time can be reduced by using a more powerful solver (such as M4RI [2]), but this has not been explored here.

Table 4. Achieved efficiency, size (in KB), and seconds taken to build *blocked* XORSAT and SAT filters with an expected 3072 elements per block, variables per clause $k = 5$, and desired false positive rate $p = \frac{1}{2^{10}}$. Desired SAT filter efficiency was set to 75% and desired XORSAT filter efficiency was set to 98%. The SAT filter hamming weight metric [31] was set to 48%. Timeout ('-') was set at one hour. Query speed (in millions of queries per second) is also given for XORSAT, SAT, and Bloom filters.

m	XORSAT filter				SAT filter						
	Build time				Build time				Query speed		
	1 Core	8 Cores	$\mathcal{E}$	Size	1 Core	8 Cores	$\mathcal{E}$	Size	XORSAT	SAT	Bloom
2^{15}	<1	<1	98%	41	336	105	43%	56	18	4	23
2^{16}	1	<1	98%	81	883	183	43%	111	18	4	23
2^{17}	2	<1	98%	163	1768	394	43%	222	18	4	23
2^{18}	5	1	98%	326	3441	723	44%	444	18	4	23
2^{19}	8	1	97%	659	-	1724	44%	887	18	4	23
2^{20}	17	2	97%	1321	-	-	-	-	18	-	22
2^{21}	33	4	97%	2646	-	-	-	-	17	-	22
2^{22}	92	12	97%	5298	-	-	-	-	13	-	20
2^{23}	186	26	97%	10601	-	-	-	-	9	-	20
2^{24}	372	52	97%	21204	-	-	-	-	11	-	20
2^{25}	751	104	96%	42416	-	-	-	-	10	-	17
2^{26}	1515	208	96%	84958	-	-	-	-	7	-	12

However, build time can be significantly reduced by building *blocked* XORSAT filters. As discussed in Sect. 2.3, first hashing elements into small blocks and then, in parallel, building one filter for each block can drastically reduce the build time of a large data set without increasing query time and only marginally reducing efficiency. Since build time is one-time work, discovering techniques for reducing build time any further is likely unnecessary.

Results in Table 3 can be used to tune blocked XORSAT filter schemes. For example, setting the block size between 2^{11} and 2^{12} and $k = 5$ will enable fast building of blocked XORSAT filters with $\mathcal{E} \approx 98\%$. Table 4 presents the build time, achieved efficiency, filter size, and query speed for blocked XORSAT and SAT filters using these sample parameters. The table also presents query speed for Bloom filters built and queried using the same data sets, false positive rate, bucketing, and element hashing scheme. Though, Bloom filters can only achieve a maximum efficiency of $\ln 2 = 69\%$, meaning that they use approximately 44% more long-term storage than an XORSAT filter for the same data at the same false positive rate.

Table 4 demonstrates that it is practical to build XORSAT filters very near the information theoretic limit while maintaining a high query speed. Since each block can be build in parallel, linear speedup is achieved and demonstrated in the results. As with SAT filters, XORSAT filter query speed can be increased by decreasing k which may in turn decrease efficiency.

The query speed of the above filter implementations begin to drop after data sets grow above 2^{20}. This is due to size of the filter overwhelming the caching mechanisms of the computer running the experiments. It may be possible to create a cache-aware implementation of XORSAT filters that increases query speed overall and removes some of the query speed variance seen in Table 4, though this has not been explored.

Efficiency also slowly drops as filters increase in size. As discussed in Sect. 2.3, since blocks may not all hold the same number of elements, it is necessary to store additional information so that the blocks can be accessed during query. This additional information must be stored as part of the filter and, hence, increases the size of the filter, decreasing efficiency.

7   Conclusions and Future Work

The XORSAT filter is the first *practical* SAT filter construction, overcoming many of the previous hurdles presented in [31]. It is a simple offline filter construction for very large data sets that can consistently achieve the efficiency bound in practice while maintaining fast queries. This new filter construction is also parameterized so that it can be easily tailored to support an application needing, for example, fast build time, fast queries, a small memory footprint, and metadata storage and retrieval.

Potential future work includes considering XORSAT filters as part of a secure search scheme [18]. This would involve tailoring the filter construction to make it secure or resistant to various attacks such as inversion and intersection, as well as many others [27,32].

Moving from disjunctive clauses to XOR clauses provides for a SAT filter with different features (ex. near perfect efficiency, fast build time, metadata support, hints of security). Hence, it is possible that SAT filters built from other constraint types could provide other common sought after filter features such as streaming (online filters), element deletion, or element counting.

References

1. Achlioptas, D.: Random satisfiability. In: Biere et al. [5], pp. 245–270
2. Albrecht, M., Bard, G.: The M4RI library (2018). https://m4ri.sagemath.org/
3. Azinović, M., Herr, D., Heim, B., Brown, E., Troyer, M.: Assessment of quantum annealing for the construction of satisfiability filters. SciPost Phys. **2**, 013 (2017). https://doi.org/10.21468/SciPostPhys.2.2.013
4. Bard, G.V.: The method of four Russians. In: Algebraic Cryptanalysis. Springer, Boston (2009). https://doi.org/10.1007/978-0-387-88757-9_9
5. Biere, A., Heule, M., van Maaren, H., Walsh, T. (eds.): Handbook of Satisfiability. Frontiers in Artificial Intelligence and Applications, vol. 185. IOS Press, Amsterdam (2009)
6. Bloom, B.H.: Space/time trade-offs in hash coding with allowable errors. Commun. ACM **13**(7), 422–426 (1970)

7. Brodnik, A., Munro, J.I.: Membership in constant time and almost-minimum space. SIAM J. Comput. **28**(5), 1627–1640 (1999)
8. Chazelle, B., Kilian, J., Rubinfeld, R., Tal, A.: The Bloomier filter: an efficient data structure for static support lookup tables. In: Proceedings of the Fifteenth Annual ACM-SIAM Symposium on Discrete Algorithms, pp. 30–39. Society for Industrial and Applied Mathematics (2004)
9. Cohen, S., Matias, Y.: Spectral Bloom filters. In: Proceedings of the 2003 ACM SIGMOD International Conference on Management of Data, pp. 241–252. ACM (2003)
10. Collet, Y.: xxHash: extremely fast hash algorithm (2017)
11. Daudé, H., Ravelomanana, V.: Random 2-XORSAT at the satisfiability threshold. In: Laber, E.S., Bornstein, C., Nogueira, L.T., Faria, L. (eds.) LATIN 2008. LNCS, vol. 4957, pp. 12–23. Springer, Heidelberg (2008). https://doi.org/10.1007/978-3-540-78773-0_2
12. Dietzfelbinger, M., Goerdt, A., Mitzenmacher, M., Montanari, A., Pagh, R., Rink, M.: Tight thresholds for cuckoo hashing via XORSAT. In: Abramsky, S., Gavoille, C., Kirchner, C., Meyer auf der Heide, F., Spirakis, P.G. (eds.) ICALP 2010. LNCS, vol. 6198, pp. 213–225. Springer, Heidelberg (2010). https://doi.org/10.1007/978-3-642-14165-2_19
13. Dietzfelbinger, M., Pagh, R.: Succinct data structures for retrieval and approximate membership (Extended Abstract). In: Aceto, L., Damgård, I., Goldberg, L.A., Halldórsson, M.M., Ingólfsdóttir, A., Walukiewicz, I. (eds.) ICALP 2008. LNCS, vol. 5125, pp. 385–396. Springer, Heidelberg (2008). https://doi.org/10.1007/978-3-540-70575-8_32
14. Douglass, A., King, A.D., Raymond, J.: Constructing SAT filters with a quantum annealer. In: Heule, M., Weaver, S. (eds.) SAT 2015. LNCS, vol. 9340, pp. 104–120. Springer, Cham (2015). https://doi.org/10.1007/978-3-319-24318-4_9
15. Erdős, P., Renyi, A.: On a classical problem of probability theory (1961)
16. Fan, B., Andersen, D.G., Kaminsky, M., Mitzenmacher, M.D.: Cuckoo filter: practically better than Bloom. In: Proceedings of the 10th ACM International Conference on emerging Networking Experiments and Technologies, pp. 75–88. ACM (2014)
17. Fang, C., Zhu, Z., Katzgraber, H.G.: NAE-SAT-based probabilistic membership filters. preprint arXiv:1801.06232 (2018)
18. Goh, E.J., et al.: Secure indexes. IACR Cryptology ePrint Archive 2004, 216 (2004)
19. Heule, M.J., van Maaren, H.: Parallel SAT solving using bit-level operations. J. Satisf. Boolean Model. Comput. **4**, 99–116 (2008)
20. Ibrahimi, M., Kanoria, Y., Kraning, M., Montanari, A.: The set of solutions of random XORSAT formulae. In: Proceedings of the Twenty-Third Annual ACM-SIAM Symposium on Discrete Algorithms, pp. 760–779. SIAM (2012)
21. Kader, A.A., Dorojevets, M.: Novel integration of Dimetheus and WalkSAT solvers for k-SAT filter construction. In: Systems, Applications and Technology Conference (LISAT 2017), pp. 1–5. IEEE (2017)
22. Krimer, E., Erez, M.: The power of $1+\alpha$ for memory-efficient Bloom filters. Internet Math. **7**(1), 28–44 (2011)
23. Mitchell, D., Selman, B., Levesque, H.: Hard and easy distributions of SAT problems. In: AAAI 1992, pp. 459–465 (1992)
24. Mitzenmacher, M.D.: Compressed Bloom filters. IEEE/ACM Trans. Netw. (TON) **10**(5), 604–612 (2002)
25. Pittel, B., Sorkin, G.B.: The satisfiability threshold for k-XORSAT. Comb. Probab. Comput. **25**(02), 236–268 (2016)

26. Porat, E.: An optimal Bloom filter replacement based on matrix solving. In: Frid, A., Morozov, A., Rybalchenko, A., Wagner, K.W. (eds.) CSR 2009. LNCS, vol. 5675, pp. 263–273. Springer, Heidelberg (2009). https://doi.org/10.1007/978-3-642-03351-3_25

27. Pouliot, D., Wright, C.V.: The shadow nemesis: inference attacks on efficiently deployable, efficiently searchable encryption. In: Proceedings of the 2016 ACM SIGSAC Conference on Computer and Communications Security, pp. 1341–1352. ACM (2016)

28. Putze, F., Sanders, P., Singler, J.: Cache-, hash-, and space-efficient Bloom filters. J. Exp. Algorithmics **14**, 4 (2009)

29. Arlazarov, V.L., Dinitz, Y.A., Kronrod, M.A., Faradzev, I.A.: On economical construction of transitive closure of an oriented graph. Dokl. Akad. Nauk SSSR **194**(3), 487 (1970)

30. Walker, A.: Filters. Master's thesis, Haverford College (2007). http://math.uchicago.edu/~akwalker/filtersFinal.pdf

31. Weaver, S.A., Ray, K.J., Marek, V.W., Mayer, A.J., Walker, A.K.: Satisfiability-based set membership filters. J. Satisf. Boolean Model. Comput. **8**, 129–148 (2014)

32. Zhang, Y., Katz, J., Papamanthou, C.: All your queries are belong to us: the power of file-injection attacks on searchable encryption. In: USENIX Security Symposium, pp. 707–720 (2016)

ALIAS: A Modular Tool for Finding Backdoors for SAT

Stepan Kochemazov and Oleg Zaikin[✉]

ISDCT SB RAS, Irkutsk, Russia
veinamond@gmail.com, zaikin.icc@gmail.com

Abstract. We present ALIAS, a modular tool aimed at finding back-doors for hard SAT instances. Here by a *backdoor* for a specific SAT solver and SAT formula we mean a set of its variables, all possible instantiations of which lead to construction of a family of subformulas with the total solving time less than that for an original formula. For a particular back-door, the tool uses the Monte-Carlo algorithm to estimate the runtime of a solver when partitioning an original problem via said backdoor. Thus, the problem of finding a backdoor is viewed as a black-box optimization problem. The tool's modular structure allows to employ state-of-the-art SAT solvers and black-box optimization heuristics. In practice, for a number of hard SAT instances, the tool made it possible to solve them much faster than using state-of-the-art multithreaded SAT-solvers.

1 Introduction

Informally, a *backdoor* is some hidden flaw in a design of a system that allows one to do something within that system that should not be possible otherwise. In the context of Constraint Satisfaction Problems (CSP) a backdoor is usually a small subset of problem variables which has a peculiar property: instantiating backdoor variables results in a subproblem that is significantly easier to solve. For the first time the concept of backdoors arose in the context of CSP in [26], where *strong backdoors* were introduced and analyzed. Their main disadvantage is that they rely on polynomial algorithms to solve simplified subproblems, and thus strong backdoors that can be used in practice are very hard to find [15,23].

In the present paper, we consider more general backdoors to SAT, that do not rely on polynomial algorithms to solve simplified subproblems. In particular, we search for such sets of variables of a considered SAT instance that all possible instantiations of backdoor variables results in a family of subproblems, for which a total solving time is less than that for an original SAT instance. It is clear that such subproblems can be solved in parallel. For a given SAT instance C, solver S and backdoor B one can effectively compute the estimation of runtime of S on a family of subproblems produced by assigning values to variables from B in C using a Monte-Carlo method. Thus there is defined a black-box pseudo-Boolean function with aforementioned inputs. Then, it is possible to use arbitrary black-box pseudo-Boolean optimization methods to traverse the search space of possible general backdoors to find one with a good estimation.

© Springer International Publishing AG, part of Springer Nature 2018
O. Beyersdorff and C. M. Wintersteiger (Eds.): SAT 2018, LNCS 10929, pp. 419–427, 2018.
https://doi.org/10.1007/978-3-319-94144-8_25

We implemented this approach in the form of *modulAr tooL for fInding bAck-doors for Sat* (ALIAS) – a convenient customizable scalable tool that can employ arbitrary incremental state-of-the-art SAT solvers and black-box optimization heuristics to search for backdoors to hard SAT instances. The found backdoor is then used to solve the corresponding instance by the same incremental solver. Thereby, ALIAS can be viewed as a tool for constructing backdoor-based divide-and-conquer parallel SAT solvers. The ALIAS tool and our benchmarks are available at https://github.com/Nauchnik/alias.

2   On Backdoors to SAT

Suppose C is a SAT instance, X is a set of its variables, and A is a polynomial algorithm. If we assign values $\alpha = (\alpha_1, \ldots, \alpha_k)$ to variables from set $B, |B| = k$, $B \subseteq X$, then the simplified SAT instance is denoted as $C[\alpha/B]$.

Definition 1 (Backdoor [26]). *A nonempty subset B of variables from C is called a backdoor in C for A if for some instantiation β of variables from B an algorithm A returns a satisfying assignment of $C[\beta/B]$.*

Note that the definition of backdoor implies only satisfiable instances and can not be easily extended to unsatisfiable ones. Also, even if backdoor is known, it is necessary to find such β that A would be able to solve a considered instance.

Definition 2 (Strong Backdoor [26]). *A nonempty subset B of variables from C is a strong backdoor in C for A if for any instantiation γ of variables from B an algorithm A returns a satisfying assignment or concludes unsatisfiability of $C[\gamma/B]$.*

For SAT instances the natural choice of polynomial algorithm A is the Unit Propagation rule (UP) [8]. A Strong Backdoor w.r.t UP is called Strong Unit Propagation Backdoor Set (SUPBS). For any SAT instance the whole set of its variables is a SUPBS (further it is called *trivial SUPBS*). If a SAT instance encodes a Boolean circuit, a set of variables encoding its input can usually serve as a SUPBS.

Compared to a backdoor a strong backdoor is much more powerful: given a strong backdoor B, one can traverse through possible instantiations of variables from B thus solving C in time $\approx 2^{|B|} \times |C|$ (here $|C|$ is the size of a SAT instance C in computer memory). However, it is unclear what to do if, for example, $|B| > 100$. Also, the problem of finding strong backdoors for SAT is particularly hard, see e.g. [15].

The main disadvantage of the notion of strong backdoor lies in polynomial complexity requirement for an algorithm used to solve constructed subproblems. The following definition in a way extends the notion of backdoors to non-polynomial algorithms. Assume that G is an arbitrary complete SAT solving algorithm.

Definition 3 (Non-deterministic Oracle Backdoor Set (NOBS) [21]). *A non-empty set B of variables from C is a Non-deterministic Oracle Backdoor Set (NOBS) w.r.t. algorithm G if the total running time of G given formulas $C[\beta/B]$, $\beta \in \{0,1\}^{|B|}$, is less than the running time of G on the original formula C.*

Without formally defining NOBS, the corresponding idea was used in a number of papers on application of SAT to cryptanalysis instances, such as [7,9,22,27]. Compared to strong backdoors, NOBS do not give a straightforward way to estimate the runtime of G for solving C using backdoor B. However, it can be done via the Monte-Carlo method [17] as follows. We treat the average runtime of G on an arbitrary subproblem $C[\gamma/B]$, $\gamma \in \{0,1\}^{|B|}$ as a random variable. The intermediate goal is to estimate its expected value. For this purpose, first, construct a random sample of size N of instantiations of variables from B:$\{\beta_1,\ldots,\beta_N\}$, $\beta_i \in \{0,1\}^{|B|}$, $i \in \{1,\ldots,N\}$. Second, measure the runtime of G on $C[\beta_i/B]$, $i \in \{1,\ldots,N\}$, denote it by $T_G(\beta_i)$. Then the runtime estimation can be computed using the formula:

$$Runtime\_Estimation(C, B, G, N) = 2^{|B|} \times \frac{1}{N} \times \sum_{i=1}^{N} T_G(\beta_i) \qquad (1)$$

Since G is a complete algorithm, theoretically, the value of *Runtime_Estimation* function can be computed for any B. Essentially, it is a blackbox function. One possible way to find a good backdoor B is for fixed C, G and N to minimize the value of *Runtime_Estimation(C, B, G, N)* by varying B. Since any backdoor B can be uniquely represented by a Boolean vector from $\{0,1\}^{|X|}$, where X is the set of variables from C, it means that the corresponding search space is $\{0,1\}^{|X|}$.

3 The ALIAS tool

Essentially, the ALIAS tool implements the blackbox optimization in the space of possible NOBS. The blackbox function in ALIAS is computed according to (1). The flowchart of the tool is presented in Fig. 1.

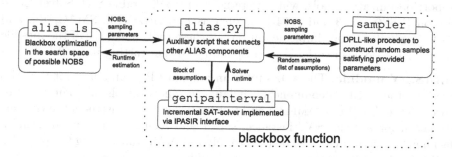

Fig. 1. ALIAS flowchart

ALIAS consists of four interconnected modules: ALIAS_LS, ALIAS.PY, SAMPLER and GENIPAINTERVAL program. The latter three implement the aforementioned blackbox function which for a given incremental SAT solver, SAT instance and NOBS computes a runtime estimation for this instance. Detailed comments are presented below.

ALIAS_LS module. Note that due to the fact that the search space of possible NOBS in the general case is extremely large, any possible way to restrict it is welcomed. Because of this, in ALIAS the search space of possible NOBS always consists of possible subsets of a SUPBS, either trivial or nontrivial.

Now, assume that a SUPBS for a considered SAT instance contains N variables. Then the search space has 2^N points, each point corresponding to some NOBS. For each NOBS we can compute the runtime estimation using (1) (for a fixed SAT solver). So the goal of ALIAS_LS is to traverse the search space towards NOBSs with minimal runtime estimations. Currently, for this purpose ALIAS_LS uses a simple optimization algorithm based on the Greedy Best First Search (GBFS) [19]. GBFS uses SUPBS as a starting point to construct a baseline runtime estimation. Then it checks all points from the neighborhood of the starting point (a set of points at Hamming distance of 1). If it finds a better point, then it starts checking its neighborhood. If all points from a neighborhood are worse than the current *best known value*, then it means that a local minimum is reached. Since the computation of (1) for an arbitrary point is quite costly, all passed points are stored in order to avoid recomputing (1) for corresponding backdoors.

The GBFS implementation in ALIAS uses two additional heuristics. First, at the beginning of the search the algorithm tries to quickly traverse the search space by removing large amount of randomly chosen variables (10 in our experiments) from the current record point at each step as long as it leads to updating the record. It often allows to quickly move from NOBS with hundreds of variables to that with dozens. The second heuristic is that when a local minimum is reached, the algorithm tries to jump from it by constructing a new starting point by permuting the current record point. The algorithm stops either if the time limit is exceeded, if the limit of jumps is reached or if the current runtime estimation is lower than the (scaled) remaining time.

On the current stage the ALIAS components are configured in a way to support optimization tools, which were used in Configurable SAT Solver Competition (CSSC) 2013 and 2014 [14], such as ParamILS [13], SMAC [12], and GGA [1]. Similar to our implementation, all these tools make use of the .pcs file that contains Boolean variables corresponding to the starting point (SUPBS).

ALIAS.PY module. The ALIAS.PY is an auxiliary Python 3.6 script that ties together other ALIAS components. It launches and controls all computations, processes the data from SAMPLER and GENIPAINTERVAL, constructs the runtime estimation for a given SAT instance, solver and NOBS, thus implementing a blackbox function. It can also be used to solve a SAT instance using the provided NOBS. In all modes ALIAS.PY can employ multiple CPU cores.

When constructing a runtime estimation, ALIAS.PY implements the Monte-Carlo method: it uses SAMPLER to construct a random sample of subproblems (in the form of assumptions for a SAT solver), then gives them in blocks of fixed size to GENIPAINTERVAL solver (by a *block* we mean a set of instantiations of backdoor variables, in form of assumptions), computes the average solving time for an arbitrary subproblem from random sample, uses it to compute runtime estimation and returns it to ALIAS_LS.

In the solving mode ALIAS.PY splits all possible instantiations of a provided NOBS variables into small blocks and feeds them to GENIPAINTERVAL until either all blocks are processed or a satisfying assignment is found.

SAMPLER module. SAMPLER is a program for generating random samples that is implemented on the basis of COMINISATPS solver [18]. Generally speaking, a random sample can be constructed in many different ways. In the most simple case for a NOBS B we can simply take N randomly generated vectors from $\{0,1\}^{|B|}$ and view them as a random sample by assigning corresponding values to variables from B. This approach was used in [9,20,22]. However, the described straightforward sampling procedure might not benefit fully from the incremental solving ability of state-of-the-art SAT solvers because the assignments of variables are too distant from each other (for example Hamming distance-wise). Thus, by default SAMPLER uses the sampling strategy proposed in [27]. Informally, it attempts to construct sequences of backdoor instantiations which are close to each other as nodes of the search tree. At the same time SAMPLER when possible puts into a random sample only subproblems that can not be solved using UP.

GENIPAINTERVAL module. The GENIPAINTERVAL program, given a CNF formula and a set of assumptions processes the latter sequentially in incremental way. To build it one needs the IPASIR API [4] and sources of some generic IPASIR-compatible incremental SAT solver. It is natural to consider only incremental solvers since the subproblems produced by instantiating NOBS variables are very similar to each other. Currently, different GENIPAINTERVAL instances running in parallel are not configured to share any information.

4 Experimental Results

In all experiments described below we employed one node of the HPC-cluster "Academician V.M. Matrosov"[1] (2 × 18-core Intel Xeon E5-2695 CPUs and 128 Gb of RAM). Each considered solver was launched on 1 node with 36 threads.

We benchmarked ALIAS against the Top 3 solvers from the SAT Competition 2017 Parallel track: SYRUP [3], PLINGELING [5] and PAINLESS-MAPLECOMSPS [10]. All these solvers are portfolio. As IPASIR-based solvers for ALIAS we used the Top 3 solvers from the SAT Competition 2017 Incremental track: ABCDSAT [6], GLUCOSE [2] and RISS [16]. The resulting parallel solvers are denoted as ALIAS-ABCDSAT, ALIAS-GLUCOSE and ALIAS-RISS.

---

[1] Irkutsk Supercomputer Center of SB RAS, http://hpc.icc.ru.

In preliminary experiments we compared the effectiveness of GBFS implementation in ALIAS_LS with that of SMAC tool [12] as blackbox heuristics for finding NOBS. For all considered instances GBFS found backdoors with better runtime estimation, thus it was used in all experiments below.

Each ALIAS solver works as follows: first ALIAS_LS is launched with a specified time limit. Once it found a good NOBS or exceeded the time limit, the best found NOBS is then used to solve the instance for the remaining time (if any) by instantiating backdoor variables and solving corresponding subproblems in parallel.

Two benchmark sets of hard SAT instances were considered. The first set consists of instances, in which a relatively small SUPBS is known. It is formed by SAT encodings of cryptanalysis instances for the alternating step generator (ASG) [11] and two its modifications, MASG and MASG0 [25]. SAT instances for these problems were taken from [27]: 10 for each of ASG-72, ASG-96, MASG-72 and MASG0-72 (40 in total). Naturally, for ASG-72, MASG-72 and MASG0-72 there is a SUPBS of 72 variables and for ASG-96 a SUPBS of 96 variables (corresponding to secret keys). Thus, ALIAS-based solvers were provided with this information. Each instance from this set has exactly one satisfying assignment.

The second benchmark set contains hard small crafted SAT instances. To construct it we first took all crafted instances with less than 500 variables from SAT Competitions 2007, 2009, 2011, 2014, 2016, 2017 and also *challenge-105.cnf* described in [24]. Then we launched SYRUP, PLINGELING and PAINLESS-MAPLECOMSPS on each of them with the time limit of 5000 s. It turned out that 33 instances were not solved in time by any solver: 7 from SAT Competition 2007, 10 from SAT Competition 2009, 9 from SAT Competition 2011, 6 from SAT Competition 2014 and also *challenge-105.cnf*. Thus, these 33 instances form the second benchmark set. For the instances from the second benchmark set the ALIAS-based solvers were given only a trivial SUPBS – the whole set of variables of a corresponding formula.

The 6 considered solvers were launched on two described sets (73 instances in total) with the time limit of 1 day. The obtained results are presented in Figs. 2a and b. Note, that 26 out of 33 instances from the second benchmark set were not solved within the time limit by any considered solver. Table 1 lists the instances from the second set, which were solved within the time limit by at least one solver. This table also contains data on found backdoors. For ASG-72, ASG-96, MASG-72 and MASG0-72 the information is presented for 1 instance out of 10, the results for other instances from the series are similar. Here $|B|$ is a size of a found backdoor, BT – time spent to find it, RE – its runtime estimation (1), ST – the solving time using the found backdoor.

On the first benchmark set the ALIAS-based solvers greatly outperform the competitors. We also tested ALIAS on ASG tests with trivial SUPBS provided (instead of much smaller nontrivial one) and it yielded much worse results. Hence, the knowledge of a nontrivial SUPBS is a big advantage. On the second set the situation is more complex: there are instances which are solved by ALIAS-based solvers but not by the competitors and vice versa. Among crafted instances only *sgen3-n240-s78945233-sat* and *sgen1-sat-250-100* are satisfiable.

Table 1. Data on found backdoors. RE, BT, ST – time in seconds.

Instance	ALIAS-GLUCOSE				ALIAS-ABCD				ALIAS-RISS			
	$|B|$	RE	BT	ST	$|B|$	RE	BT	ST	$|B|$	RE	BT	ST
ASG-72-0	23	365	412	432	15	390	1713	210	20	308	330	103
MASG-72-0	19	347	348	7	19	330	443	44	19	380	585	38
MASG0-72-0	22	417	503	361	18	1167	2195	397	22	723	882	548
ASG-96-0	26	34548	13270	9177	26	36704	22605	42300	27	37661	12872	42583
mod4-2-9-u2	29	1.4e+6	86400	-	28	1.3e+5	19059	-	31	1.8e+5	24510	-
sgen1-sat-250	31	1.7e+5	15740	63220	33	3.2e+5	9379	-	29	1.8e+6	86400	-
sgen6-1200-5	26	1045	1503	14990	26	3684	3685	8092	24	2045	2050	5431
sgen6-1320-5	28	13974	4299	40934	28	16068	4623	45897	29	23388	3281	42833
sgen6-1440-5	29	39239	11896	70725	30	62516	18079	56168	31	144779	13092	-
sgen3-n240	31	1.0e+5	5505	37590	27	17504	7120	51026	30	73406	10341	31451
challenge-105	24	6121	6234	12780	25	3414	3422	22069	26	4840	4218	22033

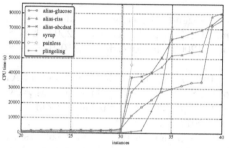
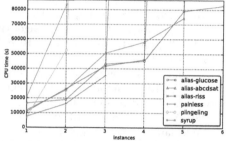

(a) ASG-72, ASG-96, MASG-72, MASG0-72   (b) Hard small crafted instances

Fig. 2. Comparison of 3 ALIAS-based solvers with the Top 3 solvers from the SAT competition 2017 parallel track

It should be noted, that strictly speaking, the blackbox optimization procedure employed in ALIAS does not guarantee that the found backdoors are really NOBS (see Sect. 2). It turned out, that for ASG-72, MASG-72 and MASG0-72 only few found backdoors are NOBS. A possible reason for this is that these instances are quite simple even for sequential solvers. Nevertheless, ALIAS-based solvers and their parallel competitors showed comparable results on them. Meanwhile, for all ASG-96 instances the found backdoors are NOBS, and here ALIAS-based solvers are clear winners. Note, that in Fig. 2a values from 31 to 40 on the x-axis correspond to the ASG-96 instances. In the second benchmark set, for *sgen6-1200-5-1* and *challenge-105* the found backdoors are indeed NOBS. For the remaining instances it was impractical to check it, because it would take up to several weeks per instance.

5   Conclusion

The experiments show that the approach to solving hard SAT instances based on sampling, while not a silver bullet, clearly has its applications. We believe that the presented ALIAS tool may be useful in the study of hard SAT instances and sometimes can shed the light on some aspects of their inner structure undetectable by state-of-the-art SAT solvers.

Acknowledgements. We thank anonymous reviewers for their insightful comments that made it possible to significantly improve the quality of presentation.

The research was funded by Russian Science Foundation (project No. 16-11-10046). Stepan Kochemazov was additionally supported by Council for Grants of the President of the Russian Federation (stipend no. SP-1829.2016.5).

References

1. Ansótegui, C., Sellmann, M., Tierney, K.: A gender-based genetic algorithm for the automatic configuration of algorithms. In: Gent, I.P. (ed.) CP 2009. LNCS, vol. 5732, pp. 142–157. Springer, Heidelberg (2009). https://doi.org/10.1007/978-3-642-04244-7_14

2. Audemard, G., Lagniez, J.-M., Simon, L.: Improving glucose for incremental SAT solving with assumptions: application to MUS extraction. In: Järvisalo, M., Van Gelder, A. (eds.) SAT 2013. LNCS, vol. 7962, pp. 309–317. Springer, Heidelberg (2013). https://doi.org/10.1007/978-3-642-39071-5_23

3. Audemard, G., Simon, L.: Lazy clause exchange policy for parallel SAT solvers. In: Sinz, C., Egly, U. (eds.) SAT 2014. LNCS, vol. 8561, pp. 197–205. Springer, Cham (2014). https://doi.org/10.1007/978-3-319-09284-3_15

4. Balyo, T., Biere, A., Iser, M., Sinz, C.: SAT race 2015. Artif. Intell. **241**, 45–65 (2016)

5. Biere, A.: CaDiCaL, Lingeling, Plingeling, Treengeling, YalSAT entering the SAT competition 2017. In: Proceedings of SAT Competition 2017, vol. B-2017-1, pp. 14–15 (2017)

6. Chen, J.: Improving abcdSAT by At-Least-One recently used clause management strategy. CoRR abs/1605.01622 (2016). http://arxiv.org/abs/1605.01622

7. Courtois, N.: Low-complexity key recovery attacks on GOST block cipher. Cryptologia **37**(1), 1–10 (2013)

8. Dowling, W.F., Gallier, J.H.: Linear-time algorithms for testing the satisfiability of propositional horn formulae. J. Log. Program. **1**(3), 267–284 (1984)

9. Eibach, T., Pilz, E., Völkel, G.: Attacking bivium using SAT solvers. In: Kleine Büning, H., Zhao, X. (eds.) SAT 2008. LNCS, vol. 4996, pp. 63–76. Springer, Heidelberg (2008). https://doi.org/10.1007/978-3-540-79719-7_7

10. Le Frioux, L., Baarir, S., Sopena, J., Kordon, F.: PaInleSS: a framework for parallel SAT solving. In: Gaspers, S., Walsh, T. (eds.) SAT 2017. LNCS, vol. 10491, pp. 233–250. Springer, Cham (2017). https://doi.org/10.1007/978-3-319-66263-3_15

11. Günther, C.G.: Alternating step generators controlled by de Bruijn sequences. In: Chaum, D., Price, W.L. (eds.) EUROCRYPT 1987. LNCS, vol. 304, pp. 5–14. Springer, Heidelberg (1988). https://doi.org/10.1007/3-540-39118-5_2

12. Hutter, F., Hoos, H.H., Leyton-Brown, K.: Sequential model-based optimization for general algorithm configuration. In: Proceedings of LION-5, pp. 507–523 (2011)

13. Hutter, F., Hoos, H.H., Leyton-Brown, K., Stützle, T.: ParamILS: an automatic algorithm configuration framework. JAIR **36**, 267–306 (2009)
14. Hutter, F., Lindauer, M., Balint, A., Bayless, S., Hoos, H., Leyton-Brown, K.: The configurable SAT solver challenge (CSSC). Artif. Intell. **243**, 1–25 (2017)
15. Kilby, P., Slaney, J.K., Thiébaux, S., Walsh, T.: Backbones and backdoors in satisfiability. AAAI **2005**, 1368–1373 (2005)
16. Manthey, N.: Towards next generation sequential and parallel SAT solvers. Constraints **20**(4), 504–505 (2015)
17. Metropolis, N., Ulam, S.: The Monte Carlo method. J. Amer. Stat. Assoc. **44**(247), 335–341 (1949)
18. Oh, C.: Between SAT and UNSAT: the fundamental difference in CDCL SAT. In: Heule, M., Weaver, S. (eds.) SAT 2015. LNCS, vol. 9340, pp. 307–323. Springer, Cham (2015). https://doi.org/10.1007/978-3-319-24318-4_23
19. Russell, S., Norvig, P.: Artificial Intelligence A Modern Approach, 3rd edn. Prentice Hall, Englewood Cliffs (2009)
20. Semenov, A., Zaikin, O.: Algorithm for finding partitionings of hard variants of Boolean satisfiability problem with application to inversion of some cryptographic functions. SpringerPlus **5**(1), 1–16 (2016)
21. Semenov, A., Zaikin, O., Otpuschennikov, I., Kochemazov, S., Ignatiev, A.: On cryptographic attacks using backdoors for SAT. In: Proceedings of the Thirty-Second AAAI Conference on Artificial Intelligence, New Orleans, Louisiana, USA, 2–7 February 2018 (2018). https://www.aaai.org/ocs/index.php/AAAI/AAAI18/paper/view/16855
22. Soos, M., Nohl, K., Castelluccia, C.: Extending SAT solvers to cryptographic problems. In: Kullmann, O. (ed.) SAT 2009. LNCS, vol. 5584, pp. 244–257. Springer, Heidelberg (2009). https://doi.org/10.1007/978-3-642-02777-2_24
23. Szeider, S.: Backdoor sets for DLL subsolvers. J. Autom. Reasoning **35**(1), 73–88 (2005)
24. Van Gelder, A., Spence, I.: Zero-one designs produce small hard SAT instances. In: Strichman, O., Szeider, S. (eds.) SAT 2010. LNCS, vol. 6175, pp. 388–397. Springer, Heidelberg (2010). https://doi.org/10.1007/978-3-642-14186-7_37
25. Wicik, R., Rachwalik, T.: Modified alternating step generators. IACR Cryptology ePrint Archive 2013, 728 (2013)
26. Williams, R., Gomes, C.P., Selman, B.: Backdoors to typical case complexity. In: IJCAI, pp. 1173–1178 (2003)
27. Zaikin, O., Kochemazov, S.: An improved SAT-based guess-and-determine attack on the alternating step generator. In: Nguyen, P., Zhou, J. (eds.) ISC 2017. LNCS, vol. 10599. Springer, Cham (2017). https://doi.org/10.1007/978-3-319-69659-1_2

PySAT: A Python Toolkit
for Prototyping with SAT Oracles

Alexey Ignatiev[1,2]($\boxtimes$), Antonio Morgado[1], and Joao Marques-Silva[1]

[1] LASIGE, Faculdade de Ciências, Universidade de Lisboa, Lisbon, Portugal
{aignatiev,ajmorgado,jpms}@ciencias.ulisboa.pt
[2] ISDCT SB RAS, Irkutsk, Russia

Abstract. Boolean satisfiability (SAT) solvers are at the core of efficient approaches for solving a vast multitude of practical problems. Moreover, albeit targeting an NP-complete problem, SAT solvers are increasingly used for tackling problems beyond NP. Despite the success of SAT in practice, modeling with SAT and more importantly implementing SAT-based problem solving solutions is often a difficult and error-prone task. This paper proposes the PySAT toolkit, which enables fast Python-based prototyping using SAT oracles and SAT-related technology. PySAT provides a simple API for working with a few state-of-the-art SAT oracles and also integrates a number of cardinality constraint encodings, all aiming at simplifying the prototyping process. Experimental results presented in the paper show that PySAT-based implementations can be as efficient as those written in a low-level language.

1 Introduction

When compared with Satisfiability Modulo Theories (SMT), Answer Set Programming (ASP) or Constraint Programming (CP), a well-known drawback of Propositional Logic (concretely, its satisfiability (SAT) problem) is the low level at which the problem constraints are represented and the low-level programmatic interface that must be used. These limitations hinder a wider adoption of SAT solvers, but in part they are also one reason for the observed performance gains that SAT-based solutions often enable. Moreover, it is generally perceived that SAT-based modeling is difficult and also error-prone. Clearly, the aforementioned alternatives, SMT, ASP and CP, also enable some sort of direct encoding to SAT, and then invoking a SAT solver, but often key aspects of the problem formulation are lost. Other approaches that directly encode problems into SAT have been considered, including NP-SPEC [10].

This paper describes PySAT, a toolkit that simplifies prototyping problem solvers with SAT solvers as oracles. Similarly to existing solutions for SMT, the prototyping language is Python, with a simple interface to an abstract SAT

This work was supported by FCT funding of post-doctoral grants SFRH/BPD/103609/2014, SFRH/BPD/120315/2016, FCT grant ABSOLV (028986/02/SAICT/2017), and LASIGE Research Unit, ref. UID/CEC/00408/2013.

O. Beyersdorff and C. M. Wintersteiger (Eds.): SAT 2018, LNCS 10929, pp. 428–437, 2018.
https://doi.org/10.1007/978-3-319-94144-8_26

solver that abstracts most details away, but also aims at compromising little in terms of performance. The paper illustrates the ease of modeling reasonably challenging problems, concretely MUS extraction, but also provides empirical evidence that the toolkit can achieve reasonably efficient implementations when compared with existing state-of-the-art tools. PySAT is open source, and it is publicly available on GitHub. Furthermore, PySAT is also readily installable as a Python package.

This paper is organized as follows. Basic definitions and notation are introduced in the next section. Section 3 describes the toolkit, its design and interface. Section 4 outlines the implementation of a deletion-based MUS extractor. Section 5 presents experimental results comparing a PySAT-based prototype of a MaxSAT algorithm compared to the state-of-the-art implementation. Section 6 overviews prior work related with PySAT. Finally, the paper concludes in Sect. 7.

2   Preliminaries

This section introduces the notation and definitions used throughout the paper. Standard propositional logic definitions apply (e.g. [9]). CNF formulas are defined over a set of propositional variables. A CNF formula F is a propositional formula represented as a conjunction of clauses, also interpreted as a set of clauses. A clause is a disjunction of literals, also interpreted as a set of literals. A literal is a variable or its complement. Throughout the paper, SAT solvers are viewed as oracles. Given a CNF formula F, a SAT oracle decides whether F is satisfiable, in which case it returns a satisfying assignment. A SAT oracle can also return an unsatisfiable core $U \subseteq F$, if F is unsatisfiable. Conflict-driven clause learning (CDCL) SAT solvers are summarized in [9].

CNF formulas are often used to model overconstrained problems, for example, the maximum satisfiability (MaxSAT) problem and the minimal unsatisfiable subset (MUS) extraction problem. In general, clauses in a CNF formula are characterized as hard, meaning that these must be satisfied, or soft, meaning that these are to be satisfied, if at all possible. A weight can be associated with each soft clause, and the goal of MaxSAT is to find an assignment to the propositional variables such that the hard clauses are satisfied, and the sum of the satisfied soft clauses is maximized. Algorithms for MaxSAT have been overviewed in [1,9,30]. Recent algorithms based on implicit hitting sets have been described in [4]. In the analysis of unsatisfiable CNF formulas, consider a given unsatisfiable CNF formula F. An MUS of F is a set of clauses $M \subseteq F$ which is both unsatisfiable and irreducible. The goal of the MUS extraction problem is to determine an MUS of a given unsatisfiable CNF formula.

3   PySAT Toolkit Description

This section describes the design and implementation of the PySAT toolkit as well as its capabilities. The toolkit aims at simplifying the work with SAT oracles. It is to be used for fast prototyping solvers and tools that target tackling practical problems and exploit the power of the state-of-the-art SAT technology.

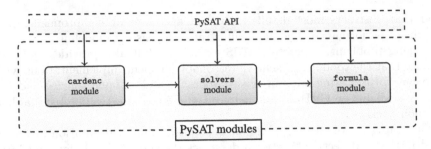

Fig. 1. PySAT toolkit and its modules.

The choice of the Python programming language was done having the following in mind. First, the language is easy-to-use and proved itself a great language for fast prototyping. This enables users to focus on implementing and improving an algorithm rather than struggling with its low-level details. Also, Python programs are typically easy to debug. Second, Python is required for installation and, thus, ready for use on almost any operating system of the POSIX family including plenty of Linux distributions, BSD and MacOS among a multitude of others. Third, the use of Python enables a user to tightly and easily integrate his/her tools with the existing technology that provides Python API, e.g. such renowned packages for scientific computing as *NumPy* [32], *SciPy* [45] and *matplotlib* [23], ILP solvers (*ILOG CPLEX* [19], *Gurobi* [18]), graph and network related libraries (e.g. *networkX* [31] and *graphviz* [17]), state-of-the-art machine learning, data analysis and mining toolkits including *scikit-learn* [44], *PyTorch* [42] and *pandas* [34], among a number of other libraries and toolkits, which find myriads of practical use cases.

3.1   PySAT Design

As the PySAT toolkit targets fast prototyping with SAT oracles, it provides interface to a number of state-of-the-art CDCL SAT solvers including MiniSat 2.2 [12,28] and its GitHub version [29], and also Glucose 3 and Glucose 4.1 [3,16]. Additionally, it also includes a couple of SAT solvers augmented with extra reasoning capabilities, namely Lingeling [7,8,21] strengthened with *Gaussian elimination* and *cardinality-based reasoning* and MiniCard 1.2 [27], which besides clauses can natively work with a special kind of constraints called *cardinality constraints* [9], i.e. constraints of the form $\sum_{i=1}^{n} l_i \circ k$ where $i, n, k \in \mathbb{N}$, each l_i is either a positive or a negative literal of a Boolean variable and $\circ \in \{<, \leq, =, \neq, \geq, >\}$. The module of the PySAT toolkit responsible for providing an API to the SAT solvers is called `solvers`.

In many cases, SAT-based problem solving requires to efficiently deal with cardinality constraints. MiniCard can handle them natively but other solvers need them to be encoded into a CNF formula. There are multiple ways to encode cardinality constraints into a set of clauses and most state-of-the-art cardinality encodings are supported by PySAT including *pairwise* and *bitwise* encodings [37], *sequential counters* [47], *sorting networks* [6], *cardinality networks* [2], *ladder* [15], *totalizer* [5], *modulo totalizer* [33], and *iterative totalizer* [22]. This functionality is provided by the second module of the toolkit, namely by the `cardenc` module.

Additionally, PySAT provides a user with an input/output interface for simplified reading and writing formulas in the DIMACS format including plain CNF, partial CNF and weighted partial CNF formulas (WCNF). This is covered by the third module of the toolkit, which is referred to as `formula`.

As a result, the toolkit has three modules, two of which are implemented as C/C++ extensions (i.e. `solvers` and `cardenc`) and one module (`formula`) is a pure Python module. The structure of the PySAT toolkit can be seen as shown in Fig. 1.

3.2 Provided Interface

Boolean variables in PySAT are represented as natural identifiers, e.g. numbers from $\mathbb{N}$. A positive (negative, resp.) literal in PySAT is assumed to be a positive (negative, resp.) integer, e.g. -1 represents a literal $\neg x_1$ while 5 represents a literal x_5. A clause is a list of literals, e.g. $[-3, -2]$ is a clause $(\neg x_3 \lor \neg x_2)$.

The `pysat.solvers` module provides an interface to SAT solvers directly as well as the abstract `Solver` class. Each SAT solver can be used in the MiniSat-like *incremental* fashion [13], i.e. with the use of *assumption literals*, and exhibits methods `add_clause()`, `solve()`, `get_model()`, and `get_core()`.[1] Using a solver incrementally can be helpful when multiple calls to the solver are needed in order to solve a problem, e.g. in MaxSAT solving or in MUS/MCS extraction and enumeration. In this case, a user needs to create a solver and feed it with a CNF formula only once while calling it multiple times with different sets of assumption literals. Observe that instead of using a solver incrementally, one can opt to create a new solver from scratch at every invocation.

```
>>> from pysat.solvers import Glucose3
>>> g = Glucose3()
>>> g.add_clause([-1, 2])
>>> g.add_clause([-2, 3])
>>>
>>> print g.solve()
True
>>> print g.get_model()
[-1, -2, 3]
>>> g.delete()
```

The `pysat.formula` module can be used for performing input/output operations when working with DIMACS formulas. This can be done using classes

---

[1] The method `get_model()` (`get_core()`, resp.) can be used if a prior SAT call was made and returned `True` (`False`, resp.). The `get_core()` method additionally assumes the SAT call was provided with a list of assumptions.


```
input  : Unsatisfiable CNF F
output : MUS M

M ← F

foreach cᵢ ∈ M do
    if not SAT(M \ {cᵢ}) then
        M ← M \ {cᵢ}
    end
end
return M
```

```
# oracle: SAT solver (initialized)
# as:      full set of assumptions
i = 0
while i < len(as):
    ts = as[:i] + as[(i + 1):]
    if oracle.solve(assumptions=ts):
        i += 1
    else:
        as = ts
return as
```

(a) Pseudo-code of deletion-based MUS extraction.   (b) Its possible implementation with PySAT.

Fig. 2. An example of a PySAT-based algorithm implementation.

CNF and WCNF of this module. CNF and WCNF objects have a list of clauses, which can be added to a SAT oracle directly. The cardenc module operates through the pysat.card interface and provides access to the atmost(), atleast(), and equals() methods (they return an object of class CNF) of the abstract class CardEnc, e.g. in the following way:

```
>>> from pysat.card import *
>>> am1 = CardEnc.atmost(lits=[1, -2, 3], encoding=EncType.pairwise)
>>> print am1.clauses
[[-1, 2], [-1, -3], [2, -3]]
>>>
>>> from pysat.solvers import Solver
>>> with Solver(name='m22', bootstrap_with=am1.clauses) as s:
...     if s.solve(assumptions=[1, 2, 3]) == False:
...         print s.get_core()
[3, 1]
```

3.3 Installation

The PySAT library can be installed from the PyPI repository [40] simply by executing the following command:

```
$ pip install python-sat
```

Alternatively, one can manually clone the library's GitHub repository [41] and compile all of its modules following the instructions of the README file.

4 Usage Example

Let us show how one can implement prototypes with the use of PySAT. Here we consider a simple deletion-based algorithm for MUS extraction [46]. Its main procedure is shown in Fig. 2a. The idea is to try to remove clauses of the formula one by one while checking the formula for unsatisfiability. Clauses that are necessary for preserving unsatisfiability comprise an MUS of the input formula and are reported as a result of the procedure. Figure 2b shows a possible PySAT-based implementation. The implementation assumes that a SAT oracle denoted by variable `oracle` is already initialized, and contains all clauses of the input formula $\mathcal{F}$. Another assumption is that each clause $c_i \in \mathcal{F}$ is augmented with a selector literal $\neg s_i$, i.e. considering clause $c_i \vee \neg s_i$. This facilitates simple activation/deactivation of clause c_i depending on the value of variable s_i. Finally, a list of assumptions `as` is assumed to contain all clause selectors, i.e. `as` $= \{s_i \,|\, c_i \in \mathcal{F}\}$. Observe that the implementation of the MUS extraction algorithm is as simple as its pseudo-code. This simplicity is intrinsic to Python programs, and enables users to think on algorithms rather than implementation details.

5   Experimenting with MaxSAT

One of the benefits provided by the PySAT toolkit is that it enables users to prototype quickly and sacrifice just a little in terms of performance. In order to confirm this claim in practice, we developed a simple (non-optimized) PySAT-based implementation of the Fu&Malik algorithm [14] for MaxSAT. The implementation is referred to as *fm.py*. The idea is to compare this implementation to the state-of-the-art MaxSAT solver *MiFuMaX* [20], which can be seen as a well thought and efficient implementation of the Fu&Malik algorithm written in C++ and available online [26]. MiFuMaX has proven its efficiency by winning the *unweighted* category in the MAX-SAT evaluation 2013 [24].

For the comparison, we chose all (i.e. unweighted and weighted) benchmarks from MaxSAT Evaluation 2017 [25]. The benchmarks suite contains 880 unweighted and 767 weighted MaxSAT instances. The experiments were performed in Ubuntu Linux on an Intel Xeon E5-2630 2.60 GHz processor with 64 GByte of memory. The time limit was set to 1800 s and the memory limit to 10 GByte for each individual process to run.

The cactus plot depicting the performance of MiFuMaX and fm.py is shown in Fig. 3. As to be expected, our simple implementation of the Fu&Malik algorithm is outperformed by MiFuMaX. However, one could expect a larger performance gap between the two implementations given the optimizations used in MiFuMaX. Observe that MiFuMaX solves 384 unweighted and 226 weighted instances while fm.py can solve 376 and 219 unweighted and weighted formulas, respectively. The performance of the two implemetations is detailed in Fig. 4. In both cases (unweighted and weighted benchmarks) MiFuMaX tends to be at most a few times faster than fm.py. Also note that even though surprising, there are instances, which are solved by fm.py more efficiently than by MiFuMaX. Overall, the performance of fm.py demonstrates that a PySAT-based implementation of a problem solving algorithm can compete with a low-level implementation of the same algorithm, provided that most of the computing work is done by the underlying SAT solver, which is often the case in practice.

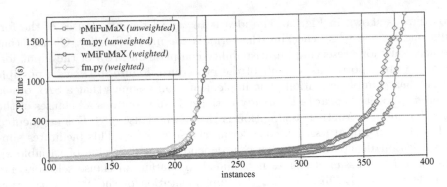

Fig. 3. Performance of fm.py and MiFuMaX on the MSE17 benchmarks.

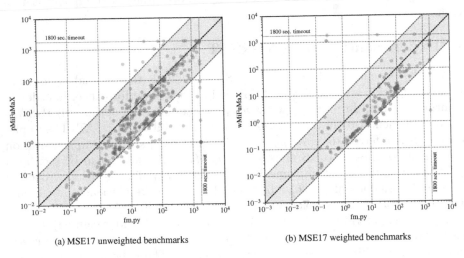

(a) MSE17 unweighted benchmarks   (b) MSE17 weighted benchmarks

Fig. 4. Detailed comparison of fm.py and MiFuMaX.

6   Related Work

A number of Python APIs for specific SAT solvers have been developed in the recent past. These include *PyMiniSolvers* [39] providing an interface to MiniSat and MiniCard, *satispy* [43] providing an API for MiniSat and lingeling, *pylgl* [38] for working with lingeling, and the Python API for CryptoMiniSat [11,48,49]. Compared to these solutions, PySAT offers a wider range of SAT solvers accessed through a unified interface, more functionality provided (e.g. unsatisfiable core and proof extraction), as well as a number of encodings of cardinality constraints. Cardinality constraints (as well as pseudo-Boolean constraints) can be alternatively manipulated using encodings provided by some other libraries. One such example is the *PBLib* library [35,36]. However, PBLib currently does not expose a Python API.

7   Conclusions

Despite the remarkable progress observed in SAT solvers for over two decades, in many settings the option of choice is often not a SAT solver, even when this might actually be the ideal solution. One reason for overlooking SAT solvers is the apparent difficulty of modeling with SAT, and of implementing basic prototypes. This paper describes PySAT, a Python toolkit that enables fast prototyping with SAT solvers. The Python interface offers (incremental) access to a blackbox exposing the basic interface of a SAT solver, but which compromises little in terms of performance. The toolkit also offers access to a number of often-used implementations of cardinality constraints. A simple implementation of a MaxSAT solver shows performance comparable with a state-of-the-art C++ implementation. The PySAT tookit is publicly available as open source from GitHub, and also as a Python package on most POSIX-compliant OSes. It is expected that the community will contribute to improving the toolkit further, with additional features, but also with proposals for improvements. Several extensions are planned. These include the integration of more SAT solvers (e.g. CryptoMiniSat and other MiniSat- and Glucose-based solvers), lower level access to the SAT solver's parameters and policies when necessary (e.g. setting preferred "polarities" of the variables), high-level support for arbitrary Boolean formulas (e.g. by Tseitin-encoding them [50] internally), and encodings of pseudo-Boolean constraints.

References

1. Ansótegui, C., Bonet, M.L., Levy, J.: SAT-based MaxSAT algorithms. Artif. Intell. **196**, 77–105 (2013)
2. Asín, R., Nieuwenhuis, R., Oliveras, A., Rodríguez-Carbonell, E.: Cardinality networks and their applications. In: Kullmann, O. (ed.) SAT 2009. LNCS, vol. 5584, pp. 167–180. Springer, Heidelberg (2009). https://doi.org/10.1007/978-3-642-02777-2_18
3. Audemard, G., Lagniez, J.-M., Simon, L.: Improving glucose for incremental SAT solving with assumptions: application to MUS extraction. In: Järvisalo, M., Van Gelder, A. (eds.) SAT 2013. LNCS, vol. 7962, pp. 309–317. Springer, Heidelberg (2013). https://doi.org/10.1007/978-3-642-39071-5_23
4. Bacchus, F., Hyttinen, A., Järvisalo, M., Saikko, P.: Reduced cost fixing in MaxSAT. In: Beck, J.C. (ed.) CP 2017. LNCS, vol. 10416, pp. 641–651. Springer, Cham (2017). https://doi.org/10.1007/978-3-319-66158-2_41
5. Bailleux, O., Boufkhad, Y.: Efficient CNF encoding of Boolean cardinality constraints. In: Rossi, F. (ed.) CP 2003. LNCS, vol. 2833, pp. 108–122. Springer, Heidelberg (2003). https://doi.org/10.1007/978-3-540-45193-8_8
6. Batcher, K.E.: Sorting networks and their applications. In: AFIPS Conference, pp. 307–314 (1968)
7. Biere, A.: Lingeling, plingeling and treengeling entering the SAT competition 2013. In: Balint, A., Belov, A., Heule, M., Järvisalo, M. (eds.) Proceedings of SAT Competition 2013, volume B-2013-1 of Department of Computer Science Series of Publications B, pp. 51–52. University of Helsinki (2013)

8. Biere, A.: Lingeling essentials, a tutorial on design and implementation aspects of the SAT solver lingeling. In: Pragmatics of SAT workshop, p. 88 (2014)

9. Biere, A., Heule, M., van Maaren, H., Walsh, T. (eds.): Handbook of Satisfiability. Frontiers in Artificial Intelligence and Applications, vol. 185. IOS Press, Amsterdam (2009)

10. Cadoli, M., Schaerf, A.: Compiling problem specifications into SAT. Artif. Intell. **162**(1–2), 89–120 (2005)

11. CryptoMiniSat. https://github.com/msoos/cryptominisat/

12. Eén, N., Sörensson, N.: An extensible SAT-solver. In: Giunchiglia, E., Tacchella, A. (eds.) SAT 2003. LNCS, vol. 2919, pp. 502–518. Springer, Heidelberg (2004). https://doi.org/10.1007/978-3-540-24605-3_37

13. Eén, N., Sörensson, N.: Temporal induction by incremental SAT solving. Electron. Notes Theor. Comput. Sci. **89**(4), 543–560 (2003)

14. Fu, Z., Malik, S.: On solving the partial MAX-SAT problem. In: Biere, A., Gomes, C.P. (eds.) SAT 2006. LNCS, vol. 4121, pp. 252–265. Springer, Heidelberg (2006). https://doi.org/10.1007/11814948_25

15. Gent, I., Nightingale, P.: A new encoding of alldifferent into SAT. In: International Workshop on Modelling and Reformulating Constraint Satisfaction Problem, pp. 95–110 (2004)

16. Glucose 3 and Glucose 4.1. http://www.labri.fr/perso/lsimon/glucose/

17. graphviz. https://www.graphviz.org/

18. Gurobi. http://www.gurobi.com/

19. IBM ILOG: CPLEX optimizer 12.7.0 (2016). http://www-01.ibm.com/software/commerce/optimization/cplex-optimizer

20. Janota, M.: MiFuMax – a literate MaxSAT solver. JSAT **9**, 83–88 (2015)

21. Lingeling bbc-9230380-160707. http://fmv.jku.at/lingeling/

22. Martins, R., Joshi, S., Manquinho, V., Lynce, I.: Incremental cardinality constraints for MaxSAT. In: O'Sullivan, B. (ed.) CP 2014. LNCS, vol. 8656, pp. 531–548. Springer, Cham (2014). https://doi.org/10.1007/978-3-319-10428-7_39

23. Matplotlib. https://matplotlib.org/

24. Eighth Max-SAT Evaluation. http://www.maxsat.udl.cat/13/

25. MaxSAT Evaluation 2017. http://mse17.cs.helsinki.fi/

26. MiFuMax. http://sat.inesc-id.pt/~mikolas/

27. MiniCard 1.2. https://github.com/liffiton/minicard/

28. MiniSat 2.2. http://minisat.se/MiniSat.html

29. MiniSat GitHub. https://github.com/niklasso/minisat/

30. Morgado, A., Heras, F., Liffiton, M.H., Planes, J., Marques-Silva, J.: Iterative and core-guided MaxSAT solving: a survey and assessment. Constraints **18**(4), 478–534 (2013)

31. networkx. https://networkx.github.io/

32. NumPy. http://www.numpy.org/

33. Ogawa, T., Liu, Y., Hasegawa, R., Koshimura, M., Fujita, H.: Modulo based CNF encoding of cardinality constraints and its application to MaxSAT solvers. In: ICTAI, pp. 9–17 (2013)

34. pandas. https://pandas.pydata.org/

35. PBLib. http://tools.computational-logic.org/content/pblib.php

36. Philipp, T., Steinke, P.: PBLib – a library for encoding pseudo-boolean constraints into CNF. In: Heule, M., Weaver, S. (eds.) SAT 2015. LNCS, vol. 9340, pp. 9–16. Springer, Cham (2015). https://doi.org/10.1007/978-3-319-24318-4_2

37. Prestwich, S.D.: CNF encodings. In: Handbook of Satisfiability, pp. 75–97 (2009)

38. pylgl. https://github.com/abfeldman/pylgl/
39. PyMiniSolvers. https://github.com/liffiton/PyMiniSolvers/
40. PyPI. https://pypi.python.org/
41. PySAT. https://pysathq.github.io/
42. Pytorch. http://pytorch.org/
43. satispy. https://github.com/netom/satispy/
44. scikit-learn. http://scikit-learn.org/
45. SciPy. https://scipy.org/
46. Silva, J.M.: Minimal unsatisfiability: models, algorithms and applications (invited paper). In: ISMVL, pp. 9–14 (2010)
47. Sinz, C.: Towards an optimal CNF encoding of Boolean cardinality constraints. In: van Beek, P. (ed.) CP 2005. LNCS, vol. 3709, pp. 827–831. Springer, Heidelberg (2005). https://doi.org/10.1007/11564751_73
48. Soos, M.: Enhanced Gaussian elimination in DPll-based SAT solvers. In: POS@SAT, pp. 2–14 (2010)
49. Soos, M., Nohl, K., Castelluccia, C.: Extending SAT solvers to cryptographic problems. In: Kullmann, O. (ed.) SAT 2009. LNCS, vol. 5584, pp. 244–257. Springer, Heidelberg (2009). https://doi.org/10.1007/978-3-642-02777-2_24
50. Tseitin, G.S.: On the complexity of derivations in the propositional calculus. Stud. Math. Math. Log. Part II, 115–125 (1968)

Constrained Image Generation Using Binarized Neural Networks with Decision Procedures

Svyatoslav Korneev[1], Nina Narodytska[2(✉)], Luca Pulina[3],
Armando Tacchella[4], Nikolaj Bjorner[5], and Mooly Sagiv[2,6]

[1] Department of Energy Resources Engineering, Stanford, USA
skorneev@stanford.edu
[2] VMware Research, Palo Alto, USA
nnarodytska@vmware.com, msagiv@acm.org
[3] Chemistry and Pharmacy Department, University of Sassari, Sassari, Italy
lpulina@uniss.it
[4] DIBRIS, University of Genoa, Genoa, Italy
armando.tacchella@unige.it
[5] Microsoft Research, Redmond, USA
nbjorner@microsoft.com
[6] Tel Aviv University, Tel Aviv, Israel

Abstract. We consider the problem of binary image generation with given properties. This problem arises in a number of practical applications, including generation of artificial porous medium for an electrode of lithium-ion batteries, for composed materials, etc. A generated image represents a porous medium and, as such, it is subject to two sets of constraints: topological constraints on the structure and process constraints on the physical process over this structure. To perform image generation we need to define a mapping from a porous medium to its physical process parameters. For a given geometry of a porous medium, this mapping can be done by solving a partial differential equation (PDE). However, embedding a PDE solver into the search procedure is computationally expensive. We use a binarized neural network to approximate a PDE solver. This allows us to encode the entire problem as a logical formula. Our main contribution is that, for the first time, we show that this problem can be tackled using decision procedures. Our experiments show that our model is able to produce random constrained images that satisfy both topological and process constraints.

1 Introduction

We consider the problem of constrained image generation of a porous medium with given properties. Porus media occur, e.g., in lithium-ion batteries and composed materials [1,2]; the problem of generating porus media with a given set of properties is relevant in practical applications of material design [3–5]. Artificial porous media are useful during the manufacturing process as they allow

© Springer International Publishing AG, part of Springer Nature 2018
O. Beyersdorff and C. M. Wintersteiger (Eds.): SAT 2018, LNCS 10929, pp. 438–449, 2018.
https://doi.org/10.1007/978-3-319-94144-8_27

the designer to synthesize new materials with predefined properties. For example, generated images can be used in designing a new porous medium for an electrode of lithium-ion batteries. It is well-known that ions macro-scale transport and reactions rates are sensitive to the topological properties of the porous medium of the electrode. Therefore, manufacturing the porous electrode with given properties allows improving the battery performance [1].

Images of porous media[1] are black and white images that represent an abstraction of the physical structure. Solid parts (or so called grains) are encoded as a set of connected black pixels; a void area is encoded a set of connected white pixels. There are two important groups of restrictions that images of a porous medium have to satisfy. The first group constitutes a set of "geometric" constraints that come from the problem domain and control the total surface area of grains. For example, an image contains two isolated solid parts. Figure 1(a) shows examples of 16×16 images from our datasets with two (the top row) and three (the bottom row) grains. The second set of restrictions comes from the physical process that is defined for the corresponding porous medium. In this paper, we consider the macro-scale transportation process that can be described by a set of dispersion coefficients depending on the transportation direction. For example, we might want to generate images that have two grains such that the dispersion coefficient along the x-axis is between 0.5 and 0.6. The dispersion coefficient is defined for the given geometry of a porous medium. It can be obtained as a numerical solution of the diffusion Partial Differential Equation (PDE). We refer to these restrictions on the parameters of the physical process as process constraints.

Fig. 1. (a) Examples of images from train sets with two and three grains; (b) Examples of images generated by a GAN on the dataset with two grains. Examples of generated images with (c) $d \in [40, 50)$, (d) $d \in [60, 70)$, and (e) $d \in [90, 100]$.

The state of the art approach to generating synthetic images is to use generative adversarial networks (GANs) [6]. However, GANs are not able learn geometric, three-dimensional perspective, and counting constraints which is a known issue with this approach [7,8]. Our experiments with GAN-generated images also reveal this problem. There are no methods that allow embedding of declarative constraints in the image generation procedure at the moment.

In this work we show that the image generation problem can be solved using decision procedures for porous media. We show that both geometric and process constraints can be encoded as a logical formula. Geometric constraints are

---

[1] Specifically, we are looking at a transitionally periodic "unit cell" of porous medium assuming that porous medium has a periodic structure [5].

encoded as a set of linear constraints. To encode process constraints, we first approximate the diffusion PDE solver with a Neural Network (NN) [9,10]. We use a special class of NN, called BNN, as these networks can be encoded as logical formulas. Process constraints are encoded as restrictions on outputs of the network. This provides us with an encoding of the image generation problem as a single logical formula. The contributions of this paper can be summarized as follows: (i) We show that constrained image generation can be encoded as a logical formula and tackled using decision procedures. (ii) We experimentally investigate a GAN-based approach to constrained image generation and analyse their advantages and disadvantages compared to the constraint-based approach. (iii) We demonstrate that our constraint-based approach is capable of generating random images that have given properties, i.e., satisfy process constraints.

2   Problem Description

We describe a constrained image generation problem. We denote $I \in \{0,1\}^{t \times t}$ an image that encodes a porous medium and $d \in \mathbb{Z}^m$ a vector of parameters of the physical process defined for this porous material. We use an image and a porous medium interchangeably to refer to I. We assume that there is a mapping function M that maps an image I to the corresponding parameters vector d, $\mathrm{M}: I \to \mathbb{Z}^m$. We denote as $C_g(I)$ the geometric constraints on the structure of the image I and as $C_p(d)$ the process constraints on the vector of parameters d. Given a set of geometric and process constraints and a mapping function M, we need to generate a random image I that satisfies C_g and C_p. Next we overview geometric and process constraints and discuss the mapping function.

The geometric constraints C_g define a topological structure of the image. For example, they can ensure that a given number of grains is present on an image and these grains do not overlap. Another type of constraints focuses on a single grain. They can restrict the shape of a grain, e.g., a convex grain, its size or position on the image. The third type of constraints are boundary constraints that ensure that the boundary of the image must be in a void area. Process constraints define restrictions on the vector of parameters. For example, we might want to generate images with $d_i^j \in [a_j, b_j]$, $j = 1, \ldots, m$.

Next we consider a mapping function M. A standard way to define M is by solving a system of partial differential equations. However, solving these PDEs is a computationally demanding task and, more importantly, it is not clear how to 'reverse' them to generate images with given properties. Hence, we take an alternative approach of approximating a PDE solver using a neural network [9, 10]. To train such an approximation, we build a training set of pairs (I_i, d_i), $i = 1, \ldots, n$, where I_i is an input image of a porous medium and d_i, obtained by solving the PDE given I, is its label. In this work, we use a special class of deep neural networks—Binarized Neural Networks (BNN) that admit an exact encoding into a logical formula. We assume that M is represented as a BNN and is given as part of input. We will elaborate on the training procedure in Sect. 5.

3   The Generative Neural Network Approach

One approach to tackle the constrained image generation problem is to use generative adversarial networks (GANs) [6,11]. GANs are successfully used to produce samples of realistic images for commonly used datasets, e.g. interior design, clothes, animals, etc. A GAN can be described as a game between the image generator that produces synthetic (fake) images and a discriminator that distinguishes between fake and real images. The cost function is defined in such a way that the generator and the discriminator aim to maximize and minimize this cost function, respectively, turning the learning process into a minimax game between these two players. Each payer is usually represented as a neural network. To apply GANs to our problem, we take a set of images $\{I_1, \ldots, I_n\}$ and pass them to the GAN. These images are samples of real images for the GAN. After the training procedure is completed, the generator network produces artificial images that look like real images. The main advantage of GANs is that it is a generic approach that can be applied to any type of images and can handle complex concepts, like animals, scenes, etc.[2] However, the main issue with this approach is that there is no way to explicitly pass declarative constraints into the training procedure. One might expect that GANs are able to learn these constraints from the set of examples. However, this is not the case at the moment, e.g., GANs cannot capture counting constraints, like four legs, two eyes, etc. [7]. Figure 1 shows examples of images that GAN produces on a dataset with two grains per image. As can be seen from these examples, GAN produces images with an arbitrary number of grains between 1 and 5 per image. In some simple cases, it is easy to filter wrong images. If we have more sophisticated constraints like convexity or size of grains, then most images will be invalid. On top of this, to take into account process constraints, we need additional restrictions on the training procedure. Overall, it is an interesting research question how to extend the GAN training procedure with physical constraints, which is beyond the scope of this paper [13]. Next we consider our approach to the image generation problem.

4   The Constraint-Based Approach

The main idea behind our approach is to encode the image generation problem as a logical formula. To do so, we need to encode all problem constraints and the mapping between an image and its label as a set of constraints. We start with constraints that encode an approximate PDE solver. We denote $[N]$ a range of numbers from 1 to N.

4.1   Approximation of a PDE Solver

One way to approximate a diffusion PDE solver is to use a neural network [9,10]. A neural network is trained on a set of binary images I_i and their labels d_i, $i = 1, \ldots, n$. During the training procedure, the networks takes an image I_i as

---

[2] GANs exhibit well-known issues with poor convergence that we did not observe as our dataset is quite simple [12].

an input and outputs its estimate of the parameter vector $\hat{d}_i$. As we have ground truth parameters d_i for each image, we can use the mean square error or absolute value error as a cost function to perform optimization [14]. In this work, we take the same approach. However, we use a special type of networks: Binarized Neural Networks (BNN). BNN is a feedforward network where weights and activations are binary [15]. It was shown in [14,16] that BNNs allow exact encoding as logical formulas, namely, they can be encoded a set of reified linear constraints over binary variables. We use BNNs as they have a relatively simple structure and decision procedures scale to reason about small and medium size networks of this type. In theory, we can use any exact encoding to represent a more general network, e.g., MILP encodings that are used to check robustness properties of neural networks [17,18]. However, the scalability of decision procedures is the main limitation in the use of more general networks. We use the ILP encoding as in [14] with a minor modification of the last layer as we have numeric outputs instead of categorical outputs. We denote ENCBNN(I, d) a logical formula that encodes BNN using reified linear constraints over Boolean variables (Sect. 4, ILP encoding [14]).

$$\text{ENCBNN}(I, d) = \left(\bigwedge_{k=1}^{q-1} \text{ENCBLK}_k(\mathbf{x}_k, \mathbf{x}_{k+1}) \right) \wedge \text{ENCO}(\mathbf{x}_q, d), \qquad (1)$$

where $x_1 = I$ is an input of the network, q is the number of layers in the network and d is the output of the network. ENCBLK$_k$ denotes encoding of the intermediate layer and ENCO denotes encoding of the last layer that maps the output of the qth layer (x_q) to the dispersion value d.

4.2   Geometric and Process Constraints

Geometric constraints can be roughly divided into three types. The first type of constraints defines the high-level structure of the image. The high-level structure of our images is defined by the number of grains present in the image. Let w be the number of grains per image. We define a grid of size $t \times t$. Figure 2(a) shows an example of a grid of size 4×4. We refer to a cell (i, j) on the grid as a pixel as this grid encodes an image of size $t \times t$. Next we define the neighbor relation on the grid. We say that a cell (h, g) is a neighbour of (i, j) if these cells share a side. For example, $(2, 3)$ is a neighbour of $(2, 4)$ as the right side of $(2, 3)$ is shared with $(2, 4)$. Let NB(i, j) be the set of neighbors of (i, j) on the gird. For example, NB$(2, 3) = \{(1, 3), (2, 2), (2, 4), (3, 3)\}$.

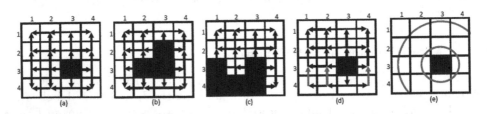

Fig. 2. Illustrative examples of additional structures used by constraint-based model.

Variables. For each cell we introduce a Boolean variable $c_{i,j,r}$, $i,j \in [t]$, $r \in [w+1]$. $c_{i,j,r} = 1$ iff the cell (i,j) belongs to the rth grain, $r = 1,\ldots,w$. Similarly, $c_{i,j,w+1} = 1$ iff the cell (i,j) represents a void area.

Each Cell Is Either a Black or White Pixel. We enforce that each cell contains either a grain or a void area.

$$\sum_{r=1}^{w+1} c_{i,j,r} = 1 \qquad j,i \in [t] \tag{2}$$

Grains Do Not Overlap. Two cells that belong to different grains cannot be neighbours.

$$c_{i,j,r} \rightarrow \neg c_{h,g,r'} \qquad (h,g) \in \mathrm{NB}(i,j), r' \in [w] \setminus \{r\} \tag{3}$$

Grains Are Connected Areas. We enforce connectivity constraints for each grain. By connectivity we mean that there is a path between two cells of the same grain using only cells that belong to this grain. Unfortunately, enforcing connectivity constraints is very expensive. Encoding the path constraint results in a prohibitively large encoding. To deal with this explosion, we restrict the space of possible grain shapes. First, we assume that we know the position of one pixel of this grain that we pick randomly. Let $s_r = (i,j)$ be a random cell, $r \in [w]$. Then we implicitly build a directed acyclic graph (DAG) G starting from this cell s_r that covers the entire grid. Each cell of a grid is a node in this graph. The node that corresponds to the cell s_r does not have incoming arcs. There are multiple ways to build a G from s_r. Figures 2(a) and (d) show two possible ways to build a G that covers a grid starting from cell $(3,3)$. We enforce that cell i,j belongs to the rth grain if its center s_r is equal to (i,j).

$$(c_{i,j,r}), \qquad s_r = (i,j), r \in [w+1]. \tag{4}$$

Next we define a parent relation in G. Let $\mathrm{PR}_G(i,j)$ be the set of parents of cell (i,j) in G. For example, $\mathrm{PR}_G(2,2) = \{(2,3),(3,2)\}$ in our example on Fig. 2(a). Given a DAG G, we can easily enforce connectivity relation w.r.t. G. The following constraint ensures that a cell (i,j) cannot belong to the rth grain if none of its parents in G belongs to the same grain.

$$\left(\textstyle\bigwedge_{(h,g)\in \mathrm{PR}_G(i,j)} \neg c_{h,g,r}\right) \rightarrow \neg c_{i,j,r}, \qquad j,i \in [t], r \in [w+1]. \tag{5}$$

Note that by enforcing connectivity constraints on the void area, we make sure that grains do not contain isolated void areas inside them.

Given a DAG G, we can generate grains of multiple shapes. For example, Fig. 2(b) shows one possible grain. However, we also lose some valid shapes that are ruled out by the choice of graph G. For example, Fig. 2(c) gives an example of a shape that is not possible to build using G in Fig. 2(a). Note that there is no path from the center $(3,3)$ of G to the cell $(3,1)$ that does not visit $(3,2)$. However, if we select a different random DAG G', e.g., Fig. 2(d), then this shape is one of the possible shapes for G'. In general, we can pick s_r and DAG randomly, so we generate a variety of shapes.

Compactness of a Grain. The second set of constraints is about restrictions on a single grain. The compactness constraint is a form of convexity constraint. We want to ensure that any two boundary points of a grain are close to each other. The reason for this constraint is that grains are unlikely to have a long snake-like appearance as solid particles tend to group together. Sometimes, we need to enforce the convexity constraint, which is an extreme case of compactness. To enforce this constraint, we again trade-off the variety of shapes and the size of the encoding. Now we assume that s_r is the center of the grain. Then we build virtual circles around this center that cover the grid. Figure 2(e) shows examples of such circles. Let $C_r(i,j) = \{C_r^1, \ldots, C_r^q\}$ be a set of circles that are built with the cell s_r as a center. The following constraint enforces that a cell that belongs to the circle C_r^v can be in the rth grain only if all cells from the inner circle C_r^{v-s} belong to the rth grain, where s is a parameter.

$$\bigvee_{c_{h,g,r} \in C_r^{v-s}} \neg c_{h,g,r} \rightarrow \neg c_{i,j,r} \qquad c_{i,j,r} \in C_r^v, v \in [q], r \in [w] \qquad (6)$$

Note that if $s = 1$ then we generate convex grains. In this case, every pixel from C_r^v has to belong to the rth grain before we can add a pixel from the circle C_r^{v+1} to this grain.

Boundary Constraints. We also have a technical constraint that all cells on the boundary of the grid must be void pixels. They are required to define boundary conditions for PDEs on generated images.

$$(c_{i,j,w+1}) \qquad j = t \vee i = t \qquad (7)$$

Connecting with BNN. We need to connect variables $c_{i,j,r}$ with the inputs of the network. We recall that an input image is a black and white image, where black pixels correspond to solid parts. Hence, if a cell belongs to a grain, i.e. $c_{i,j,r}$ is true and $r \neq w+1$, then it maps to a black pixel. Otherwise, it maps to a white pixel.

$$\begin{aligned} c_{i,j,r} &\rightarrow I_{i,j} = 1 & j,i \in [t], r \in [w], \\ c_{i,j,w+1} &\rightarrow I_{i,j} = 0 & j,i \in [t]. \end{aligned} \qquad (8)$$

Process Constraints. Process constraints are enforced on the output of the network. Given ranges $[a_i, b_i]$, $i \in [m]$ we have:

$$a_i \leq d_i \leq b_i \qquad i \in [m] \qquad (9)$$

Summary. To solve the constrained random image generation problem, we solve the conjunctions of constraint (1)–(9). Randomness comes from the random seed that is passed to the solver, a random choice of s_r and G.

5   Experiments

We conduct a set of experiments with our constraint based approach. We ran our experiments on Intel(R) Xeon(R) 3.30 GHz. We use the timeout of 600 s in all runs.

Training Procedure. For the training sets, we consider the synthetic random images of the unit cell of a periodic porous medium. For each generated image, we solve the partial differential equation to define the rate of the transport process, d. We use synthetic images because high-quality images of the natural porous medium are not available. The usage of synthetic images is a standard practice when data supply is strictly limited. We generated two datasets, D_2 with 10 K images and D_3 with 5 K images. Each image in D_2 contains two grains and each image in D_3 contains three grains. All images are black and white images of size 16 by 16. These images were labeled with dispersion coefficients along the x-axis which is a number between 0.4 and 1. We performed quantization on the dispersion coefficient value to map d into an interval of integers between 40 and 100. Intuitively, the larger the volume of the solid grains in the domain the lower the dispersion value, since the grain creates an obstacle for the transport. In the datasets, we don't use images with large enough grains to have the dispersion rate lower than 0.4 (or re-scaled to 40 for BNN), since the shape diversity of large grains is low. We use mean absolute error (MAE) to train BNN. BNN consists of three blocks with 100 neurons per inner layers and one output. The MAE is 4.2 for D_2 and 5.1 for D_3. We lose accuracy compared to non-binarized networks, e.g., MAE for the same non-binarized network is 2.5 for D_2. However, BNNs are much easier to reason about, so we work with this subclass of networks.

Image Generation. We use CPLEX and the SMT solver Z3 to solve instances produced by constraints (1)–(9). In principle, other solvers could be evaluated on these instances. The best mode for Z3 was to use an SMT core based on CDCL and a theory solver for *nested* Pseudo-Boolean and cardinality constraints. We noted that bit-blasting into sorting circuits did not scale, and Z3's theory of linear integer arithmetic was also inadequate. We considered six process constraints for d, namely, $d \in [a, b]$, $[a, b] \in \{[40, 50), \ldots, [90, 100]\}$. For each interval $[a, b]$, we generate 100 random constrained problems. The randomization comes from a random seed that is passed to the solver, the position of centers of each grain and the parameter s in the constraint (6). We used the same DAG G construction as in Fig. 2(a) in all problems.

Table 1. The number of solved instances in each interval $[a, b]$.

Solver	D_2						D_3					
	[40,50)	[50,60)	[60,70)	[70,80)	[80,90)	[90,100]	[40,50)	[50,60)	[60,70)	[70,80)	[80,90)	[90,100]
CPLEX	100	99	99	98	100	41	100	100	96	99	100	84
Z3	98	89	81	74	56	12	100	97	97	97	96	54

Table 1 shows summary of our results for CPLEX and Z3 solvers. As can be seen from this table, these instances are relatively easy for the CPLEX solver. It can solve most of them within the given timeout. The average time for D_2 is 25 s and for D_3 is 12 s with CPLEX. Z3 handles most benchmarks, but we observed it gets stuck on examples that are very easy for CPLEX, e.g.

the interval $[80, 90)$ for D_2. We hypothesize that this is due to how watch literals are tracked in a very general way on nested cardinality constraints (Z3 maintains a predicate for each nested PB constraint and refreshes the watch list whenever the predicate changes assignment), when one could instead exploit the limited way that CPLEX allows conditional constraints. The average time for the dataset D_2 is 94 s and for the dataset D_3 is 64 s with Z3.

Figures 1(c)–(e) show examples of generated images for ranges $[40, 50)$, $[60, 70)$ and $[90, 100]$ for D_2 (the top row) and D_3 (the bottom row). For the process we consider, as the value of the dispersion coefficient grows, the black area should decrease as there should be fewer grain obstacles for a flow to go through the porous medium. Indeed, images in Figs. 1(c)–(e) follow this pattern, i.e. the black area on images with $d \in [40, 50)$ is significantly larger than on images with $d \in [90, 100]$. Moreover, by construction, they satisfy geometric constraints that GANs cannot

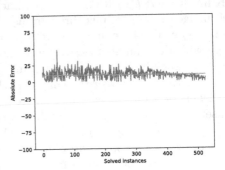

Fig. 3. The absolute error between d and its true value.

handle. For each image we generated, we run a PDE solver to compute the true value of the dispersion coefficient on this image. Then we compute the absolute error between the value of d that our model computes and the true value of the coefficient. Figure 3 shows absolute errors for all benchmarks that were solved by CPLEX. We ordered solved benchmarks by dispersion coefficient values, breaking ties arbitrarily. First, this figure shows that our model generates images with given properties. The mean absolute error is about 10 on these instances. Taking into account that BNN has MAE of 4.2 on D_2, MAE of 10 on new generated instances is a reasonable result. Ideally, we would like MAE to be zero. However, this error depends purely on the BNN we used. To reduce this error, we need to improve the accuracy of BNN as it serves as an approximator of a PDE solver. For example, we can use more binarized layers or use additional non-binarized layers. Of course, increasing the power of the network leads to computational challenges solving the corresponding logical formulas.

Limitation of the Approach. The main limitation of the presented approach is its scalability. In our experiments, we used small images that represent a unit cell of a periodic porous medium. Conceptually, our approach can handle larger images, but scalability of the underlying decision procedures becomes the main bottleneck. Unfortunately, scalability is the main limiting factor in all decision based approaches to analysis of NNs at the moment [14,17,19]. In the future, we are hoping to develop efficient domain specific decision procedures for NN analysis. The second limitation is the set of topological constraints that we can handle. Many real applications require sophisticated restrictions on the topology of solid and void areas and do not exhibit periodic structure. It is an interesting research direction to formalize these constraints and use decision procedures to generate rich topologies.

6 Related Work

There are two lines of work related. The first line of work uses constraint to enhance machine learning techniques with declarative constraints, e.g. in solving constrained clustering problems and in data mining techniques that handle domain specific constraints [20–22]. One recent example is the work of Ganji *et al.* [21] who proposed a logical model for constrained community detection. The second line of research explores embedding of domain-specific constraints in the GAN training procedure [8,13,23–25]. Work in this area is targeting various applications in physics and medicine that impose constraints, like sparsity constraints, high dynamic range requirements (e.g. when pixel intensity in an image varies by orders of magnitude), location specificity constraints (e.g. shifting pixel locations can change important image properties), etc. However, this research area is emerging and the results are still preliminary.

7 Conclusion

In this paper we considered the constrained image generation problem for a physical process. We showed that this problem can be encoded as a logical formula over Boolean variables. For small porous media, we show that the generation process is computationally feasible for modern decision procedures. There are a lot of interesting future research directions. First, the main limitation of our approach is scalability, as we cannot use large networks with a number of weights in the order of hundreds of thousands, as it is required by industrial applications. However, constraints that are used to encode, for example, binarized neural networks are mostly pseudo-Boolean constraints with unary coefficients. Hence, it would be interesting to design specialized procedures to deal with this fragment of constraints. Second, we need to investigate different types of neural networks that admit encoding into SMT or ILP. For instance, there is a lot of work on quantized networks that use a small number of bits to encode each weight, e.g. [26]. Finally, can we use similar techniques to reveal vulnerabilities in neural networks? For example, we might be able to generate constrained adversarial examples or other special types of images that expose undesired network behaviour.

References

1. Arunachalam, H., Korneev, S., Battiato, I., Onori, S.: Multiscale modeling approach to determine effective lithium-ion transport properties. In: 2017 American Control Conference (ACC), pp. 92–97, May 2017
2. Battiato, I., Tartakovsky, D.: Applicability regimes for macroscopic models of reactive transport in porous media. J. Contam. Hydrol. **120–121**, 18–26 (2011)

3. Hermann, H., Elsner, A.: Geometric models for isotropic random porous media: a review. Adv. Mater. Sci. Eng. **2014** (2014)
4. Pyrcz, M., Deutsch, C.: Geostatistical reservoir modeling (2014)
5. Hornung, U. (ed.): Homogenization and Porous Media. Interdisciplinary Applied Mathematics, vol. 6. Springer, New York (1997). https://doi.org/10.1007/978-1-4612-1920-0
6. Goodfellow, I.J., Pouget-Abadie, J., Mirza, M., Xu, B., Warde-Farley, D., Ozair, S., Courville, A.C., Bengio, Y.: Generative adversarial nets. In: Ghahramani, Z., Welling, M., Cortes, C., Lawrence, N.D., Weinberger, K.Q. (eds.) Advances in Neural Information Processing Systems 27: Annual Conference on Neural Information Processing Systems 2014, 8–13 December 2014, Montreal, Quebec, Canada, pp. 2672–2680 (2014)
7. Goodfellow, I.J.: NIPS 2016 tutorial: generative adversarial networks. CoRR abs/1701.00160 (2017)
8. Osokin, A., Chessel, A., Salas, R.E.C., Vaggi, F.: GANs for biological image synthesis. In: 2017 IEEE International Conference on Computer Vision (ICCV), pp. 2252–2261, October 2017
9. Korneev, S.: Using convolutional neural network to calculate effective properties of porous electrode. https://sccs.stanford.edu/events/sccs-winter-seminar-dr-slava-korneev
10. Arunachalam, H., Korneev, S., Battiato, I.: Using convolutional neural network to calculate effective properties of porous electrode. J. Electrochem. Soc. (2018, in preparation to submit)
11. Radford, A., Metz, L., Chintala, S.: Unsupervised representation learning with deep convolutional generative adversarial networks. CoRR abs/1511.06434 (2015)
12. Chintala, S.: How to train a GAN? Tips and tricks to make GANs work. https://github.com/soumith/ganhacks
13. Luke de Oliveira, M.P., Nachman, B.: Tips and tricks for training GANs with physics constraints. In: Workshop at the 31st Conference on Neural Information Processing Systems (NIPS), Deep Learning for Physical Sciences, December 2017
14. Narodytska, N., Kasiviswanathan, S.P., Ryzhyk, L., Sagiv, M., Walsh, T.: Verifying properties of binarized deep neural networks. CoRR abs/1709.06662 (2017)
15. Hubara, I., Courbariaux, M., Soudry, D., El-Yaniv, R., Bengio, Y.: Binarized neural networks. In: Lee, D.D., Sugiyama, M., Luxburg, U.V., Guyon, I., Garnett, R. (eds.) Advances in Neural Information Processing Systems 29, pp. 4107–4115. Curran Associates Inc., Red Hook (2016)
16. Cheng, C., Nührenberg, G., Ruess, H.: Verification of binarized neural networks. CoRR abs/1710.03107 (2017)
17. Katz, G., Barrett, C., Dill, D., Julian, K., Kochenderfer, M.: Reluplex: an efficient SMT solver for verifying deep neural networks. arXiv preprint arXiv:1702.01135 (2017)
18. Cheng, C.-H., Nührenberg, G., Ruess, H.: Maximum resilience of artificial neural networks. In: D'Souza, D., Narayan Kumar, K. (eds.) ATVA 2017. LNCS, vol. 10482, pp. 251–268. Springer, Cham (2017). https://doi.org/10.1007/978-3-319-68167-2_18
19. Huang, X., Kwiatkowska, M., Wang, S., Wu, M.: Safety verification of deep neural networks. In: Majumdar, R., Kunčak, V. (eds.) CAV 2017. LNCS, vol. 10426, pp. 3–29. Springer, Cham (2017). https://doi.org/10.1007/978-3-319-63387-9_1
20. Dao, T., Duong, K., Vrain, C.: Constrained clustering by constraint programming. Artif. Intell. **244**, 70–94 (2017)

21. Ganji, M., Bailey, J., Stuckey, P.J.: A declarative approach to constrained community detection. In: Beck, J.C. (ed.) CP 2017. LNCS, vol. 10416, pp. 477–494. Springer, Cham (2017). https://doi.org/10.1007/978-3-319-66158-2_31
22. Guns, T., Dries, A., Nijssen, S., Tack, G., Raedt, L.D.: Miningzinc: a declarative framework for constraint-based mining. Artif. Intell. **244**, 6–29 (2017)
23. Luke de Oliveira, M.P., Nachman, B.: Generative adversarial networks for simulation. In: 18th International Workshop on Advanced Computing and Analysis Techniques in Physics Research, August 2017
24. Hu, Y., Gibson, E., Lee, L.-L., Xie, W., Barratt, D.C., Vercauteren, T., Noble, J.A.: Freehand ultrasound image simulation with spatially-conditioned generative adversarial networks. In: Cardoso, M.J., et al. (eds.) CMMI/SWITCH/RAMBO -2017. LNCS, vol. 10555, pp. 105–115. Springer, Cham (2017). https://doi.org/10.1007/978-3-319-67564-0_11
25. Ravanbakhsh, S., Lanusse, F., Mandelbaum, R., Schneider, J.G., Póczos, B.: Enabling dark energy science with deep generative models of galaxy images. In: Singh, S.P., Markovitch, S. (eds.) Proceedings of the Thirty-First AAAI Conference on Artificial Intelligence, 4–9 February 2017, San Francisco, California, USA, pp. 1488–1494. AAAI Press (2017)
26. Deng, L., Jiao, P., Pei, J., Wu, Z., Li, G.: Gated XNOR networks: deep neural networks with ternary weights and activations under a unified discretization framework. CoRR abs/1705.09283 (2017)

Author Index

Printed in the United States
By Bookmasters